# Goat Meat Production and Quality

# Goat Meat Production and Quality

*Edited by*

**O. Mahgoub**

*Department of Animal & Veterinary Sciences, College of Agricultural & Marine Sciences, Sultan Qaboos University, Sultanate of Oman*

**I.T. Kadim**

*Department of Animal & Veterinary Sciences, College of Agricultral & Marine Sciences, Sultan Qaboos University, Sultanate of Oman*

*and*

**E.C. Webb**

*Department of Animal and Wildlife Sciences, University of Pretoria, South Africa*

www.cabi.org

**CABI is a trading name of CAB International**

CABI
Nosworthy Way
Wallingford
Oxfordshire OX10 8DE
UK

Tel: +44 (0)1491 832111
Fax: +44 (0)1491 833508
E-mail: cabi@cabi.org
Website: www.cabi.org

CABI
875 Massachusetts Avenue
7th Floor
Cambridge, MA 02139
USA

Tel: +1 617 395 4056
Fax: +1 617 354 6875
E-mail: cabi-nao@cabi.org



A catalogue record for this book is available from the British Library, London, UK.

**Library of Congress Cataloging-in-Publication Data**

Mahgoub, O. (Osman)
Goat meat production and quality / O. Mahgoub, I.T. Kadim, and E.C. Webb.
p. cm.
Includes bibliographical references and index.
ISBN 978-1-84593-849-9 (alk. paper)
1. Goats. 2. Goat meat. 3. Goat meat--Quality. I. Kadim, I. T. (Isam T.) II. Webb, E. C. (Edward Cottington) III. Title.

SF383.M35 2011
636.3'9--dc23

2011026537

ISBN-13: 978 1 84593 849 9

Commissioning editor: Sarah Hulbert
Editorial assistant: Gwenan Spearing
Production editor: Fiona Chippendale

Typeset by AMA Dataset, Preston, UK
Printed and bound by CPI Group (UK) Ltd, Croydon, CR0 4YY

# Contents

**Contributors** vii

**Preface** ix
*O. Mahgoub, I.T. Kadim and E.C. Webb*

1 **Overview of the Global Goat Meat Sector** 1
*O. Mahgoub, I.T. Kadim and C.D. Lu*

2 **Goat Meat Production Systems** 15
*K.W. McMillin, E.C. Webb, E.F. Donkin and F. Pinkerton*

3 **Carcass Traits of Hardy Tropical Goats** 33
*G. Alexandre and O. Mahgoub*

4 **Genetics and Breeding of Meat Goats** 52
*J.N.B. Shrestha*

5 **Reproductive Efficiency for Increased Meat Production in Goats** 119
*M. Rekik, I. Ben Salem and N. Lassoued*

6 **Nutrition of the Meat Goat** 161
*J.R. Kawas, O. Mahgoub and C.D. Lu*

7 **Growth, Development and Growth Manipulation in Goats** 196
*E.C. Webb, N.H. Casey and L. Simela*

8 **The Role of Objective and Subjective Evaluation in the Production and Marketing of Goats for Meat** 209
*B.A. McGregor*

9 **Tissue Distribution in the Goat Carcass** 231
*O. Mahgoub, I.T. Kadim and E.C. Webb*

10 **Influences of Diets on Fatty Acid Composition of Edible Tissues of Meat Goat** **250**
*J.H. Lee and G. Kannan*

11 **Mineral Composition of Goat Meat** **260**
*N.H.I. Osman and O. Mahgoub*

12 **Linear Body Measurements and Carcass Characteristics of Goats** **277**
*I.T. Kadim and O. Mahgoub*

13 **Nutritive Value and Quality Characteristics of Goat Meat** **292**
*I.T. Kadim and O. Mahgoub*

14 **Effect of Early Nutrition on Carcass and Meat Quality of Young Goats Under Milk Production Systems** **324**
*A. Argüello, N. Castro, D. Sánchez-Macías and J. Capote*

15 **Effects of Feeding System and Diet on the Body Lipid Composition of Young Goats** **337**
*P. Morand-Fehr, A. Araba, P. Bas and A. El Aich*

# Contributors

**G. Alexandre** INRA, Unité de Recherches Zootechniques, Centre Antilles-Guyane, 97170 Domaine Duclos, Petit Bourg, Guadeloupe, France; E-mail: Gisele.Alexandre@antilles.inra.fr

**A. Araba** Institut Agronomique et Vétérinaire Hassan II, Rabat Instituts, Morocco; E-mail: a.araba@iav.ac.ma

**A. Argüello** Animal Science Department, Veterinary Faculty, Las Palmas de Gran Canaria University, Arucas, 35413 Las Palmas, Spain; E-mail: aarguello@dpat.ulpgc.es

**P. Bas** UMR 791 INRA/AgroParisTech, 16 rue Claude Bernard, 75231 Paris cedex 05, France; E-mail: pierre.bas@agroparistech.fr

**I. Ben Salem** Ecole Nationale de Médecine Vétérinaire, 2020 Sidi Thabet, Tunisia; E-mail: bensalemimen@yahoo.fr

**J. Capote** Instituto Canario de Investigaciones Agrarias, La Laguna, Tenerife, Spain; E-mail: jcapote@icia.es

**N.H. Casey** Department of Animal and Wildlife Sciences, University of Pretoria, Pretoria 0002, Republic of South Africa; E-mail: nhcasey@scientia.up.ac.za

**N. Castro** Animal Science Department, Veterinary Faculty, Las Palmas de Gran Canaria University, Arucas, 35413 Las Palmas, Spain; E-mail: ncastro@dpat.ulpgc.es

**E.F. Donkin** Department of Animal and Wildlife Sciences, University of Pretoria, Pretoria 0002, Republic of South Africa; E-mail: ned.donkin@up.ac.za

**A. El Aich** Institut Agronomique et Vétérinaire Hassan II, Rabat Instituts, Morocco; E-mail: a.elaich@iav.ac.ma

**I.T. Kadim** Department of Animal & Veterinary Sciences, College of Agricultural & Marine Sciences, Sultan Qaboos University, PO Box 34, Al-Khod 123, Sultanate of Oman; E-mail: isam@squ.edu.om

**G. Kannan** Georgia Small Ruminant Research and Extension Center, Fort Valley State University, Fort Valley, GA 31088, USA; E-mail: Govindak@FVSU.EDU

**J.R. Kawas** Cuerpo Académico de Producción Animal y Recursos Naturales, Facultad de Agronomía, Universidad Autónoma de Nuevo León, Mexico; E-mail: jkawas@mnade-mexico.com

**N. Lassoued** INRA-Tunisie, Laboratoire des Productions Animales et Fourragères, Rue Hédi Karray, 2049 Ariana, Tunisia; E-mail: lassoued.narjess@iresa.agrinet.tn

**J.H. Lee** Georgia Small Ruminant Research and Extension Center, Fort Valley State University, Fort Valley, GA 31088, USA; E-mail: leej@fvsu.edu

**C.D. Lu** College of Agriculture, Forestry and Natural Resource Management, University of Hawaii, Hilo, Hawaii 96720, USA; E-mail: chrislu99@gmail.com

**O. Mahgoub** Department of Animal & Veterinary Sciences, College of Agricultural & Marine Sciences, Sultan Qaboos University, PO Box 34, Al-Khod 123, Sultanate of Oman; E-mail: osmahgob@squ.edu.om

**B.A. McGregor** Institute for Technology, Research & Innovation, Deakin University, Geelong, Victoria 3217, Australia. Formerly: Livestock Systems Group, Victorian Department of Primary Industries, Attwood, Victoria, 3049, Australia; E-mail: bmcgregor@sub.net.au

**K.W. McMillin** School of Animal Sciences, Louisiana State University Agricultural Center, Baton Rouge, Louisiana, LA 70803, USA; E-mail: KMcmillin@agcenter.lsu.edu

**P. Morand-Fehr** UMR 791 INRA/AgroParisTech, 16 rue Claude Bernard, 75231 Paris cedex 05, France; E-mail: duspiwa@agroparistech.fr

**N.H.I. Osman** Sudan Open University, Obeid Khatim Street, Arkaweet, Khartoum, Sudan; E-mail: hudaisam@yahoo.co.uk/hudaisam@gmail.com

**F. Pinkerton** Retired, Langston University, USA; E-mail: Fpinkerton@austin.rr.com

**M. Rekik** Ecole Nationale de Médecine Vétérinaire, 2020 Sidi Thabet, Tunisia; E-mail: rekik.mourad@iresa.agrinet.tn

**D. Sánchez-Macías** Animal Science Department, Veterinary Faculty, Las Palmas de Gran Canaria University, Arucas, 35413 Las Palmas, Spain; E-mail: dsanchez@becarios.ulpgc.es

**J.N.B. Shrestha** Dairy and Swine Research and Development Centre, Agriculture and Agri-Food Canada, PO Box 90, Lennoxville Station, 2000 College Street, Sherbrooke, Québec, Canada J1M 1Z3; E-mail: Jap.Shrestha@agr.gc.ca

**L. Simela** New Emerging Red Meat Producers Organization, Lynnwood, Pretoria, Republic of South Africa; E-mail: fsd@nerpo.org.za

**E.C. Webb** Department of Animal and Wildlife Sciences, University of Pretoria, Pretoria 0002, Republic of South Africa; E-mail: edward.webb@up.ac.zA/Edward.webb@up.ac.za

# Preface

*Goat Meat Production and Quality* consists of 15 chapters and was compiled with the aim of providing information on basic as well as more advanced aspects of goat meat science to the wide range of scientists and professionals concerned with goat meat research and education. It is also intended to serve as a textbook for university and college students on meat production from ruminant animals. The goat industry, which has been lagging behind other livestock industries such as cattle, sheep and pigs, would also benefit from this book. *Goat Meat Production and Quality* has been meticulously written by internationally recognized experts and includes the most recent advances in goat meat science. The contributing authors hail from various parts of the world including Africa, Asia, North and South America, Australia and Europe. The goat was one of the earliest animals to be domesticated. Over many centuries, goats have served human communities around the world to provide food, fibre and other products such as leather and manure. For decades, the goat has been regarded as a major cause of environmental destruction by overgrazing leading to desertification. However, more recently, there has been growing recognition of the goat as an animal that provides sustainable livelihoods for many people, especially in the most deprived regions of the world. Goats provide valuable animal protein at a very low cost by utilizing marginal land usually rejected by other livestock to support low-income people in Asia, Africa and other underdeveloped parts of the world. In Europe, particularly southern European countries, the goat has been used commercially for milk production and cheese making, while goat kids serve as a by-product for meat production.

In recent years, a large volume of research has been carried out on goats in various parts of the world covering a number of production aspects including: nutrition, breeding, reproduction, health, production and quality. Research findings have been published in journals and presented at scientific meetings. It is now high time to produce a comprehensive book on goat meat production and quality to benefit goat research, education and the broader goat industry. Research on goats has been carried out worldwide under different climates and in different animal production systems. The latter range from intensive dairy goat systems in Europe to extensive traditional systems in Africa and Asia. The basic elements of meat production include breeding, reproduction, nutrition and growth, which are similar across meat-producing species and are of significant economic importance. For instance, in intensive systems, the cost of animal feed would comprise almost two-thirds of production costs. Under extensive goat production systems, the cost of feeding is much lower and goat production may be a secondary activity to cropping.

Meat is the major product of the goat, as all goats can produce meat. However, goat meat is less well known to consumers compared with other meats around the world. None the less, in some regions in Africa and Asia, goat meat is preferred compared with that from other livestock. In contrast, meat from young goats (Capretto) is considered a delicacy in southern Europe and South America. With the wide movements of human populations across continents in recent times, goat meat markets have expanded to areas such as North America and Europe. The goat meat industry is not as well developed as that of other species such as beef cattle, sheep and pigs. This is mainly because societies consuming goat meat are mostly in developing countries where the meat industry is not as developed as in the developed world. None the less, goat meat has been recognized as being leaner than that from cattle, sheep and pigs. This could make it more attractive to consumers who are conscious of the health hazards of consuming meat with a high fat content. However, there are certain aspects regarding goat meat production that affect the perceived quality of goat meat. Goat meat, especially from mature male goats, has the reputation of having a strong flavour and smell. In addition, goats are often slaughtered at an older age, usually under poor pre- and post-slaughtering conditions, which may compromise meat quality.

*Goat Meat Production and Quality* will cover the most important aspects of goat meat production and quality. The chapters are essentially detailed reviews with the most up-to-date publications summarized, integrated and discussed, rather than original papers with results only from research laboratories. The book comprises two parts: goat meat production and goat meat quality. The subjects covered in the book include an extensive review of the latest situation in the global meat goat sector, followed by a description of production systems, as well as the potential of tropical goat breeds for meat production, to complement the global overview. *Goat Meat Production and Quality* also includes chapters on fundamental principles of goat production including genetics and breeding, reproduction and nutrition. It also covers areas of normal and manipulated growth and development, and carcass conformation and composition. A chapter with a more practical nature on the role of subjective and objective evaluation in the production and marketing of meat goats is also included. The section on goat meat quality addresses the nutritive value and quality characteristics of goat meat, including fatty acids and mineral composition, as well as carcass characteristics and linear body measurements. The role of nutrition in young goats, such as the use of milk replacers, and the effect of the feeding system on goat carcasses are also discussed.

We hope that *Goat Meat Production and Quality* will add to the currently published information on goats and become a useful reference on goat meat production and quality.

The editors would like to thank CABI for publishing the book and sincerely acknowledge the contributions of the authors and their collaborators. We would also like to thank everybody who supports the goat meat production and quality cause around the world, including goat owners, technical and research staff and students worldwide.

**O. Mahgoub[1], I.T. Kadim[1] and E.C. Webb[2]**
[1]Department of Animal & Veterinary Sciences, College of Agricultural & Marine Sciences, Sultan Qaboos University, Sultanate of Oman; [2]Department of Animal and Wildlife Sciences, University of Pretoria, South Africa

# 1 Overview of the Global Goat Meat Sector

O. Mahgoub[1], I.T. Kadim[1] and C.D. Lu[2]

[1]*Department of Animal & Veterinary Sciences, College of Agricultural & Marine Sciences, Sultan Qaboos University, Sultanate of Oman;* [2]*College of Agriculture, Forestry and Natural Resource Management, University of Hawaii, Hilo, Hawaii*

## 1.1 Abstract

This chapter gives an overview of the goat meat sector. It describes the current situation of the goat as a meat animal and gives an account of goat populations throughout the world and their distribution by region in different continents, as well as the countries with the highest goat populations. It also discusses the contribution of goats to the meat market. The chapter comments on the acceptability and marketability of the goat and its meat.

## 1.2 Introduction

The goat was domesticated as early as 6–7 BC, as evidenced by archaeological remains collected in western Asia. It has since played a significant socio-economic role in the evolvement of human civilization around the world. It is particularly important in the tropics and subtropics where it is used as a major source of meat, milk, fibre, skin and manure in many traditional societies. It is also used as a readily cashable source of investment. There is a tendency to keep goats as a stock of wealth and sell them proportionally less when their number rises (Seleka, 2001). Goats as mixed feeding opportunists are able to adapt to seasonal and geographical changes and utilize low-quality rangelands to produce high-quality animal protein, which is extremely important (Lu, 1987). Goats have become important livestock in arid and semi-arid regions of the world because of their characteristics of versatility in harvesting forage and their ability to survive adverse foraging conditions. As opportunistic foragers, goats are able to maintain a relatively high-quality diet, even under diverse conditions, and at times they prefer shrubs over other types of range plants (Ramirez, 1999). In Kenya, about 67% of red meat is produced in arid and semi-arid lands under pastoral production systems (Juma *et al.*, 2010). Goats are a very important component of these systems and are reported to contribute 30% of the red meat in Kenya (Ahuya *et al.*, 2005). In Africa, goats play a significant role in the production of food and provide job opportunities for women and children in rural communities (Lebbie, 2004). Small-scale goat farming is of significant benefit to families all over the world (De Vries, 2008). Goats have also been kept in developed regions ranging from the Arctic to the Alps using adapted breeds that have developed in harmony with the local climatic conditions, topography and vegetation (Dyrmundsson, 2006).

## 1.3 The Goat as a Meat-producing Animal

One of the most important products from goats is meat. The majority of goats worldwide are regarded as meat type, with only about 5% of goats being classified as dairy type (Thompson, 2006). On average, the income from dairy systems (particularly in Europe) tends to be higher than that from meat systems due to favourable market prices for milk products and a greater dependence on subsidies in the meat extensive systems (Rancourt *et al.*, 2006). However, the importance of the goat as a meat-producing animal is increasing as its meat is becoming accepted in many new markets in societies that have not previously consumed goat meat. Over the past several decades, specialized meat-goat breeds have been developed, among which the Boer goat is the most notable. The breed has been developed from a genetic pool of native African, Indian, European and Angora goats, and active selection has been carried out within the breed over the past half century. Other specialized meat-goat breeds such as Kalahari have followed the Boer goat. More recently, Tianfu goats in China have emerged as a new breed with excellent performance for meat production and reproduction efficiency, and they are easily adapted to the local environment (Wang *et al.*, 2009). Some of the factors that make the goat a successful meat-producing animal, especially under extensive systems, include: the ability to graze and utilize poor forages; the ability to walk long distances; short generation intervals and high reproductive rates; high turnover rates of investment (low-risk investment); smaller carcasses, which are conveniently marketed, preserved or consumed over short periods of time; the feasibility of herding by children and women due to the flock instinct; and their ability to stand droughts (Lebbie, 2004). In some countries, such as Korea, goat meat has been used for its medicinal properties (Son, 1999).

Historically, goats have been categorized along with sheep as small ruminants, implying similarities at physiological, functional and production levels. Because of differences in foraging behaviour, morphological characteristics and ancestry clarification, it is perhaps time to look at the goat as an independent animal sector. Devendra (1981) stated that the small ruminants sector and its potential for contribution to agricultural-based economy has been neglected for a long time. However, the economic importance of the sector and its ability to make use of available feed resources as well as the functional value as a renewable resource for poor people in less developed countries has been recognized. These less developed countries are mostly located in the tropics and subtropics and are characterized by relatively high human populations and widely constraining climates and ecosystems, with low per capita income (Devendra, 1981). Goats with their small size and high efficiency of production are ideal for small-scale village production as well as pastoral systems. Goat production systems are specialized to some extent, although they are less developed compared with those of cattle and sheep. In goat-producing countries such as South Africa, there are two distinct goat-producing sectors, the commercial sector, which keeps more mohair-producing Angora and dairy goats, and the non-commercial sector, which keeps more meat goats (Roets and Kirsten, 2005).

Domestication, which has been influenced mainly by human movement throughout history, has affected the form and functions of the goat. Goats in drier areas have acquired adaptations that allow them to thrive and produce under these harsh conditions. However, unlike sheep, goats lack fat reservoirs in their bodies that help animals in arid regions to withstand periods of nutritional stress (Iñiguez, 2004). This may give an advantage to sheep keeping, especially those of the fat-tail type, in such areas of the world. Some goats have been selected to be specialized in certain aspects of production such as milk and fibre, but almost all goat types have been used for meat production either as specialized animals or with meat as a by-product of other systems. Although humans have been involved in selecting goats for various types of production, the biodiversity among goats has been more incidental than deliberate through natural selection (Casey and Webb, 2010).

Goat production has in the past been associated with political and economic developments in various regions around the world, and this association is expected to continue in the foreseeable future. Goat production and marketing may be influenced by anticipated environmental and socio-economic challenges in the future. Climate change, consumer preference and lifestyles, as well as the global economy, are important considerations for goat production and marketing. A significant revival was observed in the small ruminant sector in central and Eastern Europe after many countries joined the European Union (EU) in 2004 (Niznikowski *et al.*, 2006). Privatization of livestock sectors including goats in central Asia was a result of dissolution of the Soviet Union. A reduction in goat populations occurred in central Asian countries except for Turkmenistan, while Kazakhstan and Kyrgyzstan experienced an increase in goat populations starting in 1998 (Iñiguez, 2004). EU economic policy reforms and its relationship with the World Trade Organization have had a great impact on the economic viability of sheep and goat production systems in northern European countries (Dyrmundsson, 2006). While the economic recession in recent years has severely affected the livestock market and resulted in a decrease in sheep numbers in developed regions such as Australia, New Zealand and Europe, the small ruminant flocks in countries with subsistence farming have increased and have become an important means of survival (Morand-Fehr *et al.*, 2004).

Historically, goats have been perceived by some as having a negative impact on the environment. However, this impact is dictated by human management. Goats can be used for biological control of weeds, which impacts positively on the environment. However, when the stocking density exceeds the carrying capacity of the rangeland, this may cause degradation and sometimes leads to desertification (El Aich and Waterhouse, 1999). Spatial, temporal and other dynamics exist between goats and the environment. For instance, in the mountainous regions of South America, a major contribution from goats is providing manure for valuable cash crops, where an element of integrated nutrient management renders the production system more sustainable, reducing the stress on the fragile range ecosystem and enabling an increase in flock size (Iñiguez, 2004). A negative environmental impact is generally more related to poorly managed, high-input and high-output systems rather than the traditional extensive production system in goats (Peacock and Sherman, 2010). It has been suggested that the greatest environmental impact is more related to intensive systems in the developed world, such as expansion of beef production in South America and intensive pig and poultry systems in Asia (Peacock and Sherman, 2010). The majority of the world's goat population is in Africa (34%) and Asia (60%), and is kept under extensive/semi-intensive low-input systems with, in most cases, a comparatively light environmental touch (Peacock and Sherman, 2010). The concept of 'sustainability' has influenced EU agricultural policy makers in recent decades. Sheep and goat farming systems in north Europe are regarded as environmentally friendly and suitable for rural development (Dyrmundsson, 2006). In developed countries, after a long period of negative image, goat farming is gaining a positive outlook (Morand-Fehr *et al.*, 2004). In Korea, goats have been regarded as environmentally friendly animals because of the concern over pollution resulting from the disposal of waste originating from other livestock species such as pigs and cattle (Son, 1999).

A sign of increasing interest in goats is evidenced by the flourishing of goat research over the past three decades. However, more goat research has been carried out in developed than in developing regions, although the majority of goats are found in the latter (Sahlu and Goetsch, 2005). Research is needed for the development of a more efficient meat production system while improving meat attributes such as carcass and meat quality in developing regions. A multidisciplinary approach has been suggested to characterize animal production systems for various interrelated production traits with special reference to parameters of interest to farmers (Morand-Fehr *et al.*, 2004; Alexandre *et al.*, 2010).

## 1.4 The Worldwide Goat Population

The world population and distribution of major red-meat-producing animals are presented in Table 1.1 as reported by the Food and Agriculture Organization of the United Nations (FAO) (FAOSTAT, 2011). The total goat population in the world in 2009 was estimated at 867,968,573 heads. This was behind cattle (1,382,241,378), sheep (1,071,274,348) and pigs (941,212,507) but was far more than camels (25,385,468). When these figures are expressed as domestic herbivore biomass (Wilson, 1984), calculated as population numbers multiplied by mean weights (18 kg for goat, 30 kg for sheep, 206 kg for cattle, etc.), goats are also ranked after cattle and sheep. A significant proportion of the world goat population is found in countries defined by the FAO as net food-importing countries (385,232,718), low-income food-deficit countries (748,264,358) and the least-developed countries (270,123,867). This accounts for 44, 86 and 31% of the total world goat population, respectively (Table 1.1).

The goat distribution by continent is presented in Fig. 1.1 (FAOSTAT, 2011). The majority of the goats in the world (60%) are found in Asia. Africa comes second with almost one-third of the world's goat population. The Americas and Europe have 4 and 2%, respectively, while Oceania has a negligible proportion. Goats are reported to represent about 30% of Africa's ruminants and produce about 17% of its meat and 12% of its milk (Lebbie, 2004). The major concentration of goats in Africa is in the sub-Saharan region with 60% of the total goat population (147 million heads) representing about 80 indigenous breeds raised under various production systems (Lebbie, 2004).

When the statistics are examined by continent, they indicate the importance of goats in developing regions, particularly in Asia and Africa. Goat numbers in Asia and Africa are comparable to those of other livestock species. In Africa, the western and eastern parts of the continent are the most populated with goats, whereas in Asia, the southern and eastern parts are the most populated (Table 1.1). Southern Europe has the highest goat population within this continent. The high goat population in this region is mainly composed of dairy goats producing milk primarily for cheese-making in France, Italy, Spain and Greece. France, Greece and Spain supply about 83% of total goat milk produced in the EU (Castel *et al.*, 2010).

**Table 1.1.** Population (number of animals) and distribution of major meat-producing animals in the world in 2009. (From FAOSTAT, 2011.)

| Region | Goats | Cattle | Sheep | Camels | Pigs |
|---|---|---|---|---|---|
| Net food-importing developing countries | 385,232,718 | 351,242,006 | 287,990,725 | 22,900,305 | 41,489,844 |
| Low-income food-deficit countries | 748,264,358 | 636,890,571 | 577,063,586 | 23,686,631 | 524,010,340 |
| Least-developed countries | 270,123,867 | 254,905,557 | 176,632,101 | 20,267,980 | 30,725,331 |
| World | 867,968,573 | 1,382,241,378 | 1,071,274,348 | 25,385,468 | 941,212,507 |
| Africa | 294,871,078 | 270,675,336 | 292,122,275 | 21,514,522 | 27,644,351 |

*Continued*

**Table 1.1.** Continued.

| Region | Goats | Cattle | Sheep | Camels | Pigs |
|---|---|---|---|---|---|
| Eastern Africa | 89,214,463 | 118,269,261 | 59,358,638 | 10,493,700 | 8,446,353 |
| Central Africa | 21,935,966 | 21,808,864 | 8,643,819 | 1,391,045 | 4,220,170 |
| Northern Africa | 61,044,200 | 51,959,200 | 108,464,500 | 5,368,000 | 57,700 |
| Southern Africa | 11,723,135 | 19,912,657 | 29,423,458 | 75 | 1,767,690 |
| Western Africa | 110,953,314 | 58,725,354 | 86,231,860 | 4,261,702 | 13,152,438 |
| Americas | 37,120,763 | 509,551,701 | 90,161,096 | | 160,293,639 |
| Northern America | 3,099,350 | 107,701,699 | 6,574,938 | | 79,548,800 |
| Central America | 8,985,947 | 46,458,797 | 8,132,118 | | 20,885,849 |
| Caribbean | 3,910,800 | 8,786,250 | 3,003,470 | | 3,792,550 |
| South America | 21,124,666 | 346,604,955 | 72,450,570 | | 56,066,440 |
| Asia | 516,660,762 | 439,175,098 | 452,629,700 | 3,863,924 | 560,425,232 |
| Central Asia | 8,163,664 | 18,988,506 | 47,215,547 | 205,888 | 1,532,911 |
| Eastern Asia | 175,830,979 | 102,623,769 | 148,015,877 | 512,050 | 472,433,849 |
| Southern Asia | 281,275,542 | 249,928,058 | 60,099,993 | 1,967,000 | 15,000,809 |
| South-Eastern Asia | 25,322,540 | 47,071,912 | 10,964,851 | | 70,577,014 |
| Western Asia | 26,068,037 | 20,562,853 | 86,333,432 | 1,178,986 | 880,649 |
| Europe | 15,911,631 | 124,222,434 | 131,222,254 | 7,022 | 187,654,883 |
| Eastern Europe | 4,542,415 | 41,948,546 | 33,969,564 | 7,022 | 53,945,250 |
| Northern Europe | 212,546 | 22,956,198 | 39,349,450 | | 23,876,624 |
| Southern Europe | 9,071,409 | 17,453,224 | 45,807,046 | | 45,095,957 |
| Western Europe | 2,085,261 | 41,864,466 | 12,096,194 | | 64,737,052 |
| Oceania | 3,404,339 | 38,616,809 | 105,139,023 | | 5,194,402 |
| Australia and New Zealand | 3,082,229 | 37,868,259 | 105,123,283 | | 2,624,502 |
| Melanesia | 280,200 | 685,500 | 15,300 | | 2,117,500 |
| Micronesia | 4,800 | 14,140 | | | 53,800 |
| Polynesia | 37,110 | 48,910 | 440 | | 398,600 |

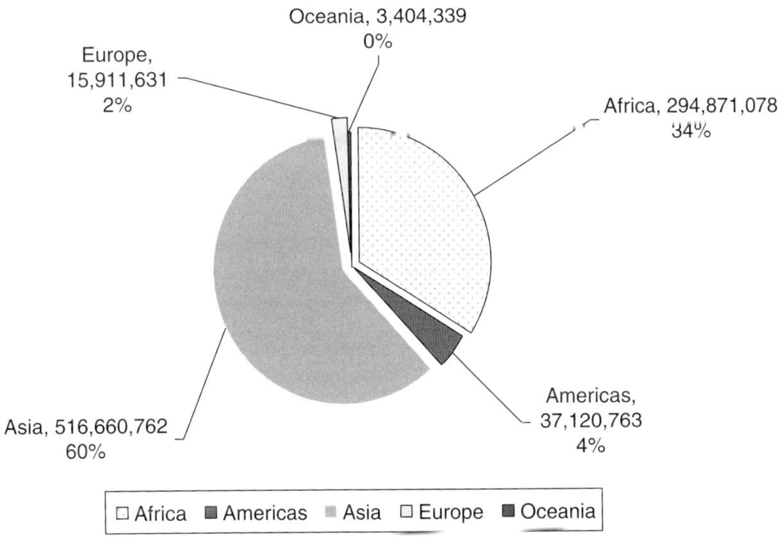


**Fig. 1.1.** Goat distribution by continent (FAOSTAT, 2011).

The goat population in this region has been maintained, and in some cases increased, compared with other livestock species. Goats have been moved to areas with reduced soil fertility that are therefore not suitable for more intensive systems such as cereal cropping or cattle ranching (Rancourt *et al.*, 2006). Spain is ranked second in terms of goat population and third in terms of goat milk yield within the EU (Castel *et al.*, 2010). The EU has 1.6% of the world's goat population but produces 13.2% of goat milk and 2.0% of goat meat (Castel *et al.*, 2010).

Trends in the numbers of goats, sheep and cattle in the world between 1980 and 2009 are shown in Fig. 1.2 (FAOSTAT, 2011). There has been a steady increase in goat numbers in the world over the past three decades. Over the same period, the numbers of sheep have declined slightly, especially during the last decade, whereas those of cattle have increased slightly. The increase in goat numbers over the past few decades has been attributed to their higher capacity to adapt to various environments, the development of goat farming in developing countries and the improved ecological image of goat farming and its products in developed countries (Morand-Fehr *et al.*, 2004).

Table 1.2 lists 18 countries with the highest goat populations in the world, with a range between 1 and 17% of the total world goat population. The top four countries are all from Asia, with 45% of the total world goat population. China and India have the largest goat populations, constituting approximately 32% of the total world goat population. With Pakistan and Bangladesh included, the Indian subcontinent has a sizeable goat population amounting to approximately 28% of the total world goat population. In Africa, Nigeria and Sudan have the highest goat populations, followed by Ethiopia, Kenya, Somalia, Niger, Tanzania, Burkina Faso and Mali.

## 1.5 Contribution of the Goat to the Meat Market

Meat is one of the major products from goats and is the major commodity income under conventional goat production systems. This includes meat from male and female young kids, castrated or intact males and culled females. Meat from the latter categories is of lower quality. Goat meat is classified according to the age of the goat and is known as capretto from young animals (weaned goats) and chevon from older goats.

The total goat meat production in 2008 was approximately 5 million t (Table 1.3) and was far less than mutton, beef and pork (8.3, 65.7 and 103 million t, respectively). Goat meat contributed approximately 2.7% of total red meat production worldwide. This could be attributed to the less developed production and marketing systems in goats compared with other species. Asia produced 70% of the world goat meat (3.5 million t), followed by Africa with 23.4% (1.2 million t), while Europe contributed only 2.5% of total world goat meat production (124,139 t). In Asia, the largest contribution came from East Asia (1,882,897 t). These figures match the pattern of distribution of live goats in the world discussed above.

Table 1.4 describes goat meat production around the world in terms of heads slaughtered, carcass weight and total production in t. The total number of goats slaughtered in 2008 was close to 400 million head around the world, with Asia contributing 279.4 million, followed by Africa with 96 million. Europe contributed 11.5 million head, with the majority in southern Europe. These numbers come from different production systems and would yield meat of various qualities. For instance, Asian (3.5 million t) and African (1.2 million t) goat meat would generally come from lower-grade carcasses produced from unfinished range animals and would have been obtained from male goats of various ages, castrated and non-castrated, as well as from older females, which are culled. Goat meat from Europe would be a by-product from milk/cheese production systems usually in the form of capretto meat from young kids. However, there are some signs of innovation in goat production systems in some parts of the world, which are targeting niche markets such as those of

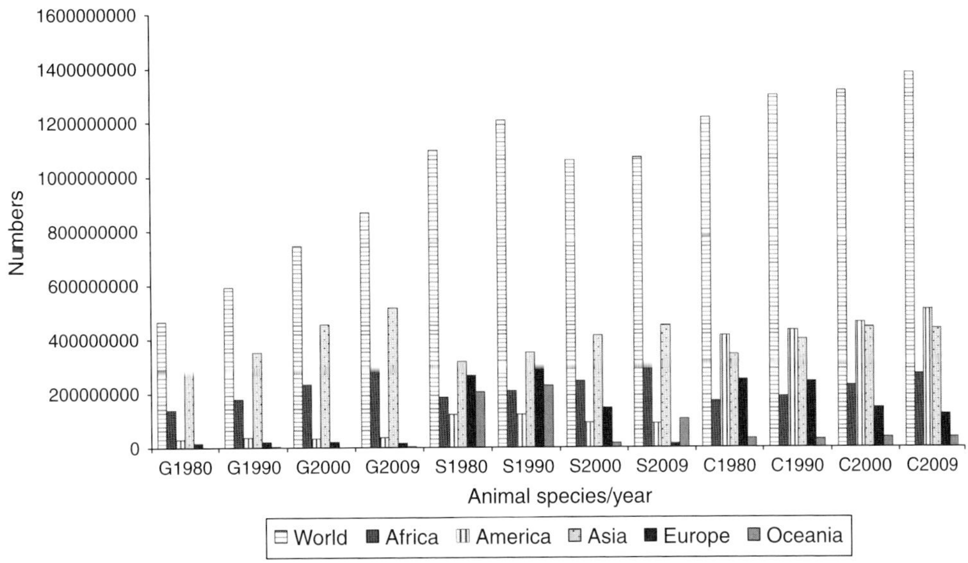


**Fig. 1.2.** Numbers of goats (G), sheep (S) and cattle (C) in the world between 1980 and 2009 (FAOSTAT, 2011).

**Table 1.2.** Countries with the highest goat population around the world in 2008 (from FAOSTAT, 2011).

| Country | Population (head) | % of total population |
|---|---|---|
| China | 149,376,747 | 17.33 |
| India | 125,732,000 | 14.59 |
| Pakistan | 56,742,000 | 6.58 |
| Bangladesh | 56,400,000 | 6.54 |
| Nigeria | 53,800,400 | 6.24 |
| Sudan | 43,100,000 | 5.00 |
| Iran, Islamic Republic of | 25,300,000 | 2.94 |
| Ethiopia | 21,884,222 | 2.54 |
| Mongolia | 19,969,400 | 2.32 |
| Indonesia | 15,805,900 | 1.83 |
| Kenya | 14,478,300 | 1.68 |
| Somalia | 12,700,000 | 1.47 |
| Niger | 12,641,352 | 1.47 |
| Tanzania, United Republic of | 12,55,0000 | 1.46 |
| Burkina Faso | 11,805,000 | 1.37 |
| Mali | 10,150,350 | 1.18 |
| Brazil | 9,500,000 | 1.10 |
| Mexico | 8,831,000 | 1.02 |

intensive goat production in southern Africa. There is some evidence that traditional goat farmers in northern Brazil have shifted from opportunistic management strategies towards more intensive ones (Primov, 1984). Meat from such improved systems would be of a higher quality and safer, and would involve better processing and packing.

**Table 1.3.** Contributions of goats to world meat production (t) in 2008 (from FAOSTAT, 2011).

| Region | Goat | Beef and buffalo | Sheep | Camel | Pig | Poultry | Turkey |
|---|---|---|---|---|---|---|---|
| World | 4,918,696 | 65,722,253 | 8,255,295 | 336,475 | 103,189,592 | 91,698,619 | 6,106,083 |
| Africa | 1,151,612 | 5,122,907 | 1,262,476 | 238,602 | 847,271 | 3,685,987 | 127,982 |
| Eastern Africa | 303,134 | 1,577,010 | 194,917 | 89,592 | 263,390 | 427,837 | 10,836 |
| Central Africa | 78,207 | 357,106 | 38,529 | 1,494 | 87,692 | 68,487 | 6 |
| Northern Africa | 257,160 | 1,284,300 | 577,768 | 92,505 | 2,538 | 1,688,000 | 112,100 |
| Southern Africa | 51,143 | 902,580 | 131,259 | | 163,137 | 1,003,250 | 5,040 |
| Western Africa | 461,967 | 1,001,911 | 320,000 | 55,011 | 330,512 | 498,413 | |
| Americas | 149,530 | 30,818,119 | 404,666 | | 18,785,952 | 40,991,598 | 4,154,655 |
| Northern America | 22,601 | 13,523,702 | 99,185 | | 12,403,329 | 21,370014 | 3,559,919 |
| Central America | 43,890 | 2,109,441 | 52,690 | | 1,337,533 | 3,334,436 | 23,945 |
| Caribbean | 12,344 | 230,473 | 10,668 | | 328,354 | 633,248 | 672 |
| South America | 70,695 | 14,954,503 | 242,122 | | 4,716,735 | 15,653,899 | 570,118 |
| Asia | 3,469,818 | 15,852,122 | 4,111,405 | 97,701 | 57,066,093 | 31,616,255 | 126,159 |
| Central Asia | 34,031 | 1,204,800 | 365,664 | 1,046 | 244,400 | 109,500 | |
| Eastern Asia | 1,882,897 | 6,993,436 | 2,051,376 | 22,860 | 49,693,424 | 17,765,043 | 3,881 |
| Southern Asia | 1,210,203 | 5,095,090 | 874,523 | 5,460 | 516,670 | 48,96,180 | 6,000 |
| South-eastern Asia | 160,957 | 1,610,222 | 68,054 | | 6,511,285 | 5,710,521 | 606 |
| Western Asia | 181,729 | 948,572 | 751,788 | 68,335 | 100,312 | 3,135,010 | 115,672 |
| Europe | 124,139 | 10,974,585 | 1,185,587 | 172 | 25,965,469 | 14,373,404 | 1,659,525 |
| Eastern Europe | 37,846 | 3,192,409 | 234,482 | 172 | 6,251,253 | 5,135,318 | 171,340 |
| Northern Europe | 612 | 1,918,068 | 426,060 | | 3,423,487 | 2,104,816 | 152,620 |
| Southern Europe | 76,356 | 2,153,502 | 380,190 | | 6,002,947 | 2,889,669 | 393,797 |
| Western Europe | 9,325 | 3,710,606 | 144,855 | | 10,287,782 | 4,243,601 | 941,768 |
| Oceania | 23,595 | 2,954,518 | 1,291,159 | | 524,806 | 1,031,373 | 37,761 |

Although goat meat supply has increased consistently over the past few decades, it still cannot meet the demand in many countries (Devendra, 2010). This has led to the slaughtering of increasing numbers of young stock with a consistent erosion of the breeding animals base, resulting in a higher meat price for goats than for other ruminants (Devendra,

**Table 1.4.** The world goat meat production in the form of slaughtered heads, yield per carcass and total production (t) in 2008 (from FAOSTAT, 2011).

| Region | Slaughtered (head) | Yield/carcass weight (kg/animal) | Production (t) |
|---|---|---|---|
| World | 398,408,444 | 123 | 4,918,696 |
| Africa | 95,748,031 | 120 | 1,151,612 |
| Eastern Africa | 27,762,100 | 109 | 000,104 |
| Middle Africa | 6,442,411 | 121 | 78,207 |
| Northern Africa | 19,995,000 | 128 | 257,160 |
| Southern Africa | 3,495,000 | 146 | 51,143 |
| Western Africa | 38,053,520 | 121 | 461,967 |
| Americas | 10,000,084 | 140 | 149,530 |
| Northern America | 796,440 | 283 | 22,601 |
| Central America | 2,601,683 | 168 | 43,890 |
| Caribbean | 897,260 | 137 | 12,344 |
| South America | 6,312,701 | 111 | 70,695 |
| Asia | 279,445,153 | 124 | 3,469,818 |
| Central Asia | 2,129,800 | 159 | 34,031 |
| Eastern Asia | 137,145,535 | 137 | 1,882,897 |
| Southern Asia | 114,819,280 | 105 | 1,210,203 |
| South-Eastern Asia | 13,475,346 | 119 | 160,957 |
| Western Asia | 11,075,192 | 153 | 181,729 |
| Europe | 11,556,540 | 107 | 124,139 |
| Eastern Europe | 2,661,773 | 142 | 37,846 |
| Northern Europe | 44,308 | 138 | 612 |
| Southern Europe | 7,852,603 | 97 | 76,356 |
| Western Europe | 997,856 | 93 | 9,325 |
| Oceania | 1,050,636 | 224 | 23,595 |
| Australia and NewZealand | 957,231 | 234 | 22,403 |
| Melanesia | 8,3900 | 124 | 1,044 |
| Micronesia | 1,600 | 183 | 29 |
| Polynesia | 7,905 | 150 | 118 |

2010). In some countries, the high demand for goat meat has also led to overslaughtering and the export of breeding females.

## 1.6 Goat Production Systems and Marketing

Goats are mostly kept under traditional, extensive and semi-intensive systems around the world. There are four major goat production systems in developing countries: rural landless, extensive, crop-baεod and rangeland-based (Devendra, 2010). The small ruminant supply is influenced by factors ranging from their population to annual rainfall and natural conditions (Seleka, 2001). Goat meat production is a commercial enterprise in only a few countries in the world including southern Africa (South Africa, Namibia and Botswana), the southern states of the USA and Mexico (Casey and Webb, 2010). Goats have an important role in self-sufficient agricultural systems in many parts of the world. This includes milk and cheese in northern countries and meat production in countries in the tropics and subtropics. Systems of goat production differ around the world. In the tropics, extensive systems prevail and depend on grazing of natural range. For instance, in Botswana, the majority of goats (97%) are raised under traditional or communal husbandry systems, whereas only 3% are raised under improved commercial production systems (Seleka, 2001). At the other extreme, in Japan, the livestock farming systems are based on concentrate feeding, which enables higher grades of meat and milk (Ozawa *et al.*, 2005). A concentrate-based goat production system is more difficult to sustain. For instance, the Japanese system is not sustainable, as only 25% of the feed is produced within Japan (Ozawa *et al.*, 2005). The current goat meat systems are not capable of supporting high-input, concentrate-feeding systems due to lack of competitive structuring and marketing competencies. A traditional intensive milk production system with meat and milk as the major commodities can be found in Spain. Under such systems, 20–40 kg kids can be marketed after spring grazing and cheese manufacturing (Castel *et al.*, 2010). Over the past few decades, this system has shifted to produce 1-month-old, 8 kg suckling kids due to legislative and economic constraints.

Production efficiency of goat-meat production under traditional subsistence systems is generally low and can be improved. Major limiting factors are high mortality and low utilization rates (sales plus home slaughter), as goats are grazed and bred under uncontrolled systems such as those in Botswana (Seleka, 2001). Kid mortality may reach up to 20% in the Inter-Andean Valley, mainly due to forage and management fluctuations and predators (Iñiguez, 2004). High mortality in goats in Africa is due to poor management practices and the occurrence of disease, whereas the low utilization rates are related to lack of appropriate and reliable marketing systems (Seleka, 2001).

In the tropics, there have been some efforts to provide public support for goat production systems to improve their efficiency. For instance, in Botswana, several support programmes from government and international organizations have been introduced for the improvement of production and marketing of small ruminants, mainly through supporting cooperatives (Seleka, 2001). In arid and semi-arid regions of the world, goats are kept within pastoralism and agropastoralism systems, which may include some subsistence cropping as well as goat keeping. This depends mainly on diversification and on subsistence activities, the herd composition and the mobility of the herds (El Aich and Waterhouse, 1999). Goat meat in these systems is destined mostly for home use or local markets, which do not require high-quality meat. In southern Africa, goats are kept by small-scale farmers, communal farmers and households (Casey and Webb, 2010). There is a trend towards improvement of small ruminant production in some countries by means other than simply increasing goat populations. These means include genetics, nutrition, housing, veterinary care and communication (Haenlein and Abdellatif, 2004). However, improvement of goats through genetics has been more evident in dairy goats by introducing established breeds such as Saanen, Alpines and Anglo-Nubian in the tropics. More recently, the Boer goat, the most characterized meat-goat breed, has started to play a role in genetic improvement of meat goats around the world (Simela and Merkel, 2008).

In developed regions, meat production from goats is not as significant. In 2006, the value of sheep and goat meat accounted for only 1% of all agricultural output and 2% of total meat consumption in six European countries (Canali, 2006). In Europe, most goats are kept for milk production. Where there is a surplus of goat milk and processed cheese, meat production poses an attractive alternative. However, in many of these countries, surplus kids are not kept for meat production and instead are culled. For instance, in Norway, goat meat contributes only about 0.3% of total meat production in the country (Asheim and Eik, 1998). This may be because goat carcasses and meat do not fit the specifications set for mutton and beef in traditional meat markets. Therefore, niche marketing of goat meat and meat products is essential in these markets.

Goat production systems in northern European countries are influenced by climatic conditions, with animals housed indoors for a significant part of the year. These systems are also supplemented in some cases with upland and mountain grazing in summer under marginal conditions and share resources with other agricultural enterprises (Dyrmundsson, 2006). The cost of production and the type and quality of products under these systems differ greatly from those of extensive pastoral tropical systems with milk being the major product and meat a by-product.

There may be an opportunity for goat meat marketing in the developing world because of the declining profitability of traditional production systems such as traditional ranching in the USA (Lupton *et al.*, 2008). A promising approach would be based on a multiproduct complementary system. Lupton *et al.* (2008) investigated a high-quality mohair-, meat- and hide-producing system from Angora goats. Special housing, feeding and watering facilities allowed the production of clean mohair due to separation from faeces and urine. Automation of feeding and watering systems reduced the labour requirements. Under feedlot and elevated floor systems, the goats grew faster than those on pasture. Goats raised under the elevated floor system produced cleaner fibres, which were suitable for hand-spinning niche markets. The net income per head was at its highest for goats raised under elevated floor systems for a niche market.

## 1.7 Acceptability of Goat Meat

Compared with mutton, beef and pork, goat meat is not as widely distributed, marketed or accepted in major world markets for various reasons. Even in countries with a high goat population, the consumption of goat meat is not as high as for meats from other species. For instance, in Kenya, beef was found to form the highest proportion of red meats consumed at home, whereas chevon and mutton were negligible (Canali Gamba, 2005). In Sudan, which is ranked among countries with a high goat population, goat meat is regarded as inferior to mutton and beef.

Acceptance of goat meat varies widely around the world. For instance, goat meat was rated by 61% of the Japanese housewives as 'not commonly eaten at home' and 'smelly', and they 'never thought about goat meat' (Ozawa *et al.*, 2005). In a survey carried out in Japan, only 16% of respondents had eaten goat meat. However, it was rated by those who consumed it as being more tender than pork (Ozawa *et al.*, 2005). On the other hand, in the developing world, goat meat is well accepted and is purchased on a small scale and in traditional markets. In certain parts of the developing world, such as the humid tropics of Africa, goat meat is the most important type of meat, as cattle and pig rearing is not practised due to disease and religious beliefs (Oludimu and Owokade, 1995).

## 1.8 Marketing of the Meat Goat

Marketing of goat meat includes marketing live goats as well as carcass and non-carcass components. Marketing of edible parts of the goat is influenced by consumer preferences due to differences in culture, religious beliefs and competition with other meats. Marketing of meat-producing livestock including goats is done in different ways around the world, with meat moving along different channels to consumers (Roets and Kirsten, 2005). Goat producers in non-commercialized farming systems use informal and less reliable marketing systems that hold many risks for producers and consumers compared with commercial systems that use well-organized systems with capital resources, a better infrastructure, institutions, legal frameworks and markets (Roets and Kirsten, 2005).

Goats are marketed in traditional subsistence markets around the developing world where most goats are found and their meat consumed. These markets are characterized by a small number of animals sold on market days by goat owners or their family members. Women are mostly responsible for goat rearing and marketing. Oludimu and Owokade (1995) reported that 95% of goat sellers in a local market in Nigeria were women. Only 20% of them were owners, while the others were apprentices, agents or relatives of the owners or were learning the trade.

Animal weight is not the only consideration for the marketing of goats and, surprisingly, is not commonly determined in traditional and subsistence markets. Goat sellers in Nigeria estimate the goat price on the basis of size, hair coat condition, age, health, meat formation and sex (Oludimu and Owokade, 1995). However, studies have revealed that the major factors that determine goat prices in traditional markets are weight, sex and age (Oludimu and Owokade, 1995). Risk, in the form of theft and mortality, is borne solely by the traders.

Some goat meat products are characterized by 'typicality', which indicates that a product is made in a specific area with specific characteristics (Rubino *et al.*, 1999). These include small goat carcasses in some countries and some types of preserved meats in Africa. The consumer's attitude and consumption behaviour can affect the growth of livestock sections (Juma *et al.*, 2010). The demand for small ruminant meat increases concomitant with increasing income or cost, as consumers perceive small ruminant meat to be a quality meat compared with beef (Juma *et al.*, 2010). Marketing of goats and their products is also affected by changes in economic trends such as globalization, which add new challenges to goat meat marketing. Globalization has initiated new demands by retailers as a result of the demand from consumers for tractability, quality and consistency of supply. Consumers prefer specialized and relatively low-volume food products (Roets and Kirsten, 2005).

The major factors that hinder goat meat marketing may be sociologically related (Morand-Fehr *et al.*, 2004). For instance, in some countries, rearing of goats or eating goat meat are related to the lower-income classes of society. These issues have to be addressed when marketing of meat and meat products from goats is considered.

In northern America, the demand for goat meat is believed to exceed the supply. Often, the demand by consumers of particular ethnic and religious groups cannot be met due to the absence of a reliable supply of goat meat, especially in urban areas of the USA. Meat goats are rarely the primary animal production enterprise in the USA, but they are becoming increasingly important contributors to the income of many producers (Glimp, 1995). Meat-goat marketing is highly unstructured in the USA, yet prices are generally higher per unit weight than other red meat-producing species (Glimp, 1995). Capretto, meat from milk-fed kids, is known to be in high demand and short supply in northern Mexico.

There is an opportunity for goat meat to be marketed in niche markets. One of these growing markets is organic meat production, with Australia and China being the most important producers (Lu *et al.*, 2010). There are no obvious differences in efficiency of production in terms of carcass yield and conformation between organic and non-organic goat meat systems (Lu *et al.*, 2010). However, consumers perceived organic goat meat to be safe and healthy, and to support production systems that emphasize animal welfare. Organic goat production contributes to the alleviation of poverty for producers who rely on marginal land for their livelihoods, while meeting the increasing demand for organic products globally. Nutritional strategies that are naturally occurring, low cost and easy to apply can improve the productive performance of goats and render organic production profitable and successful (Lu, 2011). Adequate protein intake enhances immunity and improves both resistance and resilience of the host to endoparasites and diseases. Organic goat production relies on high-forage systems; therefore, understanding plant biomass accumulation, eating behaviour, seasonal fluctuations and environmental interactions can lead to more sustainable organic production while maintaining diverse plant landscapes. Legumes are desirable for organic goat production because they contain high levels of protein and are suitable browsing for goats. Secondary plant compounds at a concentration below the level of toxicity can be utilized for disease prevention, control and treatment, and may fill the vacuum from the absence of the use of chemicals in organic goat production. In combination with other naturally occurring materials, they have the potential to improve nutrient digestion and utilization in goats. Goats are versatile at harvesting plants and able to survive under adverse foraging conditions (Lu, 1988). The tolerance of goats towards the bitterness of secondary plant compounds can play an anthelmintic role and make goats more suitable for high-forage organic production systems than other ruminant species. Rapid fetal growth associated with the physical fill of forage limits feed intake during late pregnancy and therefore presents a challenge for nutritional balance in goats under high-forage organic production systems. Understanding regulation of intake, fibre digestion and utilization can lead to nutritional balance, minimizing metabolic disorders associated with pregnancy, parturition and lactation (Lu *et al.*, 2005). The use of bioactive plants, traditional herbal or ethno-veterinary medicine may present economic and sustainable alternatives if a greater understanding of the mechanisms, interactions with other nutrients and levels of toxicity can be achieved. Chinese herbal medicines, many of which have antibacterial, antiviral and antiparasitic properties, merit further studies and verification, and may be promising in organic goat production (Lu, 2011).

There are also niche markets for young kids in Jordan, southern Turkey and southern Syria, often coupled with fattening systems because of the high value of kid meat (Iñiguez, 2004). Another niche market for goat meat is medicinal use. For instance, the Korean native black goat is used in Korea for its medicinal properties.

## References


Ahuya, C.O., Okeyo, A.M. and Peacock, C. (2005) Developmental challenges and opportunities in the goat industry: the Kenyan Experience. *Small Ruminant Research* 60, 197–206.

Alexandre, G., Gonzalez-Garcia, E., Lallo, C.H.O., Ortega-Jimenez, E., Pariacote, F., Archimede, H., Mandonnet, N. and Mahieu, M. (2010) Goat management and systems of production: global framework and study cases in the Caribbean. *Small Ruminant Research* 89, 193–206.

Asheim, L.J. and Eik, L.O. (1998) The economics of fiber and meat on Norwegian dairy goats. *Small Ruminant Research* 30, 185–190.

Canali, G. (2006) Common agricultural policy reform and its effects on sheep and goat market and rare breeds conservation. *Small Ruminant Research* 62, 207–213.

Canali Gamba, P. (2005) *Urban Domestic Consumption Patterns for Meat: Trends and Policy Implications*. Working paper. Tegemeo Institute of Agricultural Policy and Development, Egerton University, Kenya.

Casey, N.H. and Webb, E.C. (2010) Managing goat production for meat quality. *Small Ruminant Research* 89, 218–224.

Castel, J.M., Ruiz, F.A., Mena, Y. and Sanchez-Rodriguez, M. (2010) Present situation and future perspectives for goat production systems in Spain. *Small Ruminant Research* 89, 207–210.

Devendra, C. (1981) Potential of sheep and goats in less developed countries. *Journal of Animal Science* 51, 461–473.

Devendra, C. (2010) Concluding synthesis and the future for sustainable goat production. *Small Ruminant Research* 89, 125–130.

De Vries, J. (2008) Goats for the poor: some keys to successful promotion of goat production among the poor. *Small Ruminant Research* 77, 221–224.

Dyrmundsson, O.R. (2006) Sustainability of sheep and goat production in North European countries – from Arctic to the Alps. *Small Ruminant Research* 62, 151–157.

El Aich, A. and Waterhouse, A. (1999) Small ruminants in environmental conservation. *Small Ruminant Research* 34, 271–287.

FAOSTAT (2011) Food and Agricultural Organization, United Nations <http://faostat.fao.org/site/339/default.aspx>.

Glimp, H.A. (1995) Meat goat production and marketing. *Journal of Animal Science* 73, 291–295.

Haenlein, G.F.W. and Abdellatif, M.A. (2004) Trends in small ruminant husbandry and nutrition and specific reference to Egypt. *Small Ruminant Research* 51, 185–200.
Iñiguez, L. (2004) Goats in resource-poor systems in the dry environments of West Asia, Central Asia and the Inter-Andean Valleys. *Small Ruminant Research* 51, 137–144.
Juma, G.P., Ngigi, M., Baltenweck, I. and Druker, A.G. (2010) Consumer demand for sheep and goat meat in Kenya. *Small Ruminant Research* 90, 135–138.
Lebbie, S.H.B. (2004) Goats under household conditions. *Small Ruminant Research* 51, 131–136.
Lu, C.D. (1987) Implication of forage particle length on milk production in dairy goats. *Journal of Dairy Science* 70, 1411–1416.
Lu, C.D. (1988) Grazing behavior and diet selection of goats. *Small Ruminant Research* 1, 205–216.
Lu, C.D. (2011) Nutritionally related strategies for organic goat production. *Small Ruminant Research* 98, 73–82.
Lu, C.D., Kawas, J.R. and Mahgoub, O.G. (2005) Fiber digestion and utilization in goats. *Small Ruminant Research* 60, 45–52.
Lu, C.D., Gangyi, X. and Kawas, J.R. (2010) Organic goat production, processing and marketing: opportunities, challenges and outlook. *Small Ruminant Research* 89, 102–109.
Lupton, C.J., Huston, J.E., Hruska, J.W., Craddock, B.F., Pfeiffer, F.A. and Polk, W.L. (2008) Comparison of three systems for concurrent production of high quality mohair and meat from Angora male kids. *Small Ruminant Research* 74, 64–71.
Morand-Fehr, P., Boutonnet, J.P., Devendra, C., Dubeuf, J.P., Haenlein, G.F.W., Holst, P., Mowlem, L. and Capote, J. (2004) Strategy for goat farming in the 21st century. *Small Ruminant Research* 51, 175–183.
Niznikowski, R., Strzelec, E. and Popielarczyk, D. (2006) Economics and profitability of sheep and goat production under new support regimes and market conditions in Central and Eastern Europe. *Small Ruminant Research* 62, 159–165.
Oludimu, O. and Owokade, A. (1995) Goat marketing and pricing in Ile-Ife, Nigeria. *Small Ruminant Research* 17, 85–89.
Ozawa, T., Nishitani, J., Odake, S., Lopez-Villalobos, N. and Blair, H.T. (2005) Goat meat acceptance in Japan: current situation and future prospects. *Animal Science Journal* 76, 305–312.
Peacock, C. and Sherman, D.M. (2010) Sustainable goat production: some global perspectives. *Small Ruminant Research* 89, 70–80.
Primov, G. (1984) *Goat Production Within the Farming System of Smallholders of Northern Bahia, Brazil.* SR-CRSP Technical Report No. 35, University of Missouri, Columbia, USA.
Ramirez, R.G. (1999) Feed resources and feeding techniques of small ruminants under extensive management conditions. *Small Ruminant Research* 34, 215–230.
Rancourt, M., de Fois, N., Lavin, M.P., Tchakerian, E. and Vallerand, F. (2006) Mediterranean sheep and goat production: an uncertain future. *Small Ruminant Research* 62, 167–179.
Roets, M. and Kirsten, J.F. (2005) Commercialization of goat production in South Africa. *Small Ruminant Research* 60, 187–196.
Rubino, R., Morand-Fehr, P., Renieri, Peraza, C. and Sarti, F.M. (1999) Typical products of the small ruminant sector and the factors affecting their quality. *Small Ruminant Research* 34, 289–302.
Sahlu, T. and Goetsch, A.L. (2005) A foresight on goat research. *Small Ruminant Research* 60, 7–12.
Seleka, T.B. (2001) Determinants of short-run supply of small ruminants in Botswana. *Small Ruminant Research* 40, 203–214.
Simela, L. and Merkel, R. (2008) The contribution of chevon from Africa to global meat production. *Meat Science* 80, 101–109.
Son, Y.S. (1999) Production and uses of Korean native black goat. *Small Ruminant Research* 34, 303–308.
Thompson, D. (2006) *Meat Goat Breeds, Breeding Management and 4-H Market Goat Management.* Unpublished Report, Bagley, MN, USA.
Wang, D., Lu., C.D., Gangyi, X., Zhao, W. and Wang, D. (2009) Genetic diversity analysis of Tianfu goats and three relative breeds using microsatellite DNA markers. *Pacific Agriculture and Natural Resources* 1, 44–51.
Wilson, R.T. (1984) *The Camel.* Addison-Wesley Longman, London, UK.

# 2 Goat Meat Production Systems

K.W. McMillin[1], E.C. Webb[2], E.F. Donkin[2] and F. Pinkerton[3]
[1]Louisiana State University Agricultural Center, Louisiana, USA; [2]Department of Animal and Wildlife Sciences, University of Pretoria, South Africa; [3]Retired, Langston University, USA

## 2.1 Abstract

Goat production systems have been developed to adapt to the specific needs of the producer but can be grouped into several general categories. Intensive production systems are usually on smaller plots and require daily attention to management practices, while extensive systems have more reliance on goats grazing over large areas. Goats may be raised for subsistence, to generate capital, for profit or for production through communal, nomadic, transhumance and sedentary pastoralism, primitive herding and largeholder commercial farms. Small-scale goat farms are influenced by infrastructure, social systems, land tenure, economics and marketing. Production systems are dependent on climate, geography, diet, breed, reproductive capacity, diseases and goat behaviour. Goat production systems must match the desires of the producers with the many factors influencing the growth and productivity of meat goats for success of the enterprise.

Commercial meat-goat enterprises account for an overwhelming majority of the meat-goat production systems and herd sizes range from five to several thousand does. Well-managed production systems can generate weaned kid crops in excess of 130% per year, so the prolificacy, milk production and pre-weaning growth contributions by does are important in achieving this level of productivity. Goats represent a small niche in many developed countries, but they contribute significantly to the livelihoods of rural people across the globe. The different goat production systems and the factors that influence goat production are discussed in this chapter.

## 2.2 Introduction

Goats were among the earliest species of livestock to be domesticated approximately 9000 years ago. These small ruminants descended from the bezoar or wild goat (*Capra hircus*) in the hills of western Asia (Casey *et al.*, 2003). The consequence of domestication was a change in the phenotypic characteristics of wild goats, which resulted in the development of a multiplicity of goat breeds or types. These breeds or types were distributed across the world as a result of the migration and translocation of humans, usually due to changing climatic conditions and natural resources. The world goat population now consists of an enormous amount of genetic variation in morphological and production

 *Goat Meat Production and Quality* (eds O. Mahgoub, I.T. Kadim and E.C. Webb)

characteristics (Galal, 2005; Shrestha and Fahmy, 2005). According to the Food and Agriculture Organization of the United Nations (FAO, 1999), there are now about 570 breeds of goats in the goat database. The domestication of goats signalled the start of goat production in different systems and placed significant responsibility on goat keepers; hence the need to study and understand goat production systems, as well as the quality of products from animal origin.

It is important to understand the domestication of goats because this process, as well as the migration of goats to different parts of the world, influenced the type of production systems that developed over the centuries. However, in most livestock production systems, goats were regarded as easy-care animals and thus received little attention in terms of selection and breed improvement. The well-known Bezoar, Savannah and Nubian goat breeds descended from the Bezoar (*Capra aegagrus*), Markhor (*Capra aegagrus falconeri*) and Ibex (*Capra aegagrus ibex*) during the migration of nomadic pastoralists from Asia to the west (Mediterranean countries), east (Iran and Afghanistan, Turkistan, Mongolia and North China) and later via the Khyber Pass to India (Shrestha and Fahmy, 2005). Many of the later goat breeds still share similar phenotypic features, although they are kept in a variety of production systems.

Colonists from Spain and Portugal took Anglo-Nubian goats to the Americas, while those from France and the UK took similar goats to North America, Australia, New Zealand and South Africa. The development of the Boer goat in the Eastern Cape in South Africa during the 1920s is regarded as one of the cornerstones of the international goat industry because it emphasized the possibilities of selection for carcass and production characteristics in goats. Many countries are now importing more prolific breeds of goat to increase the growth characteristics and meat yields of local goats by crossbreeding and breed improvement programmes (Casey and Webb, 2010). However, goat production for meat remains a predominantly small-scale subsistence enterprise, and the benefits of such systems still surpass the disadvantages (Lebbie, 2004).

It is now estimated that 90% of the current world goat population of about 650 million occur in Asia, Africa, Central America and the Caribbean, where goat production is associated primarily with subsistence agriculture (Casey *et al.*, 2003). Asia (71%) and Africa (22%) produce most of the world's goat meat (FAO, 1999). The collapse of the goat fibre industry in Australia during the 1800s resulted in the release of goats in feral flocks and this is regarded as the origin of Australian feral goats (Werdi Pratiwi *et al.*, 2007). It is estimated that there are now between 4 and 6 million feral goats in Australia.

## 2.3 What do We Mean by a 'System' of Goat Production?

A 'system' in the context of livestock production can be considered as the integration of all the factors that influence the management of the goat and entire goat enterprise. An existing system of goat production and the way the goats are managed might be the result of the way things were created, a result of the 'balance of nature' or a result of experience after many years of 'trial and error'. In the evaluation of a system of goat production, we should also consider whether the production system is improving, deteriorating or stable, and whether it is productive and sustainable. Humans play an important role in the management of goat production systems, and management decisions often have far-reaching consequences for the animals and the environment.

Management of goat production systems will include decisions about the breed or type of goat, housing systems, breeding and feeding methods, and health management (Donkin *et al.*, 1996). All these factors will result in a system of management that determines the production system. Overall, the key factor of any goat production system lies in the competence and commitment of the owner or manager.

The success of a production system depends on the following aspects:

- Long-term sustainability
- Appropriateness for the environment
- Profitability
- Personal satisfaction.

## 2.4 Goat Production Functions and Production Systems

Goats are equally well adapted to consume and utilize leaves, grass, forbs and shoots of shrubs – goats are thus regarded as grazers and browsers (Du Plessis *et al.*, 2004). Goats also possess unique characteristics to adapt to harsh environments (Alexandre and Mandonnet, 2005). These unique characteristics enable them to thrive in a large variety of ecological regions around the world and in different production systems.

The varying levels of financial commitment, physical resources and site-specific management fall into several general types, including: (i) smallholder intensively managed; (ii) smallholder purebred; (iii) large-scale extensively managed; (iv) hobby-type for personal enjoyment; and (v) specialized herds for production of show stock or control of brushy, weedy or forest areas. Starting the goat enterprise may be the most difficult aspect because of all of the decisions that must be made. Fundamental to success in any livestock enterprise is the gathering of sufficient useful information so that wise decisions on management and production practices can be made. Failures often occur because, for example, the enterprise starts with excessive ambition or grows beyond the management ability; the numbers or kinds of goats do not match the fencing, facility, geography and climatic conditions; marketing – an important aspect of the production system – is not adequately carried out; or incorrect information is received.

Commercial meat-goat enterprises account for an overwhelming majority of the meat-goat production systems because of the desire and need for economic returns. The primary types are slaughter kid production and purebred buck and doe production. Herd sizes range from five to several thousand does, depending on the factors mentioned above. According to Casey *et al.* (2003), small-scale farmers can keep from two to three breeding does, with an average of between five and eight does and one buck, in tropical and subtropical parts of Africa. Larger commercial herds often consist of several hundred breeding does in a ratio of one buck to 25 does. Emphasis should be placed on the role of does in production because of the influences on reproduction and growth that they contribute. Well-managed production systems can generate weaned kid crops in excess of 130% per year, so the prolificacy, milk production and pre-weaning growth contributions by does are important in achieving this level of productivity.

Goats represent a small niche in many developed countries, but they contribute significantly to the livelihoods of rural people across the globe, and form a viable link in the agricultural production chain to generate an income and provide employment opportunities for women, children and the elderly. Small ruminants have a large impact on the economy and food supply of people in subtropical and tropical countries. This benefit is often not shown in national statistics because of informal trading and slaughtering.

Goats are often neglected but have served mankind with meat, milk, hair, leather and products including manure for many centuries (Webb *et al.*, 2005). It is also widely accepted that goats serve as a source of high-quality protein for rural families mainly because of their small size and affordable nature. The relatively small size of goat carcasses is beneficial for small-scale farmers because the carcass can be consumed within a short period of time and the risk of meat spoilage is reduced. A less well-known fact is that people in many rural countries consume more goat milk than cow milk. Traditional milk goats and their crosses are also used for meat production (Braker *et al.*, 2002). Many poor people and small-scale farmers use goats as a financial reserve, often due to a lack of access to financial institutions. In commercial and communal systems, goats are often used as controllers of bush encroachment. These ubiquitous animals often outnumber cattle and other production animals in many regions. Their role in sustaining agricultural production systems in many resource-poor areas clearly shows that goats are vital to economic development.

Environmentalists generally consider a sustainable livestock system to be an oxymoron (Basset and Crummey, 1993) because such systems are perceived to cause severely overgrazed hillsides or compacted tropical soils, particularly where there is individual livestock ownership but the animals are kept on communal grazing lands (Bembridge, 1987). In developed countries, livestock are often associated with human health problems that are a result of overconsumption of animal fat. Livestock are blamed for both their direct and indirect negative impact on the environment, but even livestock critics admit that, if managed correctly, livestock can play an important role in agricultural development without any negative impact on the environment. The fact is that in many countries the largest part of the country is suitable only for extensive livestock production, and in most of these regions goats are particularly well adapted.

Goats are more often poorly managed and this is attributed to their ability to survive under harsh conditions and also because most people in rural areas rear goats for their subsistence purposes to support their families (Braker *et al.*, 2002). This explains why goat farmers seldom consider the possibilities of increasing production through either cross-breeding or artificial insemination. A very important aspect in this regard is the awareness of risk by resource-poor farmers and their emphasis on minimizing it. Under such conditions, animals that are hardy against the vagaries of droughts, disease and poor management are more attractive options than more productive breeds that are vulnerable to these conditions (Braker *et al.*, 2002).

In the largest part of Africa and Asia, goats form part of subsistence farming systems, but much of the land is either under- or overutilized, leading to degradation. The misperception is that 'small scale' is often viewed as non-productive, non-commercial subsistence agriculture. However, the evidence is that small-scale agriculture has the potential to generate employment and income opportunities in rural areas irrespective of limited access to land and credit. Commercial farming for metropolitan markets is increasing in Asia, Africa and central and South America (Casey *et al.*, 2003). Security of land tenure and ownership influences the ability and willingness of farmers to adopt new technologies that will improve their production and therefore should never be underestimated. Land is a scarce resource that needs to be conserved.

Goats account for about 30% of Africa's ruminant livestock, with sub-Saharan Africa accounting for 60% of the total goat population in Africa. Goats are important in marginal agricultural land areas and play a significant role in the food chains and overall livelihoods in poor resource-households. Poverty, combined with a lack of modern agricultural skills and low-input practices, results in poor productivity due to inadequate animal nutrition, prevalence of diseases and parasites, and poor milk and meat production (Lebbie, 2004). Tick diseases may cause high mortality, which will counter high incidences of twinning (Masika and Mafu, 2004).

## 2.5 Types of Goat Production Systems

### 2.5.1 Subsistence rearing

Subsistence rearing is when livestock activity is the sole activity of the herder and his family. They raise animals to support their families and the surplus is exchanged in order to procure household necessities, with the use of money reduced to the minimum. Animals have a social role, such as being used as loans, passing them on as an inheritance, gifts and lobola ('bride price').

All these create a web of obligation of dependent relationships, which assures the cohesion of families and social group and forms the hierarchies between different groups. Capital growth comes about because of surplus herd growth, which is kept to ensure the subsistence of the stockman and his family after sales or exchanges have been made.

### 2.5.2 Rearing solely to build up capital

Farmers and people working in the tertiary sector may invest their savings in herds. The protection of their herds is entrusted to a stockman the investors are related to, or more generally to salaried or migrant share herders from pastoral ethnic groups who have left their original group.

### 2.5.3 Rearing for profit

The ultimate aim of this group is monetary gain in which techniques aim to obtain (at the best cost or least effort) animal products that are saleable at the best price.

### 2.5.4 Rearing or use of animals for animal traction

An unproductive animal may be maintained for as long as it retains its physical strength. Here, the animal is maintained for power rather than for production. Goats are not usually used for tillage or transport purposes due to their smaller size compared with other ruminant species.

## 2.6 Goat Management Systems

Goat management systems refer to the unique combination of different farming practices that are employed by different people depending on the area and type of animal involved.

### 2.6.1 Extensive system

Extensive goat management is probably the most common and popular method of managing goats. It is practised where the land is not immediately suitable for agricultural improvement or is too difficult or too costly to fence. Extensive management involves a minimum amount of labour and expense. Goat flocks tend to be relatively large due to easy access to cheap labour and relatively high returns from this system of management (Gall, 1981). Determining features are favourable climatic conditions, a short wet season, the availability of browse and few predators. Drier areas with low amounts of rainfall are more suited to goat production than cattle or sheep production as long as there are moderate temperatures. Goats prefer browse over pasture and will eat 50.5 g per kg body weight of shrubs per day (Rogosic *et al.*, 2006), so goats can be used to manage vegetative growth in overgrown or non-maintained pastures. Black locust trees and multiflora roses were found to be controlled when goats were grazed with cattle but not with cattle alone (Luginbuhl *et al.*, 2000). Predator control may be very important. Donkeys, llamas and guard dogs are used in different parts of the world where labour costs or terrain prohibit daily herding of goats by humans.

### 2.6.2 Intensive production

Kilgour and Dalton (1984) defined an intensive system as a system in which the farmer uses technology with discretion and animals are generally well managed based on good husbandry practices. In this system, goats are fed in confinement with limited access to land, for example browse is brought to them daily. It is, by definition, a system with high labour and cash input. In this system, cultivated grasses and/or by-products are fed *in situ*. Goats may be fed with cut grass with or without limited concentrates.

The behaviour of goats may be affected by the husbandry system, where lethargy and obesity may result from a lack of exercise or unbalanced diet and feet problems may arise through excessively long hooves (Kilgour and Dalton, 1984). However, these conditions can be limited by preventing inbreeding (Webb and Mamabolo, 2004). Intensive systems provide maximum protection from uncontrollable environmental factors and give complete control over the destructive aspects of the goats' feeding habits.

### 2.6.3 Tethering

Tethering is the practice where goats are confined or movement is controlled in order to prevent them from wandering and damaging neighbouring crops. In this system, goats are often pegged to rope about 3 m in length and by this they are forced to browse weeds or other undesirable plants. The disadvantage of this system is that water is only provided when goats are shifted to shelter at night. Very few or no concentrate salt or mineral licks are provided. It is only occasionally that supplements, household scraps, small quantities of grain or their by-products are given. There are some recommendations that regular changing of the tether will introduce goats to new pasture and that a running tether is preferable to a fixed one.

## 2.7 Farming Systems

### 2.7.1 Communal system

In this system, the relationship between communal goat farmers/owners and their animals is not only an economic one; social security is the main objective, not productivity (Maree and Casey, 1993). Livestock are not raised primarily for meat but as a way of capital savings and as an important source of milk and manure. Livestock are only slaughtered on special occasions such as weddings and festivals or when it is felt that they are about to die. Animals are also slaughtered for religious purposes. Normally, the land or grazing area used by the herd is communal, while individuals or one or more families own the livestock.

Livestock in many rural areas in the developing countries are regarded as wealth, and a man's social standing is measured by the size of his flock rather than money or other possessions (Poostchi, 1987). Many livestock herders use livestock for the payment of bride price (Poostchi, 1987; Maree and Casey, 1993). Here, the bridegroom's family is socially required to present animals to the bride's family before the marriage ceremony takes place. Large numbers of livestock are kept as a form of insurance against drought and famine.

The aim of most communal goat owners is an unlimited increase in the number of animals owned. Therefore, numbers and not productivity is the main objective. Each stockowner finds that he gains by maximizing the number of his animals, even though the result is deteriorating resources. Maree and Casey (1993) confirmed that each stockowner often appears to prefer to take the chance that his animals will survive, rather than reduce numbers in anticipation of deteriorating conditions.

Usually, stocking rate is not related to carrying capacity, and livestock productivity falls far below the genetic potential. The problem is that, where collective action to control stock numbers is socially and economically possible, collective effort is seldom individually attractive and hence will occur rarely unless there is deliberate intervention by external agencies. Voluntary agreement to reduce stock numbers is not in an individual's interest unless everyone has to do the same or there is local agreement or collective action. Malnutrition is the most important cause of low production rates and high mortality rates in communal systems. Often, not even the most patent and obvious need for supplementary feeding is adopted to prevent mortalities.

### 2.7.2 Communal goat production

In communal goat production systems, the focus is generally on free-ranging goats, grazing around the village, old cultivated fields or areas of regrowth or harvest (area ~ 0.5–2 ha). The goats are usually kept in pens at night. Mating is anarchic and may result in inbreeding. Animals seldom receive supplementary feeds and no records are kept of individual animals or animal performance.

## 2.8 Characteristics of Indigenous Goat Production Systems

Herd sizes vary between 2 and 120 animals and there are large variations in herd structure (buck, doe and kid ratios) (Webb *et al.*, 1998). Women and children mostly do the herding. In many cases, migrant labourers own the goats. The most important constraints of sustainable production systems include the land tenure system and poor resources, such as fencing, roads, electricity and water. It is important to get the consent of the local chief and cooperation of the extension officers before a survey is done in a communal farming area.

Bucks are often the most neglected animals in a herd, but they are expected to breed and produce offspring. Poor buck management directly affects the flock reproductive performance. Fertilization success depends on a range of factors including semen quality and mating behaviour. Testes and semen characteristics can be of great value in selecting males. Estimates of reproductive efficiency are required for goats studied in their natural habitat. Unfortunately, values obtained at experimental stations often depict 'potential' rather than 'actual' values. Some performance data of indigenous and cross-bred goats in different production systems are presented in Table 2.1.

As the fertility status of goats in communal areas is not well documented, goats were studied in Mpumalanga (Webb *et al.*, 1998). In this area, 82% of the land is entrusted to chiefs, while farmers own 18% of the land. All goats are penned at night, but the breeding management is poor. Bucks and does run together and mate as soon as puberty is reached. Mating is not controlled through the year.

### 2.8.1 Nomadic system

Nomadic systems are based on continual movement of livestock in search of grazing and water. Although nomadic systems are becoming less popular, they are more common in arid or semi-arid regions with very sparse human and animal population and where precipitation is uncertain. In other words, they are practised where the land is not immediately suitable for agricultural improvement or is too difficult or costly to fence. The harsh conditions there determine the type of animal that can be raised and the kind of movement that can be undertaken.

Nomads utilize the seasonal production of forages and the water available in different locations (El Aich and Waterhouse, 1999). The major concerns about using rangelands are overgrazing, use and management of communal property grazing lands, and drought feeding (Devendra, 2010). Forage quality and amount can vary greatly with location and among seasons. Livestock move through a series of pasture or forage systems with season. Use of water points without grazing control increases the grazing pressure on land that was previously used infrequently. Overgrazing results in rangeland degradation, and uncontrolled livestock grazing may be accelerated with drought, initiating desertification. The collective status and communal use of rangelands gives no incentives for individuals to control animal numbers or grazing pressure (El Aich and Waterhouse, 1999). The movement of nomads may also cause disagreements with settled farmers along territory borders.

The major part of the nomadic herder's food and income comes from the livestock they produce. Milk and other dairy products form the major part of their diets. Meat is used sparingly in the diet and the slaughter of animals represents a reduction in the rural or nomadic family's capital assets. Normally, the land or grazing area used by the herd is communal, while the livestock are owned by individuals or one or more families. Meanwhile, each owner attempts to keep as many animals as possible,

**Table 2.1.** Performance of indigenous and cross-bred (Thai native × Anglo-Nubian) goats raised under village and improved management systems (adapted from Milton *et al.*, 1991).

| | Village management | Improved management | |
|---|---|---|---|
| Trait | Indigenous | Indigenous | Cross-bred |
| Production (females) | | | |
| Birth weight (kg) | NA | 1.7 | 2.0 |
| Weight at: | | | |
| 3 months | 6.8 | 9.2 | 11.2 |
| 6 months | 10.0 | 12.4 | 16.1 |
| 12 months | 13.0 | 20.0 | 26.7 |
| 18 months | 17.3 | 24.1 | 32.4 |
| 24 months | 21.5 | 29.5 | 38.0 |
| Reproduction | | | |
| Kidding rate (%) | 190 | 161 | 171 |
| Pre-weaning kid mortality (%) | 29.1 | 5.0 | 6.3 |
| Annual adult mortality (%) | 7.2 | 4.7 | 1.7 |
| Body composition (males) | | | |
| No. goats | 10 | 23 | 12 |
| Dressing (%) | 45.1 | 45.7 | 45.2 |
| Saleable (%) | 70.9 | 71.4 | 71.9 |
| Muscle (%) | 70.7 | 68.4 | 66.6 |
| Bone (%) | 18.0 | 18.2 | 17.6 |
| Total fat (%) | 5.1 | 8.4 | 8.7 |
| Muscle:bone ratio | 4.0 | 3.8 | 3.8 |

NA, Not applicable.

regardless of their quality or the availability of pasture.

The low yield of grassland in arid and semi-arid areas necessitates nomadism, semi-nomadism or the development of a ranch system of farming. Poostchi (1987) found that variation in grassland utilization by livestock and the stability of the groups of people tending them included the following types of animal producers:

- Total nomadism: the owners of livestock do not have a permanent place of residence and do not practise regular crop production; their families move with the herd of livestock.
- Semi-nomadism: the livestock owners have a permanent place of residence near which supplementary crop production is practised. Farmers who live continually in a permanent settlement and who own herds, which remain in the vicinity of their place of residence, characterize partial nomadism.

The determining features in nomadism are similar to those of an extensive system – favourable climatic condition, a short wet season, the availability of grazing and browse, and few predators.

There are some disadvantages of the nomadic system:

- As illustrated by Nestel (1984), provision of veterinary services in these circumstances is a difficult task. Livestock owners or trusted members of tribal systems have to undertake first aid techniques and are depended on to report potentially serious problems. Planned programmes of vaccination and, where necessary, routine dipping or spraying of stock against ticks can be based on strategically located veterinary posts to which herds are brought on a voluntary or possibly obligatory basis when feed supplies permit.
- An essential for trade purposes in such regions is the development of a network

of stock routes with watering points and rest areas. Control posts can then be installed to ensure that animals being moved are appropriately protected against the most important disease problems they might carry or be exposed to en route. The aim is to reduce contact between groups in transit and with stock in the areas through which they pass.

#### 2.8.2 Transhumance and sedentary pastoralism

'Transhumance' is where a farmer moves his animals between two (or more) specific climatically different farms or areas according to a fixed seasonal pattern. In South Africa, livestock farmers used to have a 'summer farm' on the highveld and a 'winter farm' on the lowveld. This strategy solved the problems related to the decreasing nutritional value of the sourveld on the highveld in winter.

#### 2.8.3 Primitive herding

This system of agriculture represents a step forward and is an improvement on the system of gathering. Here, the product is the animal and the investment is labour that is needed to increase the supply of the product. Most primitive herding occurs in regions where shrubs, bunch grasses and short grasses grow and where humidity is low, rainfall unreliable and the climate arid. Goats constitute the major herds of livestock found in desert regions.

In the semi-arid and arid regions, vegetation is sparse and the amount of feed available for animals and their fodder is very limited, so the search for fodder is never-ending. The herdsmen leave their animals in one place to graze on any type of vegetation they can find until there is no more to graze; then they move to another place where the meagre supply of forage and water provides temporary feeding until the next move (Poostchi, 1987).

### 2.9 Problems Experienced by Small-scale Goat Farmers

The problems experienced by small farmers include: insecure and fragmented land rights, non-viable and small farm units, overstocking and deterioration of land and general lack of support infrastructure, water supplies, a transport network, financial support, and extension and support services. Legislative policy and institutional development have been inequitable and have aggravated the plight of the disadvantaged. These problems have contributed to low levels of production and underutilization of arable land resources, despite the relatively high agricultural potential of some of these areas. This was supported by the International Fund for Agricultural Development (IFAD, 1992), which stated that traditional 'African' land tenure systems lead to underutilization of high-potential land, thus putting pressure on marginal land, which then degrades rapidly. This has resulted in rural areas becoming more dependent on food imports. A study of commercial goat production operations in north-eastern Brazil showed that low-income families and families of average income produced the same meat product types, while cooperatives of small-size rural families produced meat of higher quality and price, and small rural families who partnered with slaughterhouses produced processed and higher-value products (Vidal and Dias, 2000).

#### 2.9.1 Infrastructure

The term 'infrastructure' in this context is usually taken to mean the existence of roads, electricity and water supplies in a district of the country. It can also be used in a more restricted sense to mean these resources at the farm level. For example, the development of an efficient grazing system is often limited by the restricted availability of water points or fencing. Farm planning should make effective use of the resources of water (e.g. rivers, dams or boreholes) to establish a suitable system of paddocks or camps, in order to allow the farmer to practise rotational grazing. The plan devised for the farm will have to be assessed to ensure that such infrastructural development is as economical as possible, and is cost-effective.

Infrastructure is also important for allowing easy access to markets for the purchase of inputs as well as for sale of livestock. An adequate infrastructure is also desirable for the well-being of the people involved in the farming. Facilities such as housing, sanitation, clean water supplies and schools can result in improved productivity of staff through better health and a stable workforce.

For the animal scientist or veterinarian, the most urgent infrastructural requirement on a farm is a well-designed set of animal-handling facilities.

### 2.9.2 People

The people who work with the animals within the farming system are often the main limitation on efficiency. The reason is usually ignorance, and this may occur at any level of management, from the owner-farmer to the stockman who works with the animals every day. In this aspect, an animal scientist, or other adviser, can have the greatest benefit, with dramatic improvements in health and productivity for very little input except time and effort to promote understanding.

The *educational* role is critically important in any technological intervention, in order that any improvements will be permanent and sustainable. The *sociology of development* and the educational processes of *extension* in a rural environment are a specialized area of study. In the end, however, people and their knowledge and skills should be our greatest resource (Donkin, 2005).

### 2.9.3 Social factors

#### *Social systems*

The social environment of any community is related to the cultural, educational and spiritual environment and conditions of the people. This is related to perceptions of our place in this life, our relationship with the environment and our relationships with other people. Change is a fact of life, even for rural people, but it is often a very slow process, and rural communities are usually ‘conservative’ by nature and resistant to change.

The reasons for this conservative approach to life are many. Some include:

- Perceptions of stability and security.
- The importance of human relationships, which are valued more than material things.
- The influence of ancestors, as well as the influence of those people who happen to be around at this time in history.
- Perceptions of economics.
- Political perceptions.
- Spiritual traditions.

Social systems must be respected, and people who are agents for development coming to the community from ‘outside’ the system need first to begin to understand the people (without arrogance) and how their social systems function. Any change will be resisted unless it is perceived to be beneficial, and unless it is perceived to be an answer to a ‘felt need’. Ultimately, the people themselves will have to make changes that might be beneficial, and they can often be assisted to do this in a controlled and planned way.

#### *Land tenure*

In Africa and many other parts of the world, traditional social systems do not include the concept of private ownership of land. This concept of land ownership is taken as a fundamental tenet in Western societies and ensures that the owner not only benefits from ownership of the land but also is directly responsible for caring for the land. In traditional societies, the land may be owned by the tribe as a whole and administered by the traditional leaders. However, in such a case, the communal ownership of land is a paradox when compared with the individual ownership of livestock. Because grazing is perceived to be 'free' and without cost, animal numbers increase to well above the sustainable carrying capacity. Deterioration of the vegetation and poor productivity of the animals are almost inevitable.

A distinction should be made between different concepts relating to communal grazing:

1. Open-access grazing. This refers to communal grazing where there is no control over animal numbers or system of usage. The environment becomes degraded.
2. Communal grazing with controls. It is possible to have communal grazing where the community agrees among themselves to limit the number of animals, and to ensure that some form of rotational grazing is practised.

Nevertheless, the communal form of land tenure is often perceived to be a major limitation in efforts to improve productivity of the natural vegetation and the animals using it as a source of feed. Land tenure is a highly sensitive political issue (Bembridge, 1987).

### Economics

Economics has sometimes been classified as a human behavioural science. It is therefore an important social factor. Money is usually the main measure of economic activity or the value placed on goods or services. In some countries, the government may have considerable control over agricultural activities through legislation. This control can bring stability in pricing and production but at the cost of reduced efficiency and reduced growth. The concept of market economics is in favour more recently and is supposed to bring greater efficiency. However, these concepts may be less applicable in terms of farming systems, because of two main factors:

1. Many farming developments are long-term by their very nature (e.g. developing a breeding herd).
2. The price of land may bear little relation to its productive potential.

Therefore, the short-term economic forces that apply to a farmer may have effects that are detrimental to long-term sustained productivity, especially to the natural vegetation.

In addition, particularly at the level of the subsistence or small-scale farmer, the usual economic criteria may not be applicable. In the context of 'household economics', the allocation of time to work or obligatory social activities may be more important than the economic (monetary) factors usually considered (Low, 1986).

### Marketing

For a farmer to enter the commercial environment, a market of some sort is required. This may depend on infrastructure (e.g. roads, communications) and on a distribution network for the products. There is no value in producing things for which there is no market (beyond subsistence requirements). Marketing is a skilled profession, and farmers may have little aptitude or experience in this aspect, as they are normally fully committed to production activities. The development of profitable marketing channels is critically important for the sustainability of farming systems. Such channels are often lacking in developing areas, contributing to overpopulation of livestock, deterioration of the environment and continuing poverty. Cattle in these areas are usually kept for reasons other than commercial gain, such as for social customs, religious ceremonies or as a store of wealth. Cattle are usually owned by the wealthier members of society, and goats are therefore more likely to be associated with those of lower social rank, who have few resources.

## 2.10 Production Systems and Natural Resources

Natural resources consist primarily of soil, climate, vegetation and animals. The land available for use may be limited in an area or in its productive capacity. Farming systems are influenced by the soil in a particular area, and soil fertility depends mainly on the geological formations from which the soil was derived. Other soil factors that need to be considered include soil type, pH and erosion.

The climate of a region or farm may vary greatly in terms of rainfall, environmental temperature and humidity. The quantity, frequency and reliability of rainfall are extremely important in planning and managing a production system, because these factors influence the type and quality of vegetation. Water supplies are important not only for the growth of natural vegetation but also as drinking water for animals. In some areas, the availability of water from boreholes or water from riverbeds for use in water points is a limiting factor.

Vegetation influences the growth, development and production characteristics of goats (Table 2.2). The available vegetation can be subdivided in two categories, natural vegetation and developed vegetation. Natural vegetation varies greatly between regions and even on different parts of the same farm, depending on soil, climate and topography. Developed vegetation includes planted forests, crops and pastures. Developed vegetation can provide a significant resource that will influence the type of production system. However, if the natural vegetation is replaced, this must be justified by improvements in productivity or improvements of the environment that will warrant the expenditure.

In general, goats will prefer browse over grasses, while body weight gains will be higher with more concentrated diets. However, goats will also usually have a higher proportion of fat in their body composition with increased energy intakes. The consumption of forages was found to be positively related to the forage dry matter content (Bateman *et al.*, 2004). The average daily gain of Spanish kid goats was higher when fed chopped lucerne compared with mixed grass hay, with no influence on carcass characteristics due to diet or sex (Wildeus *et al.*, 2007). Grazing studies with Boer × Spanish does on mixes of legumes, redroot pigweed and crabgrass grown in full sun or in pecan tree shade indicated that there was competition for moisture between the herbaceous forages and trees in periods of low rainfall, and forbs tended to survive longer in full sun mixed forages than in tree-shaded mixes, while grasses increased in the pastures over time (Goodwin *et al.*, 2002). The average daily gains of buck kid goats was higher on oak browse and rice straw than on pine browse or fermented pine browse due to the higher daily forage intakes associated with the types of browse giving the highest weight gains (Choi *et al.*, 2006). Goats prefer some tropical legume forages more than others, which affects their intake and thus body weight gain (Kanani *et al.*, 2006).

Forages often lack sufficient protein for growth. Protein supplementation of hay fed ad libitum increased total dry matter intake, decreased feed conversion, increased growth rate and gave higher proportions of lean meat percentages (Mtenga and Kitaly, 1990). Dry matter intake was similar, but average daily gain was greater for 13.8 and 22.1% crude protein in consumed dry matter than for 9.3% crude protein, although the benefit from more than 14% crude protein was minimal (Prieto *et al.*, 2000). Boer × Spanish kid goats preferred pelleted supplements over meal or liquid supplements, and supplements based on maize or soybean meal over those based on molasses or fishmeal. Consumption of dry matter was higher with stored forages than with fresh forages, with fresh cereal grain forages preferred over clover or *Brassica* spp. forages (Bateman *et al.*, 2004). The growth rate of indigenous Greek goat kids was higher when treated with an anthelmintic or supplemented with dietary protein, which would provide nutritional protection for the body against parasites (Arsenos *et al.*, 2009).

Goats kids raised in an intensive system (concentrates and enclosed area) had heavier slaughter weights than those on semi-intensive management systems (Bahia grass, millet, oats, crimson clover, browse and forb pastures), but carcass composition was not affected by the diet/management system (Johnson and McGowan, 1998). The

**Table 2.2.** Average daily gain (g) of kid meat goats with different forages and diet supplements.

| Kid goat type | Dietary treatment | | | | |
|---|---|---|---|---|---|
| | Millet pasture | Bermuda grass pasture | Bermuda grass pasture + 0.23 kg maize/head/day | Bermuda grass pasture + 0.45 kg maize/ head/day | |
| 4-month-old Boer × Spanish wethers[a] | 35 | 37 | 42 | 57 | |
| 4-month-old Spanish wethers[a] | 35 | 37 | 35 | 40 | |
| | Complete feed supplement | Intake-limited supplement | Sorghum-Sudan hay (SS) | SS with supplement | SS with intake-limited supplement |
| 5-month-old Boer ↔ Spanish does[b] | 152 | 123 | 22 | 181 | 172 |
| | Concentrate (40% protein pellets, 40% soybean hulls, 20% Bermuda grass hay) | Bahia grass pasture with 150 g/head/day protein pellets | Mimosa browse with 100 g/head/day cracked maize | | |
| Boer cross wether kid goat[c] | 117 | 49 | 90 | | |
| | 50% concentrate | 70% concentrate | 90% concentrate | | |
| Boer cross goats[d] | 97 | 103 | 90 | | |
| | Lucerne hay | 18% CP concentrate | Hay for first 45 days, then concentrate diet | | |
| 4-month-old Boer × Spanish intact males[e] | 41 | 134 | 65 | | |

[a]105-day trial (Nuti *et al.*, 2000).
[b]63-day trial (Payne *et al.*, 2006).
[c]14-week trial (Solaiman *et al.*, 2006).
[d]126-day trial (Ryan *et al.*, 2007).
[e]90-day feeding trial (Lee *et al.*, 2008).

average daily gain and feed efficiencies of goat kids were higher with increased concentrates (maize and barley) than with increased forages (lucerne hay) (Haddad, 2005), but the degree of carcass fatness with relative dietary forage:concentrate ratios was not given. Goats fed ad libitum on concentrate pellet diets and Rhodes grass hay in a feedlot grew rapidly and had decreased muscle:fat ratios with increased weight (Mahgoub *et al.*, 2005).

Boer crossbreeds had higher post-weaning average daily gains and dry matter intake than Spanish goats consuming a concentrate diet (Cameron *et al.*, 2001). Concentrates at 50% of the diet (the remainder was lucerne hay) had improved weight gain, while dry matter intake was less in Boer crossbred kid goats fed in confinement than with 70 and 90% levels of dry-rolled maize as the concentrate (Corrigan *et al.*, 2008). Feeding intensively (concentrate and leaves) or weaning at 3 months increased the weight gain and costs of the weight gain in kid goats compared with a semi-intensive system (same diet plus 8 h grazing daily) or weaning at 2 months of age (Nagpal *et al.*, 1995). Grainless diets varying in concentrate:roughage ratios under feedlot conditions affected average daily gain, but carcass traits were similar except for the dressing percentage (Sebsibe *et al.*, 2007).

Meat goats fed high energy levels have increased juiciness, tenderness and texture of goat meat, but the fat content is also increased, resulting in lower consumer acceptability of the meat than if their diet contains higher amounts of roughage (McMillin and Brock, 2005). Meat from goats grazed and fed a commercial pellet was found to be more tender and juicy than from goats fed hay and a commercial pellet feed (Carlucci *et al.*, 1998).

There are many different breeds of goats, as discussed earlier in the chapter. Some of these breeds belong to *defined* breeds, where the owners are members of breed societies, which keep records and promote the interests of the breed and the owners. However, in many parts of the world, no such organizational structures exist, and the concept of 'breed' is therefore less closely defined. Nevertheless, distinct types of goat can be identified, and there are clear differences between the types, which are usually adapted to specific areas (Table 2.3). Registered breeds usually have distinctive features (e.g. coat colour), agreed upon by the members of the breed societies, and goats not meeting these standards are not accepted. The appearance of a breed is important to establish the 'brand' of the breed, which distinguishes it from other breeds. Some breed societies specify minimum production characteristics in terms of performance, and this should be a desirable characteristic for any breed. In this way, the breed may be 'developed' to become very efficient at production. For example, the recognized dairy breeds are efficient milk producers. However, 'indigenous' types of animals may never have been consciously selected by their owners but can still have valuable characteristics such as disease resistance, which will have been established in the breed by 'survival of the fittest' over a long period of time, perhaps hundreds of years.

**Table 2.3.** Growth and gain efficiency of kid goats of different meat-goat breeds.

| | Initial growth phase | | Final growth phase | |
|---|---|---|---|---|
| Breed[a] | ADG (g/day)[b] | Efficiency[c] | ADG (g/day)[b] | Efficiency[c] |
| Alpine | 68 | 0.10 | 59 | 0.08 |
| Angora | 72 | 0.15 | 50 | 0.10 |
| Boer | 91 | 0.13 | 64 | 0.08 |
| Spanish | 62 | 0.11 | 22 | 0.04 |

[a]Wether kid goats (4 months old) were fed 75% concentrate for two consecutive 12-week periods (Urge *et al.*, 2004).
[b]ADG, Average daily gain.
[c]Efficiency was defined as weight gain per weight of feed.

Goats have adapted to different tropical climates. Goats in arid and semi-arid climates tend to be larger in size (30–50 kg) with long legs and ears and increased mobilization of fat during periods of feed shortage, and will browse over long distances due to resistance to dehydration and faecal desiccation. Goats adapted to subtropical environments are intermediate in size and have a lower water turnover. The adaptations of goats to humid and subhumid climates have resulted in small or dwarf sizes (10–25 kg) with reduced panting and evaporative cooling, low metabolic rates and reduced walking ability due to increased forage availability (Devendra, 1987).

Reproduction is among the more important aspects of goat production for meat because the small animal size results in minimal meat per animal. Profitability relies on having numerous kid goats born and raised so that marketing efficiencies and economies are obtained.

The number of young produced during the lifetime of a breeding female is of major economic importance in Africa, as indicated by Wilson (1992). The productivity of small ruminants during their lifetime can be improved by lowering the age of first parturition and by increasing litter size provided that parturition intervals are less than a year. There is a time in the life of the animal when the advantage of producing more progeny becomes an inconvenience. Parturition intervals become longer with age and litter size remains stable or is reduced. It then becomes profitable to cull the aged animals to enable a larger number of young animals to be kept in the flock.

Other aspects that need to be considered in any production system include diseases that affect low stock numbers, internal parasites, external parasites, poisonous plants and predators that occur in the region. It is important to be aware of these and to incorporate this knowledge in all management decisions. Many diseases can be controlled with modern medications and vaccines, while the use of disease-resistant or disease-tolerant animals also provides a way to ensure a high level of productivity. Other options include the strategic and scientific use of anthelmintics and management programmes such as the FAMACHA system for control of internal parasites (Bath *et al.*, 2001).

Within a goat herd, a clearly established and stable hierarchy order develops and is maintained, and the dominant and subordinate animals become more selective with available forages, while the differences in feeding become more general in nature when forages are in shorter supply. Surprisingly, the middle range of does in the hierarchy are more productive in terms of milk production, number of kids, kid birth weight and pre-weaning weights (Barroso *et al.*, 2000).

Most farming systems impose constraints on the behaviour of the livestock being farmed, but an understanding of the animal by the farmer can reduce stress and improve productivity. The following should be encouraged: evaluation of stocking density relative to animal age, size and class; separation of sexes except during breeding; prevention of isolation of individuals; avoidance of regrouping; sufficient feed trough space; minimization of human manipulation of newborn kid goats; not weaning until kid goats are 6–7 weeks of age; allowing older and more experienced goats to become herd leaders in extensive grazing systems; and daily contact between humans and goats (Miranda-de la Lama and Mattiello, 2010).

## References

Alexandre, G. and Mandonnet, N. (2005) Goat meat production in harsh environments. *Small Ruminant Research* 60, 53–66.

Arsenos, G., Fortomaris, P., Papadopoulos, E., Sotiraki, S., Stamataris, C. and Zygoyiannis, D. (2009) Growth and meat quality of kids of indigenous Greek goats (*Capra prisca*) as influenced by dietary protein and gastrointestinal nematode challenge. *Meat Science* 82, 317–323.

Barroso, F.G., Alados, C.L. and Boza. J. (2000) Social hierarchy in the domestic goat: effect on food habits and production. *Applied Animal Behaviour Science* 69, 35–53.

Basset, T.J. and Crummey, D.E. (1993) *Land African Agrarian Systems*. University of Wisconsin Press, Madison.

Bateman, H.G., White, T.W., Williams, C.C. and Alford, S. (2004) Case study: goat preference for concentrates or forages is influenced by physical and chemical characteristics of the feed. *Professional Animal Scientist* 20, 198–204.

Bath, G.F., Hansen, J.W., Krecek, R.C., van Wyk, J.A. and Vatta, A.F. (2001) *Sustainable Approaches for Managing Haemonchosis in Sheep and Goats*. FAO Animal Production and Health Paper, Final Report of FAO Technical Co-operation Project No. TCP/SAF/8821(A).

Bembridge, T.J. (1987) Aspects of cattle production in Transkei. *South African Journal of Animal Science* 17, 74–78.

Braker, M.J.E., Udo, H.M.J. and Webb, E.C. (2002) Impact of intervention objectives in goat production within subsistence farming systems in South Africa. *South African Journal of Animal Science* 32, 185–191.

Cameron, M.R., Luo, J., Sahlu, T., Hart, S.P., Coleman, S.W. and Goetsch, A.L. (2001) Growth and slaughter traits of Boer × Spanish, Boer × Angora, and Spanish goats consuming a concentrate-based diet. *Journal of Animal Science* 79, 1423–1430.

Carlucci, A., Girolami, A., Napolitano, F. and Monteleone, E. (1998) Sensory evaluation of young goat meat. *Meat Science* 50, 131–136.

Casey, N.H. and Webb. E.C. (2010) Managing goat production for meat quality. *Small Ruminant Research* 89, 218–224.

Casey, N.H., Van Niekerk, W.A. and Webb, E.C. (2003) Goat meat. In: Caballero, B., Trugo, L. and Finglass, P. (eds) *Encyclopaedia of Food Sciences and Nutrition*. Academic Press, London, pp. 2937–2944.

Choi, S.H., Choy, Y.H., Kim, Y.K. and Hur, S.N. (2006) Effects of feeding browses on growth and meat quality of Korean black goats. *Small Ruminant Research* 65, 193–199.

Corrigan, M.E., Drouillard, J.S., Loe, E.R., Depenbusch, B.E. and Quinn, M.J. (2008) Effects of concentrate level and pen configuration on performance of Boer crossbred goat kids. *Professional Animal Scientist* 24, 614–618.

Devendra, C. (1987) Herbivores in the arid and wet tropics. In: Hacker, J.B. and Ternouth, J.H. (eds) *The Nutrition of Herbivores. Proceedings of the 2nd International Symposium on the Nutrition of Herbivores*. Academic Press, New South Wales, Australia, pp. 23–46.

Devendra, C. (2010) Concluding synthesis and the future for sustainable goat production. *Small Ruminant Research* 89, 125–130.

Donkin, E.F. (2005) Sustainable livestock development in Africa: how do we help Africa to feed itself? *South African Journal of Animal Science* 6, 56–67.

Donkin, E.F., Boyazoglu, P.A., Els, H.C., MacGregor, R.G., Ramsay, K.A. and Lubout, P.C. (1996) Productivity of Saanen, South African indigenous and crossbred goats fed a complete feed: preliminary results. In: *Proceedings of the VI International Conference on Goats*, Beijing, May 1996. Volume 1, pp. 132–135.

Du Plessis, I., Van der Waal, C. and Webb, E.C. (2004) A comparison of plant form and browsing height selection of four small stock breeds – preliminary results. *South African Journal of Animal Science* 24(1), 31–34.

El Aich, A. and Waterhouse, A. (1999) Small ruminants in environmental conservation. *Small Ruminant Research* 34, 271–287.

FAO (1999) *Agricultural Statistics*. Food and Agriculture Organization, Rome, Italy.

Galal, S. (2005) Biodiversity in goats. *Small Ruminant Research* 60, 75–81.

Gall, C. (1981) *Goat Production*. Academic Press, San Diego.

Goodwin, D.J., Muir, J.P. and Wittie. R.D. (2002) *Goat Performance, Forage Selectivity and Forage Quality Dynamics in Three Cultivated Warm Season Pastures in North-central Texas*. Sheep and Goat, Wool and Mohair CPR, Texas A&M University, pp. 90–98.

Haddad, S.G. (2005) Effect of dietary forage:concentrate ratio on growth performance and carcass characteristics of growing Baladi kids. *Small Ruminant Research* 57, 43–49.

IFAD (1992) *Soil and Water Conservation in Sub-Saharan Africa – towards Sustainable Production by the Rural Poor*. International Fund for Agricultural Development, Rome, Italy.

Johnson, D.D. and McGowan. C.H. (1998) Diet/management effects on carcass attributes and meat quality of young goats. *Small Ruminant Research* 28, 93–98.

Kanani, J., Lukefahr, S.D. and Stanko. R.L. (2006) Evaluation of tropical forage legumes (*Medicago sativa, Dolichos lablab, Leucaena leucocephala* and *Desmanthus bicornutus*) for growing goats. *Small Ruminant Research* 65, 1–7.

Kilgour, R. and Dalton, C. (1984) *Livestock Behaviour – a Practical Guide.* Granada, London.

Lebbie, S.H.B. (2004) Goats under household conditions. *Small Ruminant Research* 51, 131–136.

Lee, J.H., Kouakou, B. and Kannan. G. (2008) Chemical composition and quality characteristics of chevon from goats fed three different post-weaning diets. *Small Ruminant Research* 75, 177–184.

Low, A. (1986) *Agricultural Development in Southern Africa: Farm-household Economics and the Food Crisis.* James Currey, London.

Luginbuhl, J.-M., Green, J.T. Jr, Poore, M.H. and Conrad, A.P. (2000) Use of goats to manage vegetation in cattle pastures in the Appalachian region of North Carolina. *Sheep and Goat Research Journal* 16, 124–135.

Mahgoub, O., Kadim, I.T., Al-Saqry, N.M. and Al-Busaidi, R.M. (2005) Potential of Omani Jebel Akhdar goat for meat production under feedlot conditions. *Small Ruminant Research* 56, 223–230.

Maree, C. and Casey, N.H. (1993) *Livestock Production Systems: Principles and Practice.* Agricultural Development Foundation, Brooklyn, Pretoria, South Africa.

Masika, P.J. and Mafu, J.V. (2004) Aspects of goat farming in the communal farming systems of the central Eastern Cape, South Africa. *Small Ruminant Research* 52, 161–164.

McMillin, K.W. and Brock, A.P. (2005) Production practices and processing for value-added goat meat. *Journal of Animal Science* 83 (Suppl.), E57–E68.

Milton, J.T.B., Saithanoo, S. and Praditrungwatana, P. (1991) Goat management in the Asian humid tropics. In: Saithanoo, S. and Norton, B.W. (eds) *Goat Production in the Asian Humid Tropics.* Proceedings of an international seminar held at Prince of Songkla University, Hat Yai, Thailand.

Miranda-de la Lama, G.C. and Mattiello, S. (2010) The importance of social behaviour for goat welfare in livestock farming. *Small Ruminant Research* 90, 1–10.

Mtenga, L.A. and Kitaly, A.J. (1990) Growth performance and carcass characteristics of Tanzanian goats fed *Chloris gayana* hay with different levels of protein supplement. *Small Ruminant Research* 3, 1–8.

Nagpal, A.K., Singh, D., Prasad, V.S.S. and Jain, P.C. (1995) Effect of weaning age and feeding system on growth performance and carcass traits of male kids in three breeds in India. *Small Ruminant Research* 17, 45–50.

Nestel, B. (1984) *Development of Animal Production Systems.* World Animal Science A2. Elsevier, New York.

Nuti, L., Pinkerton, F. and McMillin, K.W. (2000) Experts study benefit of corn supplement on pastured wethers. *Goat Rancher* September, 20–21.

Payne, B., Crenwelge, J., Lambert, B.D. and Muir, J.P. (2006) A self-limiting complete feed changes forage intake and animal performance of growing meat goats. *South African Journal of Animal Science* 36, 257–260.

Poostchi, I. (1987) *Rural Development and the Developing Countries.* Alger Press Limited, Oshawa.

Prieto, I., Goetsch, A.L., Banskalieva, V., Cameron, M., Puchala, R., Sahlu, T., Dawson, L.J. and Coleman, S.W. (2000) Effects of dietary protein concentration on postweaning growth of Boer crossbred and Spanish goat wethers. *Journal of Animal Science* 78, 2275–2281.

Rogosic, J., Pfister, J.A., Provenza, F.D. and Grbesa, D. (2006) Sheep and goat preference for and nutritional value of Mediterranean maquis shrubs. *Small Ruminant Research* 64, 169–179.

Ryan, S.M., Unruh, J.A., Corrigan, M.E., Drouillard, J.S. and Seyfert, M. (2007) Effect of concentrate level on carcass traits of Boer crossbred goats. *Small Ruminant Research* 73, 67–76.

Sebsibe, A., Casey, N.H., van Niekerk, W.A., Tegegne, A. and Coertze, R.J. (2007) Growth performance and carcass characteristics of three Ethiopian goat breeds fed grainless diets varying in concentrate to roughage ratios. *South African Journal of Animal Science* 37, 221–232.

Shrestha, J.N.B. and Fahmy, M.H. (2005) Breeding goats for meat production: a review. 1. Genetic resources, management and breed evaluation. *Small Ruminant Research* 58, 93–106.

Solaiman, S.G., Shoemaker, C.E. and D'Andrea, G.H. (2006) The effect of high dietary Cu on health, growth performance, and Cu status in young goats. *Small Ruminant Research* 66, 85–91.

Urge, M., Merkel, R.C., Sahlu, T., Animut, G. and Goetsch, A.L. (2004) Growth performance by Alpine, Angora, Boer and Spanish wether goats consuming 50 or 75% concentrate diets. *Small Ruminant Research* 55, 149–158.

Vidal, D.D.L. and Dias, R.P. (2000) Caprine commercialization systems and carcass sections in the northeast of Brazil. In: *Proceedings of the VIIth International Conference on Goats*, pp. 501–503.

Webb, E.C. and Mamabolo, M.J. (2004) Production and reproduction characteristics of South African indigenous goats in communal farming systems. *South African Journal of Animal Science* 34, 236–239.

Webb, E.C., Mamabolo, M.J., Du Preez, E.R. and Morris, S.D. (1998) Reproductive status of goats in communal systems in South Africa. In: Webb, E.C., Cronjé, P.B. and Donkin, E.F. (eds) *Research and Training Strategies for Goat Production Systems in South Africa.* University of Pretoria, South Africa, pp. 79–85.

Webb, E.C., Casey, N.H. and Simela, L. (2005) Goat meat quality. *Small Ruminant Research* 60, 153–166.

Werdi Pratiwi, N.M.W., Murray, P.J. and Taylor, D.G. (2007) Feral goats in Australia: a study on the quality and nutritive value of their meat. *Meat Science* 75, 168–177.

Wildeus, S., Luginbuhl, J.-M., Turner, K.E., Nutall, Y.L. and Collins, J.R. (2007) Growth and carcass characteristics in goat kids fed grass- and alfalfa-hay-based diets with limited concentrate supplementation. *Sheep and Goat Research Journal* 22, 15–19.

Wilson, R.T. (1992) Goat meat production and research in Africa and Latin America. In: *Proceedings of the Vth International Goat Conference*, New Delhi, India, pp. 458–472.

# 3 Carcass Traits of Hardy Tropical Goats

G. Alexandre[1] and O. Mahgoub[2]
[1]INRA, Unité de Recherches Zootechniques, Centre Antilles-Guyane, Guadeloupe, France; [2]Department of Animal & Veterinary Sciences, College of Agricultural & Marine Sciences, Sultan Qaboos University, Sultanate of Oman

## 3.1 Abstract

The purpose of this chapter is to review the fitness and meat production potential of hardy indigenous goats compared with modern single-purpose breeds, which are widely used in the developing tropics and need modern methods of husbandry. Meat production is discussed based on the reproductive performance of goats. The major variations in carcass traits between diverse breeds and husbandry conditions are reviewed. Case studies are presented based on the Creole goat of the Caribbean and Dhofari breed of Oman in order to support the views presented in the chapter.

## 3.2 General Considerations

The developmental pathway for tropical livestock production for many years was to upgrade or completely replace indigenous livestock with exotic breeds of supposedly higher genetic merit. Indications that this might not be the most appropriate approach were largely ignored (Wilson, 2009). Modern single-purpose breeds and modern methods of husbandry for small ruminant production have penetrated remote parts of the developing tropics. Cases of failure or success were reviewed (Kosgey *et al.*, 2006). Reasons for failure to establish highly productive enterprises in the tropics with exotic livestock include: high import costs, heavy mortality, poor fertility, reduced appetite due to high temperature and humidity, low-quality pasture, susceptibility to internal and external parasites and inadequate management skills (Ogink, 1993; Ahuya *et al.*, 2005). In an extensive review, Wilson (2009) observed that recently there has been a reverse movement towards using locally developed livestock species and breeds. These are now seen as pools of irreplaceable genetic material of unacknowledged merit and value that must not be lost but must be conserved for possible future use. The Food and Agriculture Organization (FAO) has acknowledged that there is renewed interest in native breeds, particularly to increase sustainable animal production in developing countries (FAO, 2007). Since the very first research of Rodriguez and Preston (1997), the significance of tropical indigenous breeds as valuable genetic resources has become increasingly recognized, particularly for rearing goats under harsh conditions (Ogink, 1993; Alexandre and Mandonnet, 2005; Omondi *et al.*, 2008).

In light of claims of failures in exotic breed enterprises under tropical conditions, it may be argued that production efficiency should be evaluated for whole integrated systems rather than for individual animals (Alexandre *et al.*, 2010). In that respect several elements must be taken into account:

- Many native tropical breeds, particularly goats, are not only dual-pupose but multi-purpose (Wilson, 2009). Generally, goats are defined as multifunctional animals. For instance, Peacock (1996) has listed at least 20 useful products and services from goats.
- The capacity of livestock of the tropics to utilize low-quality feed and convert it to high-quality protein for human use is essential to enable farmers to make more efficient use of available local feeds (e.g. local roughages or non-conventional feeds). This was highly recommended by Preston and Leng (1987), who called for matching livestock systems with available resources, which is even more relevant today.
- Food production systems in developing countries are moving into marginal areas for which sustainable farming systems and the most adapted livestock species and breeds have yet to emerge. These must be developed while paying due attention to the environmental, economic and equity aspects of ecosystems. Wilson (2009) noted that, in the tropics, the right animal is not always the one that produces the most.
- Moreover, preservation of biological diversity not only determines survival (Alexandre and Mandonnet, 2005; Kosgey and Okeyo, 2007) but also determines adaptation to changing environments (Hoffmann, 2010), including changes in consumer preferences (Glowatzki-Mullis *et al.*, 2008).

## 3.3 Meat Production Issues

There are still some researchers (e.g. Shrestha and Fahmy, 2007a) who support the concept of using exotic breeds within crossbreeding programmes in the tropics to promote high-yielding systems and make such programmes economically viable for local producers. Particularly when considering rapid growth and heavy carcasses, the question that frequently arises is the importance of body size (Andersen, 1978; Dickerson, 1978; Ogink, 1993). Tropical animal production systems are characterized by a large variety of breeds that differ widely in mature size combined with variable farming systems (Ogink, 1993; Wilson, 2009). To overcome what is considered as poor meat production potential, attempts have been made to increase low output levels by introducing breeds of larger size. This was based on the belief that exotic breeds are superior in terms of high yields and growth rates. The latter approach overlooks performance-limiting effects of a variety of environmental factors on specialized exotic breeds (Ogink, 1993). Over the past 70–80 years, there has been controversy about the size of animals. For example in sheep, Dickerson (1978) reported that larger or smaller body size may have important biological advantages for adaptation to climate, feed resources, predators and diseases, maternal and paternal use in crossbreeding schemes, and marketing. In hot, dry climates with sparse seasonal grazing, the genetically smaller individuals within a species presumably are better able to forage and reach final market weights and reproduce earlier than larger ones. In cattle, Andersen (1978) argued that larger genotypes normally have the highest production capacity for milk and beef but also the highest maintenance requirements. He concluded that the optimal size of cattle will depend on market demands and on climatic conditions and husbandry systems. Many researchers (e.g. Andersen, 1978; Dickerson, 1978; Ogink, 1993) were influenced by the rule of Bergmann and Allen, which states that 'in warm-blooded animals, races from warm regions are smaller than races from cold regions'.

With special reference to goats, Ogink (1993) developed one of the first extended studies on genetic size and growth. Many hypotheses on meat animals state that the variation between animals can be explained to a large extent by differences in scale, expressed by genetic size. However, while questioning the genetic size scaling theory, Ogink (1993) noted that introducing larger-sized breeds as such did not improve the biological efficiency of production. In fact, improvement of production efficiency should involve adjustment of size of the breeding female to suit a particular environment. For beef production, Dickerson (1970) noted that the primary focus should be on improvement of genetically variable functional components of performance/reproductive rate, growth rate (relative to metabolic weight) and body composition in meat animals. Dickerson (1978) recommended choosing a mature body size that is more adapted to the environment, suitable breeding systems and market factors for the species and the area of production. Dickerson (1978) concluded that 'each biological type should be evaluated under the management–marketing conditions for which it is best suited.' The qualities required in the right animal in the right place, if it is to succeed, are adaptation to local physical, nutritional and management environments (Wilson, 2009). In an analysis of growth patterns in indigenous Kambing Katjang goats and its crosses (F1, F2 and backcross) with the German Fawn goat, Tsukahara *et al.* (2008) found that backcrossed goats performed better than the pure German breeds. However, the authors stated that the fast maturing rate of backcrossed goats should be matched with the environment, which is only possible under intensive management. In smallholder farms in developing countries where management is suboptimal, backcrossed goats might be poorer in growth performance than the other crosses. Consequently, high growth rate or a heavy carcass (i.e. large body size) must not be considered as the most important trait for meat-producing animals in the tropics, as adaptive traits or fitness (*sensu lato*) characters are of paramount importance for tropical livestock (Menendez-Buxadera and Mandonnet, 2006; Wilson, 2009).

For meat output, muscle and fat partitioning must be taken into account. For instance, in a study comparing Texel and Scottish Blackface (SB) lambs of two contrasting body sizes (Lambe *et al.*, 2007), it was concluded that slaughtering SB lambs at lighter weights than Texel ones was more beneficial, because the SB lambs would not become over-fat and muscle proportions would be maintained at a higher level, although total muscle and carcass weights would be lower. SB lambs at lighter weights carry a higher proportion of total muscle in the economically more important leg region and have rounder muscle shapes in the hind leg and loin, whereas heavier lambs will deposit increasing proportions of muscle in the lower-priced thoracic area and will have a flatter muscle shape in the more valuable carcass regions.

In breeding programmes, especially in small ruminants, exotic breeds usually perform well during the active periods of projects (Ayalew *et al.*, 2003) where external supervision is available (Ahuya *et al.*, 2005). Success is usually measured in terms of increases in animal performances or in household income. There is, however, hardly ever a full cost–benefit analysis that takes into account the opportunity cost of labour and loss of land that is transformed from food to feed production. For instance, raising animals with a larger body size results in higher feeding requirements needing a larger area of land, i.e. more feeding resources. Once direct support in breeding programmes ends, it is only a matter of time before the exotic breeds disappear and are replaced by indigenous breeds (Ahuya *et al.*, 2005), sometimes with a high risk of loss of biodiversity (Ayalew *et al.*, 2003; Alexandre *et al.*, 2009). Therefore, the question that arises is how to promote the use of local breeds for meat production versus importing breeds of larger size to the tropics. This chapter aims to discuss these issues pertinent to meat production from goats in the tropics.

## 3.4 Fitness: Adaptation and Reproduction

Meat output is a product of complex traits and is dependent on numerous abiotic, biotic and socio-economic factors

(Alexandre *et al.*, 2010). One of the most prominent attributes of goats as meat producers is their high reproductive capacity, specifically under harsh conditions (Devendra and Burns, 1983; Bosman *et al.*, 1997). Another important attribute is 'fitness'. Menendez-Buxadera and Mandonnet (2006) studying genotype–environment interactions under tropical conditions developed the concept of fitness defined as follows:

> In natural conditions, the best-adapted animals will be those that utilize the available resources in the most efficient way to face the adverse environmental conditions. As all living organisms require energy to carry out their vital cycle, the surviving individuals will be those that have a metabolism more specialized in functions of resistance, adaptation, and therefore, will be the best prepared and active during the mating season. Consequently, such adapted individuals will be the greater contributors of genes to the next generation (with higher quantity of offspring produced). This property is called fitness and is the basis of the population evolution process. Fitness is the result of a group of functions and physiological properties that cumulatively offer a specific genotype a higher adaptive functional component.

According to Wilson (2009), many fitness characters of tropical livestock are poorly developed or are absent in temperate breeds These include a lower metabolic rate that generates less heat, reduced panting but more readily sweating to conserve energy, a feed intake that is less affected by high temperatures, a higher intake of poor-quality feed, higher digestibility and efficiency of feed conversion, a reduced water requirement, a greater ability to retain feed and water in the large intestine, and better resistance to parasites and some diseases.

It is well known that goats are kept under a wide range of production and farming systems throughout the tropics. This could be attributed to their high adaptive capacity (Silanikove, 2000) and also to levels of fitness. Goats are able to produce under varying and frequently unfavourable environmental conditions (Alexandre and Mandonnet, 2005). One of the most important adaptations of goats to ecological conditions is their variable body size. One of the best examples known in marginal environments is the West African Dwarf goat, which remains the only domestic species that is able to survive in its particular region of West Africa (Daramola and Adeloye, 2009). Among its physiological features are small body size and low metabolic requirements, which are important traits to enable the animal to minimize its requirements in an area or season where food resources are limited in quality and quantity. Daramola and Adeloye (2009) reported that hereditary dwarfism is common in the humid tropical zone. This complies with the rule of Bergmann and Allen, which can be interpreted as a correlation between morphological variation (body surface area) and ambient temperature (heat dissipation). Other findings reviewed by Daramola and Adeloye (2009) suggest that body size is also correlated with primary plant productivity, drought resistance, and type and quality of food. Selection pressure towards a smaller size explains the widespread dwarfism in domestic ruminants occupying the same niche. In accordance with Bergmann's rule, even non-dwarfed breeds of ruminants in the humid tropics are, in most cases, much smaller than tropical exotic breeds (Devendra and Burns, 1983). The small size of many tropical breeds is directly associated with other important traits such as early maturity, quality of products (meat, milk) and nutrient requirements for maintenance. Low per-head nutrient requirements mean that the dwarf goat fits the limited resources of small farmers or marginal grazing lands, which cannot sustain large ruminants throughout the production cycle.

Research on small or medium-sized goat breeds such as the West African Dwarf or Caribbean Creole goat indicates very high reproductive abilities (Alexandre *et al.*, 1999; Baiden, 2007; Khanum *et al.*, 2007; Mahieu *et al*,. 2008) and disease resistance (Mandonnet *et al.*, 2001, 2006; Bambou *et al.*, 2009; Chiejina *et al.*, 2009; Nnadi *et al.*, 2009), which emphasize their high fitness characteristics (Alexandre and Mandonnet, 2005; Daramola and Adeloye, 2009).

## 3.5 Comparisons of Carcass Data of Tropically Adapted and Temperate Goat Breeds

There is a considerable body of literature on goats of different breeds and that vary greatly in body sizes and functions, as well as varying systems of production (Devendra and Burns, 1983; Peacock, 1996). However, only a few objective comparative studies between these breeds are available and such comparisons may be confounded by the environmental conditions in which the goats are kept. Among goat breeds, European dairy breeds and the improved Boer goat of South Africa appear to be the fastest growing as they are large-sized breeds. The Boer goat, which is gaining popularity in many parts of the world, has a higher proportion of muscle in the carcass compared with most goat breeds. However, the data are not very conclusive because of variations in production systems (Warmington and Kirton, 1990). Extensive goat-raising systems, which are common in the tropics, markedly reduce the productive performance and carcass characteristics in the Boer goat (Almeida *et al.*, 2006).

The numerous and variable breed/system combinations and their multiple interactions have resulted in a wide variation of carcass weight, size and frame. Compilation of data from the literature has allowed us to make comparisons and reach viable conclusions. Ninety-four publications, comprising 214 comparisons, were reviewed. In order to be integrated into this database, data on carcass (weight, yield, measurements, cuts) and/or data on dissection (carcass, shoulder and leg) were obtained on different genotypes, fed various diets and slaughtered at different weight/age. More than half of the studies (52%) dealt exclusively with indigenous breeds, whereas 31% dealt with crossbreeds. The diets studied were diverse, with most of them being mixed diets (66%), i.e. composed of forage supplemented with concentrates, whereas 34% were composed of forage only. The forages used were half grass (22%), and half brush and by-products. Male growing animals represented 98% of the overall total. Generally, animals were chosen for slaughtering according to their age (57%) or weight (43%).

## 3.6 Variations in Slaughter and Carcass Weights

The distribution of slaughter weight (SW) of some published data for various goats is given in Fig. 3.1. The mean ± standard deviation (SD) was 22.0 ± 8.0 kg. Corresponding values for carcass weight (CW) were 10.4 ± 4.0 kg (Fig. 3.2). Two subsets of data were separated (Table 3.1): data obtained from papers where calculations of carcass yield were defined as dressing percentage (CW/SW, $n = 112$) and others with the yield defined as carcass output per empty body weight (CW/EBW; $n = 93$). The means were 46.1% (± 5.7%) and 52.9% (± 3.4%), respectively. Four classes of SW were discriminated in these two subsets of data (Fig. 3.2a and b). As expected, carcass yield improved with increasing SW. However, clear thresholds appeared in the values (regardless of the mode of calculation) for classes of SW higher than class 1. For lightweight kids, the dressing percentage (Fig. 3.2a) was on average 39%, while, for heavier kids (SW > 13.5–14 kg), it ranged from 48 to 50%. The corresponding values for carcass output (Fig. 3.2b) were 50% for light kids and 53–55% for heavier ones.

## 3.7 Carcass Indices and Cuts

Since the work of Fehr *et al.* (1976), very few studies have dealt with the importance of goat carcass conformation per se. Prasad and Kirton (1992) outlined the importance of carcass weight for conformation classification. These studies indicated that carcass weight/yield increases within conformation

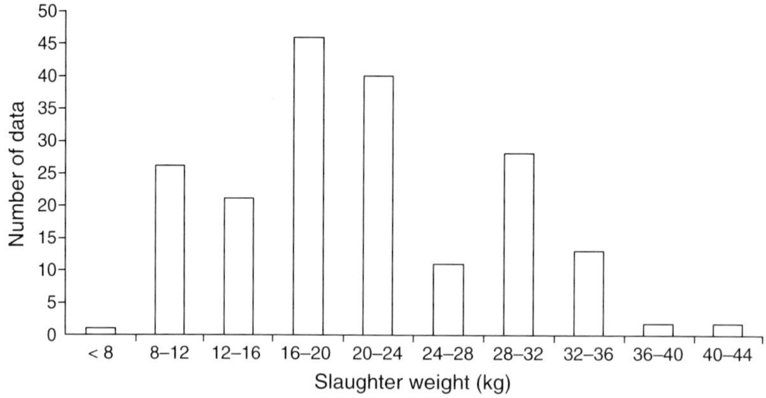


**Fig. 3.1.** Distribution of slaughter weight data.

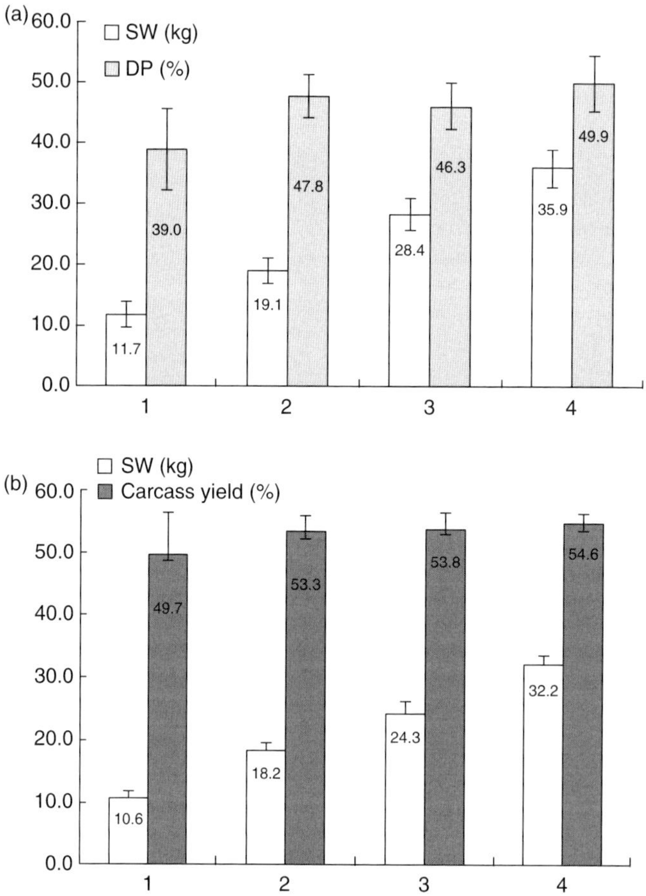


**Fig. 3.2.** Values of goat carcass yields calculated as (a) dressing percentage (DP) = carcass weight/slaughter weight (SW), and (b) carcass output = carcass weight/empty body weight, according to SW class (1–4).

**Table 3.1.** Descriptive statistics of carcass weights and yield of growing kids of different genotypes fed various feeding regimes.

| Characteristic | No. goats | Mean | SD | CV (%) | Minimum | Maximum |
|---|---|---|---|---|---|---|
| Data without calculation of EBW (36 papers) | | | | | | |
| Slaughter weight (kg) | 112 | 22.1 | 7.88 | 35.6 | 8.5 | 43.2 |
| Carcass weight (kg) | 122 | 11.0 | 4.46 | 40.5 | 3.4 | 24.0 |
| Dressing out percentage (%) | 112 | 46.1 | 5.70 | 12.4 | 27.5 | 56.6 |
| Data with calculation of EBW (30 papers) | | | | | | |
| Slaughter weight (kg) | 78 | 21.4 | 10.25 | 47.8 | 6.1 | 55.0 |
| Carcass weight (kg) | 89 | 10.1 | 4.20 | 41.4 | 2.9 | 24.5 |
| Carcass output (%) | 94 | 52.9 | 3.78 | 7.1 | 38.0 | 60.1 |

SD, Standard deviation; CV, coefficient of variation.

classes, given that these three criteria are interrelated (Fehr *et al.*, 1976; Prasad and Kirton, 1992). Oman *et al.* (1999) attempted to describe the US goat carcass conformation based on muscle shape and thickness of the leg, loin, rack and shoulder adapted from the US Department of Agriculture (USDA) sheep grading system. However, in the absence of an official US goat grading system, even in most recent work, sheep grading systems are still used for goat carcasses (Ryan *et al.*, 2007). In the French West Indies (where meat is not a by-product of milk production systems, in contrast to France, the 'mother country'), there is no official grading standard designed specifically for the French goat carcasses. Therefore, it is generally accepted that the light lamb grid employed in the Mediterranean regions within production systems with some similarities to those employed in tropical regions could be suitable for goat grading.

Given that it is difficult to compare conformation scores because of scarcity of data and differences in grading methods, different indices have been calculated:

- Carcass index: weight/length
- Carcass compactness: width/length
- Leg compactness: width/length.

It appears that there is a wide variability within the literature (Table 3.2) for carcass indices, which varied from 0.076 to 0.297 with a 32% coefficient of variation. For carcass compactness and leg compactness (Table 3.2), the difference between minimum and maximum values amounted to 160%. Although the sets of data for calculations of carcass index and carcass/leg compactness ($n = 75$ versus $n = 25$) were not similar and did not cover the same range of carcass weights, it was possible to conclude that the shape of the goat carcass differs greatly from one genotype × system to another.

The evolution of the carcass index according to the carcass weight is given in Fig. 3.3. Three groups of data were separated according to range of variation in carcass weight: light (6.8 ± 2.1 kg), medium (10.1 ± 2.9 kg) and heavy carcasses (12.3 ± 4.1 kg). Regression equations were computed (PROC REG program; SAS, 2000) to predict carcass index according to carcass weight (Table 3.3). The quadratic terms were all significant ($P < 0.05$) in the different groups, with $R^2$ approaching 0.89, 0.92 and 0.77, respectively.

An effective grading system for meat goats is needed to standardize the description and allow comparisons between breeding values and/or system differences. There is a lack of information on this specific and multifactorial concept of carcass conformation and meat yield in tropical regions. Therefore, the use of criteria such as carcass cuttability, yield of lean meat and muscularity is of paramount importance. Tatum *et al.* (1998) noted that the intrinsic value of a feeder animal is appreciated owing to its optimal proportions, at a preferred market weight, of the different carcass cuts. As for goat carcass cut

**Table 3.2.** Descriptive statistics of carcass shape and cuts of growing kids of different genotypes fed various feeding regimens.

| Characteristic | No. treatments | Mean | SD | CV (%) | Minimum | Maximum |
|---|---|---|---|---|---|---|
| Carcass index (weight/length); *n* = 21 papers | | | | | | |
| Slaughter weight (kg) | 66 | 19.3 | 7.74 | 40.2 | 6.1 | 38.2 |
| Carcass weight (kg) | 78 | 9.4 | 3.79 | 40.4 | 2.9 | 21.7 |
| Carcass index | 75 | 0.161 | 0.0517 | 32.0 | 0.076 | 0.297 |
| Carcass compactness (length/width); *n* = 11 papers; CW = 6.6 ± 2.0 kg. | | | | | | |
| Carcass compactness | 25 | 3.78 | 0.522 | 13.8 | 2.95 | 4.73 |
| Leg compactness (length/width); *n* = 12 papers; CW = 7.6 ± 3.0 kg. | | | | | | |
| Leg compactness | 26 | 1.97 | 0.239 | 12.1 | 1.46 | 2.38 |
| Shoulder (% of carcass); *n* = 18 papers; CW = 11.0 ± 4.3 kg. | | | | | | |
| Shoulder | 41 | 21.1 | 2.65 | 12.5 | 15.9 | 26.9 |
| Leg (% of carcass); *n* = 22 papers; CW = 10.6 ± 4.4 kg. | | | | | | |
| Leg | 50 | 29.6 | 5.59 | 18.9 | 22.0 | 41.9 |

SD, Standard deviation; CV, coefficient of variation; CW, carcass weight.

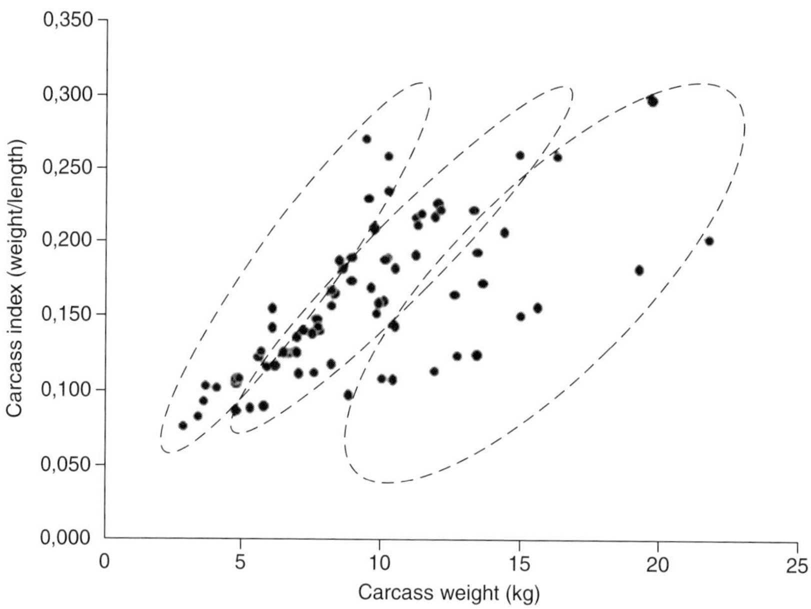


**Fig. 3.3.** Evolution of carcass index (weight/length, kg/cm) according to carcass weight (kg) in goat studies (description of three groups) (Limea, 2009).

composition, a standardized method has been developed (Colomer-Rocher *et al.*, 1987) and is increasingly being used in many countries (e.g. in Africa; Sanon *et al.*, 2008). Among the studies reporting on carcass cutting, the proportions of shoulder and leg were 21.1 ± 2.65% and 29.6 ± 5.59%, respectively (Table 3.2). Values are lower than for lambs, as demonstrated in studies comparing goat and sheep (Mahgoub and

**Table 3.3.** Regression equations to predict carcass index (weight/length; $y$) according to carcass weight (kg; $x$).

| Carcass group | $n$ | Formulae | $r^2$ | Significance |
|---|---|---|---|---|
| Total data set | 75 | $y = 0.0446x^{0.5672}$ | 0.5514 | $P > 0.05$ |
| | 75 | $y = -0.0006x^2 + 0.021x + 0.0184$ | 0.5019 | $P > 0.05$ |
| Light carcasses | 29 | $y = 0.0496e^{0.152x}$ | 0.8672 | $P < 0.01$ |
| | 29 | $y = 0.0034x^2 - 0.024x + 0.1379$ | 0.8898 | $P < 0.01$ |
| Medium weight carcasses | 30 | $y = 0.0335x^{0.7304}$ | 0.9203 | $P < 0.01$ |
| | 30 | $y = -0.002x^2 + 0.0182x + 0.0215$ | 0.8892 | $P < 0.01$ |
| Heavy carcasses | 16 | $y = 0.0446x^{0.5672}$ | 0.7550 | $P < 0.05$ |
| | 16 | $y = -0.0006x^2 + 0.021x + 0.0184$ | 0.7746 | $P < 0.05$ |

Lodge, 1996; Sheridan *et al.*, 2003; Sen *et al.*, 2004).

## 3.8 Tissue Dissection

The carcass of a meat animal is composed of varying proportions of muscle, fat and bone. Muscle, being more edible, is usually regarded as the most important carcass tissue to the consumer, while fat is more related to health issues, especially in developed countries. Based on consumer expectations, the goat carcass and goat meat are very well qualified as lean meat (Webb *et al.*, 2005). Consequently, it is important to give factual data on tissue proportions. The shoulder and leg are known to have the highest percentage of muscle within carcass cuts (Tahir *et al.*, 1994; Dhanda *et al.*, 2003a). The results of dissections of the whole carcass, shoulder and leg into lean, bone and fat (subcutaneous + intramuscular) reviewed from the literature are shown in Table 3.4. There was a great variability in the proportion of fat dissected in the carcass, the shoulder or the leg 39, 42 and 45% of coefficient of variation, respectively. This is most probably due to the effects of different feeding regimes.

Tissue partitioning is also known to differ among genotypes and according to animal age. Since the pioneering work of Hailu Hammond (1962), it has been shown that maximal growth rate is attained first by bone, then muscle and lastly by fatty tissue. Conclusions on the differences among breeds are generally inconclusive because of the effects of degrees of maturity. This was reported in sheep and cattle (Sellier *et al.*, 1992), but studies on effects of breed maturity are scarce in goats. In this species, the accretion of fat and muscle regulation and their relative body partitioning within the perspective of meat potential assessment have been insufficiently studied (e.g. Mahgoub and Lodge, 1996; Mahgoub and Lu, 1998). Two of the main reasons for this are thought to be:

1. Meat production in many cases is a secondary product of the milk or fibre sector.
2. There are not many specialized goat meat breeds.

However, some results can be highlighted from the database generated for this study. Proportions of lean vary from 60 to 65% (within treatments and cuts) and those of fat vary from 9 to 14%, which indicates a satisfactory meat potential for goat, which is generally defined as a multifunctional animal (Peacock, 1996).

In the tropics, a high proportion of bone in the carcass is considered a negative characteristic. This indicates a lack of muscularity, which characterizes most local breeds, rendering them poorly rated due to a small frame/size. An additional feature that would help in describing the meat potential of small stock is the muscle:bone ratio (Hopkins *et al.*, 1997). Many studies with sheep have shown that a fleshy sheep

**Table 3.4.** Descriptive statistics on carcass dissection of growing kids of different genotypes fed various feeding regimes.

| Characteristic | No. treatments | Mean | SD | CV (%) | Minimum | Maximum |
|---|---|---|---|---|---|---|
| Dissection of carcass (% of carcass); *n* = 27 papers | | | | | | |
| Slaughter weight (kg) | 87 | 21.9 | 8.20 | 37.6 | 6.1 | 43.2 |
| Carcass weight (kg) | 87 | 11.2 | 4.72 | 42.2 | 3.4 | 24.0 |
| Lean | 87 | 62.5 | 7.28 | 11.6 | 44.9 | 79.1 |
| Bone | 87 | 22.9 | 6.95 | 30.3 | 12.0 | 45.7 |
| Fat | 84 | 13.6 | 5.31 | 39.0 | 3.1 | 29.5 |
| Lean:bone ratio | 87 | 2.97 | 1.064 | 35.9 | 0.98 | 5.65 |
| Lean:fat ratio | 84 | 5.76 | 3.244 | 56.3 | 1.89 | 21.32 |
| Dissection of shoulder (% of shoulder); *n* = 12 papers (mean 21.2 ± 2.5%) | | | | | | |
| Lean | 28 | 60.2 | 12.88 | 21.4 | 26.0 | 71.5 |
| Bone | 28 | 22.9 | 3.28 | 14.3 | 15.0 | 31.9 |
| Fat | 26 | 12.3 | 5.18 | 42.1 | 5.8 | 16.9 |
| Lean:bone ratio | 28 | 2.68 | 0.758 | 28.3 | 1.16 | 4.77 |
| Lean:fat ratio | 26 | 5.40 | 2.696 | 49.9 | 1.29 | 12.17 |
| Dissection of leg (% of leg); *n* = 14 papers (mean 29.9 ± 2. 6%) | | | | | | |
| Lean | 37 | 65.4 | 10.03 | 15.3 | 34.8 | 78.1 |
| Bone | 37 | 24.7 | 5.19 | 21.0 | 16.1 | 38.9 |
| Fat | 28 | 9.3 | 4.20 | 45.2 | 3.4 | 17.8 |
| Lean:bone ratio | 37 | 2.80 | 0.878 | 31.4 | 1.28 | 4.72 |
| Lean:fat ratio | 28 | 10.20 | 7.645 | 74.9 | 2.54 | 27.12 |

SD, Standard deviation; CV, coefficient of variation.

breed would be graded as 'superior' for its muscularity while it could be 'inferior' for its muscle:bone ratio. For instance, Purchas *et al.* (1991) demonstrated that, although these two characteristics often change together, there are situations where differences in muscularity are not accompanied by differences in muscle:bone ratio and vice versa. In goat studies, the lean:bone ratios calculated either in the whole carcass, shoulder or leg reached values ranging from 2.68 to 2.97 (Table 3.4). As with lean:fat ratios, the values were higher and very satisfactory, in contrast to sheep. In fact, differences in levels of fatness, rather than muscle weight distribution, account for most differences in carcass grading between sheep breeds (Butler-Hog *et al.*, 1984; Laville *et al.*, 2002). Poor fat accretion and leggy conformation of goat carcasses (Colomer-Rocher *et al.*, 1992; Mahgoub and Lu, 1998; Oman *et al.*, 1999) support the assumption that these criteria are not suitable for grading goat carcasses. Goats are known to have more fat deposits, mainly in the abdominal cavity, than the carcass (Kempster, 1981; Warmington and Kirton, 1990) and therefore carcass fat parameters, as used for sheep, might not be appropriate for grading goat carcasses.

## 3.9 Case Studies

### 3.9.1 Indigenous Caribbean goats

In many countries of the Caribbean and Latin American region, the local populations of goats are frequently named 'Chèvre Créole', Creole goat or 'Criollo' (Devendra and Burns, 1983). They are derived mainly from crossbreeding between various West African, European and sometimes Indian breeds (Naves *et al.*, 2000). Over time, the native population has evolved naturally through adaptation to agroecological conditions. Goat farming systems in the French West Indies are based on the use of the Creole goat breed (Alexandre *et al.*, 1999) on grazed pastures (Mahieu *et al.*, 2008). This influences meat production aspects such as live weight, growth performance and carcass abilities. The Creole goat is known for its high weaner productivity (Mahieu *et al.*, 2008) and its genetic ability to resist gastrointestinal parasites (Mandonnet *et al.*, 2001, 2006). The Creole goat is a medium-sized meat breed with a traditional slaughter weight of 18 kg, which can be reached at 6–18 months of age, depending on the system (Mahieu *et al.*, 2008). The local goat industry in the French West Indies faces a 'fundamental' threat to its very existence that will eventually result in loss of genetic diversity (Alexandre *et al.*, 2009), a guarantee for its future. The trend of importing live exogenous animals or frozen carcasses has resulted in a gradual decline in the ratio of local production:local demand (Alexandre *et al.*, 2008). To effectively reverse this trend and to stop the anarchic crossing of native goats with exogenous breeds, a genetic improvement programme for this breed is presently under way (Alexandre and Mandonnet, 2005).

Studies have begun with Creole male goats to study carcass characteristics and meat quality in relation to feeding systems (Liméa *et al.*, 2009a) and slaughter conditions (Liméa *et al.*, 2009b). This has facilitated accumulation of a database on many of the carcass traits (yield, quality scores, carcass cuts and linear measurements). The database includes 131 intact male kids of the Creole genotype raised indoors. Two contrasting groups were discriminated according to their feeding level. The low level (LL) group ($n$ = 65) received a basal diet (green tropical forage) without concentrate, while the high level (HL) group ($n$= 66) received, in addition, a concentrate diet (280–320 g/day on average). Animals were slaughtered when they reached either their final live weight or age. Standardized procedures of carcass measuring and cutting (Colomer-Rocher *et al.*, 1987) were followed. The different carcass data obtained are shown in Table 3.5. The carcass weight and yield in the Creole goat steadily increased within ages as expected. The values (up to 30 kg carcass and 62% carcass output) were very similar to performance of other meat breeds (Table 3.1).

When fed intensively, the kids reached the slaughter weight of 22 kg 3 months sooner than their counterparts fed forage only. Comparing the HL and LL diets in kids at the same age, the increase reached 120–140% for slaughter weights and 153–179% for cold carcass weights. Undoubtedly, the proportion of fat was affected by the level of energy in the diets (Liméa *et al.*, 2009b), although the absolute values remained at a very adequate levels compared with values in the literature (Table 3.4).

The distribution of prime cuts (not tabulated) remained similar (~50%), irrespective of the conditions and attained values of the well-conformed genetic breeds reported by Dhanda *et al.* (2003a). Given that the shoulder and leg are reported to have the highest percentage of muscle (Dhanda *et al.*, 2003b) within carcass cuts, dissection of their tissues can give an assessment of the potential of the meat production from the animals. In the present set of data, the values were in favour of the leg compared with the shoulder (2–4 points more) where the values ranged from 71 to 78%. In other studies, values ranged from 68.5 to 71.4% and from 65.3 to 67.6% in the leg and shoulder, respectively. The higher muscle percentage in the Creole goat leg and shoulder is most probably due to the lower fat content (3–7%), compared with a value of 10–13% reported by Dhanda *et al.* (2003b). Similar trends were observed by comparing the Creole and New Zealand Saanen goat within a similar slaughter weight range (Colomer-Rocher *et al.*, 1992). However, these differences may have arisen not only from a higher fat but also from a higher bone percentage, which was 23–26% in the New Zealand Saanen goats compared with 18–24% in the Creole goats.

A trait that better describes the meat production potential is the muscle:bone ratio, which is particularly relevant for Creole male goats. Values of 2.8–4.2 fell within the upper range of values reported in the literature. The carcass indices followed a similar trend of slaughter weight, i.e. an increase with age and better diets. Creole

**Table 3.5.** Carcass weight, yield, cuts, measurements and dissection of Creole kids according to their feeding level (LL, low level, basal green tropical forage; and HL, high level, an additional 300–350 g concentrate/day) and their age at slaughter (from Liméa *et al.*, 2009a,b).

| Carcass characteristic | Feeding level | | | | | | |
|---|---|---|---|---|---|---|---|
| | LL | | | HL | | | |
| Age at slaughter (months) | 7 | 12 | 15 | 7 | 9 | 11 | 15 |
| Number of kids | 16 | 36 | 13 | 15 | 20 | 16 | 15 |
| Slaughter weight (kg) | 16.0 | 22.3 | 25.5 | 19.2 | 22.9 | 30.5 | 36.5 |
| Empty body weight (kg) | 10.5 | 15.6 | 18.1 | 14.8 | 18.8 | 24.4 | 29.5 |
| Hot carcass weight (kg) | 5.8 | 9.2 | 10.6 | 8.6 | 12.3 | 15.1 | 18.7 |
| Cold carcass weight (kg) | 5.5 | 8.9 | 10.3 | 8.4 | 11.7 | 14.8 | 18.4 |
| Dressing percentage (%) | 34.4 | 41.4 | 40.4 | 43.8 | 51.0 | 48.5 | 50.4 |
| Carcass output (%) | 52.8 | 56.9 | 57.0 | 57.1 | 60.0 | 60.7 | 62.4 |
| Carcass indices | | | | | | | |
| Carcass compactness | 4.025 | 3.648 | 4.007 | 4.023 | 3.391 | 3.974 | 3.909 |
| Leg compactness | 2.463 | 2.482 | 2.322 | 2.451 | 2.502 | 2.377 | 2.378 |
| Carcass index | 0.113 | 0.160 | 0.176 | 0.157 | 0.200 | 0.242 | 0.287 |
| Dissection of shoulder | | | | | | | |
| Muscle (%) | 69.1 | 72.7 | 73.7 | 71.3 | 72.8 | 74.2 | 75.1 |
| Bone (%) | 25.2 | 21.3 | 21.2 | 20.7 | 19.4 | 19 | 17.9 |
| Fat (%) | 5.6 | 6.8 | 5.1 | 7.9 | 7.75 | 6.8 | 7 |
| Muscle:bone ratio | 2.74 | 3.39 | 3.48 | 3.44 | 3.70 | 3.90 | 4.19 |
| Muscle:fat ratio | 12.34 | 10.74 | 14.45 | 9.02 | 9.41 | 10.92 | 10.73 |
| Dissection of leg | | | | | | | |
| Muscle (%) | 71.4 | 74.1 | 75.5 | 74.8 | 75.2 | 77.9 | 76.8 |
| Bone (%) | 25.5 | 22.1 | 21.2 | 21.7 | 20.8 | 18.5 | 19.3 |
| Fat (%) | 3.1 | 3.7 | 3.3 | 3.5 | 4.05 | 3.5 | 3.9 |
| Muscle:bone ratio | 2.80 | 3.37 | 3.56 | 3.45 | 3.75 | 4.21 | 3.98 |
| Muscle:fat ratio | 23.03 | 20.0 | 22.88 | 21.37 | 18.79 | 22.26 | 19.69 |

kids, when compared within a similar range of carcass weights, exhibited very good carcass indices that were similar to or even higher than the larger Boer crossbreds, contrary to the widely held belief in their inferiority in the French West Indies meat sector (Alexandre *et al.*, 2008).

The leg indices may be regarded as a similar concept to 'muscularity', a trait used for sheep by Purchas *et al.* (1991) based on femur length and the weight of surrounding muscles. Comparing our findings (not tabulated) with other available studies, our values of 0.036–0.048 are comparable to those of other breeds such as Florida kids (0.029–0.047; Pena *et al.*, 2007), Italian Jonica (0.035–0.044; Marsico *et al.*, 1993) and Canary caprine (0.022–0.053; Marichal *et al.*, 2003) but were lower than Omani breeds (0.070–0.078; Kadim *et al.*, 2003).

The small-sized Creole goat breed could be a valuable meat producer based on satisfactory carcass and leg indices, muscle:bone ratios and carcass yield and cuttability. These descriptors for carcass conformation and meat potential suggest that Creole goats, although not yet genetically improved, may be comparable to some fleshy meat breeds, and provide a prospective potential incentive for the local goat meat sector. However, more research on genetic improvement is needed for the future.

### 3.9.2 Omani native meat goats

Goats are the most important meat-producing animal in many Asian and African countries. Goats are the most numerous livestock and their meat is preferred over other meats in Oman. There are three major goat breeds of importance in Oman including Al-Jabal Al-Akhdar, Batina and Dhofari goats. The Dhofari goat contributes about 44% of goats in the country and has the smallest body size (Mahgoub, 1997a). It is a hardy breed, living mostly in the southern mountainous region of Dhofar, and resembles the Somali goat. It is used here for comparative purposes with the Creole goat of the French West Indies as it is of similar body weight range. The Creole goat has been described as a medium-sized meat breed traditionally slaughtered at 18 kg, which is reached at 6–18 months of age, depending on the system (Mahieu *et al.*, 2008).

Apart from a limited experimental genetic improvement programme in Wadi Qurayat Research Station and limited crossbreeding with Anglo-Nubian goats (Al Ojaili, 1995) and cashmere goats (M.G. El Hag, personal communication), the Dhofari goat has not been subjected to major genetic improvement. Dhofari goats are raised mainly under transhumant systems, depending mainly on natural range grazing, which is characterized by lush green seasonal grass in the south of the country for about 3–4 months. Supplementary feeding is customarily given to fattened goats and lactating does in the form of Rhodes grass hay, dates, concentrates, banana leaves, barley and household leftovers (Mahgoub and Lodge, 1996). Natural selection has been carried out on the breed mainly for survival traits.

The breed is the smallest of the local breeds with a mature weight of 30–35 kg but fattened males are usually slaughtered at 6 months of age at 18–22 kg body weight. The small size of the breed is regarded as a favourable characteristic, as the whole roasted goat carcass would comfortably fit on a tray of rice. This is important from a cultural food traditional point of view in the Arabia Gulf region.

The genetic diversity of the breed is relatively conserved, although increasing numbers of the Somali goat, which resembles the Dhofari, have been imported in the country. Under improved management, the Dhofari goat's reproductive performance was found to be above average. Al Ojaili (1995) reported a conception rate of 65%, a twinning rate of 23% and a litter size of 1.29. Dhofari goat males are reputed for having a markedly high precocity and they reach sexual maturity early in life (Mahgoub and Lu, 1998). The breed has also been reported to have better milk production performance than the other Omani goat breeds (Chesworth and Horton, 1996). This is an important meat production characteristic, as it enables dams to nurse their kids for a better growth performance.

Very few studies have been carried out on the breed to evaluate its potential for meat production under improved management systems (Mahgoub, 1997a,b; Kadim *et al.*, 2003). Some of the published data are presented in Table 3.6. Both sets of data originated from animals raised under intensive management with concentrate feed being offered besides ad libitum Rhodes grass hay feeding.

When fed intensively, kids can reach the slaughter weight of 16 kg at 5 months of age, a faster rate than that of range-fed kids. The carcass weight and yield in Dhofari goats steadily increased within ages as expected. The carcass weight of the Dhofari goat (9.2 kg at 5 months) was comparable to that of tropical breeds (Table 3.6). However, there was only an increase of 3 kg of carcass weight at the age of 17 months (Table 3.6). The values for carcass output (64.9 and 56.5% at 18 and 30 kg body weight, respectively) are comparable to those mentioned previously for Creole goats in the French West Indies, as well as for other meat breeds (Table 3.6). The dressing percentage value of 56.5% reported by Mahgoub (1997a,b) was higher than those elsewhere reported for goats but had been calculated on fasted body weight. The dressing percentage of 41.8% at 30 kg live weight (17 months) of Dhofari goats was below the average (46.1%) computed for tropical breeds (Table 3.6).

The Dhofari goat showed different indices from that of pooled data (Table 3.6) or Creole goats (Table 3.5). The value for carcass compactness was much higher for

**Table 3.6.** Carcass weight, yield, cuts, measurements and dissection of Dhofari Omani kids fed Rhodes grass hay and a concentrate (from Mahgoub, 1997a,b; Kadim *et al.*, 2003).

| Parameter | Mahgoub (1997a,b) | Kadim *et al.* (2003) |
|---|---|---|
| Number of kids | 10 | 14 |
| Age at slaughter (months) | 5 | 17 |
| Slaughter weight (kg) | 16.3 | 29.9 |
| Empty body weight (kg) | 14.3 | 22.1 |
| Hot carcass weight (kg) | 9.2 | 12.7 |
| Cold carcass weight (kg) | – | 12.5 |
| Dressing percentage (%) | 56.5 | 41.8 |
| Carcass output (%) | 64.9 | 56.6 |
| Carcass indices | | |
| Carcass compactness | | 7.48 |
| Carcass index | | 0.291 |
| Carcass dissection | | |
| Muscle (%) | 70.0 | |
| Bone (%) | 13.2 | |
| Fat (%) | 12.9 | |
| Muscle:bone ratio | 5.35 | |
| Muscle:fat ratio | 5.64 | |
| Shoulder (% of carcass) | | 43 |
| Leg (% of carcass) | | 32 |

the Dhofari goat, indicating longer carcasses. The carcass index for Dhofari was also higher than for tropical goat breeds (Table 3.6).

The prime cuts of shoulder and leg contributed to a large proportion of the carcass in the Dhofari goat (Table 3.6). The shoulder and leg are reported to have the highest percentage of muscle (Dhanda *et al.*, 2003b).

The values for the Dhofari goat carcass tissue contents (Table 3.6), which can give an indication of the potential of the meat production from the animals, differed from those of tropical breeds (Table 3.1) and Creole goats (Table 3.5). The higher muscle percentage in the Dhofari goat is due to the lower bone content (13.2%) compared with 23% for pooled data (Table 3.4), as fat contents were similar. A high bone percentage of 23–26% was reported for New Zealand Saanen goats (Colomer-Rocher *et al.*, 1992) and 18–24% for Creole goats (Liméa *et al.*, 2009a,b).

The muscle:bone ratio describes the meat production potential. Values for the Dhofari goat were much higher than those in the pooled data or for the Creole goat. This was mainly because of the lower bone content. They were above the upper range of values reported in the literature. However, the muscle:fat ratio was similar to that of the pooled data (Table 3.4).

Mahgoub and Lu (1998) compared Omani goats of large size (Batina) and small size (Dhofari) for meat production. Although Batina goats grew faster and reached the destined slaughter weights earlier, the smaller Dhofari goat had higher growth rates relative to final body weight. At 18 kg body weight, Dhofari goats had a higher dressing percentage, carcass muscle content, and carcass and non-carcass fat content but a lower bone content. There were differences in muscle distribution, with Dhofari goats having higher proportions of muscle in the proximal hind limbs and around the vertebral column, which are regarded as muscles of higher value.

Similar to the Creole goat, the small size of the Dhofari goat breed could favour it to become a valuable meat producer based on the satisfactory carcass indices, muscle:bone ratios, and carcass yield and cuttability. These descriptors for carcass conformation and meat potential suggest that Dhofari goats, although not yet genetically improved, may be comparable to some larger meat breeds, and offer a prospective potential for the local goat meat sector.

## 3.10 Conclusions

- The fitness and meat production potential of hardy indigenous does are superior compared with modern single-purpose breeds in the tropics, especially if modern husbandry techniques are not available.
- There is no universal genotype for use in all environments (Tsukahara *et al.*, 2008; Wilson, 2009).
- Indigenous hardy goat breeds such as those in the tropics should not be automatically replaced with imported large breeds for meat production under local systems.
- Despite the increasing interest in indigenous livestock, breed evaluation schemes for tropical goats are relatively scarce (Shrestha and Fahmy, 2007b).
- Production efficiency should be evaluated for the integrated system (Alexandre *et al.*, 2010), not just the individual animal.
- Analyses of published work on body sizes and/or carcass frames are somewhat inconclusive when tissue partitioning is insufficiently studied.
- Consumer expectations should be taken into account at the local sector level.

## References


Ahuya, C.O., Okeyo, A.M., Mwangi-Njuru, D.M. and Peacock, C. (2005) Developmental challenges and opportunities in the goat industry in Kenya. *Small Ruminant Research* 60, 197–206.

Alexandre, G. and Mandonnet, N. (2005) Goat meat production in harsh environments. *Small Ruminant Research* 60, 53–66.

Alexandre, G., Aumont, G., Mainaud, J.C., Fleury, J. and Naves, M. (1999) Productive performances of Guadeloupean Creole goats during the suckling period. *Small Ruminant Research* 34, 157–162.

Alexandre, G., Asselin de Beauville, S., Shitalou, E. and Zebus, M.F. (2008) An overview of the goat meat sector in Guadeloupe: conditions of production, consumer preferences, cultural functions and economic implications. *Livestock Research for Rural Development* 20 <http://www.lrrd.org/lrrd20/1/alex20014.htm>.

Alexandre, G., Leimbacher, F., Maurice, O., Domarin, D., Naves, M. and Mandonnet, N. (2009) Goat farming systems in Martinique: management and breeding strategies. *Tropical Animal Health and Production* 41, 635–644.

Alexandre, G., González-García, E., Lallo, C.H.O., Ortega-Jimenez, E., Pariacote, F., Archimède, H., Mandonnet, N. and Mahieu, M. (2010) Goat management and systems of production: global framework and study cases in the Caribbean. *Small Ruminant Research* 89, 193–206.

Almeida, A.M., Schwalbach, L.M., de Waal, H.O., Greyling, J.P.C. and Cardoso, L.A. (2006) The effect of supplementation on productive performance of Boer goat bucks fed winter veld hay. *Tropical Animal Health and Production* 38, 443–449.

Al Ojaili, A.A. (1995) Production traits in Dhofari and Dhofari × Anglo-Nubian goats in Oman. *International Journal of Animal Science* 10, 13–16.

Andersen, B.B. (1978) Animal size and efficiency, with special reference to growth and feed conversion in cattle. *Animal Production* 27, 381–391.

Ayalew, W., Rischkowsky, B., King, J.M. and Bruns, E. (2003) Crossbreds did not generate more net benefits than indigenous goats in Ethiopia. *Agricultural Systems* 76, 1137–1156.

Baiden, R.Y. (2007) Birth weight, birth type and pre-weaning survivability of West African Dwarf goats raised in the Dangme West District of the Greater Accra Region of Ghana. *Tropical Animal Health and Production* 39, 141–147.

Bambou, J.C., Gonzalez-Garcia, E., Chevrotiere, C., Arquet, R., Vachiery, N. and Mandonnet, N. (2009) Peripheral immune response in resistant and susceptible Creole kids experimentally infected with *Haemonchus contortus*. *Small Ruminant Research* 82, 34–39.

Bosman, H.G., Moll, H.A.J. and Udo, H.M.J. (1997) Measuring and interpreting the benefits of goat keeping in tropical farm systems. *Agricultural Systems* 53, 349–372.

Butler-Hog, B.W., Francombe, M.A. and Dransfield, E. (1984) Carcass and meat quality of ram and ewe lambs. *Animal Production* 39, 107–114.

Chesworth, J.M. and Horton, G.M.J. (1996) Lactation in indigenous Omani goats. *International Journal of Animal Sciences* 11, 7–12.

Chiejina, S.N, Behnke, J.M., Nnadi, P.A, Ngongeh, L.A. and Mugong, G.A. (2009) The responses of two ecotypes of Nigerian West African Dwarf goat to experimental infections with *Trypanosoma brucei* and *Haemonchus contortus*. *Small Ruminant Research* 85, 91–98.

Colomer-Rocher, F., Morand-Fehr, P. and Kirton, A.H. (1987) Standard methods and procedures for goat carcass evaluation, jointing and tissue separation. *Livestock Production Science* 17, 149–159.

Colomer-Rocher, F., Kirton, A.H., Merces, G.J.K. and Duganzich, D.M. (1992) Carcass composition of New Zealand Saanen goats slaughtered at different weights. *Small Ruminant Research* 7, 161–173.

Daramola, J.O. and Adeloye, A.A. (2009) Physiological adaptation to the humid tropics with special reference to the West African Dwarf (WAD) goat. *Tropical Animal Health and Production* 41, 1005–1016.

Devendra, C. and Burns, M. (1983) *Goat Production in the Tropics*. Commonwealth Agricultural Bureaux, London.

Dhanda, J.S., Taylor, D.G. and Murray, P.J. (2003a) Part 1. Growth, carcass and meat quality parameters of male goats: effects of genotype and liveweight at slaughter. *Small Ruminant Research* 50, 57–66.

Dhanda, J.S., Taylor, D.G. and Murray, P.J. (2003b) Part 2. Carcass composition and fatty acid profiles of adipose tissue of male goats: effects of genotype and liveweight at slaughter. *Small Ruminant Research* 50, 67–74.

Dickerson, G.E. (1978) Animal size and efficiency: basic concepts. *Animal Production* 27, 367–379.

FAO (2007) Subregional report on animal genetic resources: the Caribbean. Annex to *The State of the World's Animal Genetic Resources for Food and Agriculture*. FAO, Rome, Italy.

Fehr, P.M., Sauvant, D. and Dumont, B.L. (1976) Croissance et qualité des carcasses des chevreaux de boucherie. *2èmes Journées de la Recherche Ovine et Caprine, Croissance, Engraissement et Qualité des Carcasses d'Agneaux et de Chevreaux*. INRA-ITOVIC, pp. 166–189.

Glowatzki-Mullis, M.L., Muntwyler, J., Baumle, E. and Gaillard, C. (2008) Genetic diversity measures of Swiss goat breeds as decision-making support for conservation policy. *Small Ruminant Research* 74, 202–211.

Hailu Hammond, J. (1962) *Growth and Development of Mutton Qualities in the Sheep*. Oliver and Boyd, Edinburgh, UK.

Hoffmann, I. (2010) Climate change and the characterization, breeding and conservation of animal genetic resources. *Animal Genetics* 41 (Suppl. 1), 32–46.

Hopkins, D.L., Fogarty, N.M. and Menzies, D.J. (1997) Differences in composition, muscularity, muscle:bone ratio and cut dimensions between six lamb genotypes. *Meat Science* 45, 439–450.

Kadim, I.T., Mahgoub, O., Al-Ajmi, D.S., Al-Maqbaly, R.S., Al-Saqri, N.M. and Ritchie, A. (2003) An evaluation of the growth, carcass and meat quality characteristics of Omani goat breeds. *Meat Science* 66, 203–210.

Kempster, A.J. (1981) Fat partition and distribution in the carcasses of cattle, sheep and pigs: a review. *Meat Science* 5, 83–98.

Khanum, S.A., Hussain, M. and Kausar, R. (2007) Assessment of reproductive parameters in female Dwarf goat (*Capra hircus*) on the basis of progesterone profiles. *Animal Reproduction Science* 102, 267–275.

Kosgey, I.S. and Okeyo, A.M. (2007) Genetic improvement of small ruminants in low-input, smallholder production systems: technical and infrastructural issues. *Small Ruminant Research* 70, 76–88.

Kosgey, I.S., Baker, R.L., Udo, H.M.J. and Arendonk, J.M. (2006) Successes and failures of small ruminant breeding programs in the tropics: a review. *Small Ruminant Research* 61, 13–28.

Lambe, N.R., Navajas, E.A., McLean, K.A., Simm, G. and Bünger, L. (2007) Changes in carcass traits during growth in lambs of two contrasting breeds, measured using computer tomography. *Livestock Science* 107, 37–52.

Laville, E., Bouix, J., Sayd, T., Eychenne, F., Marcq, F., Leroy, P.L., Elsen, J.M. and Bibé, B. (2002) Carcass conformation in lambs. A study of genetic variability among breeds. *INRA, Productions Animales* 15, 53–66.

Liméa, L. (2009) Effets des conditions d'alimentation et d'abattage sur les caractéristiques de carcasse et de viande du caprin creole. PhD AgroPariTech, Ecole Doctorale Abies, 228 pp.

Liméa, L., Gobardham, J., Gravillon, G., Nepos, A. and Alexandre, G. (2009a) Growth and carcass traits of Creole goats under different pre-weaning, fattening and slaughter conditions. *Tropical Animal Health and Production* 41, 61–70.

Liméa, L., Boval, M., Mandonnet, N., Garcia, G., Archimède, H. and Alexandre, G. (2009b) Fattening performances, carcass quality and non-carcass components of indigenous Caribbean goats under varying nutritional densities. *Journal of Animal Science* 87, 3770–3781.

Mahgoub, O. (1997a) Meat production from the Omani Dhofari goat. 1. Live-weight growth and body composition. *International Journal of Animal Science* 12, 25–30.

Mahgoub, O. (1997b) Meat production from the Omani Dhofari goat. 2. Distribution of carcass tissue. *International Journal of Animal Science* 12, 31–38.

Mahgoub, O. and Lodge, G.A. (1996) Growth and body composition in meat production of Omani Batina goats. *Small Ruminant Research* 19, 233–246.

Mahgoub, O. and Lu, O.D. (1998) Growth, body composition and carcass tissue distribution in goats of large and small sizes. *Small Ruminant Research* 27, 267–278.

Mahieu, M., Archimède, H., Fleury, J., Mandonnet, N. and Alexandre, G. (2008) Intensive grazing system for small ruminants in the Tropics: the French West Indies experience and perspectives. *Small Ruminant Research* 77, 195–207.

Mandonnet, N., Aumont, G., Arquet, R., Gruner, L., Bouix, J. and Khang, J. (2001) Assessment of genetic variability of resistance to gastrointestinal parasites in Creole goats. *Journal of Animal Science* 79, 1706–1712.

Mandonnet, N., Menendez-Buxadera, A., Arquet, R., Mahieu, M., Bachand, M. and Aumont, G. (2006) Genetic variability in resistance to gastro-intestinal strongyles during early lactation in Creole goats. *Animal Science* 82, 283–287.

Marichal, A., Castro, N., Capote, J., Zamorano, M.J. and Argüello, A. (2003) Effects of live weight at slaughter (6, 10 and 25 kg) on kid carcass and meat quality. *Livestock Production Science* 83, 247–256.

Marsico, G., Vicenti, A., Centoducati, P. and Braghieri, A. (1993) Influence of weaning age on productive performance of kids slaughtered at 107 days of age. *Small Ruminant Research* 12, 321–328.

Menendez-Buxadera, A. and Mandonnet, N. (2006) The importance of the genotype–environment interaction for selection and breeding programmes in tropical conditions. *CAB Reviews: Perspectives in Agriculture, Veterinary Science, Nutrition and Natural Resources* 1, no. 026.

Naves, M., Leimbacher, F., Alexandre, G. and Mandonnet, N. (2000) Development of animal breeding strategies for the local breeds of ruminants in the French West Indies. *ICAR Technical Series* 3, 379–385.

Nnadi, P.A., Kamalu, T.N. and Onah, D.N. (2009) The effect of dietary protein on the productivity of West African Dwarf (WAD) goats infected with *Haemonchus contortus*. *Veterinary Parasitology* 161, 232–238.

Ogink, N.W.M. (1993) Genetic size and growth in goats. PhD thesis, Wageningen Agricultural University, Wageningen, the Netherlands.

Oman, J.S., Waldron, D.F., Griffin, D.B. and Savell, J.W. (1999) Effect of breed-type and feeding regimen on goat carcass traits. *Journal of Animal Science* 77, 3215–3218.

Omondi, I.A., Baltenweck, I., Drucker, A.G., Obare, G.A and Zander, K.K. (2008) Valuing goat genetic resources: a pro-poor growth strategy in the Kenyan semi-arid tropics. *Tropical Animal Health and Production* 40, 583–596.

Peacock, C. (1996) Improving goat production in the tropics. In: *A Manual for Development Workers*. Oxfam/FARM-Africa Publication, Oxford, UK.

Pena, F., Perea, J., García, A. and Acero, R. (2007) Effects of weight at slaughter and sex on the carcass characteristics of Florida suckling kids. *Meat Science* 75, 543–550.

Prasad, V.S.S. and Kirton, A.H. (1992) Evaluation and classification of live goats and their carcasses and cuts. In: Lockeshwar, R.R. (ed.) *Proceedings of the Vth International Conference on Goats*. International Academic Publishers, New Delhi, India, pp. 440–449.

Preston, T.R. and Leng, R.A. (1987) *Matching Ruminant Production Systems with Available Resources in the Tropics and Sub-tropics*. CTA, Wageningen, the Netherlands.

Purchas, R.W., Davies, A.S. and Abdullah, A.Y. (1991) An objective measure of muscularity: changes with animal growth and differences between genetic lines of Southdown sheep. *Meat Science* 30, 81–94.

Rodriguez, L. and Preston, T.R. (1997) Local feed resources and indigenous breeds: fundamental issues in integrated farming systems. *Livestock Research for Rural Development* 9 <http://www.lrrd.org/lrrd9/2/lylian92.htm>.
Ryan, S.M., Unruh, J.A., Corrigan, M.E., Drouillard, J.S. and Seyfert, M. (2007) Effects of concentrate level on carcass traits of Boer crossbred goats. *Small Ruminant Research* 73, 67–76.
Sanon, H.O., Kaboré-Zoungrana, C. and Ledin, I. (2008) Growth and carcass characteristics of male Sahelian goats fed leaves or pods of *Pterocarpus lucens* or *Acacia senegal*. *Livestock Science* 17, 192–202.
SAS (2000) SAS/STAT User's guide. In: *SAS Language Guide for Personal Computers*, Version 8.1. SAS Institute, Cary, North Carolina.
Sellier, P., Bouix, J., Renand, G. and Molénat, M. (1992) Les objectifs et les critères de selection: les aptitudes bouchères: croissance, efficacité alimentaire et qualité de la carcasse. In: *Génétique Quantitative*. Productions Animales, INRA, pp. 147–160.
Sen, A.R, Santra, A. and Karim, S.A. (2004) Carcass yield, composition and meat quality attributes of sheep and goat under semiarid conditions. *Meat Science* 66, 757–763.
Sheridan, R., Hoffman, L.C. and Ferreira, A.V. (2003) Meat quality of Boer goat kids and Mutton Merino lambs. 1. Commercial yields and chemical composition. *Animal Science* 76, 63–71.
Shrestha, J.N.B. and Fahmy, M.H. (2007a) Breeding goats for meat production. 2. Crossbreeding and formation of composite population. *Small Ruminant Research* 67, 93–112.
Shrestha, J.N.B. and Fahmy, M.H. (2007b) Breeding goats for meat production. 3. Selection and breeding strategies. *Small Ruminant Research* 67, 113–125.
Silanikove, N. (2000) The physiological basis of adaptation in goats to harsh environments. *Small Ruminant Research* 35, 181–193.
Tahir, M.A.H., Abdulla, A.H.H. and Al-Jassim, A.F. (1994) Effect of castration and weight at slaughter on carcass traits and meat quality of goat. *Indian Journal of Animal Sciences* 64, 778–782.
Tatum, J.D., Samber, J.A., Gillmore, B.R., LeValley, S.B. and Williams, F.L. (1998) Relationship of visual assessments of feeder lamb muscularity to differences in carcass yield traits. *Journal of Animal Science* 76, 774–780.
Tsukahara, Y., Chomei, Y., Oishi, K., Kahi, A.K., Panandam, J.M., Mukherjee, T.K. and Hirooka, H. (2008) Analysis of growth patterns in purebred Kambing Katjang goat and its crosses with the German Fawn. *Small Ruminant Research* 80, 8–15.
Warmington, B.G. and Kirton, A.H. (1990) Genetic and non-genetic influence on growth and carcass traits of goats. *Small Ruminant Research* 3, 147–165.
Webb, E.C., Casey, N. and Simela, L. (2005) Goat meat quality. *Small Ruminant Research* 60, 153–166.
Wilson, T.R. (2009) Fit for purpose – the right animal in the right place. *Tropical Animal Health and Production* 41, 1081–1090.

## Additional references used in the literature database

Anandan, S., Musalia, L.M., Sastry, V.R.B. and Agrawal, D.K. (2003) Carcass characteristics of goats fed ammoniated neem (*Azadirachta indica*) seed kernel cake. *Asian–Australasian Journal of Animal Sciences* 16, 1451–1454.
Anous, M.R. and Mourad, M.M. (1993) Crossbreeding effects on reproductive traits of does and growth and carcass traits of kids. *Small Ruminant Research* 12, 141–149.
Anous, M.R. and Mourad, M.M. (2001) Some carcass characteristics of Alpine kids under intensive versus semi-intensive systems of production in France. *Small Ruminant Research* 40, 193–196.
Attah, S., Okubanjo, A.O., Omojola, A.B. and Adesehinwa, A.O.K. (2004) Body and carcass linear measurements of goats slaughtered at different weights. *Livestock Research for Rural Development* 16 <http://ftp.sunet.se/wmirror/www.cipav.org.co/lrrd/lrrd16/8/atta16062.htm>.
Bayraktarolu, E.A., Akman, N. and Tuncel, E. (1988) Effects of early castration on slaughter and carcass characteristics in crossbred Saanen × Kilis goats. *Small Ruminant Research* 1, 189–194.
Cameron, M.R., Luo, J., Sahlu, T., Hart, S.P., Coleman, S.W. and Goetsch, A.L. (2001a) Growth and slaughter traits of Boer × Spanish, Boer × Angora and Spanish goats consuming a concentrate-based diet. *Journal of Animal Science* 79, 1423–1430.

Cameron, M.R., Hart, S.P., Sahlu, T., Gilchrist, C. and Coleman, S.W. (2001b) Effects of gender and age on performance and harvest traits of Boer × Spanish goats. *Journal of Applied Animal Research* 20, 141–155.

Dhanda, J.S., Taylor, D.G., McCosker, J.E. and Murray, P.J. (1999a) The influence of goat genotype on the production of Capretto and Chevon carcasses. 1. Growth and carcass characteristics. *Meat Science* 52, 355–361.

Dhanda, J.S., Taylor, D.G., McCosker, J.E. and Murray, P.J. (1999b) The influence of goat genotype on the production of Capretto and Chevon carcasses. 3. Dissected carcass composition. *Meat Science* 52, 369–374.

El-Shahat, A.A. (1993) Density values of carcass joints of local Egyptian goats. *Small Ruminant Research* 12, 221–225.

Gallo, C., Le Breton, Y., Wainnright, I. and Berkoff, M. (1996) Body and carcass composition of male and female Criollo goats in the South of Chile. *Small Ruminant Research* 23, 163–169.

Hailu Dadi, H., Woldu, T. and Lema, T. (2005) Comparison of carcass characteristics of Borana and Arsi Bale goats under different durations of feedlot management. *Livestock Research for Rural Development* 17 <http://www.lrrd.org/lrrd17/12/dadi17137.htm>.

Hogg, B.W., Mercer, G.J.K., Mortimer, B.J., Kirton, A.H. and Duganzich, D.M. (1992) Carcass and meat quality attributes of commercial goats in New Zealand. *Small Ruminant Research* 8, 243–256.

Koyuncu, M., Duru, S., Kara Uzun, Ş., Öziş, Ş. and Tuncel, E. (2007) Effect of castration on growth and carcass traits in hair goat kids under a semi-intensive system in the South-Marmara region of Turkey. *Small Ruminant Research* 72, 38–44.

Mahgoub, O., Lu, O.D, Hameed, M.S., Richie, A., Al-Halhali, A.S. and Annamalai, K. (2005) Performance of Omani goats fed diets containing various metabolise energy densities. *Small Ruminant Research* 58, 175–180.

Melaku, S. and Betsha, S. (2008) Bodyweight and carcass characteristics of Somali goats fed hay supplemented with graded levels of peanut cake and wheat bran mixture. *Tropical Animal Health and Production* 40, 553–560.

Meneses, R.R., Rojas, A.O., Flores, H.P. and Romero, O.Y. (2004) Rendimientos y composicion de canales de cabritos Criollos e hidridos cashmere. *Archivos de Zootecnia* 53, 107–110.

Mourad, M., Gbanamou, G. and Balde, I.B. (2001) Carcass characteristics of West African Dwarf goats under extensive system. *Small Ruminant Research* 42, 81–85.

Mushi, D.E., Safari, J., Mtenga, L.A., Kifaro, G.C. and Eik, L.O. (2009) Effects of concentrate levels on fattening performance, carcass and meat quality attributes of Small East African × Norwegian crossbred goats fed low quality grass hay. *Livestock Science* 124, 148–155.

Nour, A.Y.M., Thonney, M.L., Stouffer, J.R. and White, W.R.C. (1983) Changes in carcass characteristics with increasing weight of large and small cattle. *Journal of Animal Science* 57, 1154–1172.

Phengvichith, V. and Ledin, I. (2007) Effect of a diet high in energy and protein on growth, carcass characteristics and parasite resistance in goats. *Tropical Animal Health and Production* 39, 59–70.

Simela, L., Ndlovu, L.R. and Sibanda, L.M. (1999) Carcass characteristics of the marketed Matebele goat from south-western Zimbabwe. *Small Ruminant Research* 32, 173–179.

Tadesse, M. (2007) The influence of age and feeding regimen on the carcass traits of Arsi-Bale goats. *Livestock Research for Rural* Development 19 <http://www.lrrd.org/lrrd19/4/tade19047.htm>.

Tshabalala, P.A., Strydom, P.E., Webb, E.C. and de Kock, H.L. (2003) Meat quality of designated South African indigenous goat and sheep breeds. *Meat Science* 65, 563–570.

# 4 Genetics and Breeding of Meat Goats

J.N.B. Shrestha
*Dairy and Swine Research and Development Centre, Agriculture and Agri-Food Canada, Québec, Canada*

## 4.1 Abstract

The Bezoar, Savannah and Nubian types of goat that were domesticated from their respective wild Bezoar, Markhor and Ibex ancestors are the predecessors of the 1183 breeds, populations and landraces in the world. More than 10,000 years of exposure to the forces of evolution and creative human activity have contributed towards a colossal amount of variability in morphological characteristics and production performance. Despite the available diversity, goats have not benefited from scientific achievements in quantitative genetics, nutrition and disease prevention to the same extent as other livestock and poultry species. This is because goats were often a neglected species, kept in developing countries by the poor and the landless at the end of the social scale. At the same time, goats were considered responsible for soil erosion due to controversy and ignorance ascribed to their destructive nature. Only in the last three decades has their role in alleviating poverty and sustaining food production by increasing household income gained recognition. In goats, the primary source of knowledge prior to the 1970s has been from their use as an experimental animal in biomedical research. A great deal of attention continues to be directed towards dairy and pashmina goats, often impeding efforts to allocate the resources necessary to develop meat goats. Researchers have provided irrefutable evidence to confirm that climate, terrain, breed (or population, or landrace), availability of feed and grazing land, diseases, culture, economic status of the producer and government policy, which vary from country to country and from region to region within a country, significantly influence the productivity of goats. Studies of breeds and their crosses under varying management schemes, although mostly from institutional herds, have identified breeds with the necessary potential to improve efficiency of meat-goat production. Estimates of heterosis demonstrate prospects for improving vigour, reproduction and the maternal ability of the dam, as well as survival, growth, uniformity of the carcass and meat quality of the kid. Crossbreeding of complementary breeds such as Alpine, Beetal, Boer, Damascus, Jamunapari, Nubian and Saanen with indigenous goats, as well as composite populations derived from the combination of two or more breeds, has improved the productivity of goats worldwide. Consumer acceptability of meat and meat products from crossbred animals has been well established. Genetic parameter estimates for reproduction, growth, meat quality and milk yield in goats assembled from numerous studies offer theoretical promise in direct selection for efficiency of meat production. Likewise, purebred selection has benefited from a greater proportion of additive genetic variance associated with economically important production traits. As additive genetic variance is exhausted, the role of direct selection for non-additive genetic covariance among crossbred offspring needs to be exploited. The breeding of populations with as broad a genetic base as possible is therefore critical in sustaining the genetic response to selection. Improvement in meat quality and production in goats can be accomplished with a comprehensive and technically sound assessment of important production traits that have sufficient flexibility to meet diverse environmental and managerial conditions in harmony with social, religious and cultural attributes. An integral and indispensable part of breeding strategies to maximize production efficiency in other domestic livestock and poultry has been the establishment of optimal breeding objectives along with the use of multi-trait mixed animal model methodology to obtain precise estimates of genetic parameters and the prediction of breeding values of the offspring and their parents. In practice, genetic improvement of meat goats can be accomplished by a simple procedure that involves the identification, measurement, recording, selection criteria based on realistic economic values, estimation of genetic parameters and prediction of breeding values intended for pertinent morphological characteristics and production performance. In the future, novel technology based on molecular markers associated with economically important morphological characteristics and production performance could be integrated into genetic improvement of productivity in meat goats.

## 4.2 Introduction

In the last century, advances in the theory and application of quantitative genetic principles have played an important role in exploiting the biological potential of livestock and poultry utilized in the production of milk, meat, eggs, fibre and power. General principles have included breed evaluation, crossbreeding, formation of composite populations and selection criteria based on realistic economic values. Goats in general have not benefited from scientific achievements made in genetics, nutrition and husbandry to the same extent as other livestock and poultry species. Furthermore, depiction of the goat as evil in medieval ages, along with controversy and ignorance ascribed to the destructive nature of the goat species, has not helped their development. Prior to the 1970s, the use of the goat as an experimental animal in biomedical research was the primary source of scientific knowledge (Gall, 1982). The majority of research that followed, although negligible compared with other livestock and poultry species, was from developing countries. In the goat species, more attention has been and continues to be directed towards dairy and pashmina goats (Fahmy and Shrestha, 2000). The limited number of studies on genetics and the environment influencing production traits of economic importance has impeded efforts to develop the goat as a meat animal (Shelton, 1978). Only in the last three decades has the genetics of goat meat production been featured in the scientific literature (Shrestha and Fahmy, 2005, 2007a,b).

Goat numbers and production statistics and their relation to human populations as well as economic standing are important parameters necessary for the development of national policies worldwide in support of development activities. Countries have been grouped according to gross national income (GNI) per capita into low (US$995 or less), lower middle (US$996–3945), middle (US$3946–12,195) and upper middle (US$12,196 or more) according to the World Bank. Furthermore, countries are classified according to economies into emerging and developing, and advanced as described in the World Economic Outlook Database – WEO Groups and Aggregates Information of the International Monetary Fund. The number of goats worldwide (in 1980, 1990, 2000 and 2008), goat breeds (2008), human population (2008) and GNI per capita

(World Bank, http://databank.worldbank.org/) for individual countries, continents (Africa, Asia, Europe, America and Oceania), countries grouped according to GNI per capita (low, lower middle, middle and upper middle) and economies (emerging and developing, and advanced) are presented in Table 4.1. Worldwide statistics illustrating the number of goats slaughtered, average carcass weight and total meat production are presented in Table 4.2.

In 1994, the European Union produced 75,000 million t of meat from more than 11 million goats in Greece (50%), Spain (25%), Italy (10%), France (8%) and other countries (Carbo and del la Calle, 1995). Spain produced 16,000 million t of meat from 2.8 million goats of the Serrana (30%), Murciana Granadina (20%), Malaga (10%), Canary Island (6%) and Verata (4%) breeds, while 20% were crossbreds. France with 1 million goats in 1996 produced goat meat as a by-product of the dairy industry (Decoster and Berinstain-Bailly, 1996). Australia was the leading exporter of goat meat from the Angora, Cashmere, Dairy, Feral, Boer and Boer × Cashmere populations valued at about A$20 million annually (B.A. McGregor, Victoria, Australia, 2004, personal communication). Currently, the total production of goat meat in the world, 4,938,655 million t (FAOSTAT, 2010, http://faostat.fao.org/), represents only a fraction of the meat produced by other livestock: 5% of pig meat and 8% of cattle meat. Nevertheless, there has been a steady rise in the consumption of goat meat, as a result of a growing appetite for food of exotic origin and an influx of ethnic populations into the developed countries where goat meat tends to be a by-product of the dairy and fibre industry.

The first inventory of domestic animal diversity in the world listed 570 goat breeds, types, populations and landraces, of which 187 (33%) were in Europe, 146 (26%) in Asia and the Pacific region and 89 (16%) in Africa (Scherf, 2000). Goat numbers continue to increase worldwide, although the resources allotted for their development has been negligible (French, 1970; Devendra, 1998). This is apparent when goat numbers for 1980 are compared with those in 1990, 2000 and 2008: a recurring increase of 27, 61 and 86%, respectively can be seen (Table 4.1). Emerging and developing economies with 85% of the human population share 75% of all breed populations and 97% of goats worldwide, distributed mainly in the continents of Asia (59%) and Africa (34%). The ratio of goats to humans is 1:7 in emerging and developing economies in contrast to 1:5 in advanced economies. Notably goats are concentrated in some of the poorest countries with the highest human populations, demonstrating their importance to the livelihood of the poor and landless. The countries that raised the largest number of goats in 2008 were China (17%), India (15%), Pakistan and Bangladesh (7% each), Nigeria (6%), Sudan (5%), Iran and Ethiopia (3% each), and Mongolia, Indonesia and Kenya (2% each).

In emerging and developing economies, farmers with smallholdings and the landless through the ownership of goats utilize milk for domestic consumption while meat and meat products provide much-needed cash accounting for 70–80% of the total income of the household. Sources of goat meat tend to vary according to the husbandry practised, with no specific age or weight requirement for slaughter. This is because little attention is paid to quality, as much of the meat produced is either for home consumption or sold at local markets. Previously, the total number of goats, number slaughtered, average carcass weight and total goat meat production in the world (FAO, 1999) revealed that the continents of Asia (71%) and Africa (22%) produced most of the world's goat meat. These figures exclude informal slaughter and consumption of goat meat. In 2008, the number of goats slaughtered in the emerging and developing economies was 97% of goats worldwide, distributed mainly in the continents of Asia (69%) and Africa (25%), being concentrated in the dry and humid, and tropical and subtropical climates (Table 4.2). The major countries that slaughtered goats for their meat and skins were China (34%), India (12%), Bangladesh (8%), Nigeria (5%), Pakistan and Sudan (4% each), and Ethiopia,

**Table 4.1.** Countries, continents, gross national income (GNI) per capita and economies for the years indicated in relation to numbers of goats, goat breeds, human populations and GNI per capita.

| | Number of goats[a] (×1000) | | | | Breeds[b] | Human population (×1000)[c] | |
|---|---|---|---|---|---|---|---|
| Source | 1980 | 1990 | 2000 | 2008 | 2008 | 2008 | GNI per capita (US$)[d] |
| Africa | | | | | | | |
| Algeria | 2,723 | 2,472 | 3,027 | 3,751 | 8 | 34,373 | 4,260 |
| Angola | 1,270 | 1,500 | 2,150 | 2,478 | 3 | 18,021 | 3,340 |
| Bahrain | 15 | 16 | 19 | 19 | | 776 | 25,420 |
| Benin | 931 | 872 | 1,234 | 1,472 | 4 | 8,662 | 700 |
| Botswana | 638 | 2,092 | 1,900 | 1,980 | 5 | 1,921 | 6,760 |
| Burkina Faso | 3,400 | 6,580 | 9,104 | 11,805 | 8 | 15,234 | 480 |
| Burundi | 657 | 927 | 855 | 1,650 | 2 | 8,074 | 140 |
| Cameroon | 2,340 | 3,520 | 4,410 | 4,400 | 7 | 19,088 | 1,120 |
| CapeVerde | 65 | 109 | 109 | 202 | | 499 | 2,830 |
| Central African Rep. | 956 | 1,242 | 2,614 | 4,069 | 3 | 4,339 | 410 |
| Chad | 2,620 | 2,838 | 5,179 | 6,288 | 8 | 10,914 | 540 |
| Comoros | 87 | 113 | 113 | 118 | 3 | 644 | 750 |
| Congo Democratic Rep. | 2,681 | 3,850 | 4,131 | 4,046 | 7 | 64,257 | 150 |
| Congo, Rep. of the | 159 | 278 | 280 | 295 | 1 | 3,615 | 1,810 |
| Côte d'Ivoire | 900 | 888 | 1,116 | 1,282 | 1 | 20,591 | 980 |
| Djibouti | 545 | 502 | 511 | 512 | 2 | 849 | 1,210 |
| Egypt | 1,451 | 2,400 | 3,425 | 4,473 | 10 | 81,527 | 1,800 |
| Equatorial Guinea | 7 | 8 | 9 | 9 | 1 | 659 | 14,980 |
| Eritrea | | | 1,700 | 1,730 | 6 | 4,927 | 300 |
| Ethiopia | | | 8,598 | 21,884 | 26 | 80,713 | 280 |
| Ethiopia PDR | 17,180 | 17,200 | | | | | |
| Gabon | 78 | 80 | 91 | 92 | 1 | 1,448 | 7,320 |
| Gambia | 162 | 180 | 145 | 374 | 1 | 1,660 | 400 |
| Gaza Strip | | | | | | 1,527 | |
| Ghana | 1,934 | 2,019 | 3,077 | 4,405 | 2 | 23,350 | 680 |
| Guinea | 405 | 525 | 1,008 | 1,696 | 1 | 9,833 | 350 |
| Guinea Bissau | 100 | 208 | 325 | 393 | 1 | 1,575 | 250 |
| Kenya | 8,000 | 10,186 | 9,923 | 14,478 | 10 | 38,765 | 730 |
| Lesotho | 784 | 844 | 830 | 917 | 5 | 2,049 | 1,060 |
| Liberia | 200 | 230 | 220 | 285 | 1 | 3,793 | 170 |
| Libyan Arab Jamahiriya | 1,500 | 1,100 | 1,263 | 2,500 | 1 | 6,294 | 12,380 |
| Madagascar | 1,438 | 1,256 | 1,033 | 1,260 | 3 | 19,111 | 420 |
| Malawi | 650 | 853 | 1,689 | 3,106 | 7 | 14,846 | 260 |
| Mali | 6,750 | 6,086 | 7,087 | 10,150 | 9 | 12,705 | 610 |
| Mauritania | 2,597 | 3,400 | 5,087 | 5,610 | 8 | 3,215 | 980 |
| Mauritius | 70 | 95 | 73 | 26 | 3 | 1,269 | 6,720 |
| Mayotte | | | | | | 191 | |
| Morocco | 6,154 | 5,335 | 4,931 | 5,178 | 7 | 31,606 | 2,520 |
| Mozambique | 335 | 2,000 | 4,900 | 4,325 | 7 | 22,383 | 380 |
| Namibia | 1,917 | 1,860 | 1,850 | 2,100 | 9 | 2,130 | 4,210 |
| Niger | 9,132 | 6,240 | 9,327 | 12,641 | 3 | 14,704 | 330 |

*Continued*

**Table 4.1.** Continued.

| Source | Number of goats[a] (×1000) | | | | Breeds[b] | Human population (×1000)[c] | GNI per capita (US$)[d] |
|---|---|---|---|---|---|---|---|
| | 1980 | 1990 | 2000 | 2008 | 2008 | 2008 | |
| Nigeria | 11,297 | 23,321 | 42,500 | 53,800 | 9 | 151,212 | 1,170 |
| Palestine, Occupied Tr. | | | 309 | 322 | | | |
| Qatar | 56 | 98 | 178 | 160 | | 1,281 | |
| Réunion | 33 | 31 | 38 | 40 | 1 | | |
| Rwanda | 885 | 1,075 | 757 | 1,736 | 7 | 9,721 | 410 |
| Saint Helena | 2 | 1 | 1 | 1 | | 7 | |
| Sao Tome and Principe | 4 | 4 | 5 | 5 | 5 | 160 | 1,020 |
| Senegal | 973 | 2,552 | 3,879 | 4,471 | 2 | 12,211 | 980 |
| Seychelles | 4 | 5 | 5 | 5 | 4 | 87 | 10,530 |
| Sierra Leone | 136 | 149 | 200 | 540 | 1 | 5,560 | 320 |
| Somalia | 17,000 | 18,500 | 12,300 | 12,700 | 8 | 8,926 | |
| South Africa | 5,794 | 6,100 | 6,706 | 6,529 | 15 | 48,793 | 5,870 |
| Sudan | 12,748 | 15,277 | 38,548 | 43,100 | 11 | 41,348 | 1,120 |
| Swaziland | 303 | 298 | 422 | 276 | 2 | 1,168 | 2,560 |
| Tanzania, United Rep. | 5,662 | 8,526 | 11,889 | 12,550 | 13 | 42,484 | 460 |
| Togo | 516 | 2,043 | 1,425 | 1,508 | 4 | 6,459 | 410 |
| Tunisia | 922 | 1,279 | 1,448 | 1,496 | | 10,327 | 3,540 |
| Uganda | 2,544 | 4,710 | 6,396 | 8,523 | 7 | 31,657 | 420 |
| West Bank | | | | | | 3,937 | |
| West Sahara | 147 | 162 | 172 | 173 | | 394 | |
| Zambia | 258 | 534 | 1,249 | 2,000 | 5 | 12,620 | 960 |
| Zimbabwe | 982 | 2,540 | 2,950 | 3,100 | 6 | 12,463 | |
| Asia | | | | | | | |
| Afghanistan | 2,850 | 3,350 | 7,300 | 6,386 | 6 | 29,021 | 370 |
| Armenia | | | 43 | 39 | 4 | 3,077 | 3,350 |
| Azerbaijan, Rep. | | | 494 | 587 | 2 | 8,680 | 3,830 |
| Bangladesh | 9,208 | 21,031 | 34,100 | 56,400 | 3 | 160,000 | |
| Bhutan | 16 | 37 | 31 | 30 | 1 | 687 | 1,770 |
| Brunei Darussalam | 1 | 3 | 3 | 3 | | 392 | |
| Cambodia | | | | | 1 | 14,562 | 630 |
| China | 80,762 | 98,313 | 148,401 | 149,377 | 62 | 1,324,655 | 3,060 |
| Cyprus | 220 | 208 | 346 | 368 | 5 | 862 | 26,940 |
| Dhekelia | | | | | | 16 | |
| Georgia | | | 80 | 83 | 5 | 4,307 | 2,450 |
| Hong Kong | | | | | | 6,978 | 31,420 |
| India | 86,900 | 113,200 | 123,533 | 125,732 | 39 | 1,139,965 | 1,080 |
| Indonesia | 7,691 | 11,298 | 12,566 | 15,147 | 14 | 227,345 | 2,010 |
| Iran, Islamic Rep. | 17,358 | 24,748 | 25,757 | 25,300 | 10 | 71,956 | 4,120 |
| Iraq | 2,080 | 1,550 | 1,300 | 1,475 | 6 | 30,711 | 2,060 |
| Israel | 145 | 120 | 62 | 90 | 4 | 7,309 | 24,710 |
| Japan | 67 | 35 | 35 | 15 | 4 | 127,704 | 37,930 |
| Jordan | 453 | 600 | 461 | 1,083 | 7 | 5,812 | 3,520 |
| Kazakhstan | | | 931 | 2,610 | 5 | 15,674 | 6,140 |
| Korea Dem Rep. | 490 | 650 | 276 | 3,441 | | 23,819 | |

*Continued*

**Table 4.1.** Continued.

| Source | Number of goats[a] (×1000) | | | | Breeds[b] | Human population (×1000)[c] | GNI per capita (US$)[d] |
|---|---|---|---|---|---|---|---|
| | 1980 | 1990 | 2000 | 2008 | 2008 | 2008 | |
| Korea, Rep. | 201 | 211 | 445 | 266 | 7 | 48,607 | 21,570 |
| Kuwait | 273 | 40 | 153 | 160 | | 2,728 | |
| Kyrgyzstan | | | 543 | 873 | 8 | 5,278 | 790 |
| Lao People's Dem. Rep. | 49 | 139 | 122 | 289 | 3 | 6,205 | 750 |
| Lebanon | 444 | 435 | 417 | 450 | 2 | 4,194 | 6,860 |
| Macau | | | | | | 526 | |
| Malaysia | 342 | 331 | 238 | 285 | 11 | 27,014 | 7,250 |
| Maldives | | | | | 2 | 305 | 3,690 |
| Mongolia | 4,715 | 4,959 | 11,034 | 19,969 | 9 | 2,641 | 1,670 |
| Myanmar | 610 | 1,036 | 1,392 | 2,624 | 3 | 49,563 | |
| Nepal | 4,650 | 5,324 | 6,325 | 8,136 | 11 | 28,810 | 400 |
| Oman | 630 | 720 | 979 | 1,620 | 6 | 2,785 | 17,890 |
| Pakistan | 24,953 | 35,446 | 47,426 | 56,742 | 36 | 166,112 | 950 |
| Philippines | 2,960 | 4,790 | 6,245 | 4,174 | 9 | 90,348 | 1,700 |
| Saudi Arabia | 2,240 | 3,406 | 2,462 | 2,200 | 4 | 24,807 | 17,700 |
| Singapore | 0.8 | 0.2 | 0.5 | 0.6 | | 4,839 | 37,650 |
| Sri Lanka | 493 | 522 | 495 | 377 | 7 | 20,156 | 1,780 |
| Syrian Arab Rep. | 1,026 | 1,000 | 1,050 | 1,579 | 4 | 19,748 | 2,150 |
| Taiwan | | | | | | 22,921 | |
| Tajikistan | | | 706 | 1,424 | 8 | 6,836 | 600 |
| Thailand | 56 | 121 | 144 | 374 | 2 | 65,493 | 3,670 |
| Timor-Leste | 26 | 97 | 75 | 137 | | 1,098 | 2,460 |
| Turkey | 18,775 | 11,942 | 7,774 | 6,286 | 11 | 73,914 | 8,890 |
| Turkmenistan | | | 500 | 900 | 3 | 5,044 | 2,760 |
| United Arab Emirates | 342 | 657 | 1,279 | 1,570 | | 4,485 | |
| Uzbekistan | | | 886 | 2,000 | 7 | 27,314 | 910 |
| Vietnam | 173 | 372 | 544 | 1,484 | 9 | 86,211 | 910 |
| Yemen | 2,898 | 5,333 | 6,918 | 8,708 | 8 | 22,917 | 960 |
| Europe | | | | | | | |
| Albania | 811 | 1,144 | 1,104 | 820 | 20 | 3,143 | 3,840 |
| Austria | 35 | 36 | 72 | 60 | 13 | 8,337 | 46,350 |
| Belarus | | | 58 | 72 | | 9,681 | 5,380 |
| Belgium | | | 16 | 31 | 8 | 10,708 | 45,010 |
| Belgium Luxembourg | 6 | 9 | | | | | |
| Bosnia and Herzegovina | | | 98 | 70 | 1 | 3,773 | 4,530 |
| Bulgaria | 433 | 433 | 1,046 | 495 | | 7,623 | 5,390 |
| Croatia | | | 79 | 84 | 5 | 4,434 | 13,580 |
| Czech Rep. | | | 32 | 17 | 6 | 10,424 | 16,670 |
| Czechoslovakia | 63 | 50 | | | | | |
| Denmark | | | | | 3 | 5,493 | 58,550 |
| Estonia | | | 3 | 4 | 1 | 1,340 | 14,410 |
| Finland | 2 | 4 | 9 | 6 | 1 | 5,313 | 47,630 |
| France | 1,125 | 1,226 | 1,211 | 1,224 | 14 | 62,279 | 42,190 |
| Germany | 61 | 90 | 135 | 190 | 22 | 82,110 | 42,800 |

*Continued*

**Table 4.1.** Continued.

| Source | Number of goats[a] (×1000) 1980 | 1990 | 2000 | 2008 | Breeds[b] 2008 | Human population (×1000)[c] 2008 | GNI per capita (US$)[d] |
|---|---|---|---|---|---|---|---|
| Greece | 4,532 | 5,348 | 5,614 | 5,346 | 7 | 11,237 | 27,650 |
| Hungary | 15 | 16 | 189 | 67 | 11 | 10,038 | 12,800 |
| Iceland | 0.2 | 0.3 | 0.4 | 0.6 | 1 | 317 | 49,360 |
| Ireland | | | 8 | 9 | 8 | 4,426 | 49,480 |
| Italy | 978 | 1,246 | 1,397 | 920 | 54 | 58,832 | 35,230 |
| Latvia | | | 8 | 13 | 3 | 2,266 | 11,940 |
| Liechtenstein | 0.1 | 0.2 | 0.3 | 0.3 | | 36 | 113,210 |
| Lithuania | | | 25 | 20 | 6 | 3,358 | 11,890 |
| Luxembourg | | | 1 | 3 | | 489 | 74,890 |
| Malta | 6 | 6 | 5 | 6 | 1 | 412 | |
| Moldova, Rep. | | | 99 | 99 | 2 | 3,633 | 1,500 |
| Netherlands | 30 | 61 | 179 | 390 | 14 | 16,445 | 48,990 |
| Norway | 85 | 89 | 76 | 70 | 2 | 4,768 | 86,670 |
| Poland | | | 190 | 136 | 7 | 38,126 | 11,820 |
| Portugal | 733 | 857 | 630 | 496 | 6 | 10,622 | 20,540 |
| Romania | 375 | 1,017 | 558 | 865 | 2 | 21,514 | 8,290 |
| Russian Federation | | | 2,148 | 2,213 | 15 | 141,950 | 9,650 |
| Serbia | | | | 154 | 0.5 | 7,350 | 5,520 |
| Serbia and Montenegro | | | 241 | | | | |
| Slovakia | | | 51 | 37 | 3 | 5,406 | 16,590 |
| Slovenia | | | 15 | 28 | 5 | 2,021 | 24,280 |
| Spain | 2,100 | 3,780 | 2,627 | 2,959 | 24 | 45,556 | 31,630 |
| Sweden | | | | | 3 | 9,220 | 52,460 |
| Switzerland | 80 | 68 | 62 | 81 | 11 | 7,648 | 56,370 |
| Ukraine | | | 825 | 645 | 3 | 46,258 | 3,210 |
| UK | | 114 | 77 | 95 | 12 | 60,944 | 46,150 |
| USSR | 5,824 | 6,562 | | | | | |
| Former Yug. Rep. Macedonia | | | 65 | 133 | 2 | 2,041 | 4,120 |
| America | | | | | | | |
| Antigua and Barbuda | 12 | 12 | 34 | 37 | 6 | 87 | 13,020 |
| Argentina | 3,000 | 3,300 | 3,490 | 4,250 | 13 | 39,883 | 7,190 |
| Aruba | | 3 | | | | 105 | |
| Bahamas | 13 | 14 | 14 | 15 | | 338 | |
| Barbados | 15 | 4 | 5 | 5 | 5 | 255 | 520 |
| Belize | 1 | 0.1 | 0.1 | 0.2 | | 322 | 3,740 |
| Bermuda | 0.5 | 0.6 | 0.3 | 0.4 | | 64 | |
| Bolivia | 2,007 | 1,445 | 1,714 | 1,979 | 5 | 9,694 | 1,460 |
| Brazil | 8,326 | 11,895 | 9,347 | 9,355 | 21 | 191,972 | 7,490 |
| British Virgin Islands | 12 | 10 | 10 | 10 | | | |
| Canada | 22 | 27 | 30 | 30 | 11 | 33,311 | 43,490 |
| Cayman Islands | 0.3 | 0.3 | 0.3 | 2 | | 54 | |
| Chile | 600 | 600 | 740 | 740 | 8 | 16,804 | 9,470 |

*Continued*

**Table 4.1.** Continued.

| Source | Number of goats[a] (×1000) | | | | Breeds[b] | Human population (×1000)[c] | GNI per capita (US$)[d] |
|---|---|---|---|---|---|---|---|
| | 1980 | 1990 | 2000 | 2008 | 2008 | 2008 | |
| Colombia | 645 | 959 | 1,185 | 1,200 | 4 | 45,012 | 4,610 |
| Costa Rica | 1 | 2 | 3 | 5 | 5 | 4,519 | 6,060 |
| Cuba | 99 | 90 | 715 | 1,134 | 7 | 11,205 | |
| Dominica | 6 | 10 | 10 | 10 | 6 | 73 | 4,790 |
| Dominican Rep. | 451 | 550 | 178 | 190 | 0.7 | 9,953 | 4,330 |
| Ecuador | 257 | 311 | 280 | 150 | 1 | 13,481 | 3,700 |
| El Salvador | 14 | 15 | 13 | 11 | 5 | 6,134 | 3,460 |
| French Guiana | 1 | 1 | 1 | 1 | | | |
| Grenada | 13 | 10 | 7 | 7 | 3 | 104 | 5,870 |
| Guadeloupe | 64 | 69 | 34 | 48 | 2 | | |
| Guatemala | 72 | 105 | 111 | 101 | 5 | 13,686 | 2,680 |
| Guyana | 70 | 78 | 79 | 79 | 3 | 763 | 1,450 |
| Haiti | 1,000 | 1,110 | 1,942 | 1,910 | 5 | 9,876 | |
| Honduras | 24 | 26 | 31 | 25 | 7 | 7,319 | |
| Jamaica | 380 | 440 | 440 | 440 | 7 | 2,687 | 4,800 |
| Martinique | 9 | 28 | 14 | 14 | 0.1 | | |
| Mexico | 9,638 | 10,439 | 8,704 | 8,831 | 12 | 106,350 | 10,000 |
| Montserrat | 6 | 7 | 7 | 7 | 1 | | |
| Netherlands | 21 | 14 | 14 | 14 | 1 | 195 | |
| Nicaragua | 6 | 6 | 7 | 7 | 5 | 5,667 | 1,050 |
| Panama | 6 | 5 | 5 | 6 | 4 | 3,399 | 6,290 |
| Paraguay | 115 | 148 | 123 | 130 | 6 | 6,238 | 2,140 |
| Peru | 1,699 | 1,722 | 2,023 | 1,904 | 4 | 28,837 | 3,990 |
| Puerto Rico | 25 | 21 | 9 | 3 | 1 | 3,955 | |
| St Kitts and Nevis | 10 | 10 | 14 | 9 | 6 | 49 | 11,210 |
| St Lucia | 10 | 12 | 10 | 9 | 5 | 170 | 5,430 |
| St Pierre and Miquelon | | 0.04 | | | | | |
| St Vincent and Grenadines | 4 | 6 | 6 | | 1 | 109 | 5,130 |
| Suriname | 6 | 10 | 7 | 4 | 2 | 515 | 4,760 |
| Trinidad and Tobago | 50 | 57 | 58 | 60 | 6 | 1,333 | 15,580 |
| USA | 1,400 | 1,900 | 2,300 | 3,118 | 16 | 304,375 | 48,190 |
| Uruguay | 12 | 14 | 15 | 17 | 5 | 3,334 | 8,020 |
| US Virgin Islands | 6 | 4 | 4 | 4 | | 110 | |
| Venezuela, Bolivar Rep. | 1,338 | 1,650 | 1,205 | 1,415 | 7 | 27,935 | 9,170 |
| Oceania | | | | | | | |
| Australia | 65 | 1,630 | 1,905 | 3,000 | 14 | 21,432 | 41,890 |
| Cook Islands | 3 | 5 | 3 | 1 | 3 | | |
| Fiji | 110 | 170 | 241 | 250 | 1 | 844 | 4,060 |
| French Polynesia | 12 | 14 | 17 | 17 | | 266 | |
| Guam | 0.5 | 0.6 | 0.7 | 0.7 | | 176 | |
| Micronesia, Fed. States | | | 4 | 4 | | 110 | 2,460 |
| New Caledonia | 8 | 17 | 10 | 8 | | 247 | |

*Continued*

**Table 4.1.** Continued.

| Source | Number of goats[a] (×1000) | | | | Breeds[b] | Human population (×1000)[c] | GNI per capita (US$)[d] |
|---|---|---|---|---|---|---|---|
| | 1980 | 1990 | 2000 | 2008 | 2008 | 2008 | |
| New Zealand | 53 | 1,063 | 183 | 96 | 4 | 4,269 | 26,830 |
| Pacific Islands Trust Tr. | 4 | 4 | | | | | |
| Papua New Guinea | 2 | 2 | 2 | 3 | 1 | 6,577 | 1,090 |
| Tonga | 13 | 16 | 13 | 13 | 2 | 104 | 3,240 |
| Tuvalu | 0.04 | | | | | | |
| Vanuatu | 11 | 11 | 12 | 19 | 1 | 234 | 2,490 |
| Wallis and Futuna | 7 | 7 | 7 | 7 | | | |
| Continents | | | | | | | |
| Africa | 141,520 | 177,766 | 235,999 | 296,606 | 289 | 995,514 | 2,650 |
| Asia | 273,757 | 351,366 | 454,592 | 509,224 | 341 | 3,999,738 | 7,484 |
| Europe | 17,294 | 22,157 | 18,765 | 17,860 | 301 | 733,912 | 34,083 |
| America | 31,464 | 37,139 | 34,937 | 37,286 | 220 | 910,363 | 8,904 |
| Oceania | 288 | 2,939 | 2,396 | 3,418 | 32 | 35,275 | 8,314 |
| GNI (US$/year: 2008)[c] | | | | | | | |
| Low (≤$995) | 90,579 | 120,051 | 175,035 | 230,442 | 267 | 898,530 | 551 |
| Lower middle ($996–3,945) | 228,622 | 294,461 | 410,439 | 442,279 | 303 | 3,440,275 | 2,342 |
| Middle ($3,946–12,195) | 75,206 | 83,412 | 80,404 | 82,230 | 246 | 1,003,648 | 6,863 |
| Upper middle (≥$12,196) | 16,403 | 23,449 | 22,615 | 25,547 | 323 | 1,036,540 | 40,429 |
| Economies[d] | | | | | | | |
| Emerging and developing | 451,954 | 572,791 | 728,516 | 844,768 | 889 | 5,678,394 | 4,136 |
| Advanced | 12,003 | 18,170 | 17,524 | 18,953 | 283 | 984,189 | 40,962 |
| Total | 464,323 | 591,367 | 746,689 | 864,394 | 1,183 | 6,674,801 | 12,687 |

[a]DAD-IS: http://dad.fao.org/.
[b]FAOSTAT: http://faostat.fao.org/.
[c]World Bank: http://databank.worldbank.org/.
[d]International Monetary Fund: http://www.imf.org/.

Iran and Indonesia (2% each). The ranking of countries based on total goat meat produced was as follows: China (38%), India (10%), Nigeria (6%), Pakistan (5%), Bangladesh and Sudan (4% each) and Iran (2%). The following countries each share 1% of the total goat meat production in the world: Indonesia, Ethiopia, Greece, Philippines, Niger, Mongolia, Kenya, Nepal, Mexico, Mali, Somalia, Afghanistan, Turkey, South Africa, Tanzania, Burkina Faso, Yemen, Brazil, Uganda and Saudi Arabia.

In the years following World War II, more emphasis was placed on improving efficiency to encourage the production of cheaper commodities to meet the growing demand of the increasing number of affluent families worldwide. This has increased

**Table 4.2.** Countries, continents, gross national income (GNI) per capita[a] and economies[b] in relation to the number of goats slaughtered, average carcass weight and total meat production[c].

| Source | Goats slaughtered (×1000) | | | | Average carcass weight (kg) | | | | Total meat production (million t) | | | |
|---|---|---|---|---|---|---|---|---|---|---|---|---|
| | 1980 | 1990 | 2000 | 2008 | 1980 | 1990 | 2000 | 2008 | 1980 | 1990 | 2000 | 2008 |
| Africa | | | | | | | | | | | | |
| Algeria | 926 | 841 | 1,230 | 1,410 | 10 | 10 | 10 | 10 | 9,259 | 8,405 | 12,300 | 14,100 |
| Angola | 380 | 450 | 645 | 743 | 9 | 9 | 15 | 15 | 3,420 | 4,050 | 9,675 | 11,149 |
| Bahrain | 60 | 25 | 10 | 380 | 15 | 15 | 15 | 15 | 900 | 375 | 150 | 5,700 |
| Benin | 279 | 305 | 408 | 491 | 10 | 10 | 10 | 10 | 2,792 | 3,050 | 4,076 | 4,914 |
| Botswana | 219 | 460 | 440 | 460 | 12 | 12 | 12 | 12 | 2,630 | 5,520 | 5,280 | 5,520 |
| Burkina Faso | 1,100 | 2,105 | 2,913 | 3,736 | 7 | 8 | 8 | 8 | 7,700 | 17,053 | 23,594 | 30,259 |
| Burundi | 197 | 370 | 324 | 610 | 10 | 10 | 9 | 10 | 1,971 | 3,700 | 2,850 | 6,100 |
| Cameroon | 702 | 1,254 | 1,570 | 1,570 | 10 | 10 | 10 | 10 | 7,020 | 12,540 | 15,700 | 15,700 |
| Cape Verde | 23 | 39 | 45 | 85 | 10 | 10 | 10 | 10 | 230 | 394 | 450 | 846 |
| Central African Rep. | 170 | 267 | 530 | 809 | 16 | 16 | 19 | 19 | 2,720 | 4,278 | 10,000 | 14,800 |
| Chad | 700 | 760 | 1,557 | 1,970 | 13 | 12 | 12 | 12 | 9,100 | 8,740 | 18,684 | 23,640 |
| Comoros | 27 | 35 | 35 | 38 | 10 | 10 | 10 | 10 | 270 | 350 | 350 | 380 |
| Congo Dem. Rep. | 736 | 1,492 | 1,638 | 1,602 | 10 | 11 | 12 | 11 | 7,300 | 17,097 | 19,000 | 17,753 |
| Congo Rep. | 48 | 83 | 85 | 87 | 9 | 9 | 9 | 9 | 430 | 749 | 760 | 783 |
| Côte d'Ivoire | 470 | 395 | 314 | 370 | 10 | 10 | 9 | 9 | 4,700 | 3,950 | 2,919 | 3,471 |
| Djibouti | 220 | 178 | 188 | 188 | 13 | 13 | 13 | 13 | 2,744 | 2,222 | 2,350 | 2,350 |
| Egypt | 1,176 | 1,528 | 1,351 | 970 | 18 | 18 | 19 | 19 | 21,000 | 27,500 | 25,000 | 18,000 |
| Equatorial Guinea | 2 | 3 | 4 | 4 | 11 | 11 | 11 | 11 | 22 | 31 | 42 | 44 |
| Eritrea | | | 680 | 683 | | | 9 | 9 | | | 5,800 | 5,800 |
| Ethiopia | | | 3 007 | 7,600 | | | 9 | 9 | | | 25,560 | 64,600 |
| Ethiopia PDR | 6,800 | 7,850 | | | | | 9 | 8 | | | 57,800 | 66,700 |
| Gabon | 21 | 23 | 27 | 28 | 10 | 10 | 10 | 10 | 210 | 225 | 270 | 280 |
| Gambia | 49 | 54 | 36 | 94 | 11 | 11 | 11 | 11 | 534 | 592 | 394 | 1,028 |
| Ghana | 480 | 505 | 769 | 1,035 | 10 | 10 | 13 | 13 | 4,560 | 4,794 | 10,150 | 13,662 |
| Guinea | 84 | 118 | 398 | 678 | 16 | 16 | 11 | 12 | 1,344 | 1,887 | 4,459 | 8,426 |

*Continued*

**Table 4.2.** Continued.

| Source | Goats slaughtered (× 1000) | | | | Average carcass weight (kg) | | | | Total meat production (million t) | | | |
|---|---|---|---|---|---|---|---|---|---|---|---|---|
| | 1980 | 1990 | 2000 | 2008 | 1980 | 1990 | 2000 | 2008 | 1980 | 1990 | 2000 | 2008 |
| Guinea-Bissau | 55 | 62 | 98 | 121 | 9 | 9 | 9 | 9 | 495 | 561 | 877 | 1,095 |
| Kenya | 1,600 | 2,800 | 2,800 | 4,200 | 11 | 11 | 11 | 11 | 17,600 | 30,800 | 30,800 | 46,200 |
| Lesotho | 260 | 330 | 245 | 225 | 8 | 8 | 9 | 9 | 2,080 | 2,640 | 2,205 | 2,025 |
| Liberia | 63 | 74 | 74 | 89 | 9 | 9 | 9 | 9 | 567 | 666 | 666 | 800 |
| Libyan Arab Jamahiriya | 420 | 350 | 400 | 790 | 12 | 15 | 15 | 15 | 5,050 | 5,250 | 6,000 | 11,850 |
| Madagascar | 486 | 420 | 351 | 425 | 15 | 15 | 15 | 15 | 7,291 | 6,300 | 5,265 | 6,375 |
| Malawi | 195 | 256 | 600 | 1,630 | 12 | 12 | 12 | 12 | 2,341 | 3,072 | 7,200 | 19,557 |
| Mali | 1,900 | 1,666 | 1,976 | 3,045 | 13 | 14 | 14 | 14 | 24,700 | 23,321 | 27,665 | 42,631 |
| Mauritania | 505 | 521 | 830 | 970 | 15 | 15 | 15 | 15 | 7,575 | 7,815 | 12,450 | 14,550 |
| Morocco | 1,850 | 1,550 | 1,550 | 1,980 | 10 | 14 | 14 | 11 | 19,000 | 21,500 | 22,000 | 22,000 |
| Mozambique | 134 | 800 | 2,050 | 1,770 | 12 | 12 | 12 | 12 | 1,608 | 9,600 | 24,600 | 21,240 |
| Namibia | 380 | 370 | 400 | 329 | 12 | 12 | 12 | 12 | 4,560 | 4,440 | 4,800 | 3,840 |
| Niger | 3,020 | 2,048 | 2,986 | 4,400 | 12 | 12 | 12 | 12 | 36,240 | 24,576 | 35,832 | 52,800 |
| Nigeria | 4,650 | 9,500 | 17,420 | 21,318 | 13 | 13 | 13 | 13 | 59,054 | 120,649 | 221,234 | 270,742 |
| Palestine, Occupied Tr. | | | 184 | 230 | | | 21 | 22 | | | 3,946 | 4,945 |
| Qatar | 21 | 24 | 48 | 46 | 14 | 14 | 14 | 14 | 294 | 341 | 669 | 644 |
| Réunion | 5 | 7 | 8 | 9 | 12 | 11 | 11 | 11 | 61 | 76 | 86 | 94 |
| Rwanda | 270 | 341 | 242 | 555 | 11 | 11 | 11 | 11 | 2,970 | 3,751 | 2,662 | 6,105 |
| Sao Tome and Principe | 1 | 2 | 2 | 2 | 11 | 11 | 11 | 11 | 13 | 16 | 18 | 19 |
| Senegal | 337 | 895 | 1,009 | 1,219 | 10 | 10 | 10 | 11 | 3,370 | 8,950 | 10,422 | 13,736 |
| Seychelles | 1 | 2 | 2 | 2 | 11 | 11 | 11 | 11 | 13 | 17 | 20 | 20 |
| Sierra Leone | 34 | 37 | 47 | 128 | 9 | 9 | 9 | 9 | 306 | 333 | 423 | 1,152 |
| Somalia | 3,108 | 4,190 | 2,500 | 3,250 | 13 | 13 | 13 | 13 | 40,403 | 54,469 | 32,500 | 42,250 |
| South Africa | 1,820 | 2,100 | 2,240 | 2,310 | 16 | 16 | 16 | 16 | 28,400 | 34,600 | 36,000 | 37,190 |
| Sudan | 3,382 | 2,691 | 9,077 | 14,530 | 13 | 13 | 13 | 13 | 43,966 | 34,987 | 118,000 | 188,900 |
| Swaziland | 145 | 115 | 160 | 103 | 18 | 18 | 18 | 18 | 2,610 | 2,070 | 2,880 | 1,854 |

| | | | | | | | | | | | | |
|---|---|---|---|---|---|---|---|---|---|---|---|---|
| Tanzania, United Rep. | 1,278 | 1,790 | 2,450 | 2,550 | 12 | 12 | 12 | 12 | 15,332 | 21,484 | 29,400 | 30,600 |
| Togo | 129 | 520 | 399 | 420 | 9 | 9 | 9 | 9 | 1,161 | 4,680 | 3,591 | 3,777 |
| Tunisia | 489 | 744 | 850 | 870 | 6 | 10 | 11 | 12 | 3,130 | 7,300 | 9,200 | 10,500 |
| Uganda | 890 | 1,500 | 2,050 | 2,400 | 12 | 12 | 12 | 12 | 10,680 | 18,000 | 24,600 | 29,000 |
| West Sahara | 42 | 50 | 54 | 55 | 12 | 12 | 12 | 12 | 504 | 600 | 648 | 660 |
| Zambia | 77 | 160 | 387 | 640 | 12 | 12 | 12 | 12 | 927 | 1,922 | 4,644 | 7,680 |
| Zimbabwe | 255 | 870 | 1,100 | 1,147 | 12 | 12 | 12 | 12 | 3,063 | 10,440 | 13,200 | 13,764 |
| Asia | | | | | | | | | | | | |
| Afghanistan | 1,380 | 1,540 | 3,700 | 3,200 | 13 | 13 | 13 | 13 | 17,940 | 20,020 | 48,100 | 41,600 |
| Bangladesh | 4,604 | 10,516 | 18,400 | 30,000 | 5 | 7 | 7 | 7 | 23,673 | 73,639 | 129,000 | 210,000 |
| Bhutan | 8 | 16 | 17 | 15 | 9 | 9 | 9 | 9 | 74 | 139 | 153 | 135 |
| Brunei Darussalam | 0.1 | 1 | 3 | 4 | 10 | 10 | 10 | 10 | 1 | 11 | 25 | 35 |
| China | 18,841 | 44,042 | 96,816 | 133,340 | 11 | 12 | 13 | 14 | 200,500 | 520,571 | 1,211,688 | 1,828,316 |
| Cyprus | 167 | 185 | 243 | 166 | 13 | 23 | 26 | 23 | 2,200 | 4,200 | 6,300 | 3,874 |
| India | 30,240 | 43,000 | 43,931 | 47,775 | 10 | 10 | 10 | 10 | 302,400 | 430,000 | 469,000 | 477,750 |
| Indonesia | 3,630 | 5,830 | 6,900 | 6,602 | 10 | 10 | 10 | 10 | 36,300 | 58,300 | 44,890 | 66,027 |
| Iran, Islamic Rep. | 5,200 | 7,100 | 7,821 | 7,550 | 14 | 14 | 14 | 14 | 72,800 | 99,500 | 109,500 | 106,000 |
| Iraq | 760 | 652 | 590 | 610 | 12 | 12 | 12 | 24 | 9,120 | 7,827 | 7,080 | 14,335 |
| Israel | 70 | 70 | 150 | 247 | 16 | 16 | 16 | 15 | 1,120 | 1,120 | 2,400 | 3,702 |
| Jordan | 136 | 81 | 138 | 314 | 15 | 23 | 11 | 13 | 1,980 | 1,900 | 1,568 | 4,210 |
| Kazakhstan | | | 278 | 1,319 | | | 14 | 16 | | | 3,900 | 20,631 |
| Korea Dem. Rep. | 147 | 195 | 680 | 940 | 15 | 15 | 15 | 15 | 2,205 | 2,925 | 10,200 | 14,100 |
| Korea Rep. of | 56 | 64 | 185 | 105 | 15 | 15 | 15 | 15 | 840 | 960 | 2,775 | 1,575 |
| Kuwait | 130 | 10 | 43 | 45 | 13 | 13 | 13 | 13 | 1,690 | 130 | 559 | 585 |
| Kyrgyzstan | | | 200 | 420 | | | 18 | 18 | | | 3,656 | 7,365 |
| Lao People's Dem. Rep. | 8 | 24 | 31 | 80 | 14 | 14 | 14 | 14 | 105 | 331 | 429 | 1,120 |
| Lebanon | 204 | 182 | 140 | 228 | 18 | 18 | 18 | 18 | 3,678 | 3,276 | 2,520 | 4,098 |
| Malaysia | 85 | 52 | 66 | 94 | 9 | 9 | 9 | 9 | 775 | 472 | 602 | 855 |
| Mongolia | 1,600 | 1,551 | 2,402 | 3,044 | 15 | 16 | 12 | 14 | 24,000 | 24,700 | 30,000 | 47,510 |
| Myanmar | 216 | 527 | 907 | 2,504 | 10 | 10 | 10 | 9 | 2,155 | 5,271 | 9,070 | 23,700 |
| Nepal | 2,611 | 2,609 | 3,268 | 4,240 | 9 | 11 | 11 | 11 | 23,500 | 28,896 | 36,930 | 46,188 |

*Continued*

**Table 4.2.** Continued.

| Source | Goats slaughtered (×1000) | | | | Average carcass weight (kg) | | | | Total meat production (million t) | | | |
|---|---|---|---|---|---|---|---|---|---|---|---|---|
| | 1980 | 1990 | 2000 | 2008 | 1980 | 1990 | 2000 | 2008 | 1980 | 1990 | 2000 | 2008 |
| Oman | 140 | 179 | 190 | 550 | 25 | 25 | 25 | 43 | 3,500 | 4,475 | 4,750 | 23,500 |
| Pakistan | 16,310 | 19,141 | 18,760 | 15,350 | 10 | 15 | 17 | 17 | 157,000 | 296,000 | 310,000 | 261,000 |
| Philippines | 1,184 | 1,915 | 3,151 | 3,037 | 12 | 14 | 11 | 17 | 14,000 | 26,704 | 33,721 | 53,152 |
| Saudi Arabia | 379 | 1,627 | 2,079 | 1,730 | 13 | 14 | 11 | 14 | 5,000 | 22,900 | 22,200 | 24,900 |
| Singapore | 0.5 | 1 | 0.6 | 1 | 15 | 11 | 12 | 12 | 7 | 16 | 7 | 11 |
| Sri Lanka | 137 | 83 | 85 | 58 | 20 | 20 | 20 | 20 | 2,730 | 1,640 | 1,690 | 1,160 |
| Syrian Arab Rep. | 397 | 362 | 281 | 338 | 17 | 17 | 17 | 25 | 6,544 | 5,976 | 4,633 | 8,444 |
| Thailand | 17 | 36 | 43 | 93 | 15 | 15 | 15 | 15 | 250 | 541 | 649 | 1,395 |
| Timor-Leste | 8 | 29 | 23 | 42 | 10 | 10 | 10 | 10 | 80 | 290 | 225 | 420 |
| Turkey | 3,830 | 4,204 | 3,400 | 2,700 | 14 | 16 | 16 | 15 | 52,600 | 66,000 | 53,000 | 41,600 |
| Turkmenistan | | | 300 | 413 | | | 15 | 16 | | | 4,500 | 6,400 |
| United Arab Emirates | 155 | 270 | 1,400 | 950 | 16 | 16 | 16 | 16 | 2,480 | 4,320 | 22,400 | 15,200 |
| Vietnam | 87 | 186 | 320 | 754 | 15 | 15 | 15 | 15 | 1,305 | 2,790 | 4,800 | 11,310 |
| Yemen | 1,350 | 1,766 | 2,254 | 2,958 | 10 | 10 | 10 | 10 | 13,500 | 17,660 | 22,540 | 29,579 |
| Europe | | | | | | | | | | | | |
| Albania | 332 | 377 | 900 | 725 | 12 | 11 | 8 | 8 | 3,900 | 4,000 | 7,200 | 5,800 |
| Austria | 37 | 31 | 66 | 45 | 11 | 9 | 12 | 13 | 410 | 280 | 799 | 583 |
| Belgium | | | 3 | 2 | | | 20 | 21 | | | 56 | 45 |
| Bulgaria | 299 | 278 | 600 | 463 | 11 | 11 | 13 | 11 | 3,301 | 2,974 | 7,500 | 5,115 |
| Croatia | | | 13 | 12 | | | 18 | 18 | | | 225 | 216 |
| Czech Rep. | | | 34 | 16 | | | 9 | 8 | | | 300 | 119 |
| Czechoslovakia | 37 | 34 | | | 11 | 12 | | | 405 | 411 | | |
| Estonia | | | | 1 | | | | 20 | | | | 24 |
| France | 1,126 | 1,298 | 847 | 846 | 7 | 7 | 8 | 8 | 8,000 | 9,104 | 6,600 | 7,100 |
| Germany | 11 | 9 | 17 | 26 | 9 | 18 | 18 | 18 | 100 | 165 | 306 | 478 |
| Greece | 4,114 | 5,252 | 4,458 | 5,174 | 10 | 10 | 10 | 11 | 41,984 | 50,715 | 44,200 | 55,500 |
| Hungary | | 5 | 25 | 21 | | 13 | 13 | 14 | | 65 | 325 | 305 |
| Italy | 689 | 557 | 453 | 311 | 7 | 7 | 8 | 8 | 5,110 | 3,972 | 3,687 | 2,372 |
| Lithuania | | | 20 | 20 | | | 29 | 15 | | | 559 | 300 |
| Malta | 2 | 1 | 4 | 3 | 20 | 17 | 10 | 11 | 31 | 12 | 40 | 35 |

| | | | | | | | | | | | | |
|---|---|---|---|---|---|---|---|---|---|---|---|---|
| Netherlands | 27 | 69 | 23 | 50 | 14 | 13 | 13 | 13 | 381 | 897 | 305 | 650 |
| Norway | 26 | 28 | 21 | 24 | 11 | 13 | 12 | 13 | 286 | 356 | 257 | 306 |
| Portugal | 560 | 460 | 256 | 143 | 7 | 6 | 8 | 6 | 3,844 | 2,911 | 2,105 | 889 |
| Romania | 312 | 858 | 490 | 542 | 13 | 8 | 8 | 8 | 3,900 | 7,000 | 3,966 | 4,154 |
| Russian Federation | | | 1,000 | 977 | | | 20 | 18 | | | 20,316 | 17,923 |
| Serbia and Montenegro | | | 41 | | | | 18 | | | | | |
| Slovakia | | | 23 | 35 | | | 8 | 8 | | | | 298 |
| Slovenia | | | 15 | 24 | | | 13 | 16 | | | 200 | 329 |
| Spain | 1,416 | 2,164 | 1,951 | 1,220 | 10 | 8 | 8 | 8 | 13,682 | 16,417 | 16,488 | 9,253 |
| Switzerland | 50 | 46 | 33 | 32 | 13 | 13 | 15 | 17 | 648 | 576 | 505 | 536 |
| Ukraine | | | 619 | 543 | | | 13 | 16 | | | 8,000 | 8,750 |
| USSR | 1,960 | 2,375 | | | 15 | 16 | | | 30,000 | 38,000 | | |
| Americas | | | | | | | | | | | | |
| Antigua and Barbuda | 4 | 4 | 11 | 13 | 10 | 10 | 10 | 10 | 36 | 39 | 105 | 125 |
| Argentina | 950 | 1,000 | 1,364 | 1,500 | 7 | 7 | 7 | 7 | 6,270 | 6,600 | 9,002 | 9,900 |
| Bahamas | 5 | 5 | 5 | 6 | 12 | 12 | 12 | 12 | 55 | 57 | 62 | 72 |
| Barbados | 6 | 2 | 2 | 2 | 14 | 14 | 14 | 14 | 81 | 23 | 25 | 28 |
| Belize | 0.4 | 0.05 | 0.06 | 0.08 | 12 | 12 | 12 | 14 | 4 | | | |
| Bermuda | 0.2 | 0.3 | 0.1 | 0.1 | 12 | 12 | 12 | 12 | 2 | 3 | 1 | 1 |
| Bolivia | 480 | 389 | 524 | 525 | 11 | 11 | 11 | 11 | 5,280 | 4,279 | 5,764 | 5,775 |
| Brazil | 2,000 | 3,000 | 2,550 | 2,570 | 11 | 11 | 11 | 11 | 22,800 | 34,200 | 29,300 | 29,450 |
| British Virgin Islands | 3 | 3 | 3 | 3 | 12 | 12 | 12 | 12 | 40 | 36 | 36 | 36 |
| Chile | 240 | 240 | 297 | 320 | 18 | 18 | 18 | 18 | 4,320 | 4,320 | 5,346 | 5,760 |
| Colombia | 158 | 259 | 425 | 435 | 15 | 16 | 16 | 16 | 2,425 | 4,202 | 6,670 | 6,850 |
| Costa Rica | 0.5 | 0.7 | 1 | 0.8 | 12 | 12 | 12 | 12 | 5 | 7 | 12 | 9 |
| Cuba | 30 | 55 | 118 | 306 | 12 | 12 | 14 | 13 | 354 | 660 | 1,651 | 3,850 |
| Dominica | 2 | 3 | 4 | 4 | 12 | 12 | 12 | 12 | 25 | 40 | 42 | 42 |
| Dominican Rep. | 135 | 110 | 60 | 60 | 12 | 12 | 12 | 15 | 1,620 | 1,320 | 720 | 720 |
| Ecuador | 88 | 105 | 35 | 42 | 15 | 15 | 29 | 30 | 1,320 | 1,575 | 1,023 | 1,299 |
| El Salvador | 6 | 6 | 5 | 6 | 15 | 15 | 15 | 15 | 82 | 87 | 81 | 88 |
| French Guiana | 0.4 | 0.5 | 0.6 | 0.7 | 10 | 10 | 10 | 10 | 3 | 4 | 6 | 7 |
| Grenada | 2 | 3 | 2 | 2 | 10 | 11 | 11 | 12 | 20 | 32 | 24 | 25 |

*Continued*

**Table 4.2.** Continued.

| Source | Goats slaughtered (× 1000) | | | | Average carcass weight (kg) | | | | Total meat production (million t) | | | |
|---|---|---|---|---|---|---|---|---|---|---|---|---|
| | 1980 | 1990 | 2000 | 2008 | 1980 | 1990 | 2000 | 2008 | 1980 | 1990 | 2000 | 2008 |
| Guadeloupe | 23 | 24 | 14 | 11 | 11 | 13 | 13 | 15 | 250 | 312 | 180 | 165 |
| Guatemala | 28 | 30 | 31 | 29 | 15 | 15 | 15 | 15 | 420 | 450 | 465 | 435 |
| Guyana | 19 | 25 | 26 | 27 | 10 | 10 | 10 | 10 | 190 | 250 | 260 | 270 |
| Haiti | 220 | 244 | 432 | 400 | 15 | 15 | 15 | 15 | 3,300 | 3,660 | 6,480 | 6,000 |
| Honduras | 11 | 10 | 13 | 10 | 15 | 15 | 15 | 15 | 166 | 156 | 187 | 148 |
| Jamaica | 33 | 35 | 40 | 41 | 14 | 13 | 14 | 16 | 450 | 450 | 551 | 681 |
| Martinique | 2 | 8 | 7 | 6 | 11 | 12 | 13 | 13 | 20 | 100 | 83 | 80 |
| Mexico | 2,696 | 2,777 | 2,510 | 2,550 | 11 | 13 | 15 | 17 | 30,305 | 36,102 | 38,760 | 43,128 |
| Montserrat | 1 | 2 | 2 | 2 | 12 | 12 | 12 | 12 | 15 | 20 | 25 | 25 |
| Netherlands Antilles | 5 | 13 | 4 | 4 | 9 | 9 | 9 | 9 | 47 | 113 | 35 | 38 |
| Nicaragua | 2 | 2 | 3 | 2 | 16 | 16 | 16 | 19 | 30 | 27 | 46 | 33 |
| Paraguay | 62 | 69 | 74 | 95 | 10 | 10 | 10 | 10 | 619 | 693 | 739 | 950 |
| Peru | 742 | 748 | 542 | 499 | 12 | 12 | 13 | 13 | 9,100 | 8,750 | 6,963 | 6,447 |
| Puerto Rico | 23 | 24 | 0.3 | 1 | 11 | 11 | 11 | 12 | 262 | 254 | 3 | 16 |
| St Kitts and Nevis | 4 | 4 | 5 | 1 | 14 | 14 | 14 | 13 | 49 | 54 | 70 | 13 |
| St Lucia | 4 | 5 | 4 | 3 | 14 | 14 | 14 | 14 | 49 | 63 | 51 | 47 |
| St Vincent and Grenadines | 1 | 2 | 2 | 3 | | | | | 12 | 24 | 23 | 30 |
| Suriname | 3 | 4 | 2 | 1 | 10 | 10 | 10 | 10 | 30 | 39 | 20 | 13 |
| Trinidad and Tobago | 24 | 25 | 25 | 27 | 15 | 15 | 15 | 15 | 352 | 375 | 375 | 405 |
| USA | | | | 796 | | | | 28 | | | | 22,600 |
| US Virgin Islands | 2 | 2 | 2 | 2 | 11 | 11 | 11 | 11 | 26 | 19 | 19 | 20 |
| Venezuela | 502 | 656 | 407 | 480 | 11 | 8 | 17 | 12 | 5,457 | 5,119 | 7,051 | 5,870 |
| Oceania | | | | | | | | | | | | |
| Australia | 15 | 315 | 425 | 650 | 25 | 25 | 25 | 25 | 375 | 7,875 | 10,625 | 16,250 |
| Cook Islands | 0.6 | 0.6 | 0.6 | 0.2 | 15 | 15 | 15 | 15 | 8 | 8 | 9 | 2 |
| Fiji | 33 | 56 | 73 | 74 | 14 | 12 | 13 | 13 | 451 | 648 | 934 | 983 |
| French Polynesia | 4 | 4 | 5 | 5 | 15 | 15 | 15 | 15 | 54 | 63 | 75 | 75 |
| Guam | 0.1 | 0.2 | 0.2 | 0.3 | 36 | 35 | 43 | 50 | 6 | 7 | 10 | 15 |

| | | | | | | | | | | | | |
|---|---|---|---|---|---|---|---|---|---|---|---|---|
| Micronesia, Fed. States | | | 1 | 1 | | | 11 | 11 | | | 13 | 14 |
| New Caledonia | 4 | 9 | 2 | 2 | 12 | 12 | 12 | 12 | 48 | 108 | 22 | 22 |
| New Zealand | 55 | 162 | 115 | 107 | 17 | 12 | 11 | 11 | 950 | 1,998 | 1,285 | 1,153 |
| Pacific Islands Tr. | 1 | 1 | | | 11 | 11 | | | 13 | [illegible]3 | | |
| Papua New Guinea | 0.4 | 0.6 | 0.6 | 0.8 | 15 | 15 | 15 | 15 | 6 | 8 | 8 | 11 |
| Tonga | 3 | 2 | 2 | 2 | 15 | 15 | 15 | 15 | 42 | 30 | 24 | 26 |
| Vanuatu | 2 | 2 | 3 | 3 | 10 | 10 | 10 | 10 | 23 | 23 | 26 | 28 |
| Wallis and Futuna | 1 | 1 | 1 | 1 | 15 | 15 | 15 | 15 | 15 | 15 | 15 | 15 |
| Continents | | | | | | | | | | | | |
| Africa | 42,841 | 56,219 | 74,213 | 98,339 | 12 | 12 | 12 | 12 | 496,620 | 663,928 | 916,923 | 1,193,211 |
| Asia | 93,936 | 147,781 | 218,358 | 271,255 | 13 | 14 | 14 | 15 | 983,648 | 1,729,326 | 2,593,215 | 3,386,936 |
| Europe | 10,997 | 13,843 | 11,911 | 11,257 | 11 | 11 | 13 | 13 | 115,982 | 137,857 | 124,870 | 121,080 |
| Americas | 8,515 | 9,893 | 9,550 | 10,788 | 12 | 12 | 13 | 14 | 95,891 | 114,514 | 122,256 | 151,451 |
| Oceania | 119 | 554 | 628 | 846 | 17 | 16 | 17 | 17 | 1,391 | 10,795 | 13,046 | 18,594 |
| GNI (US$/year: 2,008)[a] | | | | | | | | | | | | |
| Low (≤$995) | 37,017 | 45,564 | 59,491 | 71,273 | 12 | 12 | 12 | 12 | 389,585 | 597,04[illegible] | 775,413 | 890,326 |
| Lower middle (US$996–3,945) | 71,316 | 117,079 | 190,657 | 240,750 | 12 | 13 | 13 | 14 | 774,757 | 1,326,783 | 2,263,105 | 3,078,049 |
| Middle (US$3,946–12,195) | 20,817 | 25,395 | 26,464 | 26,975 | 12 | 12 | 13 | 13 | 265,534 | 334,579 | 367,013 | 371,700 |
| Upper middle (≥US$12,196) | 9,451 | 12,936 | 12,081 | 13,552 | 13 | 14 | 14 | 15 | 94,873 | 135,22[illegible] | 133,724 | 194,746 |
| Economies[b] | | | | | | | | | | | | |
| Emerging and developing | 147,828 | 217,388 | 305,044 | 382,125 | 12 | 12 | 13 | 13 | 1,612,309 | 2,552,53[illegible] | 3,665,519 | 4,737,344 |
| Advanced | 8,462 | 10,754 | 9,329 | 10,027 | 13 | 14 | 14 | 14 | 80,449 | 102,13[illegible] | 99,592 | 127,712 |
| Total | 156,407 | 228,291 | 3[illegible]4,659 | 392,484 | 12 | 13 | 13 | 13 | 1,694,132 | 2,656,42 | 3,770,310 | 4,871,272 |

[a]World Bank, http://databank.worldbank.org/.
[b]International Monetary Fund, http://www.imf.org/.
[c]FAOSTAT, http://faostat.fao.org/.

**Table 4.3.** Goat breeds in the world by speciality, climate-environment and country of origin. (From Shrestha and Fahmy, 2005, 2007a.)

| Breed/speciality | Country | Breed/speciality | Country |
|---|---|---|---|
| MEAT | | *Tropical dry* | |
| *Tropical very dry* | | Barbari | India |
| Sudan Desert | Sudan | Beetal | India |
| *Tropical dry* | | Dera Din Panah | Pakistan |
| Barbari | Pakistan | Granadina | Spain |
| Black Bengal | Indian subcontinent | Jamunapari | India |
| Bugri | Pakistan | Jhakrana | India |
| Cutchi | India | Kamori | Pakistan |
| Damani | Pakistan | Nigerian Dwarf | West Africa |
| Ganjam | India | Nubian | Sudan |
| Kaghani | Pakistan | Surti | India |
| Kail | Pakistan | Zaraiby | Egypt |
| Lehri | Pakistan | *Tropical humid* | |
| Marwari | India | Malabar | India |
| Nigerian Dwarf | West Africa | *Subtropical dry* | |
| Nubian | Sudan, UK | Damascus | Syria and Lebanon |
| Osmanabadi | India | *Subtropical mild* | |
| Patteri | Pakistan | Alpine | Switzerland |
| Sangamaneri | India | Fawn | Belgium and Germany |
| Sirohi | India | LaMancha | USA |
| Tapri | Pakistan | Oberhasli | Switzerland |
| *Tropical humid* | | Rove | France |
| Fijian | Fiji | Saanen | Switzerland |
| Katjang | Indonesia | Toggenburg | Switzerland |
| Katukachchiya | Sri Lanka | PROLIFIC | |
| *Subtropical dry* | | *Tropical very dry* | |
| Boer | South Africa | Sudan Desert | Sudan |
| *Subtropical humid* | | *Tropical dry* | |
| Angora | Turkey | Barbari | India |
| Banjiao | China | Black Bengal | Indian subcontinent |
| Chengdu Ma | China | *Tropical humid* | |
| Du An | China | Malabar | India |
| Fuquing | China | West African Dwarf | West Africa |
| Guizhou White | China | *Subtropical dry* | |
| Haimen | China | Boer | South Africa |
| Huai | China | *Subtropical humid* | |
| Kheri | Nepal | Ma'tou | China |
| Leizhou | China | *Tropical and subtropical dry* | |
| Longlin | China | Criollo | South America |
| Ma'tou | China | PASHMINA (Cashmere) | |
| Okinawa | Japan | *High mountains cold* | |
| Shanzi White | China | Chyangra | Nepal |
| Terai | Nepal | Kashmiri | Central Asia |
| *Subtropical mild* | | Singhal | Nepal |
| Rove | France | MOHAIR | |
| Spanish | UK | *Subtropical humid* | |
| *Mountain humid* | | Angora | Turkey |
| Khasi | India | SKINS | |
| MILK | | *Tropical dry* | |
| *Tropical very dry* | | Black Bengal | Indian subcontinent |
| Black Bedouin | Israel and Egypt | *Tropical humid* | |
| | | Maradi | Niger and Nigeria |
| | | Mubende | Uganda |

production efficiency, uniformity of carcasses and consumer acceptability, while adding value to meat and meat products. Crossbreeding of the dairy (Alpine, Damascus, Jamunapari, Saanen and Toggenburg) and meat (Beetal, Boer and Nubian) breeds with indigenous breed populations in many parts of the world has produced kids suitable for meat production (Shrestha and Fahmy, 2005, 2007a). Meat goats represent the majority of goats in the tropics and include those not specialized in the production of milk, pashmina (cashmere), mohair and skins (Table 4.3). There is evidence that goat producers have benefited from the application of breeding strategies that have helped reduce costs associated with goat meat production (Quartermain, 1991). Estimates of genetic parameters and the genetic response to selection for economically important traits in goats corroborate theoretical promise for improving the efficiency of meat production. The use of mixed animal model methodology (Henderson and Quaas, 1976), which has contributed to genetic improvement in other livestock and poultry, is also applicable in the breeding of meat goats. Thus, predicting breeding values of parents and their offspring from the simultaneous evaluation of multiple traits associated with improving production efficiency has considerable potential. The subject of this chapter is the application of quantitative genetic principles and skilled practices of the breeder, along with complementary goat genetic resources worldwide, to achieve rapid and permanent improvement of morphological characteristics and production performance that are of economic importance to meat production.

## 4.3 Domestication and Dissemination of Goat Genetic Resources

Domestication has contributed to physical changes in the wild goat species during their adaptation in captivity while being made suitable for farming. This is an ongoing process of artificial selection and skilful breeding practice in an environment provided by the farmer. Archaeological studies of bones uncovered adjacent to human settlements in areas adjoining Iran, Turkey, Syria, Jordan, Pakistan, Israel and the Palestinian Territories, along with radiocarbon dating, suggest that prehistoric communities as early as 8500 BC (the end of the Mesolithic period) domesticated herds of goats and flocks of sheep. Three predominant types, the Bezoar, Savannah and Nubian goats, evolved from their respective wild ancestors, the Bezoar or Pasang (*Capra aegagrus aegagrus*), Markhol (*C. aegagrus falconeri*) and Ibex (*C. aegagrus ibex*) goats. According to Herre (1958), the Bezoar type may have been the first domesticated goat. The expansion of agriculture and the stock raising culture led to the movement of nomadic and semi-nomadic pastoralists with their goats in westerly and easterly directions from the west Asian centre of civilization. Goats travelled from Iran and Afghanistan through Turkistan to Mongolia arriving at North and South China, which was later known as the 'silk road', and to the Hainan Island and the western plain of Taiwan. The Tsinghai and Mongolian goats were established in the Tibetan plateau sometime during this era. Goats indigenous to central and eastern China may have originated from crosses between Mongolian goats and the meat goat from south China.

The expansion of the Aryan Empire in the 2nd century BC contributed to the migration of goats through the Khyber Pass into the Indian subcontinent. In the 6th century BC, the thriving trade between the Indian subcontinent and south-eastern Asia along the maritime route to Burma, Thailand, Malaysia and Indonesia, and from South China into south Asia was responsible for the introduction of goats to these regions. It has been suggested that the present population of indigenous goats in Thailand, Malaysia and Indonesia may have been derived from crosses of Jamunapari and Anglo-Nubian goats (Devendra and Nozawa, 1976). Kambing Katjang or Pea goats indigenous to Malaysia and Indonesia are morphologically similar to goats in the Philippines, Taiwan and the Islands of south-west Japan, and resemble Black Bengal and south China goats.

The rise and the expansion of the Roman Empire were characterized by European goats with large body size (Bökönyi, 1974). It has been suggested that hornless goats first appeared during this period. Following the demise of the Empire, body size decreased in the 3rd to 6th centuries, remaining small during the Middle Ages to the 14th century. In the 5th century BC, goats with scimitar-shaped horns, which originated in Egypt, migrated into areas adjoining Syria and Palestine. Migration of goats from north to south in East Africa and east to west in North Africa was followed by subsequent dispersal across the African continent. Currently, Savannah-type goats are widespread across the African continent where the majority of animals are raised for their meat and milk by pastoralists. Nubian-type goats raised by sedentary agriculturists in North Africa possibly originated from goats in India or Iran that subsequently migrated into Syria and Egypt. It has been suggested that Anglo-Nubian goats were derived from crosses between prick-eared goats indigenous to the UK and Nubian-type Zaraibi, Chitral and Jamunpari goats from Africa and India.

Goats indigenous to Europe accompanied voyages from Spain and Portugal to the Americas; from France and England into North America; from England to Australia, New Zealand and South Africa; and from Africa to the Caribbean countries. Consequently, large populations of feral goats were established in Australia and New Zealand. In the USA, Spanish goats account for 20% of the population and produce most of the goat meat in the country. These goats, of Mexican origin, are distinct from the Angora or dairy breeds. Natural selection and contribution from the Nubian (syn. Anglo-Nubian) or Toggenburg breeds have resulted in a variety of morphological characteristics.

### 4.3.1 Concept of breed

In the early 18th century, Robert Blakewell, known as the father of animal breeding, developed the concept of combining stocks with similar morphological characteristics, and established *Coates' Herdbook for Shorthorn Cattle and Herd Societies in England.* This was followed by registration of animal ancestry and the maintenance of herd books to restrict introductions contributing to the development of a large number of pure breeds (Lush, 1945). Dairy goats from Switzerland were introduced into England leading to the development of the Saanen, Toggenburg and Alpine breeds. Purebred goats also accompanied voyages, which contributed to their introduction into the colonies where a large number of goat herds currently remain. The rapid expansion of the goat numbers in the newly established colonies could be attributed to favourable climatic conditions, an abundance of vegetative growth and the absence of communicable diseases as well as predators. Domestication and the history of goats have been described in comprehensive reviews on the subject (Harris, 1962; Epstein, 1971; Bökönyi, 1974; Devendra and Nozawa, 1976; Nozawa, 1991).

In the mid-20th century, Mason (1951) published the first edition of *A World Dictionary of Livestock Breeds, Types and Varieties.* This has been complemented with further details on distribution, morphological characteristics and production performance of goats in China (Phillips *et al.*, 1945; Epstein, 1969; Cheng, 1984; Ying, 1995), India (Acharya, 1982), Pakistan (Hasnain, 1985), Turkey (Yalçin, 1986) and Europe (Simon and Buchenauer, 1993). This was followed by the development of the Global Animal Genetic Data Bank (Simon, 1990) and the publication of the Animal Genetic Resource Information by the Food and Agriculture Organization of the United Nations (FAO). Centuries of evolution and skilful breeding practice have established specific genetic combinations contributing to the development of a colossal amount of goat genetic resources distributed worldwide across all ecosystems. These goats have adapted to varying production environments (Mason and Maule, 1960; Gall, 1982, 1996; Galal, 1987; Simon, 1990; Simon and Buchenauer, 1993; Mason, 1996; Scherf, 2000; Morand-Fehr *et al.*, 2004) and exhibit diversity in performance providing clear evidence of their potential for genetic improvement of productivity.

### 4.3.2 Classification of indigenous goats

There are many goats indigenous to the tropical, subtropical and high mountain climate exposed to humid, very dry, dry, and humid and cold environments. The absence of herd books in many developing countries has led to the classification of goats according to morphological characteristics distinct from other populations in the vicinity. Goats have been described according to specific local names, distribution and primary product. In India, the Council of Scientific and Industrial Research (CSIR, 1970) classified goats into the following breed groups according to their primary product: 34 for meat, 12 for milk, eight for prolificacy and three each for pashmina and skin production. Cashmere-like goats were distributed in the trans-Himalayan mountain range, milk goats in the northern dry regions and meat goats in the Deccan Plateau and areas adjoining the Bay of Bengal. Goats in India were further classified into 20 breeds: Sirohi, Marwari, Beetal, Jhakrana, Barbari, Jamunapari, Mehsana, Gohilwadi, Zalawadi, Kutchi and Surti in the north-western arid and semi-arid regions; Sangamneri, Malabari, Osmanabadi and Kannaiadu in the southern peninsular region; Ganjam and Bengal in the eastern region; and Gaddi, Changthangi and Chigu in the northern temperate region. In Pakistan, goats were classified on the basis of hair and smooth coat characteristics into 25 breeds: Kajli, Khurassani and Lehri of Baluchistan; Damani, Gaddi and Kaghani of the North-west Frontier province; Beetal, Nachi, Dera Din Panah and Teddy of Punjab; Barbari, Chappar, Kamori and Sind Desi of Sind; Baltistani, Jararkheil, Kohai Ghizer and Piamiri of northern areas; and Beiari, Buchi, Desi, Kooti, Labri, Pothohari and Shurri of Azad Kashmir. In China, meat goats were classified into 12 breeds: Banjiao, Chengdu Ma, Du An, Fuqing, Guizhou White, Haimen, Huai, Leizhou, Longlin, White goat, Matou and Nanjiang yellow. Further details on issues relating to the conservation of goat genetic resources have been described in detail in the textbook *Goat Science and Production* and a comprehensive review on the subject (Shrestha and Galal, 2010, 2011).

## 4.4 Divergent Production Environments

In general, meat goats are raised in small and large multi-species herds under nomadism and semi-nomadism in a range of agro-pastoral production systems, arid rangelands and within the perimeter of the farm, or in the vicinity of smallholder farming systems. The divergent management practices are traditional, socially tolerant and in harmony with the natural vegetation and prevailing environment but make efficient use of labour surplus to household requirements with sustained productivity. The resource-poor households not only lack the necessary finance to purchase concentrates but also fail to appreciate the scientific logic behind the concept of fattening goats to improve meat quality. The practice of fattening livestock prior to slaughter by providing high-energy concentrate diets in feedlots, which is common in other meat animals, does not exist in goats.

### 4.4.1 Asia

Goats are often tethered in stubble fields or browse along roadsides, adjoining forests and community pastures, while frequently being supplemented with cut-and-carry fodder. In general, women and children in rural areas are responsible for daily feeding and husbandry requirements. Around urban areas, goats browse on vegetation surrounding irrigated fields or verdure along property boundaries and vacant land but remain secure and always within the proximity of the owner. During the summer months, the availability and quality of forage and shrubs diminish in adjoining villages and areas of the city. Thus, goats need to travel a greater distance for foraging or are supplemented with agricultural by-products from industrial waste produced during food processing, along with vegetable toppings from kitchen waste.

In countries bordering the Himalayan mountain range, goats raised in large multispecies herds benefit from an abundant supply of natural vegetation, which is the primary source of diet in transhumance systems (Shrestha *et al.*, 1998). Nomadic pastoralists practise the age-old tradition of ascending in the spring to the high mountain pastures of indigenous grasses, mountain vegetation and shrubs, and descending in autumn into populated areas adjoining croplands or irrigated fields. During gestation and lactation, supplementary feed to meet the dietary nutrient requirements comes from green and stored fodder, stubble from recently harvested fields, and concentrates or locally available agricultural by-products. Feed is the major constraint to the number of animals that can be maintained during the winter months. There is also competition from humans due to the limitation of concentrates available in the region. The practice of slaughtering large numbers of goats and sheep in the autumn coincides with cultural practices and religious rituals that occur during the harvesting of crops.

In the Tibetan plateau, large herds of Tsinghai and Mongolian goats are raised for milk, meat, skin and hair by nomadic pastoralists under adverse environmental conditions of low rainfall, hot summers and cold winters under a transhumance system. In the late 19th century, the mountain and community pastures began to experience rampant overgrazing because of increasing numbers of livestock continuing to share a fixed acreage of land. The transhumance system of management is under threat from loss of pasture, restricted access to National Parks, overgrazing, predators, a toll for grazing, diseases and rural exodus of the human population. Consequently, an increasing number of sedentary herds graze around irrigated land and rely on locally available feed and feed by-products for subsistence.

In South and South-east Asia, there have been attempts to increase goat meat production by integrating the raising of goats with crop production and farming systems. Devendra (1988) has described the integration of goat meat production and crop farming systems based on wheat, rice, maize, oilseeds, cash crops, vegetables, cut-and-carry fodders, coconuts, oil palm and rubber plantations for various stocking densities owned by small farmers, the landless, labourers and peasants. In the Philippines, a 300-goat herd kept for meat production along with fish culture of Nile tilapia (*Oreochromis niloticus*) at a stocking density of 10,000/ha resulted in the harvest of fish weighing 78 g at 120 days (Libunao, 1990). In South China, including Hainan Island and the western plain of Taiwan, small herds of goats continue to thrive around well-irrigated river basins.

### 4.4.2 Africa and the Middle East

In the Sahel region of Africa, nomadic tribes move with their goats in search of vegetation around oases, waterholes and adjoining irrigated lands to provide milk and meat for domestic consumption. Each household maintains three to four goats that roam freely, producing meat with minimum labour and input costs. This form of traditional husbandry results in a 30% greater return over investment. In the south-eastern region of Nigeria, an average household has six West African Dwarf goats raised under intensive management, while women participate as meat retailers (Chidebelu and Ngo, 1998). Details and description of meat-goat production and marketing in Asia, Africa and Latin America, as well as consumer preference, carcass quality and composition, and the influence of season, climate, breed, sex, age and nutrition have been reviewed (Devendra and Burns, 1983; Devendra, 1988; Upton, 1988; Wilson, 1992).

In the Mediterranean and northern Europe, goats are kept with the primary objective of producing milk for cheese-making. In Tunisia, Algeria, Morocco, Crooco, Spain, Italy and France, significant amounts of meat and meat products are sold in the domestic market within an industry that varies among countries (Le Jaouen, 1997). In Tunisia, sedentary goat keepers of the oases of Nefta and Tozeur usually keep five does in smallholdings together with sheep or dairy cattle to provide income for the household from the sale of meat while the milk is used for human consumption (Rekik *et al.*, 1996). This production system appears to be highly efficient, despite a low return from a negligible investment of capital and labour. In 1991, Algeria had 2.5 million goats, which were located mostly in the eastern and southern regions of the country and raised together with multiple species in herds of 20–200. Larger herds of more than 50 goats were common in the mountains of the Kabylie region of central Algeria. In general, indigenous goats of the Barber breed were kept for meat, whereas those of the Arabia breed provided both meat and milk. In 1985, approximately 20,000 million t of goat meat was produced in Morocco, demonstrating the importance of goats to the economy. In 1989, drought, along with the implementation of a regressive legislation to protect forests, decreased the goat population dramatically, from 8 million in 1960 to 5 million in 1999. In France, the endangered Rove goat raised under a transhumance system is being promoted to produce meat from feed based on wild flora. This concept is based on the idea of safe keeping of the forest diversity while protecting employment in areas where rural exodus is rampant. Details and descriptions of the herding systems including transhumance and sedentary herds, breed, vegetation, grazing and production management, as well as socio-economic and marketing aspects of goat-meat production and processing in Europe, have been reviewed (Flamant and Morand-Fehr, 1990; Morand-Fehr *et al.*, 1992; El-Aich *et al.*, 1995; d'Avant, 2001).

### 4.4.3 The Americas

In the eastern USA, demand for goat meat is from the increasing population of ethnic origin and the growing appreciation for food of exotic origin. This has contributed to research on silvopastoral production systems based on a high proportion of browse in diets composed of herbage from several species of the genus *Paulownia* (Mueller *et al.*, 2001). In the north and west central regions of Argentina, 50,152 farms have 3.7 million Criolo (syn. Creole) and Anglo-Nubian goats on pasture for meat production (Angel-Neelem and Nellem, 1998). In Mexico, Galina *et al.* (2000) reported that goats grazing on semi-arid woody brush (*Caducifolio espinoso*) rangeland were more profitable when supplemented with a complex catalytic feed containing non-protein nitrogen compared with those fed traditional rations balanced by feeding supplements. In Guadeloupe, Creole meat goats on a semi-intensive production system based on breeding cycles of 8 months and grazing on fertilized and irrigated pastures of *Digitaria decumbens*, weaned 1.4 kg of kid weight/ha (Alexandre *et al.*, 1997, 1999). Creole goats attained 90% fertility, a prolificacy of 2.25 and 78% viability to weaning, doubling their productivity. The authors concluded that meat production in the humid tropics should be assessed in relation to agronomic sources of variation that included grazing systems, animal management and ingestion levels other than incidence of gastrointestinal parasitism.

## 4.5 Breed Evaluation

Important information pertinent to the genetic improvement of morphological characteristics and production performance of goats maintained under specific management conditions is available from worldwide studies, which include performances of pure- and crossbred goats kept under varying management practices across all ecosystems. Worldwide, 28% of goat breeds and populations either have not been reported or are of unknown origin. Compared with other ruminant livestock, this represents the largest proportion (Galal, 2005). The problem can be attributed to difficulties encountered in collecting census data from remote regions of the world. Only a fraction of the 1183 goat breed populations worldwide as reported by DAD-IS (2010, http://dad.fao.org/) have been subjected to evaluation. The majority of studies are based on a small number of offspring derived from few sires. These studies are often confounded with feeding and management practices that are not consistent with smallholder farming systems in the country. In developing countries, income from carcass weight should not be considered as the only saleable product for meat goats because non-carcass components have economic and religious significance. While the reproductive rate of the doe and kid survival are considered important, there is no conscious effort to improve growth rates and meat quality. This is because increased numbers of animals in the household are indicators of wealth and status in the community.

Haenlein (1996) reported that improved goat breeds, such as the Alpine breeds from continental Europe, recognized for their mature body size and higher milk yield when crossed with the Anglo-Nubian, Boer and Jamunapari breeds, have demonstrated potential for increasing meat and milk production. Galal (2005) assessed the productivity of goat breeds worldwide and concluded that breeds in Europe were more prolific with a higher milking ability and heavier body weight compared with those in Latin America. Breeds in Africa, Asia, the Pacific and Near East regions, and North America were intermediate, whereas those in the Caribbean were smallest (Table 4.4). Meat breeds in India have a mature weight ranging from 19–37 kg at 15–18 months of age in contrast to large breeds that weigh 58–60 kg, while dwarf breeds weigh 15–25 kg at the same age (Taneja, 1982).

Evidence has been presented that suggests the presence of significant diversity among breeds for economically important traits associated with goat meat production. In India, meat goats such as the Sirohi, Kannaiada, Black Bengal and Assam Hill breeds and their crosses, as well as the Jamunpari, Beetal, Barbari and Black Bengal breeds, were evaluated for growth, feed conversion and carcass characteristics (Acharya, 1988). Also included in the study were growth rate and milk production of crossbred offspring from bucks of the Alpine, Saanen and Anglo-Nubian breeds and does of the Malabari, Beetal, Assam Hill, Black Bengal and Jamunapari breeds. At the National Dairy Research Institute in India, Alpine × Beetal and Saanen × Beetal produced crossbred offspring that exceeded their indigenous parental breeds in body weight and dressing percentage (Acharya *et al.*, 1982). The Jamunapari breed characterized with large body size weighed significantly more at maturity than the Beetal, Barbari, Jhakrana, Sirohi and Marwari breeds (Taneja, 1982). At the Haryana Agricultural University in India, Alpine × Beetal and Anglo-Nubian × Beetal kids fed individually to 5 months on feedlot exceeded their purebred contemporaries in body weight, hot carcass weight and dressing percentage. Bucks of the Jamunapari and Beetal breeds crossed with does of the Black Bengal, Beetal and Sirohi breeds produced offspring that were more productive. In addition, Jamunapari-sired Black Bengal does produced offspring that weighed significantly more than their smaller parent. The Beetal breed, known to be highly fecund and better adapted when crossed with does of the smaller Sirohi breed, produced large goats for meat production (Misra *et al.*, 1980).

In Nepal, Jamunapari-, Barbari- and Beetal-sired indigenous does produced kids that grew rapidly to 6 months, weighing more than their indigenous contemporaries; however, at 12 months of age, the indigenous goats were heavier (Sainju *et al.*, 1994). Crossbred kids from Jamunapari × Khari and Saanen × indigenous goats weighed more than their parental breeds at 15 weeks of age by 3.4–3.5 and 5.9–13 kg, respectively. In Malaysia, crossbreeding of the German Fawn breed with the indigenous Katjang goats produced large difference in growth rate (Hirooka *et al.*, 1997). In China, the productivity of the Banjiao, Chengdu Ma, Du An, Fuqing, Guizhou White, Haimen, Huai, Leizhou, Longlin, White goat, Matou and Nanjiang yellow breeds and their crosses, considered as meat goats, has been assessed (Jiang, 1982, 1995). In Japan,

**Table 4.4.** Diversity in the production performance of goats by region of the world (from Galal, 2005).

| Performance | Region | Sex | Goats[a] | Mean ± SD | Minimum | Maximum |
|---|---|---|---|---|---|---|
| Body weight (kg) | Africa | Male | 32 | 53 ± 37 | 20 | 130 |
| | | Female | 32 | 39 ± 22 | 20 | 94 |
| | Asia and Pacific | Male | 106 | 43 ± 17 | 16 | 130 |
| | | Female | 106 | 32 ± 12 | 14 | 100 |
| | Europe | Male | 123 | 66 ± 13 | 35 | 120 |
| | | Female | 124 | 49 ± 12 | 24 | 120 |
| | Latin America and Caribbean | Male | 10 | 40 ± 14 | 15 | 70 |
| | Near East | Male | 48 | 44 ± 14 | 17 | 75 |
| | | Female | 47 | 33 ± 8 | 15 | 50 |
| | North America | Male | 5 | 41 ± 21 | 22 | 67 |
| | | Female | 4 | 35 ± 26 | 13 | 60 |
| Litter size (kid) | Africa | | 17 | 1.4 ± 0.33 | 1 | 2.1 |
| | Asia and Pacific | | 79 | 1.4 ± 0.37 | 1 | 2.9 |
| | Europe | | 28 | 1.6 ± 0.35 | 1 | 2.2 |
| | Latin America and Caribbean | | 11 | 1.4 ± 0.3 | 1.1 | 2.0 |
| | Near East | | 27 | 1.6 ± 0.43 | 1.1 | 2.5 |
| | North America | | 4 | 1.0 | 1 | 1 |
| Milk yield (kg/lactation) | Africa | | 3 | 126 ± 116 | 50 | 500 |
| | Asia and Pacific | | 63 | 136 ± 109 | 16 | 550 |
| | Europe | | 41 | 299 ± 225 | 40 | 775 |
| | Latin America and Caribbean | | 2 | 63 ± 4 | 60 | 65 |
| | Near East | | 32 | 150 ± 97 | 35 | 460 |

SD, Standard deviation.
[a]Number of goat breeds with records.

Shinjo and Toma (1985) reported that indigenous Okinawa meat goats were more prolific with year-round kidding than the Japanese Saanen goats, whereas fecundity was lower.

In the Bahamas and Fiji, indigenous goats sired by the Anglo-Nubian breed produced kids with superior body conformation and larger body size, demonstrating an advantage in favour of the exotic sire breed (Wilson *et al.*, 1980; Hussain *et al.*, 1983). In Mexico, indigenous Mexican goats sired by the Nubian and Saanen breeds resulted in kids that were 8% heavier in post-kidding body weight compared with those sired by the Toggenburg breed (Montaldo *et al.*, 1995). Also, litter weights of kids sired by the Alpine breed were 18% heavier at birth than those sired by the Granadina breed. The performance of goats indigenous to Africa such as the Bornu White, Red Sokoto, Small East African, Sudanese Desert and West African Dwarf breeds has been described in detail (Quartermain, 1991). In

Egypt, kids of Alpine × Rove surpassed their parental breeds for weight gain from 30 to 90 days of age, carcass yield, width and shape but produced more internal fat (Mourad and Anous, 1998).

There has been a wealth of information on breed evaluation published in the second half of the last century, demonstrating the opportunity to identify breeds with potential genetic merit for increased goat meat production. These studies are comprised of crossbreeding the Damascus with the Anglo-Nubian and Jamunapari in Oman; the Katjang with the Saanen, Anglo-Nubian, British Alpine and Jamunapari in Malaysia; the Boer with the Small East African in Kenya and Malawi; the Creole with the Saanen and the Barbados in Puerto Rico, and with the Nubian, French Alpine and Toggenburg in Venezuela; the Anatolian Black with the Saanen and Maltese, and the Saanen with the Sardinian and the Kilis in Turkey; and the Jamunapari with indigenous goats in Sri Lanka. Further evaluations encompass the Alpine, Anglo-Nubian and Saanen breeds with the Beetal goat in India and Fiji; the Jamunapari, Saanen, Barbari and Beetal breeds with indigenous goats in Nepal; the German Fawn breed and Katjang goats in Malaysia; the Alpine and Rove breeds in Egypt; the Japanese Saanen breed with Okinawa meat goats in Japan; the Nubian, Saanen, Toggenburg, Alpine and Granadina breeds with indigenous goats in Mexico; and the Boer breed with the Spanish and Angora, and the Nubian and Alpine breeds in the USA.

A number of studies have reported that offspring derived from breeds with genetic potential for increased growth rate have a requirement for a high-quality diet, signifying the importance of breed × nutrition interaction. Glimp (1995) suggested that increased productivity could be achieved by providing better nutrition for crossbred kids from any improved breed. In India, kids of Beetal × Sirohi raised extensively on range failed to demonstrate any advantage in body weight at 12 months of age over those of the Sirohi breed. When kids were raised on feedlot, performance was significantly superior, whereas mortality, feed efficiency and dressing percentage were similar. The advantage of crossbred offspring under an abundance of feed has also been reported in Fiji, where stall-fed Alpine × Beetal, Anglo-Nubian × Beetal and Saanen × Beetal goats not only grew rapidly but had heavier pre-slaughter weight and hot carcass weight to 5 months of age.

In the USA, kids of the Boer breed fed a diet based on high-quality forage weighed more than Spanish kids, whereas the feeding of diets based on medium-quality forage resulted in a reduction or reversal of the advantage in growth rate (Blackburn, 1995). In another study, Mislevy *et al.* (2000) was not able to demonstrate any association with type and quality of grass when Spanish × Boer kids foraged Argentine, Paraguay-22 and Tifton-9 Bahia grasses and Florico stargrass. Boer × Spanish and Boer × Angora kids fed a concentrate-based diet resulted in similar early post-weaning growth rates, whereas Boer crosses exceeded Spanish crosses (Cameron *et al.*, 2001). Boer × Alpine kids gained substantially more with better feed conversion at 31 weeks than at 50 weeks of age, suggesting that growth rate tended to decrease as kids grew older (Luo *et al.*, 2000). Kids slaughtered at a constant age showed that a Boer × Spanish cross compared with the Spanish breed produced heavier body weight and hot carcass weight, whereas Boer × Spanish and Spanish × Angora were similar, and the Angora breed was significantly lighter (Oman *et al.*, 2000). Breeds did not vary in lean colour, surface discoloration, overall appearance or off-odour, although kids of a Spanish × Angora cross tended to have a higher proportion of lean and less fat than those of the Angora breed, suggesting crossbreeding as an option to improve carcass characteristics.

## 4.6 Crossbreeding

Progressive breeders made use of crossbreeding long before the importance of heterosis was well recognized by the scientific community, recommended by specialists and willingly accepted for commercial production. Previously, breeding strategies in livestock and poultry were patterned somewhat after methods employed by plant breeders to produce hybrid maize. Commercial producers not only benefited from outstanding genetic merit in the parental breeds but also profited from heterosis prevalent among complementary breed crosses. There has been irrefutable evidence from researchers worldwide to suggest that crossbred offspring obtained in a systematic manner grew more rapidly, were more fecund and survived better than their purebred contemporaries.

Studies with pigs and sheep have confirmed that heterosis improves the maternal environment and enhances milk production in the crossbred female parent, as well as the growth performance of their offspring (Shrestha, 1973; Shrestha and Heaney, 2003, 2004). Multiparous domestic mammals, which include meat goats, benefit from the maternal environment provided during gestation and the nursing of their offspring. In industrial nations, the practical importance of heterosis in improving efficiency of production for livestock and poultry has been widely appreciated by commercial producers. Therefore, breeders should opt for crossbreeding to maximize production efficiency for the commercial production of goat meat.

Earlier studies using single crosses derived from two divergent inbred populations authenticated the benefits from heterosis. Unlike plants, the concept of inbred lines is not practical in the livestock species. This is because substantial loss in productivity can result from lower inherent potential for reproduction in the inbred parents, along with an increase in inbreeding and susceptibility to diseases in their offspring. In livestock and poultry species, approaches to crossbreeding that have been widely practised are contingent on the number of parental breeds or populations, and the order of mating among specialized pure breeds and their crosses. These include a number of approaches, such as specific or rotational crosses involving multiple breeds, backcrossing to the male parent, and the combination of specific breeds and rotational crosses.

The breakdown of desirable combinations of segregating alleles inherited from many of the parental breeds during crossbreeding or the development of synthetic populations could lead to a loss of desirable morphological characteristics and production performance (Dickerson, 1969a,b; Kinghorn, 1980). This may be attributed to the inter-breed recombination among non-allelic genes (epistasis) decreasing the proportion of retained heterosis, both direct and maternal. In addition to recombination loss, selection over subsequent generations increases the loss of within-breed variability, decreases the effective population size and hastens the rate of inbreeding. Research results on recombination loss in the parents and their offspring have often been conflicting because precise estimates require a large number of breeds and their crosses independent of environmental influence. Shrestha (2010) has discussed in detail heterosis retention and inter-breed recombination among non-allelic genes associated with crossbreeding and synthesis of breeds.

In practice, breeders are faced with the difficulty of simultaneously improving a number of economically important characteristics to increase efficiency of goat meat production. Therefore, genetic and environment components associated with production traits of economic importance need to be considered. There have been several reports suggesting substantial improvement in the productivity of offspring from crosses between exotic breeds and indigenous goats, leading to the myth that exotic crosses are more productive under prevailing conditions and requirements (Shrestha, 1998). In developing countries, this approach has contributed to successive generations of unintentional crossbreeding, which has often resulted in upgrading indigenous goats to resemble the exotic breed. On the other hand, the lack of adequate nutrition and husbandry has contributed to loss of adaptability, fecundity and disease resistance, while the genetic base has been reduced by repeated use of the same parents to the detriment of performance in the crossbred population. The success or failure of the breeding programme to a large extent depends on socio-economic values, fiscal constraints, religious rituals, responsiveness to indigenous knowledge and the traditional skills acquired by the producer.

Of the large number of breeds worldwide, some have considerable merit for meat production while others have shown promise in crossbred combinations (Garcõa and Gall, 1981; Devendra, 1982, 1988; Gall, 1982, 1996; Devendra and Burns, 1983). The lack of inherent potential for lean muscle yield among established or indigenous goats demonstrates prospects for complementing with exotic breeds that excel in genetic merit. Potential sources of exotic breeds have demonstrated considerable promise in improving efficiency of production when used for crossbreeding and development of composite populations. Some of the popular goats in the Asia-Pacific region with potential genetic merit are the Barbari, Beetal, Damani, Daira Deen Panah, Jamunapari and Kamori breeds for milk and meat; the Black Bengal, Fijian, Kambing, Katjang, Kheri, Marwari, Ma'tou, Nubian, Sirohi, Teddy and Terai breeds for meat; the Chyangra, Kashmiri and Singhal breeds for meat and pashmina; and the Black Bengal, Malabari, Barbari and Ma'tou breeds for fecundity (Shrestha and Fahmy, 2005). The Alpine, Saanen and Toggenburg dairy breeds from continental Europe and the Boer breed from South Africa have considerable merit for use in goat meat production.

In each country, the sources of animals selected for crossbreeding need to be assessed carefully. If breeds must be imported, consultation with breed associations or societies as well as producer organizations before making any decision on their importation is essential. It is important to note that breeders, besides having raised goats for a number of years, are privy to a great deal of information on the goats' genetic background, health status, behaviour and previously available knowledge on their performance. Breeding animals may be introduced into the population chosen either as a source of divergent genetic resources or for their inherent potential without jeopardizing the animal health status of the country. There is always the possibility of a disadvantage associated with the imported breed, due to restrictions associated with the test environment and inadequacy of nutrients necessary for expression of the full inherent potential in the crossbred offspring. The problem is more pronounced with incidence of mortality. This is because goat production tends to be concentrated in marginal lands that are not suitable for crossbred offspring with higher nutritional requirements.

Reproductive technology, which includes collection, processing, storage and use of fresh and frozen buck semen and embryos from does, has achieved success in controlling the spread of diseases. Currently, semen and embryos can be transported with relative ease while minimizing the risk of introducing diseases. This avenue provides an opportunity to sample a large number of unrelated goats and establish a foundation herd consisting of as broad a genetic base as possible. Achieving acceptable fertility will reduce associated costs incurred during the importation of exotic breeds and their propagation. The choice of animals within breeds selected for crossbreeding may depend largely on previously acquired knowledge and the availability of a healthy breeding stock of appropriate age. Experience has shown that fiscal constraints usually result in the introduction of a limited number of unrelated animals, mostly bucks. Details and descriptions of findings from studies worldwide presented in the following sections substantiate the advantage of crossbreeding complementary breeds to maximize production efficiency for commercial production of goat meat.

### 4.6.1 Backcrosses

A common practice among goat breeders in developing countries is to produce purebred offspring necessary for herd replacement from about one-third of the superior female parents, while the remaining two-thirds are crossed with bucks of an alternative breed to produce two-breed-cross offspring. Selected two-breed-cross does are mated to bucks of the sire breed to produce backcross offspring for market. The operational advantage lies in keeping a single breed in the farm or household and purchasing only bucks of an alternative breed to produce both the single- and backcross offspring. In this procedure, there is an advantage from heterosis in maternal performance and survival of the crossbred female parent, along with growth rate and survival of their offspring. The drawback is associated with the inability to benefit from the full complement of heterosis, which can result in lower performance of the backcross than that of a single cross. In addition, the genetic superiority associated with parental breeds may not be optimal. It is important to note that backcross combinations involving exotic breeds may result in offspring that may not be well adapted to their new environment unless there is provision for adequate nutrition and husbandry.

A number of studies worldwide revealed that productivity of crossbred kids exceeded that of their contemporary purebred parents. In India, Saanen × (Saanen × Malabari) compared with Saanen × Malabari goats weighed –5 to 17% from birth to 6 months of age (Acharya, 1982). Similarly, Angora × (Angora × Sangamaneri) compared with Angora × Sangamaneri goats weighed –11 to 4% from birth to 12 months of age (Table 4.5). In contrast, Taneja (1982) reported that Beetal × (Beetal × Sirohi) compared with Beetal × Sirohi goats weighed 7–19% less from birth to 9 months of age (Table 4.6).

In the USA, Gebrelul *et al.* (1994) reported that the Alpine and Nubian breeds and their single crosses varied in body weight at birth, weaning and 6 months by 4–11, 8–17 and 3–5% of their dam breed, respectively, while mortality to 90 days increased substantially (Table 4.7). Corresponding estimates for backcross kids were –3 to 6, 8–16 and –1 to 16%, respectively, while mortality to 90 days was variable. Estimates of direct heterosis for body weight at birth, weaning and 6 months of age were 0.24, 1.9, 0.6 kg, respectively, and for kid mortality was 4%, whereas maternal heterosis estimates for body weight were 0.16, 1 and 1.4 kg, respectively, and for kid mortality was 3%. Maternal heterosis had a significant influence on weaning weight, demonstrating the advantage of using crossbred dams for goat meat production. In order to reduce the high mortality in crossbred offspring, management issues need to be addressed.

In Egypt, Barki × (Zaraibi × Barki) and Barki × (Damascus × Barki) goats exceeded the Barki breed in productivity (Abdelsalam *et al.*, 1994). Barki × (Zaraibi × Barki) compared with Zaraibi × Barki goats weighed –15 to 11% from birth to market, and produced 21% less total 16-week milk yield. Similarly Barki × (Damascus × Barki) compared with Zaraibi × Barki goats weighed 12–4% less from birth to market, and produced 5% more total 16-week milk yield (Table 4.8). At the same time, the Zaraibi breed was more productive than the Damascus breed (Abdelsalam *et al.*, 2000). Does of the Barki breed compared with those of Zaraibi × Barki and Damascus × Barki produced 47 and 71% less total 16-week milk yield than the Barki breed. In addition, does of the Barki breed compared with those of Barki × (Zaraibi × Barki) and Barki × (Damascus × Barki) produced 16 and 54% less total 16-week milk yield. These results suggest that crossbred does with a higher proportion of the Barki breed produced less milk from a shorter lactation period.

In Sri Lanka, the Boer, Jamunapari and Kottukachchiya breeds and 50% Boer cross weighed 3, 2.8, 2.2 and 2.7 kg, respectively at birth (Premasundeba *et al.*, 1998). Backcrossing with the Boer breed resulted in offspring with heavier birth weight, while *inter se* mating of Boer crosses did not decrease body weight.

In Thailand, the Thai breed, Anglo-Nubian × Thai and Anglo-Nubian × (Anglo-Nubian × Thai) slaughtered at 6, 11 or 14 months of age produced carcasses with similar yield and percentage of fat and bone (Pralomkarn *et al.*, 1995). Correspondingly, Anglo-Nubian × (Anglo-Nubian × Thai) compared with the Thai breed and Anglo-Nubian × Thai produced lower lean muscle by 1 and 4%, respectively.

**Table 4.5.** Means (± SEM) for body weight at birth (kg) and at various ages, age at first kidding, kidding interval, service period and litter size for various goat breeds and crosses[a] (from Acharya *et al.*, 1982 and Acharya, 1988).

| Sire breed | Beetal | Alpine | Saanen | Malabari | Alpine | Saanen | Saanen | Sangamaneri | Angora | Angora |
|---|---|---|---|---|---|---|---|---|---|---|
| Dam breed | Beetal | Beetal | Beetal | Malabari | Malabari | Malabari | Saanen × Malabari | Sangamaneri | Sangamaneri | Angora × Sangamaneri |
| Body weight at: | | | | | | | | | | |
| Birth | 2.9 ± 0.05 | 3.2 ± 0.05 | 3.4 ± 0.06 | 1.7 ± 0.02 | 1.9 ± 0.05 | 2.3 ± 0.02 | 2.7 ± 0.11 | 1.9 ± 0.00 | 2.1 ± 0.02 | 2.2 ± 0.07 |
| 3 months | 7.7 ± 0.11 | 10.3 ± 0.13 | 10.4 ± 0.20 | 5.7 ± 0.12 | 6.3 ± 0.16 | 6.1 ± 0.11 | 5.9 ± 0.26 | 7.3 ± 0.28 | 7.3 ± 0.06 | 8.2 ± 0.09 |
| 6 months | 12.2 ± 0.21 | 13.8 ± 0.20 | 14.3 ± 0.53 | 9.3 ± 0.19 | 8.9 ± 0.36 | 10.2 ± 0.25 | 9.7 ± 0.52 | 10.6 ± 0.39 | 10.6 ± 0.09 | 11.2 ± 0.11 |
| 9 months | | | | 11.1 ± 0.19 | 12.2 ± 0.57 | 13.3 ± 0.20 | | 13.5 ± 0.14 | 13.0 ± 0.15 | 13.4 ± 0.17 |
| 12 months | 21.8 ± 0.8 | 40.1 ± 1.8 | 26.9 ± 1.4 | 15.2 ± 0.4 | 17.6 ± 1.2 | 17.8 ± 0.5 | | 17.3 ± 2.2 | 16.0 ± 0.2 | 15.4 ± 0.2 |
| Age of first kidding (days) | 534 ± 34 | 495 ± 14 | 546 ± 19 | 700 ± 23 | 685 ± 27 | 585 ± 28 | | | | |
| Kidding interval (days) | 313 ± 7 | 323 ± 5 | 300 ± 11 | 295 ± 10 | 329 ± 20 | 407 ± 23 | | | | |
| Service period (days) | 173 ± 5 | 201 ± 9 | | 141 ± 9 | 184 ± 19 | 250 ± 63 | | | | |
| Litter size (%) | | | | | | | | | | |
| Single | 41 | 59 | 67 | 55 | 65 | 31 | | | | |
| Twin | 51 | 36 | 29 | 41 | | 63 | | | | |
| Triplet | 9 | 5 | 5 | 4 | 35 | 6 | | | | |

| Sire breed | Alpine | Anglo Nubian | Beetal | Alpine | Saanen | Sangamaneri | Angora |
|---|---|---|---|---|---|---|---|
| Dam breed | Beetal | Beetal | Beetal | Beetal | Beetal | Sangamaneri | Sangamaneri |
| Slaughter | | | | | | | |
| Age (months) | 5 | 5 | 9 | 9 | 9 | 9 | 9 |
| Live weight (months) | 14.1 ± 1.70 | 23.9 ± 3.40 | 15.2 ± 0.65 | 18.3 ± 0.41 | 18.8 ± 0.81 | 11.5 ± 1.46 | 12.9 ± 0.50 |
| Hot carcass (%) | 5.3 ± 0.83 | 9.3 ± 1.48 | 7.7 ± 0.30 | 9.5 ± 0.36 | 9.4 ± 0.44 | 5.1 ± 0.16 | 4.4 ± 0.42 |
| Dressing percentage | 38 | 39 | 50 | 52 | 50 | 44 | 38 |

**Table 4.6.** Means (± SE) for body weight of single born kids at birth (kg) and at various ages by breed and their crosses[a].

| Breed | | Body weight (kg)[b] at: | | | | | | | | | |
|---|---|---|---|---|---|---|---|---|---|---|---|
| Sire | Dam | No. kids | Birth | No. kids | 3 months | No. kids | 6 months | No. kids | 9 months | No. kids | 12 months |
| 1978–1979 | | | | | | | | | | | |
| Sirohi | Sirohi | 309 | 2.8 ± 0.02 (0) | 288 | 9.9 ± 0.12 (0) | 190 | 13.6 ± 0.17 (0) | 167 | 17.1 ± 0.18 (0) | 164 | 21.3 ± 0.18 (0) |
| Beetal | Sirohi | 261 | 3.1 ± 0.04 (11) | 244 | 10.3 ± 0.13 (4) | 163 | 14.3 ± 0.19 (5) | 142 | 17.1 ± 0.19 (0) | 138 | 22.3 ± 0.20 (5) |
| Year: 1980 | | | | | | | | | | | |
| Sirohi | Sirohi | 57 | 2.9 ± 0.04 (0) | 55 | 9.6 ± 0.29 (0) | 34 | 12.5 ± 0.33 (0) | 39 | 17.0 ± 0.41 (0) | | |
| Beetal | Sirohi | 36 | 3.1 ± 0.06 (7) | 33 | 10.4 ± 0.43 (8) | 22 | 13.3 ± 0.44 (6) | 21 | 17.6 ± 0.52 (4) | | |
| Beetal | Beetal × Sirohi | 24 | 2.9 ± 0.08 (–6) | 19 | 8.4 ± 0.46 (–19) | 13 | 12.4 ± 0.47 (–7) | 10 | 16.2 ± 0.53 (–8) | | |

| Breed | | Body weight[b] (kg) at: | | | | | | | | | |
|---|---|---|---|---|---|---|---|---|---|---|---|
| Sire | Dam | No. kids | Birth | No. kids | 1 month | No. kids | 2 months | No. kids | 3 months | No. kids | 4 months |
| Black Bengal | Black Bengal | 65 | 1.2 ± 0.03 (0) | 44 | 2.1 ± 0.10 (0) | 41 | 3.0 ± 0.16 (0) | 29 | 3.8 ± 0.30 (0) | 24 | 4.4 ± 0.37 (0) |
| Jamunapari | Black Bengal | 135 | 1.4 ± 0.04 (17) | 64 | 2.5 ± 0.12 (19) | 56 | 3.3 ± 0.14 (10) | 38 | 4.4 ± 0.21 (16) | 27 | 5.5 ± 0.40 (25) |

| Breed | | Body weight[c] (kg) at: | | | | |
|---|---|---|---|---|---|---|
| Sire | Dam | No. kids | Birth | 3 months | 6 months | 12 months |
| Black Bengal | Black Bengal | 131 | 1.8c ± 0.07 (0) | 7.6ab ± 0.28 (0) | 11.4b ± 0.40 (0) | 16.5b ± 0.55 (0) |
| Beetal | Beetal | 274 | 2.5a ± 0.04 (0) | 8.6a ± 0.18 (0) | 12.9a ± 0.23 (0) | 19.0a ± 0.34 (0) |
| Beetal | Black Bengal | 115 | 1.9c ± 0.07 (6) | 7.9ab ± 0.30 (4) | 12.4ab ± 0.43 (9) | 18.7ab ± 0.58 (13) |
| Black Bengal | Beetal | 96 | 2.3b ± 0.06 (–8) | 7.2b ± 0.28 (–16) | 12.7ab ± 0.40 (–2) | 18.2ab ± 0.54 (–4) |

*Continued*

**Table 4.6.** Continued.

| Breed | | Reproductive traits | | | | |
|---|---|---|---|---|---|---|
| Sire | Dam | No. does | Age at first conception (days) | Age at first kidding (days) | First service period (days) | First kidding interval (days) |
| Black Bengal | Black Bengal | 91 | 307a ± 12 (0) | 451b ± 13 (0) | 151b ± 3 (0) | 294b ± 3 (0) |
| Beetal | Black Bengal | 241 | 395a ± 12 (29) | 543a ± 12 (20) | 167a ± 3 (11) | 314a ± 3 (7) |

[a]The percentage deviation from the dam breed is shown in parentheses.
[b]Taneja (1982).
[c]Means within a column and class not followed by the same letter differ significantly ($P < 0.05$).

**Table 4.7.** Means (± SE) for body weight (BW) at birth and at various ages, and mortality to 90 days by breed and their crosses[a].

| Breed | | | Body weight (kg)[b] at: | | | | | |
|---|---|---|---|---|---|---|---|---|
| Sire | Dam | No. kids born | Birth (kg) | No. kids weaned | 10–12 weeks (kg) | No. kids marketed | 3 months (kg) | Mortality to 90 days (%) |
| Alpine | Alpine | 89 | 2.7 ± 0.08 (0) | 68 | 13.2 ± 0.4 (0) | 54 | 19.9 ± 0.7 (0) | 1.8 ± 2 (0) |
| Nubian | Nubian | 47 | 2.6 ± 0.11 (0) | 24 | 12.7 ± 0.7 (0) | 17 | 18.3 ± 1.1 (0) | 4.5 ± 4 (0) |
| Alpine | Nubian | 48 | 2.7 ± 0.11 (4) | 33 | 13.7 ± 0.6 (8) | 27 | 18.8 ± 0.9 (3) | 10.9 ± 4 (142) |
| Nubian | Alpine | 79 | 3.0 ± 0.09 (11) | 62 | 15.5 ± 0.4 (17) | 47 | 20.9 ± 0.7 (5) | 5.3 ± 3 (194) |
| Alpine | Alpine × Nubian | 25 | 2.8 ± 0.16 (4) | 16 | 15.9 ± 0.9 (16) | 13 | 20.3 ± 1.3 (8) | 13.1 ± 5 (20) |
| Alpine | Nubian × Alpine | 57 | 2.9 ± 0.11 (–3) | 41 | 14.8 ± 0.5 (–5) | 29 | 20.7 ± 0.8 (–1) | 5.2 ± 3 (–2) |
| Nubian | Alpine × Nubian | 26 | 2.8 ± 0.15 (4) | 14 | 13.8 ± 0.9 (1) | 11 | 21.9 ± 1.3 (16) | 11.4 ± 5 (5) |
| Nubian | Nubian × Alpine | 24 | 3.2 ± 0.16 (6) | 17 | 14.2 ± 0.8 (–8) | 13 | 20.6 ± 1.3 (–1) | 11.2 ± 5 (111) |

| Breed | | | | Body weight (kg)[c] at: | | |
|---|---|---|---|---|---|---|
| Sire | Dam | No. does | Litter size at birth | No. kids | Birth | 90 days |
| Spanish | Spanish | 38 | 1.46a ± 0.1 (0) | 38 | 2.8a ± 0.1 (0) | 13.6a ± 0.4 (0) |
| Boer × Spanish | Spanish | 31 | 1.40a ± 0.1 (–4) | 31 | 3.1b ± 0.1 (11) | 15.8b ± 0.5 (16) |

[a]The percentage deviation from the dam breed is shown in parentheses.
[b]Gebrelul *et al.* (1994).
[c]Lopez-Perez *et al.* (1998).

**Table 4.8.** Means (± SE) for body weight (BW) at birth and at various ages, and reproductive traits by breed and their crosses[a].

| Breed | | Body weight[b] (kg) at: | | | | | |
|---|---|---|---|---|---|---|---|
| Sire | Dam | No. kids | BW at birth | No. kids | 4 months | No. kids | 9 months |
| 1982–1985 | | | | | | | |
| East African | East African | 189 | 2.0 ± 0.06 (0) | 95 | 8.8 ± 0.33 (0) | 21 | 11.9 ± 0.64 (0) |
| Toggenburg | East African | 218 | 2.2 ± 0.06 (10) | 98 | 9.7 ± 0.33 (10) | 31 | 14.2 ± 0.64 (19) |
| Anglo-Nubian | East African | 98 | 2.2 ± 0.06 (10) | 55 | 9.5 ± 0.33 (8) | 20 | 13.0 ± 0.64 (9) |
| Galla | Galla | 75 | 2.6 ± 0.06 (0) | 43 | 9.1 ± 0.33 (0) | 19 | 13.1 ± 0.64 (0) |
| Toggenburg | Galla | 139 | 2.6 ± 0.06 (0) | 57 | 10.2 ± 0.33 (12) | 26 | 13.9 ± 0.64 (6) |
| Anglo-Nubian | Galla | 91 | 2.8 ± 0.06 (8) | 44 | 10.4 ± 0.33 (14) | 25 | 15.1 ± 0.64 (15) |

| Breed | | Body weight[c] at: | | | | | Productivity[d] | | | |
|---|---|---|---|---|---|---|---|---|---|---|
| Sire | Dam | No. kids | Birth (kg) | 56 days (kg) | Market (kg) | Mortality to 56 days (%) | No. does | Litter weight at weaning (kg) | Milk intake per 1 kg gain | Total 16 weeks' milk yield (kg) |
| Barki | Barki | 146 | 2.1 (0) | 6.2 (0) | 12.6 (0) | 24 (0) | 47 | 7.6 ± 0.78 (0) | 9.17 ± 0.45 (0) | 80.0 (0) |
| Zaraibi | Zaraibi | 309 | 2.1 (0) | 6.3 (0) | 10.9 (0) | 31 (0) | 38 | 9.2 ± 0.78 (0) | 7.43 ± 0.45 (0) | 116.5 (0) |
| Damascus | Damascus | 225 | 3.1 (0) | 9.1 (0) | 17.5 (0) | 35 (0) | 73 | 11.9 ± 0.78 (0) | 6.78 ± 0.45 (0) | 146.0 (0) |
| Zaraibi | Barki | 52 | 2.6 (24) | 7.1 (15) | 13.2 (5) | 25 (4) | 19 | 10.2 ± 0.78 (29) | 7.70 ± 0.45 (–16) | 117.7 (47) |
| Damascus | Barki | 80 | 2.6 (24) | 7.8 (26) | 15.6 (24) | 33 (38) | 20 | 10.4 ± 0.78 (33) | 7.70 ± 0.45 (–16) | 136.8 (71) |
| Barki | Zaraibi × Barki | 32 | 2.2 (–15) | 7.5 (6) | 14.7 (11) | 34 (36) | 16 | 7.9 ± 0.78 (–22) | 8.73 ± 0.45 (13) | 93.0 (–21) |
| Barki | Damascus × Barki | 66 | 2.3 (–12) | 7.5 (–4) | 14.8 (–5) | 21 (–36) | 29 | 10.4 ± 0.78 (–1) | 8.20 ± 0.45 (6) | 123.0 (5) |

[a]The percentage deviation from the dam breed is shown in parentheses.
[b]Ruvuna *et al.* (1988).
[c]Abdelsalam *et al.* (1994).
[d]Abdelsalam *et al.* (2000).

In Taiwan, Nubian × Taiwan, Nubian × (Nubian × Taiwan) and Nubian × [Nubian × (Nubian × Taiwan)] varied in performance with fertility of 84, 72, 73 and 100%, a prolificacy of 1.6, 1.8, 1.9 and 1.6, and average daily gains of 97, 124, 99 and 88 g, respectively, while food efficiencies were similar (Wen *et al.*, 1997). The Nubian breed was lower in performance than their crosses, with fertility of 77% and a higher prolificacy of 2.1. Kids from the Nubian and Taiwan breeds and Nubian × Taiwan breeds varied in performance, with carcass fat contents of 10.5, 7.1 and 8.3%, respectively.

### 4.6.2 Rotational crosses

These involve the mating of selected crossbred female offspring derived from two or more breeds to bucks of a pure breed not used in the previous breeding, in succession (Figs 4.1 and 4.2), while kids surplus to breeding requirements are marketed. There is a drawback that can result from large differences in the performance of offspring due to the changing proportion of parental breeds in successive breeding. This contributes to lack of uniformity in the marketing of meat and meat products to the consumer. In theory, the performance of the two-breed rotational cross is expected to be lower by one-third of the difference between the single cross and the average of the two parental breeds, which corresponds to the reduction in heterozygosity. Furthermore, the benefit from maternal versus individual performance from the use of all breeds in succession is not the same as in specific crosses. In a three-breed rotational cross, the performance may decrease by one-seventh, the difference between the average performance of three single crosses and the average of the three parental breeds. Further loss in maternal and individual performance can occur from inter-breed recombination in the gametes of the offspring associated with genetic components transmitted from the dam and maternal grand dam.

### 4.6.3 Specific breed crosses

In theory, the genetic basis of the average superiority of the three-breed crosses over constituent single crosses can be explained by heterosis in maternal performance and its interaction with transmitted and maternal deviations in performance of the purebreds. These have been described to include epistasis and the average level of inbreeding characteristic of the breed together with greater average heterozygosity in the offspring (Dickerson, 1969a,b; Hill, 1972a,b). The production of specific three-breed-cross offspring involves crossing bucks of a meat-type sire breed usually recognized for superior growth rate and meat quality with single-cross does derived from fecund-type dam breeds that excel in reproductive rate and maternal ability of the female parent. One of the two dam breeds may be indigenous in order to benefit from availability, adaptability, mothering ability and survival, while the other is a complementary, highly fecund dam breed. In practice, facilities and resource requirements to maintain three breeds can be a serious drawback. It is, however, possible to maintain only one fecund-type breed or indigenous goat population and to purchase bucks of an alternative breed to produce the crossbred female parents. These crossbred females are mated to a meat-type sire breed resulting in terminal-cross offspring. The specific three-breed-cross kids produced will benefit from rapid growth rate, carcass quality and increased uniformity in the marketing of meat and meat products to the consumer.

Usually, goat breeders produce a two-breed-cross female parent by mating two-thirds of does in the herd to bucks of an alternative dam breed. Purebred goats for herd replacement are retained from the remaining one-third of does. In practice, the female parent, probably from a breed population indigenous to the region, can be raised within the farm, while male parents with potential for either growth or fecundity may be purchased from reputable breeders. Galal (1987) proposed a breeding strategy that involves mating a proportion

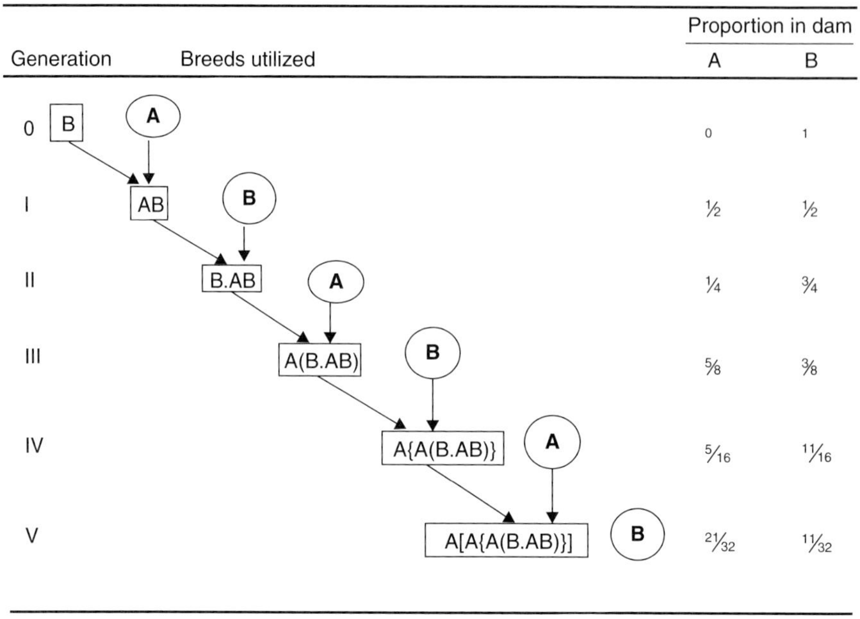


**Fig. 4.1.** Rotational crossbreeding based on two breeds (A and B).

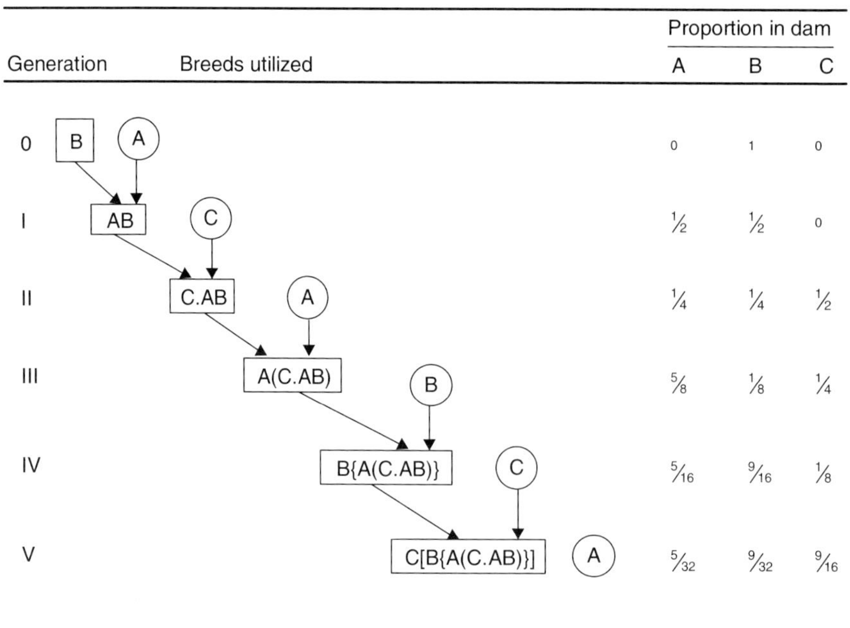


**Fig. 4.2.** Rotational crossbreeding based on three breeds (A, B and C).

of indigenous goats to produce purebred goats for herd replacement. The remaining indigenous goats are crossed with bucks of an improved dairy breed to produce crossbred does for milk production. Finally, the mating of crossbred does to bucks of a meat-type breed produces specific three-breed-cross kids with rapid growth rate deemed suitable for meat production.

In India, Acharya (1988) summarized the performance of the Beetal, Malabari and Sangamaneri breeds and their crosses sired by the Alpine, Saanen and Angora breeds and concluded that crossbred offspring sired by dairy breeds tended to exceed the female parent in body weight from birth to 12 months of age, age at first kidding, kidding interval, service period, litter size and weight, age and weight at slaughter, hot carcass weight and dressing percentage (Table 4.5). According to Taneja (1982), the Beetal and Sirohi breeds with marginal differences in mature weight produced Beetal × Sirohi and Sirohi kids that were similar in average body weight from birth to 12 months of age, feed efficiency and dressing percentage at 6–7 months of age, whereas overall mortalities were 3.5–11.9 and 2.4–10.1%, respectively (Table 4.6). Misra (1983) evaluated Jamunapari × Black Bengal, Beetal × Black Bengal and Black Bengal kids for body weight and carcass traits. The author concluded that the larger Jamunapari and Beetal breeds had potential merit for improving meat production when crossed with small and medium-sized goats in India. Black Bengal and Beetal × Black Bengal goats had a similar age at first conception and, in the latter, age at first kidding, first service period and first kidding interval were significantly longer (Singh *et al.*, 2000). The Beetal breed exceeded the Black Bengal breed, as well as their reciprocal crosses, in body weight at birth, 3, 6 and 12 months of age, resulting in heterosis estimates of –2, –7, 3 and 4%, respectively (Singh *et al.*, 2002). These findings further support the use of bucks from breeds characterized by larger body size and mature weight to produce heavier crossbred offspring.

In China, Boer × Xuhuai crossbreds weighed more at birth, 2, 6 and 12 months of age than the Xuhuai breed. Consequently, Zhou *et al.* (2001) recommended Boer-sired offspring for meat production.

In Kenya, indigenous East African and Galla goats and their crossbreds sired by the Toggenburg and Anglo-Nubian breeds were evaluated for body weight from birth to 9 months of age, including their growth rates (Table 4.8). Ruvuna *et al.* (1988) ranked East African goats as the lightest, Anglo-Nubian × Galla as the heaviest and Galla, Toggenburg × East African, Toggenburg × Galla and Anglo-Nubian × East African as intermediate. Toggenburg- and Anglo-Nubian-sired East African and Galla does exceeded the dam breed in body weight at birth by 10 and 0–8%, respectively, and at 9 months by 9–19 and 6–15%, respectively. In another study, Ruvuna *et al.* (1992) reported that kids born to Galla does compared with those from East African does produced 2.3 kg heavier live weight, 3.2% greater hot and chilled carcasses at 14.7 months, 0.9% more internal fat, 1.4% more lean at 7.2 months and 2% more lean at 14.7 months. Toggenburg × Galla goats slaughtered at 2 years of age compared with those of Toggenburg × East African, Anglo-Nubian × East African and Anglo-Nubian × Galla produced significantly more carcass lean content by 3.1, 3.4 and 3.8%, respectively. Kids from Toggenburg- and Anglo-Nubian-sired does had a similar growth rate, slaughter weight and carcass composition, while Galla does exceeded East African does in producing heavier and leaner kids.

In Bangladesh, the Black Bengal, Barbari and Anglo-Nubian breeds and Barbari × Black Bengal breeds weighed 1.4, 2.2, 3.2 and 1.6 kg at birth, 7.7, 11.5, 19 and 10.6 kg at 6 months, and 11.3, 21.9, 31.9 and 16.5 kg at 12 months of age, respectively (Mia *et al.*, 1993). The body weight of Barbari × Black Bengal goats was intermediate to their parental breeds with heterosis estimates of –11% at birth, 10% at 6 months and –1% at 12 months of age.

In Oman, male kids of Anglo-Nubian × Dhofari compared with the Dhofari breed were significantly heavier from birth to 1 year of age, with greater wither height and body length (Al-Ojaili, 1995). The author concluded that crossbreeding does of the Dhofari breed with bucks of a temperate breed demonstrated potential for increasing meat production.

In Egypt, the Alpine breed compared with the Rove breed was more fecund with a lighter body weight and similar carcass traits. The estimate of heterosis for body weight increased from 7% at 10 days to 27% at 210 days for female kids, 12% at 90 days of age for male kids, and –9 to 4% for lean yield, width and shape of carcass and internal fat (Anous and Mourad, 1993). Heterosis estimates for length, width and conformation of gigot were not significant. The authors concluded that crossbreeding the Alpine breed with the Rove breed produced does that were more prolific, while their kids grew rapidly with wide and compact carcass but more internal fat. Heterosis estimates of Zaraibi × Barki goats for body weight at birth, 56 days and market age were 24, 14 and 12%, respectively, and for mortality to 90 days was –9%, whereas those for Damascus × Barki goats were 0, 2 and 4%, respectively, with 12% for mortality to 90 days (Table 4.8). The authors concluded that Zaraibi × Barki and Damascus × Barki were intermediate between those of their parental breeds for body weight at birth, 56 days and market age (Abdelsalam *et al.*, 1994). Barki goats sired by the Damascus breed were more productive than those sired by the Zaraibi breed, resulting in more milk from a longer lactation, heavier litter weights and better milk conversion ratio, while contributing towards improved performance of kids (Abdelsalam *et al.*, 2000). At the same time, the crossbreds had an advantage over the Barki breed by producing more milk and heavier litters.

In the UK, castrated male kids of the British Saanen breed were heavier than Boer × British Saanen kids at 8 weeks of age and when slaughtered at 28, 33 and 38 kg live weight, whereas the latter were heavier than the Anglo-Nubian breed (Gibb *et al.*, 1993). The Anglo-Nubian breed compared with the British Saanen breed resulted in kids with heavier carcasses (hot, cold and empty), more muscle and less subcutaneous and intermuscular fat at each of the three slaughter weights. The reduction in intermuscular fat was presumed to be detrimental to eating quality. Boer × British Saanen compared with the British Saanen breed slaughtered at 28 or 33 kg live weight produced carcasses with proportionately more intermuscular fat. Those slaughtered at 38 kg live weight showed a little increase in internal and intermuscular fat deposition, resulting in carcasses with slightly more subcutaneous fat and lower overall fat content. The authors recommended mating bucks of the Boer breed to a proportion of does of the British Saanen breed to produce kids for market. Dairy-goat producers were encouraged to grow kids to a live weight of 38 kg or more for slaughter in order to take advantage of heavier carcasses without incurring excessive fat.

In the USA, Johnson *et al.* (1995) reported that Nubian × Florida native compared with Florida native and Spanish × Florida native goats weighed significantly more at slaughter (22 versus 19 and 19.3 kg, respectively) with a heavier carcass weight (10.9 versus 9.5 and 9.6 kg, respectively) and larger rib eye area (8.3 versus 7.42 and 7.11 cm$^2$, respectively), whereas Florida native compared with Nubian × Florida native and Spanish × Florida native goats were significantly lower in the percentage of bone (20.4 versus 21.7 and 21.2%, respectively) and had more fat than Nubian × Florida native (11.6 versus 9.3%). Boer × Spanish goats compared with the Spanish breed weighed more at birth and 90-day weaning by 0.26 and 0.67 kg, respectively, and at 8 months (either on pasture or feedlot by 2.1 and 4.1 kg, respectively), whereas feed consumption, subcutaneous backfat thickness and longissimus muscle area adjusted for carcass weight were similar (Waldron *et al.*, 1995).

Feedlot trials revealed that Boer × Spanish compared with Spanish goats resulted in kids that had significantly heavier live and carcass weights, as well as greater actual and adjusted fat thickness, carcass conformation score and leg circumference (Oman *et al.*, 1999). Only actual and adjusted fat thickness and carcass conformation remained significant when adjusted for live weight. Again, kids on feedlot compared with those on pasture were significantly heavier in live and carcass weights, with more fat and lean but less bone as a percentage of carcass weight. In another study, the Angora breed compared with the Spanish breed, Boer × Spanish and Spanish × Angora resulted in kids that had significantly lighter live and hot carcass weights, smaller longissimus muscle area and leg circumference, a lower percentage of lean and a higher percentage of fat (Oman *et al.*, 2000). In addition, kids of Boer × Spanish goats exceeded those of the Spanish breed in live and hot carcass weights and carcass conformation score ($P < 0.05$) but were similar to Spanish × Angora goats. In general, the Spanish breed and Boer × Spanish compared with the Angora breed and Spanish × Angora breeds produced carcasses with a higher percentage of lean and a lower percentage of fat. Again, Spanish × Angora compared with the Angora breed produced carcasses with more lean and less fat in most side and primal cuts (Table 4.9). There was no difference among breeds and their crosses for lean colour, surface discolouration, overall performance and off-odour. The authors concluded in favour of crossbreeding due to an advantage associated with large body size, rapid growth rate and carcasses with a greater proportion of lean.

In Canada, preliminary evaluation revealed that bucks of the Boer breed compared with those of the Alpine breed resulted in kids that were 9% heavier at birth, whereas the Alpine breed (Alpine × Saanen and Boer × Alpine) kids had similar body weights at weaning and 160 days (Goonewardene *et al.*, 1998). There was a significant breed of dam effect for body weights, while the Alpine breed and crosses with Saanen does had a similar dressing percentage and rib eye muscle areas. The authors suggested that research based on a larger sample size was necessary to verify why carcass traits of Boer-sired kids failed to demonstrate an advantage over Alpine-sired kids.

In a Northern Mexican farm, Alpine, Granadina, Nubian, Saanen and Toggenburg goats raised in a stall-fed system were grouped according to the proportion of exotic breed into high-grade ($\geq ^7/_8$) and low-grade ($< ^7/_8$) goats (Sánchez *et al.*, 1994). The high- and low-grade Alpine, Saanen and Toggenburg goats compared with Nubian goats were similar in birth weight (Table 4.10). In another study under stall-feeding conditions, high- and low-grade Alpine, Saanen and Toggenburg goats compared with local Mexican goats produced more milk from a longer lactation period, while increasing their efficiency, litter size and weight (Montaldo *et al.*, 1995). The authors suggested the need to estimate heterosis for economically important traits based on further studies of crossbred goats raised under varying goat production systems. Potential merit consistent with bioeconomic efficiency indices of performance for stall-fed systems resulted in Montaldo and Meza (1999) proposing crossbreeding local goats in Mexico with the Granadina and Nubian breeds for meat production, and with the Alpine and Saanen breeds for both meat and milk production.

In Australia, capretto (6–10 kg carcass of pink meat from suckling kids) production from Boer × Saanen (BS) and Saanen × Feral (SF) compared with Feral (FF) and Saanen × Angora (SA) goats resulted in significantly higher daily gain (165 and 162 versus 128 g/day), reducing the time to reach market weight (77 and 83 versus 99 and 101 days), and longer carcasses (49 and 49.9 versus 46.7 and 47.9 cm), while Boer × Angora (BA) goats (19 g/day, 88 days and 38.4 cm) were similar or lower (Dhanda *et al.*, 1999a). Also, SA produced 0.9% more kidney and pelvic fat, and 1% more omental fat measured as a ratio of empty body weight, while BA compared with other crossbreeds and FF produced 2.6 cm more subcutaneous fat. Chevon (16–22 kg carcass from older kids) production from BS and SF compared with BA and FF resulted in a significantly higher average daily gain (140 and 130 versus 106 and 95 g/day), reducing the time required to reach market weight (238 and 257 versus 282 and

**Table 4.9.** Least squares means (± SE) for carcass yield and quality measures of Boer × Spanish, Spanish, Spanish × Angora and Angora goats on range and feedlot.

| | Range | | Feedlot | | | |
|---|---|---|---|---|---|---|
| Source | Boer × Spanish[a] | Spanish[a] | Boer × Spanish[a,b] | Spanish[a,b] | Spanish × Angora[b] | Angora[b] |
| No. of kids | 12 | 12 | 12 | 12 | 6 | 6 |
| Live weight (kg) | 20.5z ± 1.4 | 18.4z ± 1.4 | 38.2a x ± 1.4 | 33.5b,y ± 1.4 | 36.5ab ± 2.0 | 28.0c ± 2.0 |
| Hot carcass weight (kg) | 10.0z ± 0.7 | 8.8z ± 0.7 | 21.7a,x ± 0.7 | 19.2b,y ± 0.7 | 20.1ab ± 1.0 | 14.5c ± 1.0 |
| Longissimus muscle area ($cm^2$) | 6.3y ± 0.7 | 5.3y ± 0.7 | 12.5a,x ± 0.7 | 11.5a,x ± 0.7 | 11.5a ± 0.5 | 9.3b ± 0.5 |
| 12th rib (cm) | | | | | | |
| Actual fat thickness | 0.03y ± 0.01 | 0.03y ± 0.01 | 0.12a,x ± 0.01 | 0.07b,x ± 0.01 | 0.13a ± 0.02 | 0.12a ± 0.02 |
| Adjusted fat thickness | 0.04z ± 0.02 | 0.04z ± 0.02 | 0.16ab,x ± 0.02 | 0.11b,y ± 0.02 | 0.23a ± 0.03 | 0.22a ± 0.03 |
| Body wall thickness (cm) | 0.62y ± 0.09 | 0.53y ± 0.09 | 1.32x ± 0.09 | 1.40x ± 0.09 | 1.55 ± 0.12 | 1.40 ± 0.12 |
| Carcass conformation score[c] | 3.3z ± 0.8 | 1.8z ± 0.8 | 11.4a,x ± 0.8 | 8.3b,y ± 0.8 | 10.7ab ± 1.1 | 9.0ab ± 1.1 |
| Carcass length (cm) | 92y ± 1.1 | 90y ± 1.1 | 107a,x ± 1.1 | 105a,x ± 1.1 | 103b ± 1.6 | 94c ± 1.6 |
| Leg circumference (cm) | 44z ± 0.6 | 43z ± 0.6 | 55a,x ± 0.6 | 53ab,y ± 0.6 | 53a ± 0.9 | 48b ± 0.9 |
| Lean maturity score[d] | 1.4x ± 0.15 | 1.4x ± 0.15 | 1.4x ± 0.15 | 1.5x ± 0.15 | 1.9 ± 0.21 | 1.4 ± 0.21 |
| Skeletal maturity score[d] | 1.4y ± 0.08 | 1.5y ± 0.08 | 1.7x ± 0.08 | 1.7x ± 0.08 | 1.5 ± 0.12 | 1.5 ± 0.12 |
| Marbling score[e] | 1.7y ± 0.23 | 1.8y ± 0.23 | 3.4ab,x ± 0.23 | 3.1b,x ± 0.23 | 4.1a ± 0.32 | 4.1a ± 0.32 |
| Frank streaking score[e] | 2.0y ± 0.17 | 1.8y ± 0.17 | 3.6b,x ± 0.17 | 3.4b,x ± 0.17 | 4.3a ± 0.24 | 4.2ab ± 0.24 |
| Buckiness score[f] | 1.6y ± 0.3 | 1.3y ± 0.3 | 4.4a,x ± 0.3 | 4.0a,x ± 0.3 | 4.8a ± 0.4 | 3.2b ± 0.4 |

[a]Oman *et al.* (1999); [b]Oman *et al.* (2000).
[c]Means based on a 15-point descriptive scale (1 = very angular, narrow and thin and 15 = extremely thick and bulging).
[d]Means based on the USDA (1992) skeletal and lean maturity score for lambs, where 1 = $A^{00}$ and 2 = $B^{CO}$.
[e]Means based on the USDA (1992) marbling and flank streaking scores, where 1 = practically devoid$^{00}$ and 5 = modest$^{00}$.
[f]Means based on a 5-point scale where 1 = no buckiness and 5 = extreme buckiness.
a,b,c or x,y,z Means within a row not followed by the same letter differ ($P < 0.05$).

**Table 4.10.** Means (± SE) for birth weight, total milk production, lactation length, post-kidding body weight, efficiency, and litter size and weight[a].

| Genetic group | No. kids[a] | Birth weight (kg) | No. does[b] | Total milk production (kg) | Lactation length (d) | Post-kidding body weight (kg) | Efficiency (kg)[c] | Litter at birth | |
|---|---|---|---|---|---|---|---|---|---|
| | | | | | | | | Size | Weight (kg) |
| Local | | | 30 | 299c±34 | 288c ± 11 | 42.7c ± 1.3 | 7.2bc ± 0.7 | 1.72ab ± 0.11 | 5.2ab ± 0.4 |
| Low-grade | | | | | | | | | |
| Alpine | 290 | 3.2ac ± 0.04 | 91 | 459ab ± 20 | 251bc ± 7 | 43.2bc ± 0.8 | 10.5a ± 0.4 | 1.73ab ± 0.07 | 5.5ab ± 0.2 |
| Granadina | 192 | 2.9c ± 0.05 | 52 | 353c ± 26 | 237bc ± 9 | 43.5bc ± 1.0 | 8.3bc ± 0.6 | 1.57ab ± 0.08 | 4.6ab ± 0.3 |
| Nubian | 355 | 3.2ef ± 0.03 | 105 | 370c ± 19 | 230c ± 7 | 46.3ab ± 0.8 | 8.0bc ± 0.4 | 1.77a ± 0.06 | 5.5ab ± 0.2 |
| Saanen | 240 | 3.3ab ± 0.04 | 71 | 428bc ± 23 | 244bc ± 8 | 43.5bc ± 0.9 | 9.9ab ± 0.5 | 1.79ab ± 0.07 | 5.6ab ± 0.2 |
| Toggenburg | 180 | 3.1ef ± 0.05 | 54 | 422bc ± 26 | 259abc ± 9 | 41.9c ± 1.0 | 10.0ab ± 0.4 | 1.53ab ± 0.08 | 4.8ab ± 0.3 |
| High-grade | | | | | | | | | |
| Alpine | 871 | 3.3b ± 0.03 | 330 | 469ab ± 13 | 270ab ± 4 | 44.6bc ± 0.5 | 10.6a ± 0.3 | 1.65ab ± 0.04 | 5.6a ± 0.1 |
| Granadina | 601 | 2.7d ± 0.03 | 136 | 370c ± 16 | 230c ± 5 | 43.6bc ± 0.6 | 8.4b ± 0.3 | 1.70ab ± 0.05 | 4.8b ± 0.2 |
| Nubian | 1062 | 3.1f ± 0.02 | 130 | 339c ± 15 | 228c ± 5 | 48.1a ± 0.6 | 7.0c ± 0.3 | 1.69ab ± 0.05 | 5.3ab ± 0.2 |
| Saanen | 454 | 3.3ab ± 0.03 | 130 | 513a ± 16 | 283a ± 5 | 45.9ab ± 0.6 | 11.2a ± 0.3 | 1.61ab ± 0.05 | 5.4ab ± 0.2 |
| Toggenburg | 378 | 3.3b ± 0.03 | 165 | 450b ± 16 | 263ab ± 5 | 42.7c ± 0.6 | 10.6a ± 0.3 | 1.51b ± 0.05 | 5.1ab ± 0.2 |

[a]Sánchez *et al.* (1994); [b]Montaldo *et al.* (1995).
[c]Total milk production as a ratio of post-kidding body weight
a,b,c,d,e,f, Means within a column not followed by the same letter differ ($P < 0.05$).

295 days) and longer carcasses (62.1 and 60.3 versus 54.9 and 56.4 cm). Also, SF produced 1.3% more kidney and pelvic fat, and 2.1% more omental fat. Capretto production from BS compared with other crossbreeds and FF resulted in significantly paler longissimus muscle, a desirable quality with a mean objective score of 1.4 on a scale of 1–5, chromameter values of 53.6, 8.6 and 6.2 for lightness ($L^*$), redness ($a^*$) and yellowness ($b^*$) (Hunter scale), and a fibre optic probe value of 53 (Dhanda *et al.*, 1999b). In addition, BS compared with SA resulted in significantly more muscle in the flank (57.7 versus 52.1%) and ribs, and compared with FF and SA leg length (64.3 versus 68.8 and 67.7%) was lower (Dhanda *et al.*, 1999c). Chevon production from FF compared with other crosses resulted in 43–73% more muscle ($P<0.05$) and various primal cuts in carcasses, while BA and SF compared with FF produced more subcutaneous fat (6.2 and 6 versus 4.5%) and intermuscular fat (8.2 and 7 versus 5.5%). Capretto production in the carcass side from SA compared with BS and SF resulted in more subcutaneous fat (6.6 versus 4.1 and 4.2%) and intermuscular fat (5.3 versus 3.3 and 3%), whereas BS resulted in a bone content of 23–38% ($P < 0.05$) and various primal cuts. Capretto and chevon production from FF compared with other crossbreeds resulted in higher muscle to bone ratios of 2.7:1 and 3.6:1, respectively. Concurrently, BA had a longissimus thoracis muscle with 7.2% more extractable fat content compared with 5% for BS and 3.2% for SF, while the proportion of individual fatty acids varied significantly among the crossbreeds and FF (Dhanda *et al.*, 1999d). The authors concluded that growth rate, carcass characteristics and composition were in favour of crossbred goats with no important influence on meat quality and chemical composition.

In Norway, Asheim and Eik (1998) proposed an alternative source of revenue to cashmere production for a specialized dairy goat farm. This approach would require kidding in the month of May and raising offspring for slaughter at about 8 months of age or 16 kg live weight. Surplus milk in summer and a growing niche market for goat meat were expected to increase farm family earnings by 18.5% annually.

Modelling based on economic and genetic parameters has been helpful in developing crossbreeding strategies for the raising of livestock and poultry. Bett *et al.* (2011) used bioeconomic models to study profit and the economic values of four breeding groups: purebreds (indigenous goats and German Alpine breed), single crosses, backcrosses, and second backcrosses with the objective of defining breeding objectives to determine optimum crossbreeding levels for goats in smallholder production systems. The authors concluded that crossbred goats derived from backcrossing would increase the profitability of the smallholder production in Kenya, whereas crossbreeding to a higher proportion of the German Alpine breed with second backcrossing was not recommended.

## 4.7 Composite Populations

An integral part of breeding approaches employed by livestock and poultry breeders for commercial production of eggs, meat, and fibre has been the concept of composite populations (Figs 4.3 and 4.4). This is because it may be possible to achieve the same level of performance in a single composite population with a much simpler breeding structure compared with the crossbreeding of two or more breeds. The operational advantage arises from the reduced risk of introducing diseases because all replacements are produced within the farm. There is also a lower resource requirement from having to manage a single population with no requirement to purchase breeding stock. A large number of newly developed breeds worldwide have been described by Mason (1996) and Gall (1996). However, the more recent methods differ from the older principally in their intensity and deliberate application of a greater store of scientific knowledge on qualitative genetics and complementing

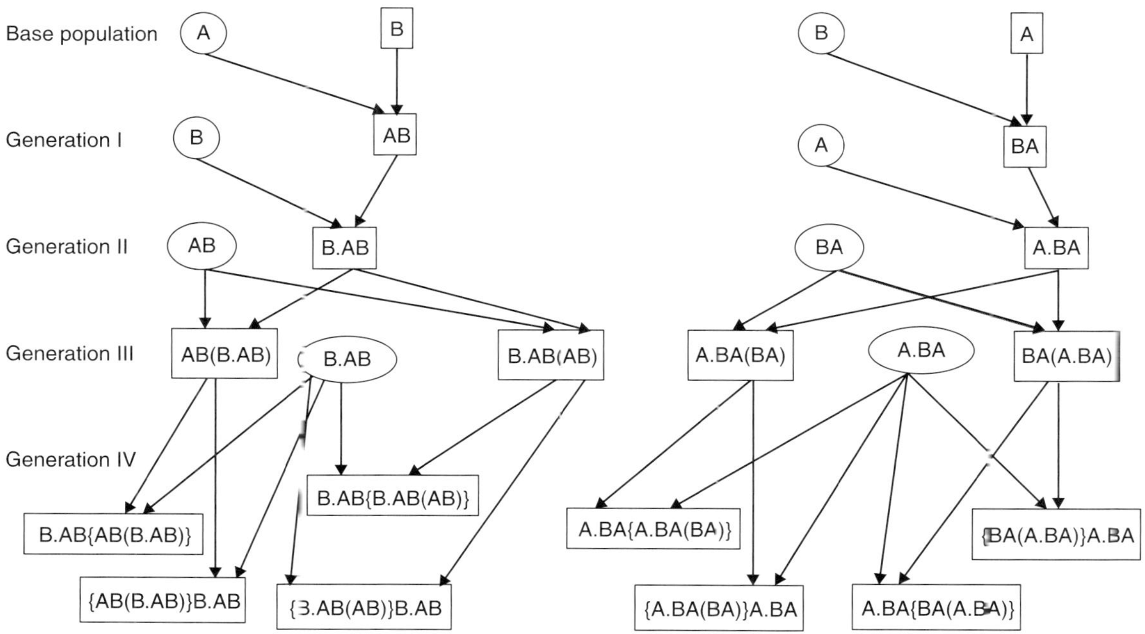


**Fig. 4.3.** Flow chart showing crossbred combinations of two breeds (A and B) leading to the development of a composite population (Shrestha, 2005).

| Parents | Sire | Dam | Sire | Dam | Sire | Dam | Sire | Dam | Sire | Dam | Sire | Dam |
|---|---|---|---|---|---|---|---|---|---|---|---|---|
| Base population | A | B | B | A | A | C | C | A | B | C | C | B |
| Generation I | C | AB | C | BA | B | AC | B | CA | A | BC | A | CB |
| | AB | C | BA | C | AC | B | CA | B | BC | A | CB | A |
| Generation II[a] | AB | C.AB | BA | C.BA | AC | B.AC | CA | B.CA | BC | A.BC | CB | A.CB |
| | C.AB | AB | C.BA | BA | B.AC | AC | B.CA | CA | A.BC | BC | A.CB | CB |
| | AB | AB.C | BA | BA.C | AC | AC.B | CA | CA.B | BC | BC.A | CB | CB.A |
| | AB.C | AB | BA.C | BA | AC.B | AC | CA.B | CA | BC.A | BC | CB.A | CB |
| Generation III[b] | C.AB | AB(C.AB) | C.BA | BA(C.BA) | B.AC | AC(B.AC) | B.CA | CA(B.CA) | A.BC | BC(A.BC) | A.CB | CB(A.CB) |
| | AB(C.AB) | C.AB | BA(C.BA) | C.BA | AC(B.AC) | B.AC | CA(B.CA) | B.CA | BC(A.BC) | A.BC | CB(A.CB) | A.CB |
| | C.AB | (C.AB)AB | C.BA | (C.BA)BA | B.AC | (B.AC)AC | B.CA | (B.CA)CA | A.BC | (A.BC)BC | A.CB | (A.CB)CB |
| | (C.AB)AB | C.AB | (C.BA)BA | C.BA | (B.AC)AC | B.AC | (B.CA)CA | B.CA | (A.BC)BC | A.BC | (A.CB)CB | A.CB |
| | AB.C | AB(AB.C) | BA.C | BA(BA.C) | AC.B | AC(AC.B) | CA.B | CA(CA.B) | BC.A | BC(BC.A) | CB.A | CB(CB.A) |
| | AB(AB.C) | AB.C | BA(BA.C) | BA.C | AC(AC.B) | AC.B | CA(CA.B) | CA.B | BC(BC.A) | BC.A | CB(CB.A) | CB.A |
| | AB.C | (AB.C)AB | BA.C | (BA.C)BA | AC.B | (AC.B)AC | CA.B | (CA.B)AC | BC.A | (BC.A)BC | CB.A | (CB.A)CB |
| | (AB.C)AB | AB.C | (BA.C)BA | BA.C | (AC.B)AC | AC.B | (CA.B)AC | CA.B | (BC.A)BC | BC.A | (CB.A)CB | CB.A |
| | Offspring | | Offspring | | Offspring | | Offspring | | Offspring | | Offspring | |
| Generation IV | C.AB{AB(C.AB)} | | C.BA{BA(C.AB)} | | B.AC{AC(B.AC)} | | B.CA{CA(B.CA)} | | A.BC{BC(A.BC)} | | A.CB{CB(A.CB)} | |
| | {AB(C.AB)}C.AB | | {BA(C.AB)}C.BA | | {AC(B.AC)}B.AC | | {CA(B.CA)}B.CA | | {BC(A.BC)}A.BC | | {CB(A.CB)}A.CB | |
| | {(C.AB)AB}C.AB | | {(C.BA)BA}C.BA | | {(B.AC)AC}B.AC | | {(B.CA)CA}B.CA | | {(A.BC)BC}A.BC | | {(A.CB)CB}A.CB | |
| | C.AB {(C.AB)AB} | | C.BA{(C.BA)BA} | | B.AC{(B.AC)AC} | | B.CA{(B.CA)CA} | | A.BC{(A.BC)BC} | | A.CB{(A.CB)CB} | |
| | AB.C{AB(AB.C)} | | BA.C{BA(BA.C)} | | AC.B{AC(AC.B)} | | CA.B{CA(CA.B)} | | BC.A{BC(BC.A)} | | CB.A{CB(CB.A)} | |
| | {AB(AB.C)}AB.C | | {BA(BA.C)}BA.C | | {AC(AC.B)}AC.B | | {CA(CA.B)}CA.B | | {BC(BC.A)}BC.A | | {CB(CB.A)} CB.A | |
| | AB.C{(AB.C)AB} | | BA.C{(BA.C)BA} | | AC.B{(AC.B)AC} | | CA.B{(CA.B)AC} | | BC.A{(BC.A)BC} | | CB.A{(CB.A)CB} | |
| | {(AB.C)AB}AB.C | | {(BA.C)BA}BA.C | | {(AC.B)AC}AC.B | | {(CA.B)AC}CA.B | | {(BC.A)BC}BC.A | | {(CB.A)CB}CB.A | |

**Fig. 4.4.** Flow chart showing crossbred combinations of three breeds (A, B and C) leading to the development of a composite population (Shrestha, 2005). [a]Crossbred parents with genetic contribution from two or three breeds. [b]Crossbred parents with genetic contribution from three breeds.

husbandry requirements (Winters, 1953, 1954; Dickerson, 1969a; Lopez-Fanjul, 1974; Shrestha and Heaney, 2003, 2004).

In theory, a newly developed breed is expected to have a lower genetic potential for performance from a reduced level of heterozygosity compared with a specific or rotational cross involving the same number of breeds (Shrestha, 1973). This is because a single population is not capable of exploiting breed difference in maternal versus individual performance. The expected loss in heterozygosity can be reduced by increasing the number of breeds assembled to three, four or more, retaining two-thirds, three-quarters or more of the average of single crosses among constituent breeds. Furthermore, there is the possibility of minimizing loss in heterosis arising from the rearrangement of genetic combinations in the development of a composite population.

In developing countries, the practice of indiscriminate crossbreeding of the indigenous population with imported breeds has led to breed replacement, and often the development of composite populations with no record of their genetic background. This point is illustrated in Algeria where the Muria and Maltese breeds were originally imported with the intention of crossbreeding indigenous Berber, Makatia and Arabia goats (Taferrant *et al.*, 1995). In the following years (1967–87), further introduction of the Alpine, Toggenburg and Saanen breeds for crossbreeding in the Kabylie region resulted in the development of composite populations. In Korea, Germany, the Russian Federation, India, Australia and New Zealand, grading up to European dairy breeds with considerable potential for improving productivity has been widespread. In Korea, Saanen bucks have been mated to a number of indigenous goats (Lee *et al.*, 1974). In the Russian Federation, the grading up of local goats has been achieved by crossbreeding through the use of semen from the Toggenburg- and Saanen-derived breeds (Orekhov, 1980). In India, the grading up of the Deccani and Gaddi breeds was based on the use of the Angora breed, which excels in mohair production (Shrestha, 2005). In Australia and New Zealand, the productivity of feral goats improved with the introduction of the Angora breed.

The interbreeding of a number of indigenous goat populations located in isolated regions has resulted in the development of small meat-type breeds in Sri Lanka, India, Papua New Guinea and Fiji; various Criollo (syn. Creole) breeds in Latin America and West Indies; Spanish goats in the south-west USA; and Sem-Raça Definida goats in Brazil (Quartermain, 1991). Some of the newly developed goat breeds, such as Anglo-Nubian, Boer, French Alpine, Killis, La Mancha and Prenakan Etawah, have considerable genetic potential for increased productivity (Gall, 1996). Crossbreeding with breeds that have demonstrated potential genetic merit has resulted in the development of 80 composite breed populations in 37 countries (Table 4.11).

In the early 1920s, South African farmers from the Eastern Cape crossbred indigenous goats kept by the Hottentot and Bantu tribe with imported Nubian and Indian goats, resulting in the development of the 'Boer breed' (Skinner, 1972). In the following years, recurrent selection for size and conformation had a significant influence in the development of this breed. The history, origin and characteristics of the Boer breed in South Africa have been reviewed (Malan, 2000). The author reported advantages of 210% fecundity, 29 kg weaning weight at 120 days, productivity to 10 years of age, the ability to raise twins, increased lean muscle yield, hardiness, adaptability and resistance to diseases. According to Greyling (2000), the Boer breed has a reproductive potential for early maturity with puberty at 28–31 kg body weight, an extended breeding season and rebreeding within 55 days of kidding. Furthermore, Boer goats have exhibited oestrus in all months of the year, although the frequency was <50% in breeding cycles of 9 months and <30% in 4-month breeding cycles with a 62-day mean interval from kidding to conception. Boer goats are also known predominantly as browsers, demonstrating potential for use in controlling shrubs and bush (Erasmus, 2000). Australia, Canada, China, France, Germany, Israel, New Zealand, Sri

**Table 4.11.** Composite breed populations in the world, number of foundation breeds and year of origin or recognition (from Shrestha, 2005).

| Country | Composite breed | No. foundation breeds[a] | Country | Composite breed | No. foundation breeds[a] |
|---|---|---|---|---|---|
| Australia | Cashgora | 2 | Kyrgyzstan | Kirgiz | 2 |
| Brazil | Branca sertaneja | 2 | Malaysia | Jermasia | 2 |
| | Parda sertaneja | 2 | Mongolia | Gobi Wool goat | 2 |
| | SRD | 2 | | Unjuul | 2 (1982) |
| Bulgaria | Bulgarian White Dairy | 2 | | Uuliin Bor | 2 (1991) |
| China | Guanzhong Dairy | 2 (1940) | Morocco | Fnideq | 2 |
| | Hailun | 3 | Mozambique | Pafuri | 2 (1928) |
| | Hongtong | 2 | Netherlands | Dutch Pied | 2 |
| | Laoshan Dairy | 2 (1919) | | Dutch Toggenburg | 2 |
| | Nanjiang Yellow | 2 (1960) | | Dutch White | 2 |
| Cyprus | Peratiki | 2 | Nigeria | Savannah Brown | 2+ |
| Denmark | Danish Landrace | 3 | Norway | Norwegian | 5 |
| Fiji | Fiji | 3 | New Zealand | Kiko | 2 |
| France | French Alpine | 2 (1930)[b] | Pakistan | Beiari | 2 |
| Germany | German Improved Fawn | 2 (1928)[b] | | Buchi | 2 |
| | German Improved White | 2 (1928)[b] | | Jattal | 2 |
| Hungary | Hungarian Improved | 2+ | | Pak Angora | 2 |
| India | Indian Mohair | 3 (1973) | | Shurri | 2 |
| | Malabari | 2 | | Sind Desi | 2 |
| | Ramdhan | 2 | Romania | Banat White | 3 |
| Indonesia | Peranakan Etawah | 2 | Russia | Altai Mountain | 2 (1982) |
| Israel | Israeli Saanen | 2 (1932) | | Angora-Don | 2 |
| | Yaez | 2+ | | Dagestan White | 2 |
| Italy | Aquila | 4 | | Don-Kirgiz cross | 2 |
| | Benevento | 4 | | Russian White | 2 (1905) |
| | Campobasso | 4 | South Africa | Boer | 2 (1959)[b] |
| | Ionica | 2 (1981)[b] | Spain | Barreña | 3 |
| | Potenza | 3 | | Murcia-Granada | 2 (1980)[b] |
| Kazakhstan | Soviet Mohair | 2 (1962) | | Murcian | 2 (1933)[b] |
| Kenya | Kenya Dual-Purpose | 4 | Tajikistan | Soviet Mohair | 2 (1962) |

[a]Year of origin or year recognized.
[b]Year breeds society, association or stud book was established.

Lanka and the USA, as well as many other countries, have imported this breed to meet the growing demand for increased production of goat meat. Nevertheless, claims stating that Boer goats have resistance to specific diseases need to be substantiated.

## 4.8 Selection

Morphological characteristics and production performance considered by the breeder to be of economic importance for improving productivity are numerous. Studies in many goat breeds and their crosses fed diets that vary in nutritive value and raised under conditions of sedentary, nomadic and semi-nomadic management substantiate the importance of environmental influences on performance (Guha *et al.*, 1968; Moulick and Syrstad, 1970; Mavrogenis *et al.*, 1984; Nicoll, 1985; Bakshi *et al.*, 1986; Herrera *et al.*, 1987; Gebrelul *et al.*, 1994; Gerstmayr and Horst, 1995). The general statement of environment influencing many of the economically important traits may be summarized as follows: bucks are heavier than does; single-born kids are heavier than twins, which are heavier than triplets; and body weight tends to increase with age of the dam up to 5–6 years of age, declining at older ages. In addition, location, year of birth, season of kidding, parity and weight at slaughter, as well as diet, have an important influence on productivity. The incidence of mortality has been reported to be lower in goats raised under range conditions and in arid and semi-arid regions of the country that are well drained with sandy soil compared with those in hot, humid environments. In contrast, mortality is higher in confinement and stall feeding as a result of the concentration of animals and diseases. The influence of environment on performance has been described in detail in a number of publications and is important but is beyond the scope of the present review.

In theory, any breeding plan that is most efficient for one performance trait may be less efficient for another. Also, increasing the number of traits considered in the selection criteria reduces the annual increment of genetic response to selection for each trait. There is agreement among researchers in the scientific community that performance traits known to be highly heritable benefit from mass selection by exploiting a greater proportion of additive genetic variation. Genetic gains realized from selection for performance traits of economic importance, which are primarily associated with additive genetic variance, are permanent and cumulative. At the same time, lowly heritable traits benefit more from family selection that could make the most of non-additive genetic variation.

An important issue in goat breeding is how selection should be exploited for the recurrent improvement of performance traits considered to be of economic importance for commercial production of goat meat. Years of intensive selection in populations derived from narrow genetic bases could eventually exhaust additive genetic variance available for selection. Breeding populations with large effective numbers of parents comprised of sufficient genetic variability for economically important traits allow increased intensity of selection while reducing the rate of inbreeding. There is evidence to suggest that selection will eventually exhaust additive genetic variation in the population when a greater proportion of genes influencing a quantitative trait is additive and a smaller proportion is non-additive. Possibly, breeding populations today may already have experienced a reduction in reproductive rate and performance traits. Therefore, it is pertinent that ways and means of exploiting variance associated with dominance and epistasis are explored. When overdominance is present, even in a few loci, the value of progeny testing and family selection for improving performance could become inadequate.

In animal breeding, the practical importance of heterosis has long been appreciated, and meat from crossbred animals is well accepted by the consumer. Consequently, in developed countries, the majority of the meat animals marketed today are of crossbred origin. Researchers have provided irrefutable evidence to suggest that livestock and poultry species can benefit from selection, both within and among populations, without tremendous costs in time and facilities compared with the development of inbred lines. Breeders must rely on the genetic superiority of parents for any improvement of crossbred progeny performance. This leads to an important issue – whether selection for commercial production should be based on performance within the purebred populations or their crossbred progeny. There is evidence to suggest theoretical promise in direct selection for maximum performance of crosses among complementary populations based on progeny tests of individuals and families in test-cross mating (Comstock *et al.*, 1949; Comstock, 1961). Selection based on crossbred progeny performance could possibly achieve further gains in productivity by exploiting additive genetic covariance between genotypes of both parental populations.

Research results suggest that genetic parameters (Tables 4.12–4.14) for weight gain and meat quality are low to moderately heritable and in the desired direction, demonstrating considerable potential for genetic improvement of meat goats (Acharya, 1982, 1988; Taneja, 1982). At the same time, the influence of environment on performance is of a positive nature. Prediction of genetic merit of meat goats based on a comprehensive and technically sound assessment of performance can be relatively simple to implement in terms of fiscal constraints and resource requirements for labour and time. This includes measuring and recording performance consisting of reproduction, growth and survival followed by estimated breeding values for each trait. Estimated breeding values are a function of the additive genetic variation among performance traits transmitted from one generation to the next (heritability) and their corresponding associations (genetic, phenotypic and environmental).

The estimate of genetic parameters for performance traits of economic importance in the several breeds and their crosses with higher levels of accuracy under the prevailing environment is of vital importance for the selection of meat goats. Estimates of genetic parameters in many goat breeds worldwide, including large populations of the Angora, Boer and Saanen breeds, have been summarized in Table 4.13. In Angora goats, the heritability estimates for body weight from weaning to yearling ranged from 0.23 to 0.59, while corresponding estimates of phenotypic and genetic correlations ranged from 0 to 0.38 and from 0.28 to 0.58, respectively (Nicoll, 1985). In Saanen goats, heritability estimates for body weight from 1 to 7 months of age ranged from 0.48 to 0.63, while corresponding estimates of phenotypic and genetic correlations ranged from 0.41 to 0.95 and from 0.36 to 0.95, respectively (Ricordeau *et al.*, 1972). In Damascus goats, phenotypic and genetic correlations between body weight at birth and 70-day weaning were 0.43 and 0.34, between body weight at birth and 140 days were 0.71 and 0.82, and between body weight at 70 and 140 days were 0.71 and 0.82, respectively (Mavrogenis *et al.*, 1984). In Ganjam goats, phenotypic and genetic correlations between body weight at birth and 6 months were 0.33 and 0.92, between body weight at birth and 12 months were 0.03 and –0.35, and between body weight at 6 and 12 months were 0.50 and 0.14, respectively (Madeli and Patro, 1984). Pooled estimates of genetic parameters from six studies, which included Alpine × Beetal, Beetal, Sirohi, Black Bengal, Jamunapari and Indigenous × exotic goats resulted in heritability estimates for body weight at birth, 3, 6, 9 and 12 months of age of 0.11, 0.22, 0.43, 0.33 and 0.41, respectively (Acharya, 1988). Corresponding estimates of phenotypic and genetic correlations were large and in the positive direction. Phenotypic and genetic correlations between body weight at birth and 3 months of age in Black Bengal goats were 0.51 and 0.78, in Jamunapari goats were 0.47 and 0.56, and in Beetal × Black Bengal goats were 0.51 and 0.49, respectively (Singh *et al.*, 1991). In local Assam goats and their crosses with the Beetal breed, heritability estimates for body weight at 6, 9 and 12 months of age were moderate, while phenotypic and genetic correlations were also moderate and in the desirable direction (Nahardeka *et al.*, 2001).

**Table 4.12.** Estimates of heritabilities and repeatabilities (± SE) for age at first kidding, litter size, multiple births, daily gain from birth to 150 days and bone content in the whole carcass.

| Breed | Heritability | Repeatability | Reference |
|---|---|---|---|
| Age at first kidding | | | |
| Alpine × Beetal | 0.56 ± 0.08 | | Nagpal and Chawla (1987a,b) |
| Beetal | 0.48 ± 0.09 | | |
| Litter size | | | |
| Beetal | 0.15 | | Amble *et al.* (1964) |
| Beetal, Black Bengal | 0.09 ± 0.25 | | |
| Alpine × African common | 0.02 | | Mourad (1994) |
| Multiple births | | | |
| Egyptian Baladi | 0.25 | 0.29 | Tantawy and Ahmed (1960) |
| Beetal | 0.15 | 0.22 | Amble *et al.* (1964) |
| Black Bengal | 0.09 | 0.15 | Moulick *et al.* (1966) |
| Black Bengal | 0.17 ± 0.20 | | Ali (1983) |
| Daily gain (birth to 150 days) | | | |
| West African Dwarf | 0.38–0.63 | 0.21–0.38 | Ebozoje and Ngere (1995) |
| Bone content in whole carcass | | | |
| Black Bengal, Beetal × Black Bengal, Jamunapari × Black Bengal | 0.71 ± 0.30 | | Singh and Yadava (1997) |

Estimates from a number of studies have shown direct and maternal heritability for body weight range from low to moderate, while genetic correlations between direct and maternal effects were low to moderate in magnitude, varying from negative to positive in direction (summarized in Table 4.14). Procedures developed by Dickerson (1969a) and Kinghorn (1980, 1983) resulted in similar estimates. In Turkish Angora goats, estimates of direct and maternal heritability for body weight at birth and 100 days, yearling fleece weight, litter size and total weight of kids at 100 days were low, while genetic correlations between the direct and maternal effects for body weight at birth and weaning varied from 0.18 to 0.91, for yearling and adult fleece weight varied from –0.91 to –0.95, for litter weight was –0.07, and for kid weight at 100 days was –0.98 (Gerstmayr *et al.*, 1989, 1992; Gerstmayr and Horst, 1995). In Canada, field records on 11 breed types consisting of Alpine, Angora, Nubian, Saanen, Toggenburg, non-descript goats and their crosses were assessed (Nadarajah and Burnside, 1994). Corresponding estimates of direct and maternal heritabilities for body weight at birth were 0.14 and 0.17, and at 60-day weaning were 0.26 and 0.01, respectively. In Angora goats, direct and maternal heritability estimates for body weight were 0.29 and 0.09, respectively, while phenotypic and genetic correlation between body weight and greasy fleece weight were 0.57 and 0.56, respectively (Snyman and Olivier, 1996). In Boer goats, estimates of direct and maternal heritability for body weight at birth and weaning were in the range of 0.16–0.33 and 0.05–0.60, respectively, while genetic correlations between the direct and maternal effects of body weight were small to moderate and in the desired direction (Schoeman *et al.*, 1997). Corresponding estimates for the Adelaide herd were smaller compared with the combination of Adelaide and Omatjenna herds. In local Malaysian goats and their crosses with the German Fawn breed, estimates of direct and maternal heritability for body weight at birth, 3 and 6 months of age, as well as genetic correlations between the direct and maternal effects, were low to moderate and in the desired direction

**Table 4.13.** Estimates of heritabilities (± SEM) for body weights from birth to 12 months.

| Breed | Body weight (kg) at: Birth | 3 months | 4 months | 6 months | 9 months | 12 months | Reference |
|---|---|---|---|---|---|---|---|
| Black Bengal | 0.07 ± 0.01 | – | 0.15 ± 0.04 | – | 0.21 ± 0.21 | 0.32 ± 0.08 | Guha *et al.* (1968) |
| Saanen | 0.63 (1 month) | 0.48 | 0.49 (5 months) | 0.49 (7 months) | – | – | Ricordeau *et al.* (1972) |
| Sirohi | 0.29 ± 0.16 | 0.11 ± 0.12 | – | 0.32 ± 0.18 | – | – | Misra (1983) |
| Black Bengal | 0.40 ± 0.24 | 0.09 ± 0.18 (4 weeks) | 0.25 ± 0.56 (wean) | – | – | – | Ali (1983) |
| Damascus | 0.31 ± 0.08 | 0.27 ± 0.07 (70 d) | 0.24 ± 0.07 (140 d) | – | – | – | Mavrogenis *et al.* (1984) |
| Osmanabadi | 0.10 ± 0.03 | 0.75 ± 0.03 | – | 0.66 ± 0.6 | 0.02 ± 0.01 | – | Siddiqui *et al.* (1981) |
| Ganjam | 0.19 ± 0.14 | – | – | 0.34 ± 0.23 | – | 0.36 ± 0.26 | Madeli and Patro (1984) |
| Angora | – | 0.59 ± 0.38 (wean) | – | 0.23 ± 0.14 | – | 0.40 ± 0.28 | Nicoll (1985) |
| Sirohi, Beetal × Sirohi | – | 0.59 ± 0.38 | – | – | – | – | CIRG (1986) |
| Indigenous × Exotic | 0.30 ± 0.11 | 0.17 ± 0.23 (2 months) | 0.10 ± 0.11 | 0.08 ± 0.10 | – | – | de Souza *et al.* (1987) |
| Alpine × Beetal | 0.27 ± 0.03 | 0.93 ± 0.12 | – | 0.71 ± 0.10 | 1.00 ± 0.14 | 0.60 ± 0.09 | Nagpal and Chawla (1987a) |
| Beetal | 0.24 ± 0.03 | – | – | 0.43 ± 0.10 | 0.86 ± 0.17 | 0.40 ± 0.16 | Nagpal and Chawla (1987a) |
| Black Bengal | 0.16 ± 0.12 | 0.16 ± 0.15 | – | – | – | – | Singh *et al.* (1991) |
| Jamunapari | 0.55 ± 0.18 | 0.42 ± 0.18 | – | – | – | – | Singh *et al.* (1991) |
| Beetal × Black Bengali | 0.12 ± 0.12 | 0.09 ± 0.15 | – | – | – | – | Singh *et al.* (1991) |
| Teddy | 0.05 ± 0.02 | – | – | 0.10 ± 0.01 (weaning) | – | – | Tahir *et al.* (1995) |
| Jamunapari | 0.46 ± 0.15 | 0.43 ± 0.15 | – | 0.25 ± 0.13 | 0.13 ± 0.10 | 0.13 ± 0.17 | Misra (1995) |
| South African Angora | – | – | – | – | 0.29 ± 0.06 (8–9 months) | – | Snyman and Olivier (1996) |
| Boer | 0.18 ± 0.04 | – | 0.19 ± 0.05 (5 mo) | – | – | – | Niekerk *et al.* (1996) |
| Criollo × imported | 0.15 | – | 0.08 | 0.22 | – | – | Garcia *et al.* (1996) |
| Angora | 0.25 | 0.25 (1 mo) | 0.25 (2 mo) | – | – | – | Hermiz, *et al.* (1997) |
| Boer at Omatjenne | 0.36 ± 0.14 | – | 0.60 ± 0.12 (4 mo) | 0.60 ± 0.17 | 0.40 ± 0.18 | 0.36 ± 0.19 | Schoeman *et al.* (1997) |
| Jamunapari | – | 0.30 | – | 0.51 | 0.23 | 0.31 | Roy *et al.* (1997) |
| German Fawn, Katjang | 0.34 | 0.18 | – | 0.30 | 0.24 | – | Hirooka *et al.* (1997) |
| African, French Alpine cross | 0.68 ± 0.14 | – | 0.49 ± 0.16 | 0.47 ± 0.14 | 0.43 ± 0.16 | – | Mourad and Anous (1998) |
| Assam local, Beetal cross | – | – | – | 0.26 ± 0.12 | 0.16 ± 0.11 | 0.31 ± 0.21 | Nahardeka *et al.* (2001) |

Superscripts in parentheses represents the age when weight was measured, if different from column heading.

**Table 4.14.** Estimates of direct and maternal heritability (± SE) and heritability (± SE) of feed conversion.

| Breed | Heritability | | Correlation | Reference(s) |
|---|---|---|---|---|
| | Direct | Maternal | (direct × maternal) | |
| Turkish Angora | | | | Gerstmayr *et al.* (1989, 1992); Gerstmayr and Horst (1995) |
| Birth weight | 0.02 | 0.10 | 0.18 | |
| 100-day weaning weight | 0.03 | 0.10 | 0.91 | |
| Yearling fleece weight | 0.06 | 0.04 | –0.91 | |
| Adult fleece weight | 0.01 | 0.02 | –0.95 | |
| Litter size | 0.06 | 0.04 | –0.07 | |
| Total weight of kids at 100 days | 0.06 | 0.04 | –0.98 | |
| Boer in the Adelaide and Omatjenna herds | | | | Schoeman *et al.* (1997) |
| Birth weight | 0.33 ± 0.07 | 0.36 ± 0.14 | | |
| Weaning weight | 0.27 ± 0.09 | 0.60 ± 0.12 | | |
| Boer in the Adelaide herd | | | | |
| Birth weight | 0.16 ± 0.06 | 0.14 ± 0.04 | –0.31 | |
| Weaning weight | 0.18 ± 0.05 | 0.05 ± 0.03 | –0.15 | |
| Malaysian local goats and their German Fawn crosses[a] | | | | Hirooka *et al.* (1997) |
| Birth weight | 0.16 (0.17) | 0.24 (0.24) | 0.19 (0.14) | |
| 3-month weight | 0.07 (0.07) | 0.11 (0.12) | 0.47 (0.41) | |
| 6-month weight | 0.18 (0.21) | 0.12 (0.14) | 0.25 (0.07) | |
| 9-month weight | 0.18 (0.16) | 0.12 (0.09) | 0.0 (0.31) | |

| Breed | Trait | Heritability | Reference |
|---|---|---|---|
| Sirohi; Sirohi × Beetal | Dry matter | 0.23 ± 0.42 | Misra (1983) |
| | Total digestible nutrients | 0.04 ± 0.42 | |
| Barbari | 6–9 months | 0.12 ± 0.10 | Khan and Singh (1995) |
| | 9–12 months | 0.06 ± 0.10 | |
| | 3–12 months | 0.17 ± 0.11 | |

[a]Estimates are based on Dickerson (1969a), while those in parentheses are based on Kinghorn (1980, 1983).

(Hirooka *et al.*, 1997). In Saudi Aradi and Damascus breeds and their crosses, estimates of direct additive genetic effects were significantly greater and in favour of the Damascus breed in the Jouf and Qassim experiments (Table 4.15), for body weight from birth to 24 weeks by 12–32 and 17–34%, respectively, and for daily gains from birth to 24 weeks in 4-week intervals by 14–31 and 12–37%, respectively (Khalil *et al.*, 2010). Corresponding estimates of direct heterosis were significant for growth traits in the Jouf and Qassim experiments, ranging from 0.16 to 1.39 kg, and from 0.31 to 1.56 kg, respectively. Maternal heterosis estimates were generally favourable in the Jouf experiment for body weight by 2–11% and for daily gains by 2–8%, suggesting that crossbred

**Table 4.15.** Estimates (± SE) for direct additive effect, direct and maternal heterosis for growth traits, carcass and meat composition of Saudi Aradi (A) and Damascus (D) goats at breeding stations at Jouf and Qassim (from Khalil *et al.*, 2010)[a].

| Source | Direct additive effect ($D^I = D^I_A - D^I_D$) | | Direct heterosis | | Maternal heterosis |
|---|---|---|---|---|---|
| | Jouf | Qassim | Jouf | Qassim | Jouf |
| Growth traits | | | | | |
| Weight (kg) at: | | | | | |
| Birth | –0.4 ± 0.06 (–12) | –0.9 ± 0.16 (–25) | 0.16 ± 0.05 (5) | 0.31 ± 0.14 (9) | 0.12 ± 0.05 (4) |
| 4 weeks | –2.5 ± 0.08 (–29) | –2.7 ± 0.48 (–34) | 0.76 ± 0.06 (9) | 0.52 ± 0.21 (7) | 0.93 ± 0.08 (11) |
| 8 weeks | –3.8 ± 0.12 (–29) | –3.3 ± 0.82 (–26) | 0.93 ± 0.07 (7) | 1.39 ± 1.04 (11) | 0.33 ± 0.08 (3)[NS] |
| 12 weeks | –5.0 ± 0.13 (–32) | –2.7 ± 1.09 (–17) | 1.20 ± 0.09 (8) | 1.48 ± 0.48 (10) | 0.38 ± 0.07 (2)[NS] |
| 16 weeks | –5.6 ± 0.15 (–30) | –3.8 ± 1.51 (–21) | 0.79 ± 0.10 (4) | 1.44 ± 0.69 (8) | 0.45 ± 0.12 (2)[NS] |
| 20 weeks | –6.2 ± 0.15 (–28) | –4.8 ± 1.54 (–23) | 0.55 ± 0.10 (2)[NS] | 1.80 ± 0.81 (9) | 0.46 ± 0.20 (4) |
| 24 weeks | –4.2 ± 0.19 (–17) | –4.9 ± 1.64 (–21) | 1.39 ± 0.13 (6) | 1.56 ± 0.46 (7) | 0.98 ± 0.23 (4) |
| Daily gain (g) | | | | | |
| 0–4 weeks | –47 ± 7.1 (–27) | –52 ± 4.1 (–34) | 12 ± 3.0 (7) | 14 ± 2.1 (9) | 15 ± 6.1 (8) |
| 4–8 weeks | –21 ± 6.2 (–14) | –36 ± 8.2 (–24) | 7 ± 2.0 (4) | 14 ± 5.2 (9) | 5 ± 8.2 (3)[NS] |
| 8–12 weeks | –39 ± 6.3 (–31) | –54 ± 5.2 (–37) | 8 ± 3.1 (6) | 17 ± 4.1 (12) | 5 ± 2.3 (4) |
| 12–16 weeks | –24 ± 6.1 (–17) | –26 ± 4.1 (–21) | 7 ± 2.2 (5) | 15 ± 6.1 (12) | 16 ± 5.2 (7) |
| 16–20 weeks | –47 ± 4.2 (–25) | –34 ± 2.2 (–26) | 6 ± 2.9 (3)[NS] | 19 ± 2.2 (15) | 5 ± 9.1 (2)[NS] |
| 20–24 weeks | –40 ± 3.1 (–23) | –16 ± 4.1 (–12) | 9 ± 3.2 (5) | 16 ± 1.1 (12) | 14 ± 6.2 (6) |

| | | |
|---|---|---|
| Carcass traits | | |
| Pre-slaughter weight (kg) | −6.3 ± 0.6 (−22) | 1.4 ± 0.34 (5) |
| Hot carcass weight (kg) | −7.1 ± 1.2 (−49) | 0.8 ± 0.13 (6) |
| Dressing percentage | −5.5 ± 1.4 (−12) | 1.2 ± 0.32 (2)[NS] |
| Weight of carcass components (g) | | |
| Head | −0.48 ± 0.05 (−24) | 125 ± 90 (6) |
| Skin | −0.08 ± 0.22 (−4)[NS] | 93 ± 230 (4) |
| Viscera | −0.01 ± 0.02 (−1)[NS] | 126 ± 136 (1)[NS] |
| Heart | −27 ± 11 (−21) | 62 ± 12 (4) |
| Liver | −128 ± 18 (−25) | 13 ± 34 (0.5)[NS] |
| Lungs | −52 ± 8 (−15) | 12 ± 52 (0.2)[NS] |
| Kidneys | 14 ± 4 (15) | 16 ± 3 (6) |
| Spleen | −17 ± 2 (−30) | 12 ± 92 (3)[NS] |
| Meat chemical compositon (%) | | |
| Moisture | 0.2 ± 0.12 (0.2)[NS] | −0.6 ± 0.23 (−0.8)[NS] |
| Crude protein | 0.4 ± 0.56 (0.5) | 0.7 ± 0.56 (0.9)[NS] |
| Ether extract | 0.5 ± 0.38 (3)[NS] | −0.5 ± 0.64 (−2.5)[NS] |
| Ash content | 0.1 ± 0.23 (2)[NS] | −0.3 ± 0.05 (−6.1) |

[a]The percentage deviation from the average of the purebreds is shown in parentheses.
All estimates are significant except where indicated as NS ($P > 0.05$).

dams were marginally superior to their purebred contemporaries in growth. In the Qassim experiment, estimates of additive genetic effects for carcass traits were significant and in favour of the Damascus breed; in contrast, chemical composition of meat traits were in favour of the Saudi Aradi breed. Estimates of direct heterosis were positive and in the desired direction for pre-slaughter weight, hot carcass weight and weight of head, skin, heart and kidneys. The wider range of variation in these estimates warrants further studies to establish the nature of the genetic relationship between direct and maternal effects associated with performance traits of economic importance. The difficulty in obtaining consistent estimates of genetic correlation arises from the inability to obtain large amounts of data necessary to increase precision.

An important constraint for the improvement of economically important performance traits in meat goats is the absence of precise estimates of genetic parameters and response to selection that would have been possible with large data sets and long-term studies such as those in developed countries for other livestock and poultry species. Genetic parameter estimates and their standard errors derived from different procedures using the same data set tend to vary (Henderson, 1953; Dickerson, 1969a; Patterson and Thompson, 1971; Rao, 1971; Hemmerle and Hartley, 1973; Henderson, 1984; Meyer, 1989, 1991; Gilmour *et al.*, 1995; van Tassell and van Vleck, 1996). The pooling of additive genetic variance estimates from pure breeds and their crosses may be questionable in terms of scientific merit. This is because additive genetic variance within breeds and additive genetic covariance among breeds are considered alike. Nevertheless, in practice, additive genetic variance and covariance components are combined to reduce the sampling variance associated with genetic parameter estimates.

There is the opportunity to increase the body weight of goats by exploiting additive genetic variability from genetically diverse breed populations based on the use of individual and family selection. This is because European and Boer breeds known to exceed 100 kg in body weight (Warmington and Kirton, 1990) have considerable potential for use in crossbreeding and the formation of composite populations to facilitate the commercial production of meat from goats (Shrestha 2005; Shrestha and Fahmy, 2007a).

In Barbari, Beetal, Osmanabadi, Malabari and Bengal breeds, estimates of heritability for reproductive trait, although negligible in magnitude, were positive. These estimates are indicative of prospects for improvement of reproduction. Despite the potential for genetic improvement, breeding strategies have often ignored the value of year-round kidding. According to Devendra and Burns (1983), the income and profitability associated with the commercial production of goat meat to a large extent depends on the reproductive rate of does. The authors suggested that fecundity should be included as a selection criterion for goat meat production, along with body weight and meat quality. Correspondingly, any measure of body weight close to market age would benefit from including reproductive rate, maternal ability and resistance to diseases in the selection criteria. An alternative approach that could be considered is to include body weight at 6 months or older along with age at kidding to augment the number and weight of kids marketed.

In Bangladesh, the Black Bengal breed selected for body weight at 6 months of age exceeded the random-bred Black Bengal and Jamunapari × Black Bengal in body weight per doe at birth and 60 days (Amin *et al.*, 2001). Concurrently, no detrimental influence on fertility, age and weight at first oestrus, litter size and kidding interval was observed. In the same study, crossbreeding did not offer any advantage as a result of delayed sexual maturity and post-partum service period, leading the authors to favour selection over crossbreeding. In Black Bengal and Jamunapari × Black Bengal goats, chilled carcass weight, loin eye area and edible body parts were similar, leading Das *et al.* (2001) to favour selection for commercial chevon production. According to Acharya (1988), economically important traits that need to be considered in the selection criteria for improvement of goat meat in India are growth, survival, feed conversion and carcass yield for kids, along with reproduction, age at first kidding, milk yield, kidding interval and longevity for mature goats.

In Mexico, the average generation interval of the Granadina, Nubian, Saanen, Alpine and Toggenburg breeds was 3.5, 3.6, 3.7, 3.8 and 4.3 years, respectively. The average generation interval between sire and offspring was 3.4–3.5 years, and between dam and offspring was 4–4.1 years (Meza *et al.*, 1994). Goats are known to kid year-round, demonstrating the opportunity to accelerate the genetic response to selection by minimizing generation interval.

Goats carry a substantial amount of abdominal and intestinal fat, which could be lowered, eliminating costs associated with producing unwanted fat. Reducing fat not only improves feed conversion but also promotes consumer acceptance. Despite the lack of precise estimates of genetic parameters for meat quality in goats, there is an opportunity for improvement. Aids to selection such as the use of ultrasonic devices to measure back fat and loin muscle area in live animals or dissecting individual carcass cuts to permit the separation of fat, meat and bone from a sample of slaughtered siblings have achieved considerable success in improving lean muscle yield (Stouffer, 1060). Furthermore, technology for scanning of live animals has improved noticeably over the years, making it possible to obtain a clear image of muscle area at a fraction of the cost of marketing the animal. Ultrasonically measured longissimus muscle area and depth in Alpine goats have demonstrated potential merit for use in genetic improvement of muscling (Stanford *et al.*, 1995). In Australian cashmere goats, boneless meat yield has been predicted from live weight and hot carcass weight (McGregor, 1990). In addition, Angora kids fed a high-quality energy diet and slaughtered at 5 months of age produced meat of commercially acceptable quality (McGregor, 1996).

## 4.9 Genetic Improvement Programmes

One of the most challenging issues for goat meat production is to implement genetic improvement programmes for the resource-poor goat keeper in developing countries. Lack of adequate resources, provision of essential services and facilities that extend across households over great distances as well as serious concerns over their feasibility have caused a great deal of difficulty in the delivery of performance testing programmes across countries. There is reason to believe that biological, cultural, statistical, social, economic and management aspects are necessary components of the decision-making process. This is because goats reared in large herds under transhumance systems along with many small herds around urban areas are closely associated with the environment, religious rituals and sustainable development. The improvement of meat goats must include a comprehensive and technically sound assessment with sufficient flexibility to meet the needs of diverse environmental and managerial conditions. Therefore, a breeding strategy for improvement of meat goats should include more than a prediction of breeding values for economically important production traits.

It is also crucial to have a clear understanding of the nature of the underlying quantitative genetic principles involved in exploiting genetic resources for genetic improvement. This is particularly important when examining possible benefits from crossbreeding against selection in terms of genetic gain in performance, improved vigour, uniform animal products and a decline in inbreeding depression against expenses incurred during breed evaluation. The negative association between the production of fibre and meat when nutrition is a limiting factor suggests that fibre production occurs at the expense of meat (Shelton, 1998). Under such conditions, productivity could be enhanced with provision for supplementary feeding.

Developing breeding strategies for improvement of economically important performance traits associated with goat meat production may rely on the diversity of goat genetic resources worldwide, along with evaluation of breeds and their crosses (Shrestha and Fahmy, 2005). The genetic base should be as broad as possible to provide sufficient genetic variation in performance traits to sustain the genetic response to selection over a number of generations. One should also consider the prior selection history of the base population for economically important production traits, the relative magnitude of non-additive to additive genetic effects and the possible role of stress environments during the choice of sire and dam breeds. Likewise, the decision to continue with either selection or crossbreeding must take into consideration the time and resource requirements necessary for breed evaluation and the benefit achieved from genetic gain. In the short run, crossbreeding has an advantage whereas selection of replacement parents based on genetic merit of the individual accelerates the genetic response, enhancing performance with either crossbreeding or composite breed formation. In this context, parameter estimates in a number of meat breeds and their crosses for traits including body weight at various ages, age at kidding, kidding interval, service period, litter size, live weight at slaughter, hot carcass weight, dressing percentage and feed conversion have been assembled from numerous studies and are presented in Tables 4.12–4.15.

Goat breeders may employ simple means of identifying, measuring and recording performance on the farm for use in the genetic evaluation of their breeding stock. In general, the genetic potential of goats under uniform feeding and management conditions are evaluated on the basis of growth rate of their offspring or in combination with reproduction of their dam. Furthermore, bucks and does are evaluated on the basis of productivity and the contemporary performance of their kids. In some instances, bucks from a number of adjoining farms are evaluated for their performance under more uniform environments in test stations located at a convenient distance from herds. In practice, bucks and does with superior genetic merit are often certified by a government agency or breed organization based on their performance and are sold at auctions across the country.

National breeding programmes for the evaluation of genetic merit in breeding animals usually adopt mixed animal model methodologies to determine estimated breeding values for multiple traits. The evaluation of parents along with their offspring has achieved considerable success in other livestock and poultry species, demonstrating potential merit in meat-goat production. Precise estimates of genetic parameters, performance records of parents and their siblings, pedigree relationships among animals and performance across herds have improved the accuracy of predicting breeding values. Information processing along with software, lowered costs of processing, memory and storage requirements has made the application of mixed model methodology feasible for breeding meat goats. Likewise, goats with sufficient genetic merit should rapidly be disseminated among herds under sedentary, nomadic and semi-nomadic management across the country to multiply genetic gain for the production of goat meat.

India, with one of the world's largest goat populations, has no breeding programme for improvement of productivity of goats (Acharya, 1988). In practice, central and state government farms maintain small herds of goats, while animals surplus to breeding requirements, usually unproven in terms of their genetic merit, are sold as potential breeding stock to breeders. It has been suggested that goat meat production could be improved in India by increasing reproduction and kid survival without increasing the total number of breeding animals in the country. At the same time, China, with one of the largest goat populations in the world, has no organized breeding activity for improvement of goats across the country (Wenxiu, 1988). The prospects for West African Dwarf goat production in the south-eastern region of Nigeria appear uncertain. Nevertheless, it has been suggested that women producers can benefit from vertical integration of goat production from the producer to the market.

The past few decades have seen a rise in goat improvement programmes around the world, although mostly for dairy and fibre production (Devendra, 1988; Wilson, 1992; El-Aich *et al.*, 1995; Delfosse and Jaouen, 1998). In South Africa, the National Small Stock Improvement Scheme provides the stud and commercial farms with the breeding values of all the animals assessed based on their performance records and pedigree information. The organization not only maintains breed standards but also encourages breeders to improve production performance. In Canada, Ontario has implemented breeding programmes for genetic improvement of meat goats (Nadarajah and Burnside, 1994). The USA is in the process of developing breeding programmes for meat goats.

Evaluation of production performance based on available records of the individual goat along with those of their parents should be adequate for the identification of goats with potential genetic merit. These include body weight at birth and 6 months of age, dams' litter size and milk yield, survival, dressing percentage, hot carcass weight, fertility, prolificacy and mortality. Breed organizations or governments could possibly encourage breeders to establish a hierarchical structure, which might include nucleus, multiplier and commercial goat breeders, or group breeding schemes similar to those available for other livestock species. In India, there are proposed Group Breeding Schemes with nucleus herds comprised of selected bucks and does for producing breeding stock (Acharya, 1988). In Malaysia, accelerating genetic gain based on open-nucleus herds along with embryo transfer has been proposed for milk and meat goats (Mukherjee *et al.*, 1996). An alternative would be to establish an open-nucleus structure where the nucleus herd produces bucks for the replacement male parent in commercial herds, while both the nucleus and commercial herds contribute replacement females for the nucleus herd. This approach will reduce inbreeding (James, 1977), as well as the lag associated with the transfer of genetic gain from the nucleus to the commercial herds.

In the absence of genetic evaluation of morphological characteristics and production performance based on the application of mixed model methodologies, breeders can rely on simple measures for genetic improvement similar to those used to determine ewe productivity (Shrestha *et al.*, 2002), which may include body weight of kids at weaning or an older age adjusted to a male equivalent basis. The efficient use of a selection index is based on the emphasis placed on several production traits relative to their economic value necessary to optimize breeding objectives. Prevailing markets, either in the region or across the country, could help in determining costs and benefits associated with meat, milk and fibre production in order to derive appropriate weighting factors for economically important production traits. An aggregate breeding value derived from a combination of relative economic weights and production traits will be meaningful for improving meat goats. Further details on breeding and genetics for goat production have been explained in the textbook *Goat Science and Production* (Shrestha and Crow, 2010).

## 4.10 Future Considerations

Prediction of performance of breed crosses from the evaluation of the performance of all possible two or more breed cross combinations can be useful to predict the performance of the optimum cross approaching maximum efficiency of production. Unfortunately, when more than two breeds are considered, the number of possible two-breed-cross combinations that need to be evaluated can be cumbersome. Even the testing of two- and three-breed crosses alone, neglecting reciprocals, can be a formidable task, and for a large number of breeds it soon becomes impossible. Carmon (1960) predicted the performance of specific three-breed cross, i.e. A × (B × C), where A, B and C are the parental breeds from the average of (A × B) and (A × C). This includes the two possible combinations of the breeds being crossed with the exception of (B × C), which is the female parent. Similarly, the performance of a four-breed cross, i.e.

(A × B) × (C × D), where A, B, C and D are the parental breeds, can be predicted from the average of (A × C), (A × D), (B × C) and (B × D) crosses. These crosses represent all possible two-breed crosses among the four parental breeds with the exception of (A × B) and (C × D), which are the male and female parents. The success in the prediction of performance of the multiple breed crosses depends largely on the nature of genetic variability associated with the economic trait in the two-breed cross. Preferably, management and environmental conditions should be alike.

The combination of rotational and specific crossbreeding may be based on mating of fecund-type crossbred does derived from a two or more breed rotation to bucks of an unrelated meat-type sire breed. In theory, an advantage in performance associated with two-thirds of the maximum maternal heterosis as well as individual heterosis can be achieved. The drawback is the need to follow a complex breeding protocol, which requires a large number of animals from diverse breeding populations. In practice, the replacement females for rotational crossbreeding and crossbred does are usually raised within the premises of the farm and derived from about one-third of the crossbred offspring. The remaining two-thirds of the crossbred offspring are mated to bucks of a meat-type breed purchased from highly reputed breeders to produce specific breed cross offspring that are marketed.

The use of crossbred males as sires has received little attention in animal breeding. In theory, crossbred male parents could capitalize on possible hybrid vigour in terms of libido and sperm production, as well as realized paternal heterosis from optimization of gene frequency for categorical or composite traits in the livestock species. According to Notter (1987), the use of crossbred males may be optimum at the production system level, e.g. net effect of heterosis for male fertility, non-linearity between individual traits, optimized gene frequencies of sire and dam combinations and overall profit. One would expect crossbred bucks to exhibit increased vigour that may be more appropriate in environments stressful to purebreds due to unfavourable climate, grazing condition and diseases. This is because purebred bucks may not be able to express their full genetic potential under extensive and harsh environments, habitats typical of a large proportion of goats worldwide. In practice, it is possible to raise one fecund-type goat breed within the farm and to purchase crossbred bucks from meat-type breeds for breeding.

Specific three-breed-cross offspring can also be produced by mating a crossbred buck based on two meat-type sire breeds with a purebred doe of a fecund-type breed. A drawback can occur from the lower performance associated with the absence of maternal heterosis and favourable genes in the crossbred dams including the rearrangement of genetic combinations between the chromosomes of crossbred parents. In a preliminary study, (Boer × Spanish) × Spanish compared with the Spanish breed weighed more at birth and 90 days by 11 and 16%, respectively (Table 4.7), whereas prolificacy was similar (Lopez-Perez *et al.*, 1998).

Specific four-breed-cross offspring may be produced by mating does derived from two fecund-type dam breeds with bucks from two meat-type sire breeds. In theory, this approach could capitalize on the full potential of maternal, paternal and individual heterosis. In this procedure, goats from one or two parental breeds and their crosses may be kept within the farm premises to produce does to be female parents. Crossbred bucks of alternative breeds, as required, may be purchased to be male parents. Despite the drawback of having to maintain more than two breeds for crossbreeding, there may be potential merit from increased productivity in commercial goat meat production. However, no research results are available on goat meat production based on the use of three- and four-breed-cross offspring.

The use of hill and lowland breeds of sheep in the UK demonstrates regional segmentation of production to support complex and efficient crossbreeding. In developing countries, efficient meat-goat production from the crossbred combination of indigenous goats with productive exotic breeds is possible (Ruvuna *et al.*, 1992). The authors proposed producing specific breed combinations to complement breed differences in carcass merit with segmentation of varying production systems and established markets for meat and meat products. In remote or tribal communities, there is the opportunity to produce crossbred does for use, as a terminal cross to be sold for immediate cash, or under intensive production in areas adjoining urban markets. Furthermore, the infrastructure for the transportation and marketing of extra kids from crossbreeding is crucial.

Breeders of goats of nomadic and seminomadic origin that receive a substantial portion of their income from meat and meat products are not aware of the scientific accomplishments in the field of genetics, nutrition and husbandry that have helped other livestock and poultry become more productive. Emerging and developing economies with 85% of the human population, 75% of the breed population and 97% of goats worldwide could benefit from increased productivity of meat goats raised by the poor and landless at the end of the social scale that are not only illiterate but also lack the necessary husbandry skills and finances. The difficulty of establishing breed improvement programmes in developing countries lies in establishing production criteria in the absence of an industry structure with no specific advantage from rapid growth rate, early maturity and favourable meat quality (Norman, 1991). Large numbers of bucks, possibly with potential genetic merit, continue to be sacrificed for religious rituals in accordance with the tradition and culture of the region. According to Purohit (1982), private or cooperative ventures to increase goat meat production in the villages of India could benefit from periodic culling of less productive animals followed by the feeding of high-energy diets in feedlots to culled animals starting one month before slaughter. Developing breeding strategies for goats raised under sedentary, nomadic and seminomadic management presents a real challenge. Adding value to meat and meat products, as well as recognizing their significant role in maintaining cultural heritage and sustaining biodiversity, can contribute to alleviation of poverty.

Alpine, Saanen and Toggenburg dairy breeds have benefited from research directed towards their genetic improvement (Flamant and Morand-Fehr, 1982). Evidence presented from a number of studies support the use of dairy breeds of goats from continental Europe for crossbreeding with indigenous goats to produce offspring with a rapid growth rate and potential for increased milk production. Besides producing more milk for sale, there is the opportunity for meat-type breeds to sire crossbred does to produce terminal-cross offspring for meat production.

The influx of ethnic populations in developed countries and the growing appetite for food of exotic origin have resulted in increased demand for goat meat. Goat meat from domesticated feral populations in New Zealand and Australia (Horton and Dawson, 1987; Kelly, 1988; B.A. McGregor, personal communication) is being exported to help meet the demand from increased consumption. This is indicative of potential for import replacement by promoting commercial production of meat from goats. Physical, chemical and sensory properties of meat from kids of Boer and Cashmere breeds and their crosses compared with those from lambs were of less intense flavour, tenderness and juiciness, while panellists accepted curries and patties made from goat meat (Swan *et al.*, 1998), demonstrating the opportunity to market more meat to consumers in developed countries. Nevertheless, goat meat continues to be a by-product of the dairy and fibre industries in developed countries (Decoster and Berinstain-Bailly, 1996; Asheim and Eik, 1998).

The application of quantitative genetic methodologies in combination with advances in molecular methodologies could accelerate genetic progress in selection, while developing more efficient means for commercial production of meat from goats. Constraints due to the seasonality of production and labour shortages may be avoided with controlled reproduction based on induction and synchronization of oestrus, fixed-time artificial insemination, year-round breeding based on 6-month breeding cycles, the use of meat-type sire and fecund-type dam breeds and crossbreeds, fattening and marketing activities. Green and stored forage and concentrates along with vitamin and mineral supplements to supply the necessary dietary nutrients to lactating goats, feeding high-energy diets to promote rapid growth rates for kids surplus to breeding requirements for slaughter at 90 days or earlier at 40–45 kg body weight, milking does following a suckling period of 3 weeks or less, milk replacers and solid feed to earlier-weaned kids, and the prevention of mastitis, abortions and parasite infestations provided in herd health programmes can contribute towards increased production efficiency.

Despite only a few objective evaluations of goats raised independent of range conditions (Warmington and Kirton, 1990), complementing specific breed combinations to exploit breed differences in carcass merit with segmentation of varying production systems and established markets for meat and meat products (Ruvuna *et al.*, 1992) has considerable merit in increasing the efficiency of meat-goat production in developing countries. The application of innovative breeding technologies can facilitate genetic improvement in large herds due to the ease in recording of performance and pedigree, uniform feeding and management, proximity to markets and the ability to deliver goods and services to the breeder on a low-cost basis. Nevertheless, one should examine the relative costs in terms of capital expenditure, labour, selection, crossbreeding, breed formation and recurrent crossbreeding to exploit heterosis before agreeing on a particular strategy for commercial production of meat goats.

The Winrock International (USA), International Development Research Centre (Canada) and International Livestock Research Institute (Kenya) have promoted meat-goat production in developing countries. Programmes supporting the production of goat meat in Africa (Upton, 1988), South-east Asia (Jalaludin *et al.*, 1992) and Africa and Latin America (Wilson, 1992) and the performance of meat-goat breeds in Asia, as well as strategies to achieve genetic improvement, have been published (Devendra, 1988).

## 4.11 Conclusions

Challenges to attain the biological ceiling of goats for meat production reveal the complexity of implementing innovative breeding technologies that involve the application of quantitative genetic methodologies for improvement of meat goats. Wilson (1968) estimated the biological ceiling at seven offspring per lambing and a potential mean lambing interval of 6 months, possibly similar for goats. The availability of 1183 breed populations worldwide along with previously documented information on their background, adaptability, productivity and husbandry need to be critically assessed. Furthermore, meat goats lack precise estimates of genetic parameters based on large data sets as well as selection studies to corroborate research results. Constraints, limitations and social and cultural attributes may include diseases, diet, climatic conditions, natural vegetation and terrain, labour surplus to household requirements, the availability of trained personnel and equipment, religious rituals, economic reality and proximity to markets. Smallholders under sedentary, nomadic and semi-nomadic management must rely on breeding animals with sufficient genetic merit to improve the efficiency of goat meat production, and benefit from crossbreeding and composite breed development. Evidence to suggest that additive genetic variance may have been exhausted for performance traits of economic importance, although not currently likely, would warrant family selection and progeny testing to achieve a genetic response. There is overwhelming evidence to support the genetic improvement of goat meat by crossbreeding complementary breeds, the formation of composite populations and selection for economically important performance traits. At the same time, genetic improvement will depend on how widely animals with potential genetic merit can be disseminated rapidly across commercial herds.

## References

Abdelsalam, M.M., Haider, A.E., Aboul-Naga, A.M., El-Kimary, I.S. and Eissa, M. (1994) Improving performance of desert Barki kids by crossing with Zaraibi and Damascus goats. *Egyptian Journal of Animal Production* 31, 85–97.

Abdelsalam, M.M., Eissa, M., Maharm, G. and Haider, A.I. (2000) Improving the productivity of Barki goat by crossbreeding with Damascus or Zaraibi breeds. *Alexandria Journal of Agriculture Research* 45, 33–42.

Acharya, R.M. (1982) *Sheep and Goat Breeds of India*. FAO Animal Production and Health Paper 30, Food and Agriculture Organization of the United Nations, Rome, Italy.

Acharya, R.M. (1988) Goat breeding and meat production. In: Devendra, C. (ed.) *Goat Meat Production in Asia*. International Development Research Corporation, Ottawa, Canada, pp. 14–29.

Acharya, R.M., Misra, R.K. and Patil, V.K. (1982) *Breeding Strategy for Goats in India*. Indian Council of Agricultural Research, New Delhi, India.

Alexandre, G., Aumont, G., Fleury, J., Coppry, O., Mulciba, P. and Nepos, A. (1997) Semi-intensive production of meat goats in a tropical area. *Productions Animales* 10, 43–53.

Alexandre, G., Aumont, G., Mandonnet, N. and Naves, M. (1999) The Creole goat of Guadeloupe (French West Indies): an important genetic resource for the humid tropics. *Animal Genetic Resource Information* 26, 45–55.

Ali, S.Z. (1983) Heritability estimates for birth weight, 4-week weight, weaning weight and multiple birth in Black Bengal goats of Bangladesh. *Indian Veterinary Journal* 60, 118–121.

Al-Ojaili, A.A. (1995) Production traits in Dhofari and Dhofari × Anglo-Nubian goats in Oman. *International Journal of Animal Sciences* 10, 13–16.

Amble, V.N., Khandekar, M. and Garg, J.N. (1964) *Statistical Studies on Breeding Data of Beetal Goats*. ICAR Research Series No. 38, Indian Council of Agricultural Research, New Delhi, India.

Amin, M.R., Husain, S.S. and Islam, A.B.M.M. (2001) Reproductive peculiarities and litter weight in different genetic groups of Black Bengal does. *Asian-Australasian Journal of Animal Sciences* 14, 297–301.

Angel-Neelem, M. and Nellem, M.A. (1998) Goat production in Argentina. Example of the Santiago del Estero province. *Capricorne* 11, 2 and 11–16.

Anous, M.R. and Mourad, M.M. (1993) Crossbreeding effects on reproductive traits of does and growth and carcass traits of kids. *Small Ruminant Research* 12, 141–149.

Asheim, L.J. and Eik, L.O. (1998) The economics of fibre and meat on Norwegian dairy goats. *Small Ruminant Research* 30, 185–190.

Bakshi, S.A., Patil, V.K. and Jagtap, D.Z. (1986) Non-genetic factors affecting number of kids per kidding in Angora and their crossbreds. *Indian Veterinary Journal* 63, 659–663.

Bett, R.C., Kosgey, I.S., Kahi, A.K. and Peters, K.J. (2011) Definition of breeding objectives and optimum crossbreeding levels for goats in the smallholder production systems. *Small Ruminant Research* 96, 16–24..

Blackburn, H.D. (1995) Comparison of performance of Boer and Spanish goats in two US locations. *Journal of Animal Science* 73, 302–309.

Bökönyi, S. (1974) *History of Domestic Mammals in Central and Eastern Europe*. Akadémiai Kiadó, Budapest, Hungary.

Cameron, M.R., Luo, J., Sahlu, T., Hart, S.P., Coleman, S.W. and Goetsch, A.L. (2001) Growth and slaughter traits of Boer × Spanish, Boer × Angora, and Spanish goats consuming a concentrate-based diet. *Journal of Animal Science* 79, 1423–1430.

Carbo, C.B. and del la Calle, J.R.C. (1995) Population, production and marketing of goats in the European Union and Spain. *Mundo Ganadero* 6, 11 and 73–80.

Carmon, J.L. (1960) A comparison of several crossbreeding systems for the prediction of crossbred performance. *Georgia Agricultural Experiment Station Technical Bulletin* 19, 22.

Cheng, P. (1984) *Livestock Breeds of China*. FAO Animal Production and Health Paper 46, Food and Agriculture Organization of the United Nations, Rome, Italy.

Chidebelu and Ngo, N.M. (1998) The economics of goat production in Southeastern Nigeria: implications for the future. *Nigerian Journal of Animal Production* 25, 2 and 93–99.

CIRG (1986) *Annual Report*. Central Institute for Research on Goats, Makhdoom, Farah, UP, India.

Comstock, R.E. (1961) Reciprocal recurrent selection with reference to swine breeding. In: *24th Annual Report of the Regional Swine Breeding Laboratory*, Ames, Iowa.

Comstock, R.E., Robinson, H.F. and Harvey, P.H. (1949) A breeding procedure designed to make use of both general and specific combining ability. *Agronomy Journal* 41, 360–367.

CSIR (1970) *The Wealth of India. Raw Materials*, Vol. VI. *Supplement on Livestock Including Poultry.* Public Information Directorate, Council of Scientific and Industrial Research, New Delhi, India.

Das, S., Husain, S.S., Hoque, M.A. and Amin, M.R. (2001) Genetic variation and correlation studies of some carcass traits in goats. *Asian-Australasian Journal of Animal Sciences* 14, 905–909.

d'Avant, F.P. (2001) On a report regarding the Rove goat in province. *Animal Genetic Resource Information* 29, 61–67.

Decoster, A., and Berinstain-Bailly, C. (1996) Goat production in France. *Capricorne* 9, 3 (in French).

Delfosse, C. and Jaouen, J.C. de (1998) From zoology to zootechnics: the development of goat selection to the 20th century. *Ethnozootechnie* 63, 101–112.

Devendra, C. (1982) Breed differences in productivity in goats. In: Maijala, K. (ed.) *World Animal Science. B8. Disciplinary Approach.* Elsevier, Amsterdam, the Netherlands, pp. 431–440.

de Souza, F.J., Fereira de Miranda, J.J., Machado, F.H.F. and Fernandes, A.A.O. (1987) Heritability estimates of phenotypic, genetic and environmental correlations at different ages of goats in Central Zone of Ceara state. In: *Proceedings of the 4th International Conference on Goats*, Vol. 2, Empresa Brasileira de Tequisa Agropecuaria, Brasilia, Brazil, p. 1335.

Devendra, C. (1988) *Goat Meat Production in Asia.* Proceedings of a workshop held in Tando Jam, Pakistan. International Development Research Centre, Ottawa, Canada.

Devendra, C. (1998) Indigenous goat genetic resources: potential importance in sustainable agriculture. In: Shrestha, J.N.B. (ed.) *Proceedings of the 4th Global Conference on Conservation Domestic Animal Genetic Resources*, Nepal Agricultural Research Council, Lalitpur, Nepal, and Rare Breeds International, Shropshire, UK, pp. 16–22.

Devendra, C. and Burns, M. (1983) *Goat Production in the Tropics.* CAB International, Wallingford, UK.

Devendra, C. and Nozawa, K. (1976) Goats in South East Asia, their status and production. *Zeitschrift für Tierzuchtun und Zuchtungsbiologie* 93, 101–120.

Dhanda, J.S., Taylor, D.G., McCosker, J.E. and Murray, P.J. (1999a) The influence of goat genotype on the production of Capretto and Chevon carcasses. 1. Growth and carcass characteristics. *Meat Science* 52, 355–361.

Dhanda, J.S., Taylor, D.G., Murray, P.J. and McCosker, J.E. (1999b) The influence of goat genotype on the production of Capretto and Chevon carcasses. 2. Meat quality. *Meat Science* 52, 363–367.

Dhanda, J.S., Taylor, D.G., McCosker, J.E. and Murray, P.J. (1999c) The influence of goat genotype on the production of Capretto and Chevon carcasses. 3. Dissected carcass composition. *Meat Science* 52, 369–374.

Dhanda, J.S., Taylor, D.G., Murray, P.J. and McCosker, J.E. (1999d) The influence of goat genotype on the production of Capretto and Chevon carcasses. 4. Chemical composition of muscle and fatty acid profiles of adipose tissue. *Meat Science* 52, 375–379.

Dickerson, G. (1969a) Techniques for research in quantitative animal genetics. In: *Techniques and Procedures in Animal Production Research.* American Society of Animal Science, Q Corporation, NY, USA, pp. 36–79.

Dickerson, G. (1969b) Experimental approaches in utilizing breed resources. *Animal Breeding Abstracts* 37, 191–202.

Ebozoje, M.O. and Ngere, L.O. (1995) Genetic analysis of preweaning growth in West African Dwarf goats and their halfbreds. *International Journal of Animal Sciences* 10, 247–251.

El-Aich, A., Landau, S., Bourbouze, A., Rubino, R. and Morand-Fehr, P. (1995) *Goat Production Systems in the Mediterranean.* Proceedings of FAO-CIHEAM Working Group on Goat Production. EAAP Publication No. 71, Wageningen Press, Wageningen, the Netherlands.

Epstein, H. (1969) *Domestic Animals of China.* Commonwealth Agricultural Bureau, Farnham Royal, UK.

Epstein, H. (1971) *The Origin of the Domestic Animals of Africa*, Vol. II. Africana Publishing Corporation, New York/London/Munich.

Erasmus, J.A. (2000) Adaptation to various environments and resistance to disease of the improved Boer goat. *Small Ruminant Research* 36, 179–187.

Fahmy, M.H. and Shrestha, J.N.B. (2000) Genetics for the improvement of goat meat production. In: *Proceedings of the 7th International Conference on Goats*, France, pp. 187–190.

FAO (1999) *Production Handbook*, Vol. 53. Food and Agriculture Organization of the United Nations, Rome, Italy.

Flamant, J.C. and Morand-Fehr, P. (1982) Milk production in sheep and goats. In: Coop, I.E. (ed.) *World Animal Science. C. Production-system Approach. 1. Sheep and Goat Production.* Elsevier, Amsterdam, the Netherlands, pp. 275–295.

Flamant, J.C. and Morand-Fehr, P. (1990) Mediterranean systems of animal production. In: Galaty, J.G. and Johnson, D.L. (eds) *The World of Pastoralism: Herding Systems in Comparative Perspective.* Guilford Press, London, UK, pp. 353–393.

French, M.H. (1970) *Observations on the Goat.* FAO Agricultural Studies No. 80, Food and Agriculture Organization of the United Nations, Rome, Italy.

Galal, E.S.E. (1987) Biological aspects of increasing production. In: *Proceedings of the 4th International Conference on Goats*, Empresa Brasileira de Tequisa Agropecuaria, Brasilia, Brazil, pp. 55–74.

Galal, E.S.E. (2005) Biodiversity in goats. *Small Ruminant Research* 60, 75–81.

Galina, M.A., Guerrero, M., Serrano, G., Morales, R. and Haenlein, G.F.W. (2000) Effect of complex catalytic supplementation with non-protein nitrogen on the ruminal ecosystem of growing goats pasturing on shrub land in Mexico. *Small Ruminant Research* 36, 33–42.

Gall, C. (1982) Breed differences in adaptation of goats. In: Maijala, K. (ed.) *World Animal Science. B8. Disciplinary Approach.* Elsevier, Amsterdam, the Netherlands, pp. 413–429.

Gall, C. (1996) *Goat Breeds of the World.* Technical Centre for Agricultural and Rural Cooperation, Wageningen, the Netherlands.

Garcia, B.O., Garcia, B.E., Bravo, J. and Kennedy, B. (1996) Analysis of a crossbreeding trial using Criollo and imported goats. 1. Growth of kids. *Revista de la Facultad de Agronomia, Universidad del Zulia* 13, 395–415.

Garcõa, O. and Gall, C. (1981) Goats in the dry tropics. In: Gall, C. (ed.) *Goat Production.* Academic Press, London, UK, pp. 515–556.

Gebrelul, S., Sartin, L.S. III and Iheanacho, M. (1994) Genetic and non-genetic effects on the growth and mortality of Alpine, Nubian and crossbred kids. *Small Ruminant Research* 13, 169–176.

Gerstmayr, S. and Horst, P. (1995) Estimates of performance traits in Turkish Angora goats. *Small Ruminant Research* 16, 141–157.

Gerstmayr, S., Haussmann, H. and Schlote, W. (1989) Estimation of variance components with an animal model including maternal effects. In: *Proceedings of the 40th Annual Meeting of the EAAP*, Dublin, Ireland, 27–31 August.

Gerstmayr, S., Haussmann, H. and Schlote, W. (1992) Estimation of variance components for filial and maternal effects in Turkish Angora goats. *Journal of Animal Breeding and Genetics* 109, 252–263.

Gibb, M.J., Cook, J.E. and Treacher, T.T. (1993) Performance of British Saanen, Boer × British Saanen and Anglo-Nubian castrated male kids from 8 weeks to slaughter at 28, 33 or 38 kg live weight. *Animal Production* 57, 263–271.

Gilmour, A.R., Thompson, R. and Cullis, B.R. (1995) Average information REML: an efficient algorithm for variance parameter estimation in linear mixed models. *Biometrics* 51, 1440–1450.

Glimp, H.A. (1995) Meat goat production and marketing. *Journal of Animal Science* 73, 291–295.

Goonewardene, L.A., Day, P.A., Patrick, N., Scheer, H.D., Patrick, D. and Suleiman, A. (1998) A preliminary evaluation of growth and carcass traits in Alpine and Boer goat crosses. *Canadian Journal of Animal Science* 78, 229–232.

Greyling, J.P.C. (2000) Reproduction traits in the Boer goat doe. *Small Ruminant Research* 36, 171–177.

Guha, H., Gupta, S., Mukherjee, A.K., Moulick, S.K. and Bhattacharya, S. (1968) Some causes of variation in growth rates of Black Bengal goats. *Indian Journal of Veterinary Science and Animal Husbandry* 38, 269–278.

Haenlein, G.F.W. (1996) Status and prospects of the dairy goat industry in the USA. *Journal of Animal Science* 74, 1173–1181.

Harris, D.R. (1962) The distribution and ancestry of the domestic goat. *Proceedings of the Linnean Society of London* 173, 79–91.

Hasnain, H.U. (1985) *Sheep and Goats in Pakistan.* FAO Animal Production and Health Paper No. 56, Food and Agriculture Organization of the United Nations, Rome, Italy.

Hemmerle, W.J. and Hartley, H.O. (1973) Computing maximum likelihood estimates for the mixed AOV model using the W-transformation. *Technometrics* 15, 819–831.

Henderson, C.R. (1953) Estimation of variance and covariance components. *Biometrics* 9, 226–252.

Henderson, C.R. (1984) *Application of Linear Models in Animal Breeding.* University Press, Guelph, Ontario, Canada.

Henderson, C.R. and Quaas, R.L. (1976) Multiple trait evaluation using relatives' records. *Journal of Animal Science* 43, 1188–1197.

Hermiz, H.N., Al-Amily, H.J. and Assak, E.A. (1997) Some genetic and non-genetic parameters for pre-weaning growth traits in Angora goats. *Agricultural Science* 24, 182–186.

Herre, W. (1958) Abstammung und Domestikationi der Haustiere. In: Hammond, J., Johansson, I. and Haring, F. (eds) *Handbuch der Tierzuchtung*, Vol. I. Paul Parey, Hamburg/Berlin, Germany, pp. 1–58.
Herrera, C.M., Sanchez, F. and Torres-Hernandez, G. (1987) Genetic and environmental factors affecting preweaning traits in goats. I. Breed and environmental factors affecting birth weight, one-month weight, and average daily gain. In: *Proceedings of the 4th International Conference on Goats*, Vol. 2. Brasileira de Tequisa Agropecuaria, Brasilia, Brazil, pp. 1320–1321.
Hill, W.G. (1972a) Estimation of genetic change. I. General theory and design of control populations. *Animal Breeding Abstracts* 40, 1–16.
Hill, W.G. (1972b) Estimation of genetic change. II. Experimental evaluation of control populations. *Animal Breeding Abstracts* 40, 193–213.
Hirooka, H., Mukherjee, T.K., Panandam, J.M. and Horst, P. (1997) Genetic parameters for growth performance of the Malayasian local goats and their crossbreds with the German (improved) Fawn goats. *Journal of Animal Breeding and Genetics* 114, 191–199.
Horton, C.T. and Dawson, J.E. (1987) Livestock diversification in New Zealand involving goats for meat and dairy production. *Proceedings of New Zealand Society of Animal Production* 47, 125–126.
Hussain, M.Z., Naidu, R., Tuvuki, I. and Singh, R. (1983) Goat production and development in Fiji. *World Animal Review* 48, 25–38.
Jalaludin, S., Ho, Y.W., Abdullah, N. and Kudo, H. (1992) Strategies for animal improvement in southeast Asia. Proceedings of the 25th International Symposium Tropical Agriculture Research, Tsukuba, Japan, 24–25 September, 1991. *Tropical Agriculture Research Series* 25, 67–76.
James, J.W. (1977) Open nucleus breeding systems. *Animal Production* 24, 287–305.
Jiang, Y. (1982) Goat breeds and ecological characteristics in China. *Journal of Beijing Agricultural University* 8, 9–12.
Jiang, Y. (1995) Chinese meat goat breeds and their crosses. *Animal Genetic Resource Information* 15, 71–81.
Johnson, D.D., McGowan, C.H., Nurse, G. and Anous, M.R. (1995) Breed type and sex effects on carcass traits, composition and tenderness of young goats. *Small Ruminant Research* 17, 57–63.
Kelly, R. (1988) Potential for development of the goat fibre and meat industries. *Journal of Agriculture, Western Australia* 29, 35–40.
Khalil, M.H., Mohamed, K.M., Al-Saef, A.M., Zeitoun, M.M. and El-Zarei, M.F. (2010) Crossbreeding components for growth, carcass and meat composition traits in crossing Saudi Aradi with Damascus goats. *Small Ruminant Research* 94, 10–16.
Khan, B.U. and Singh, S.K. (1995) Genetics of feed conversion efficiency in Barbari goats. *Small Ruminant Research* 15, 283–285.
Kinghorn, B. (1980) The expression of 'recombination loss' in quantitative traits. *Zeitschrift für Tierzuchtun und Zuchtungsbiologie* 97, 138–143.
Kinghorn, B. (1983) Genetic effects in crossbreeding. III. Epistasis loss in crossbred mice. *Zeitschrift für Tierzuchtun und Zuchtungsbiologie* 100, 209–222.
Lee, K.W., Chai, K.S., Tak, T.Y. and Sul, D.S. (1974) Improvement of Korean native goats grading-up with Saanen. II. Change on milk performance of crossbreds between Native Korean goats and Saanen. *Research Report Office of Rural Development Suweon Livestock* 16, 7–14.
Le Jaouen, J.C. (1997) Goat production in Europe: the first European meeting. *Chevre* 220, 13–15.
Libunao, L.P. (1990) Goat/fish integrated farming in the Philippines. *Ambio* 19, 8 and 408–410.
Lopez-Fanjul, C. (1974) Selection from crossbred populations. *Animal Breeding Abstracts* 42, 403–416.
Lopez-Perez, D., Lukefahr, S.D. and Waldron, D.F. (1998) Comparison of crossbred Boer × Spanish and purebred Spanish breed-types for kid growth and litter size traits. *Sheep and Goat Research Journal* 14, 144–147.
Luo, J., Sahlu, T. and Goetsch, A.L. (2000) Growth and carcass traits of Boer × Alpine goats slaughtered at the ages of 31 and 50 weeks. *Journal of Animal Feed Science* 9, 309–324.
Lush, J.L. (1945) *Animal Breeding Plans.* Iowa State University Press, Ames, Iowa, USA.
Madeli, U.C. and Patro, B.N. (1984) Heritability and correlations among body weights at different ages in Ganjam goats. *Indian Veterinary Journal* 61, 233–235.
Malan, S.W. (2000) The improved Boer goat. *Small Ruminant Research* 36, 165–170.
Mason, I.L. (1951) *A World Dictionary of Livestock Breeds, Types and Varieties*, 1st edn. Commonwealth Agricultural Bureaux, Bucks, UK.
Mason, I. L. (1996) *A World Dictionary of Livestock Breeds, Types and Varieties*, 4th edn. CAB International, Wallingford, UK.

Mason, I.L. and Maule, J.P. (1960) *The Indigenous Livestock of Eastern and Southern Africa.* Commonwealth Agriculture Bureaux, Bucks, UK.

Mavrogenis, G.A., Constantinou, A. and Louca, A. (1984) Environmental and genetic cause of variation in production traits of Damascus goats. I. Pre-weaning and Post-weaning growth. *Animal Production* 38, 91–98.

McGregor, B.A. (1990) Boneless meat yields and prediction equations from carcass parameters of Australian cashmere goats. *Small Ruminant Research* 3, 465–473.

McGregor, B.A. (1996) Carcass quality and commercial acceptance of Angora goat kids fed supplementary energy and slaughtered at 5 months of age. *Animal Production in Australia* 21, 135–138.

Meyer, K. (1989) Restricted maximum likelihood to estimate variance components for animal models with several random effects using a derivative-free algorithm. *Genetics Selection Evolution* 21, 317–340.

Meyer, K. (1991) Estimating variances and covariances for multivariate animal models by restricted maximum likelihood. *Genetics Selection Evolution* 23, 67–83.

Meza, H.C., Montaldo, V.H. and Sanchez, G.F.F. (1994) Generation interval of five goat breeds in northern Mexico. *Revista Latinoamericana de Pequenos Rumiantes* 1, 120–126.

Mia, M.M., Ali, A. and Howlider, M.A.R. (1993) Growth performance of Black Bengal, Barbari, Anglo-Nubian and Barbari × Black Bengal goats. *Indian Journal of Animal Sciences* 63, 1214–1215.

Mislevy, P., Martin, F.G. and Neilson, J.T. (2000) Selectivity of tropical grasses by Spanish × Boer goats. *Proceedings of the Soil and Crop Science Society of Florida* 59, 77–81.

Misra, R.K. (1983) Genetic analysis of growth, survivability, efficiency of feed conversion and carcass yield and composition in Sirohi (goat) and its crosses with Beetal. PhD thesis, University of Kurukshetra, Haryana, India.

Misra, R.K. (1995) Genetic improvement of feed-lot traits in goat. *Indian Journal of Animal Sciences* 10, 331–335.

Misra, R.K., Arora, C.L. and Acharya, R.M. (1980) Breed characterization of Sirohi goat. *Indian Journal of Animal Sciences* 50, 717–720.

Montaldo, H. and Meza, C. (1999) Genetic goat resources in Mexico: bio-economical efficiency of local and specialised genotypes. *Wool Technology and Sheep Breeding* 47, 184–198.

Montaldo, H., Juárez, A., Berruecos, J.M. and Sánchez, F. (1995) Performance of local goats and their backcrosses with several breeds in Mexico. *Small Ruminant Research* 16, 97–105

Morand-Fehr, O., Falagan, A., Rubino, R., Mowlem, A. and Skjevdal, T. (1992) Goat meat production and processing in Europe. In: *Proceedings of the 5th International Conference on Goats*, Vol. 2. New Delhi, India, pp. 473–486.

Morand-Fehr, P., Boutonnet, J.P., Devendra, C., Dubeuf, J.P., Haenlein, G.F.W., Holst, P., Mowlem, L. and Capote, J. (2004) Strategy for goat farming in the 21st century. *Small Ruminant Research* 51, 175–183.

Moulick, S.K. and Syrstad, O. (1970) Genetic and environmental cause of variation in birth weight of Black Bengal goats. *Journal of Agricultural Science* 74, 409–414.

Moulick, S.K., Guha, H., Gupta, S., Mitra, D.K. and Bhattacharya, S. (1966) Factors affecting multiple birth in Black Bengal goats. *Indian Journal of Veterinary Science and Animal Husbandry* 36, 154–163.

Mourad, M. (1994) Estimation of genetic and phenotypic parameters of some reproductive traits of African common goats in Rwanda. *Small Ruminant Research* 15, 67–71.

Mourad, M. and Anous, M.R. (1998) Estimates of genetic and phenotypic parameters of some growth traits in common African and Alpine crossbred goats. *Small Ruminant Research* 27, 197–202.

Mueller, J.P., Luginbuhl, J.M. and Bergmann, B.A. (2001) Establishment and early growth characteristics of six Paulownia genotypes for goat browse in Raleigh, NC, USA. *Agroforestry Systems* 52, 63–72.

Mukherjee, T.K., Horst, P. and Mathur, P.K. (1996) Current status and future perspectives of genetic improvement of goats in Asia. In: *Proceedings of the 6th International Conference on Goats*, Vol. 1. China, pp. 110–118.

Nadarajah, K. and Burnside, E.B. (1994) Genetic and environmental effects on growth of meat goats in Ontario. *Proceedings of the 5th World Congress on Genetics Applied to Livestock Production* 18, 190–193.

Nagpal, S. and Chawla, D.S. (1987a) Estimates of genetic parameters of body weights of Alpine × Beetal goats. In: *Proceedings of the 4th International Conference on Goats*, Vol. 2. Empresa Brasileira de Tequisa Agropecuaria, Brasilia, Brazil, p. 1312.

Nagpal, S. and Chawla, D.S. (1987b) Genetic estimates of body weights of Beetal goats. In: *Proceedings of the 4th International Conference on Goats*, Vol. 2. Empresa Brasileira de Tequisa Agropecuaria, Brasilia, Brazil, p. 1313.

Nahardeka, N., Das, D., Roy, T.C., Goswami, R.N., Das, G.C., Gogoi, P.K. and Das, B. (2001) Studies on body weights of Assam local goats and their crosses with Beetal. *Indian Veterinary Journal* 78, 811–814.

Nicoll, G.B. (1985) Estimates of environmental effects and some genetic parameters for weaning weight and fleece weights of young Angora goats. *Proceedings of New Zealand Society of Animal Production* 45, 217–219.

Niekerk, M.M., van Schoeman, S.J., Botha, M.E., Casey, N. and van Niekerk, M.M. (1996) Heritability estimates for preweaning growth traits in the Adelaide Boer goat flock. *South African Journal of Animal Science* 26, 6–10.

Norman, G.A. (1991) The potential of meat from goat. In: Lawrie, R.A. (ed.) *Developments in Meat Science*, Vol. 5. Elsevier Applied Science, London, UK, pp. 57–87.

Notter, D.R. (1987) The crossbred sire: theory. *Journal of Animal Science* 65, 99–109.

Nozawa, K. (1991) Domestication and history of goats. In: Maijala, K. (ed.) *World Animal Science. B8. Disciplinary Approach.* Elsevier, Amsterdam, the Netherlands, pp. 391–404.

Oman, J.S., Waldron, D.F., Griffin, D.B. and Savell, J.W. (1999) Effect of breed-type and feeding regimen on goat carcass traits. *Journal of Animal Science* 77, 3215–3218.

Oman, J.S., Waldron, D.F., Griffin, D.B. and Savell, J.W. (2000) Carcass traits and retail display-life of chops from different goat breed types. *Journal of Animal Science* 78, 1262–1266.

Orekhov, A.A. (1980) Goat breeds and their distribution in the USSR. *Ovtsevodstvo* 3, 37–38 (in Russian).

Patterson, H.D. and Thompson, R. (1971) Recovery of inter-block information when block sizes are unequal. *Biometrika* 58, 545–554.

Phillips, R.W., Johnson, R.S. and Moyer, R.T. (1945) *The Livestock of China.* US Government Printing Office, Washington, DC, USA.

Pralomkarn, W., Ngamponsai, W., Choldumrongkul, S., Kochapakdee, S. and Lawpetchara, A. (1995) Effects of age and sex on body composition of Thai native and crossbred goats. *Asian-Australasian Journal of Animal Sciences* 8, 255–261.

Premasundeba, A.S., Ravindran, V., de Silva, G.P.L., Jeyalingavatkani, S. and de Silva, G.P.L. (1998) Crossbreeding trails with Boer goats in Sri Lanka: effects on the birth weights of kids. *Tropenlandwirt* 99, 43–48.

Purohit, K. (1982) Village and small holder systems. In: Coop, I.E. (ed.) *World Animal Science. C. Production-system Approach. 1. Sheep and Goat Production.* Elsevier, Amsterdam, the Netherlands, pp. 459–480.

Quartermain, A.R. (1991) Evaluation and utilization of goat breeds. In: Maijala, K. (ed.) *World Animal Science. B8. Disciplinary Approach.* Elsevier, Amsterdam, the Netherlands, pp. 451–469.

Rao, C.R. (1971) Minimum variance quadratic unbiased estimation of variance components. *Journal of Multivariate Analysis* 1, 445–446.

Rekik, M., Gharbi, M. and Dhib, C. (1996) Potential for goat production in the oasis system: the case of Jerid. *Méditerranéennes* 7, 1 and 39–42.

Ricordeau, G., Poujardieu, B. and Bouillon, J. (1972) Genetic parameters of growth in young Saanen female goats in a testing station. *Annales de Génétique de Sélection animale* 4, 477–480.

Roy, R., Saxena, V.K., Singh, S.K. and Khan, B.U. (1997) Genetic analysis of body weights at different ages in Jamunapari goats. *Indian Journal of Animal Sciences* 67, 337–339.

Ruvuna, F., Cartwright, T.C., Blackburn, H., Okeyo, M. and Chema, S. (1988) Gestation length, birth weight and growth rates of pure-bred indigenous goats and their crosses in Kenya. *Journal of Agricultural Science* 111, 363–368.

Ruvuna, F., Taylor, J.F., Okeyo, M., Wanyoike, M. and Ahuya, C. (1992) Effects of breed and castration on slaughter weight and carcass composition of goats. *Small Ruminant Research* 7, 175–183.

Sainju, A.P., Shrestha, H.R. and Neopane, S.P. (1994) Goat improvement program. In: Shrestha, J.N.B. (ed.) *Proceedings of the First National Workshop on Animal Genetic Resources Conservation and Genetic Improvement of Domestic Animals in Nepal.* Nepal Agricultural Research Council, Nepal, pp. 89–93.

Sánchez, F., Montaldo, H. and Juárez, A. (1994) Environmental and genetic effects on birth weight in graded-up goat kids. *Canadian Journal of Animal Science* 74, 397–400.

Scherf, B.D. (2000) *World Watch List of Domestic Animal Diversity*, 3rd edn. Food and Agriculture Organization of the United Nations, Rome, Italy.

Schoeman, S.J., Els, J.F. and Van-Niekerk, M.M. (1997) Variance components of early growth traits in the Boer goat. *Small Ruminant Research* 26, 15–20.

Shelton, M. (1978) Reproduction and breeding of goats. *Journal of Dairy Science* 61, 994–1010.
Shelton, M. (1998) A review of the nutritional efficiency of fibre production in goats and sheep and the relationship of fiber and meat production. *Sheep and Goat Research Journal* 14, 214–223.
Shinjo, A. and Toma, M. (1985) Breeding season and litter size in Japanese Saanen and Okinawa meat goats. *Japanese Journal of Zootechnical Science* 55, 377–380.
Shrestha, H.K., Ghimire, S.C., Rasali, D.P., Joshi, H.D. and Karki, N.P.S. (1998) Characterization of indigenous goat populations in the western hills of Nepal. In: Shrestha, J.N.B. (ed.) *Proceedings of the 4th Global Conference on Conserversation Domestic Animal Genetic Resources.* Nepal Agricultural Research Council, Lalitpur, Nepal, and Rare Breeds International, Shropshire, UK, pp. 228–230.
Shrestha, J.N.B. (1973) *Evaluation of breeding schemes employed for the production of crossbred market pigs.* PhD thesis, University of Minnesota, St Paul, Minnesota, USA.
Shrestha, J.N.B. (1998) Wealth generation through conservation of animal genetic resources in Nepal. In: Shrestha, J.N.B. (ed.) *Proceedings of the 4th Global Conference on Conserversation Domestic Animal Genetic Resources.* Nepal Agricultural Research Council, Lalitpur, Nepal, and Rare Breeds International, Shropshire, UK, pp. 3–9.
Shrestha, J.N.B. (2005) Conserving domestic animal diversity among composite populations. *Small Ruminant Research* 56, 3–20.
Shrestha, J.N.B. (2010) Heterosis retention and inter-breed recombination among non-allelic genes associated with crossbreeding and synthesis of breeds. *Egyptian Journal of Sheep and Goat Science* 5, 35–82.
Shrestha, J.N.B. and Crow, G.H. (2010) Breeding and genetics. In: Solaiman, S. (ed.) *Goat Science and Production.* Wiley-Blackwell, Ames, IA, USA, pp. 55–76.
Shrestha, J.N.B. and Fahmy, M.H. (2005) Breeding goats for meat production: a review. 1. Genetic resources, management and breed evaluation. *Small Ruminant Research* 58, 93–106.
Shrestha, J.N.B. and Fahmy, M.H. (2007a) Breeding goats for meat production: a review (2) Crossbreeding and formation of composite population. *Small Ruminant Research* 67, 93–112.
Shrestha, J.N.B. and Fahmy, M.H. (2007b) Breeding goats for meat production: a review. 3. Selection and breeding strategies. *Small Ruminant Research* 67, 113–125.
Shrestha, J.N.B. and Galal, E.S.E. (2010) Conservation of goat genetic resources. In: Solaiman, S. (ed.) *Goat Science and Production.* Wiley-Blackwell, Ames, IA, USA, pp. 39–53.
Shrestha, J.N.B. and Galal, E.S.E. (2011) Conservation of goat genetic resources. *Small Ruminant Research* (in press).
Shrestha, J.N.B. and Heaney, D.P. (2003) Review of Canadian, Outaouais and Rideau Arcott breeds of sheep. 1. Development and characterization. *Small Ruminant Research* 49, 79–96.
Shrestha, J.N.B. and Heaney, D.P. (2004) Review of Canadian, Outaouais and Rideau Arcott breeds of sheep. 2. Crossbreeding, registration and subsequent release to the Canadian sheep industry. *Small Ruminant Research* 55, 1–13.
Shrestha, J.N.B., Boylan, W.J. and Rempel, W.E. (2002) Evaluation of breeds of sheep and their crosses based on ewe productivity indices. *Small Ruminant Research* 46, 89–96.
Siddiqui, M.F., Bonde, H.S., Rotte, S.G. and Deshpande, K.S. (1981) Studies on some growth attributes of Osmanabadi goat kids. *Journal of Agricultural Science* 97, 747–749.
Simon, D.L. (1990) The Global Animal Genetic Data Bank. In: *FAO Animal Production and Health Paper No. 80.* Food and Agriculture Organization of the United Nations, Rome, Italy, pp. 153–166.
Simon, D.L. and Buchenauer, D. (1993) *Genetic Diversity of European Livestock Breeds.* EAAP Publication No. 66, Wageningen Press, Wageningen, the Netherlands.
Singh, D.K. and Yadava, R. (1997) Genetic studies on proportion of bone in different carcass cuts in Black Bengal and its crossbreds with Jamunapari and Beetal goats. *Indian Journal of Animal Sciences* 67, 996–999.
Singh, D.K., Singh, C.S. and Mishra, H.R. (1991) Factors affecting growth of Black Bengal and its crossbreds with Jamunapari and Beetal goats. *Indian Journal of Animal Sciences* 61, 1101–1105.
Singh, D.K., Singh, D.K., Singh, N.S. and Singh, L.B. (2000) Genetic studies on reproductive trait of goats. *Indian Journal of Animal Sciences* 70, 1255–1257.
Singh, S., Rana, Z.S. and Dalal, D.S. (2002) Genetic and non-genetic factors affecting growth performance in goats. *Indian Journal Animal Research* 36, 12–16.
Skinner, J.D. (1972) Utilization of the Boer goat for intensive animal production. *Tropical Animal Health and Production* 4, 120–128.
Snyman, M.A. and Olivier, J.J. (1996) Genetic parameters for body weight, fleece weight and fibre diameter in South African Angora goats. *Livestock Production Science* 47, 1–6.

Stanford, K., McAllister, T.A., MacDougall, M. and Bailey, D.R.C. (1995) Use of ultrasound for the prediction of carcass characteristics in Alpine goats. *Small Ruminant Research* 15, 195–201.
Stouffer, J.R. (1969) Techniques for the estimation of the composition of meat animals. In: *Techniques and Procedures in Animal Production Research.* American Society of Animal Science, Q Corporation, NY, USA, pp. 207–219.
Swan, J.E., Esguerra, C.M. and Farouk, M.M. (1998) Some physical, chemical and sensory properties of chevon products from three New Zealand goat breeds. *Small Ruminant Research* 28, 273–280.
Taferrant, H., Ben-Youcef, M.T. and Khemici, E. (1995) Goat production systems in Algeria and particularly in the Kabylie region. In: El-Aich, A., Landau, S., Bourbouze, A., Rubino, R. and Morand-Fehr, P. (eds) *Goat Production Systems in the Mediterranean.* EAAP Publication No. 71, Wageningen Press, Wageningen, the Netherlands, pp. 184–201.
Tahir, M., Younas, M., Raza, S.H., Lateef, M., Iqbal, A. and Raza, P.N. (1995) A study on estimation of heritability of birth weight and weaning weight of Teddy goats kept under Pakistani conditions. *Asian-Australasian Journal of Animal Sciences* 8, 595–596.
Taneja, G.C. (1982) Breeding goats for meat production. In: *Proceedings of the 3rd Internatonal Conference on Goats*, Tucson, AZ, USA, pp. 27–30.
Tantawy, A.O. and Ahmed, I.A. (1960) Studies in Egyptian Baladi goats. I. Frequency of multiple births and sex rates. *Empire Journal of Experimental Agriulture* 28, 74–82.
Upton, M. (1988) Goat improvement program. In: Smith, O.B. and Bosman, H.G. (eds) *Proceedings of a Workshop at the University of Ife*, Ife, Nigeria, pp. 11–22.
USDA (1992) *United States Standards for Grades of Lambs, Yearling Mutton, and Mutton Carcasses.* Livestock and Seed Division, Agricultural Marketing Service, USDA, Washington, DC, USA.
van Tassell, C.P. and van Vleck, L.D. (1996) Multiple-trait Gibbs sampler for animal models: flexible program for Bayesian and likelihood based (co) variance component inference. *Journal of Animal Science* 74, 2586–2597.
Waldron, D.F., Huston, J.E., Thompson, P., Willingham, T.D., Oman, J.S. and Savell, J.W. (1995) Growth rate, feed consumption and carcass measurements of Spanish and Boer × Spanish goats. *Journal of Animal Science* 73 (Suppl. 1), 253.
Warmington, B.G. and Kirton, A.H. (1990) Genetic and non-genetic influences on growth and carcass traits of goats. *Small Ruminant Research* 3, 147–165.
Wen, S.H., Su, A.K., Hsieh, R.C., Yan, S.S., Wu, J.S. and Chang, H.J. (1997) Breeding meat goats: upgrading Taiwan Native goats with Nubian goats. *Journal of Taiwan Livestock Research* 30, 3 and 231–236.
Wenxiu, H. (1988) Goat meat production in China. In: Devenda, C. (ed.) *Goat Meat Production in Asia.* Proceedings of a workshop held in Tando Jam, Pakistan. International Development Research Centre, Ottawa, Canada, pp. 119–124.
Wilson, I.L., Katsigianis, T.S., Dorsett, A.A., Cathopoulis, T.E., Greaves, A.G. and Baylor, J.E. (1980) Performance of native and Anglo-Nubian crosses and observation on improved pastures for goats in the Bahamas. *Tropical Agriculture, Trinidad* 57, 183–190.
Wilson, P.N. (1968) Biological ceilings and economic efficiencies for the production of animal protein, AD 2000. *Chemistry and Industry* 27, 899–902.
Wilson, R.T. (1992) Goat meat production and research in Africa and Latin America. In: *Proceedings of the 5th International Conference on Goats*, Vol. 2, New Delhi, India, pp. 458–472.
Winters, L.M. (1953) Research in sheep breeding. *Minnesota Farm Home Science* 10, 9–11.
Winters, L.M. (1954) *Animal Breeding*, 5th edn. John Wiley & Sons, New York.
Yalçin, B.C. (1986) *Sheep and Goats in Turkey.* FAO Animal Production and Health Paper No. 60, Food and Agriculture Organization of the United Nations, Rome, Italy.
Ying, J. (1995) Chinese meat goat breeds and their crosses. *Animal Genetic Resource Information* 15, 71–81.
Zhou, G.S., Chen, Q.L. and Zhao, K.F. (2001) Hybridization test of Boer goat with Xuhuai goat. *Chinese Journal of Animal Science* 37, 40–41 (in Chinese).

# 5 Reproductive Efficiency for Increased Meat Production in Goats

M. Rekik[1], I. Ben Salem[1] and N. Lassoued[2]

[1]*Ecole Nationale de Médecine Vétérinaire, Tunisia;* [2]*INRA-Tunisie, Laboratoire des Productions Animales et Fourragères, Ariana, Tunisia*

## 5.1 Introduction

Reproduction efficiency in female goats (does) is determined by many different processes. These processes include, for example, early initiation of reproductive life, length of the breeding season, cyclic activity, ovulation rate, fertilization rate and post-partum anoestrous period. All these reproductive traits depend on numerous components: genotype, environment and husbandry factors. In goats, and for small ruminants in general, the most important factor affecting flock efficiency is reproduction. Increasing reproduction is the most important way of improving meat production. This chapter addresses ways of improving reproductive efficiency in meat goats by reviewing data from highly specialized and less intensive production systems. However, it is first important to describe the main features of the reproductive biology of meat goats.

## 5.2 Doe Reproductive Biology

### 5.2.1 Onset of puberty

As in other livestock species, puberty is an important reproductive trait in goats that will determine the age at first kidding and therefore initiates the start of the reproductive career of the animal. Puberty can be defined in several ways. In the female, it is commonly defined as being the age at which oestrus is first detected and is followed by the establishment of a functional corpus luteum and a characteristic cyclic ovarian activity in the non-pregnant animal. Sexual maturity is different from puberty and occurs later. In young female sheep and goats, sexual maturity, which indicates acquisition of a full reproductive potential (maturation of the hypothalamic–pituitary axis, oestrous expression, embryo survival), is reached at a later age than puberty (Drymundsson, 1983). In this respect, one major feature of sexual immaturity in Creole goats is the dissociation between oestrus and ovulation at puberty (Delgadillo *et al.*, 1997). Half of the first detected oestruses are not accompanied by ovulation and 36% of the first ovulations are silent.

The occurrence of puberty in goats is similar to ewe lambs and fits within the model that classifies chronologically the events leading to the first luteotropic hormone (LH) surge in young female sheep (Foster and Ryan, 1981). At puberty, the response of the hypothalamic–pituitary axis to inhibition by oestradiol is considerably reduced. Basal secretions of LH therefore

 *Goat Meat Production and Quality* (eds O. Mahgoub, I.T. Kadim and E.C. Webb)

increase as a result of the acceleration in the rate of LH discharges, thus resulting in one or more of the follicles developing towards the pre-ovulatory stage. This is the first follicular phase, which is associated with a steady increase in the concentration of oestradiol, which eventually elicits the first pre-ovulatory LH surge.

There are several factors that could play a role in the attainment of puberty in the female goat. These include season of birth, season of weaning, body weight, nutritional status and presence of bucks. The effect of the presence of bucks on puberty will be reported later in the chapter related to induction of puberty in young females.

Under field conditions, some of the previous factors, such as season of birth and season of weaning and also body weight and nutritional status, act in complex interactions, which makes objective assessment of the individual effects on the attainment of puberty very difficult. These factors seem to involve a breed effect, probably with two major types of genotypes: early and late sexually maturing breeds.

In the young doe, body weight is of critical importance to the attainment of puberty. In general, breeding in goats should be delayed until the animal has attained 60–75% of its mature body weight (Smith, 1980). The mean body weight at puberty of Boer goat does has been set at 30.6 kg for animals on a high-energy diet and 27.5 kg on a lower-energy diet (Greyling, 1988). However, there are reports (Attwood, 2007) that Boer goats in South Africa can reach puberty at 18 kg body weight. It has also been documented that delayed attainment of puberty in temperate breeds of goats under tropical conditions is explained by the low growth rates of animals of these breeds under unfavourable management conditions in the tropics. The effect of inadequate nutrition hampering normal growth of the animals could be mediated through adverse effects on pituitary function.

Puberty is also influenced by the season of birth of the kids. The recorded mean age at the onset of puberty in Boer goat does is 191.1 and 157.2 days for does born during August (late winter) and January (midsummer), respectively (Greyling, 2000). In northern Mexico, Creole does born in January demonstrate their first oestrus at an age of 250 days and on average at an age of 172 days for females born in August or December (Delgadillo and Malpaux, 1996). In both the previous studies, the authors concluded that season of birth is the main cue for the onset of puberty and that observed differences could not be ascribed to body weight, level of nutrition or the buck effect.

Boer kids weaned in April (during the normal breeding season in the southern hemisphere) exhibited oestrus significantly earlier than those weaned in December (outside the natural season) (Greyling, 1996). Animals weaned during the natural breeding season maintained higher LH levels compared with those weaned outside the normal breeding season, indicating greater pituitary activity during the breeding season.

A compilation of the age at puberty for some breeds of meat-producing goats is presented in Table 5.1.

### 5.2.2 Seasonality of reproduction

The seasonal character of sexual activity in small ruminants has long been known. Among the factors controlling seasonal reproduction and birth seasonality, the change in day length is the most important component. Long days are reported to be inhibitory and short days stimulatory to sexual activity (Karsch *et al.*, 1984). Indeed, the most reliable environmental cue to set kidding at the optimal time of year is the seasonal cycle of day length, which does not vary from one year to another and therefore allows predictability. The photoperiodic information is perceived by the retina and transmitted to the pineal gland where this signal modulates the rhythm of melatonin secretion (Legan and Karsch, 1983; Karsch *et al.*, 1984).

The majority of goat breeds show seasonality in reproduction activities (Chemineau *et al.*, 1992). For goats bred

**Table 5.1.** Age at puberty for some meat-producing breeds of goats.

| Breed | Country | Age at puberty (days) | Specific information | Reference |
|---|---|---|---|---|
| Boer | South Africa | 191 | Born during seasonal anoestrus | Greyling (2000) |
| Boer | South Africa | 157 | Born during the breeding season | Greyling (2000) |
| Creole | Mexico | 172 (128–204) | Born between August and December | Delgadillo *et al.* (1997) |
| Black Bengal | NS | 196 | | Bhattacharrya *et al.* (1984) |
| Baladi | Syria | 180–210 | | Kassem (2005) |
| Damascus | Cyprus | 220–270 | | Mavrogenis (2005) |
| Matou | China | 108 (79–216) | Based on first oestrus | Moaeen-ud-Din *et al.* (2008) |

NS, Not specified.

under temperate latitudes (above 40°N), the female experiences a period of anoestrus from the beginning of spring to late summer, with no behavioural or ovarian activity. In the northern hemisphere, the breeding season starts in the summer and autumn months and ends in winter. In contrast, tropical breeds do not appear to be seasonal breeders (Devendra and Burns, 1983).

In Argentina, based on the monthly distribution of parturition, Creole goats reared under range conditions were classified as non-seasonal breeders (Molina *et al.*, 1997). Conversely, other results (Rivera *et al.*, 2003) based on the systematic recording of oestrus and ovulations indicate that Argentine Creole goats, maintained under natural photoperiod and controlled nutritional levels, are seasonal breeders. Transitions between seasons were gradual, with maximum and minimum levels of sexual activity concentrated at about the winter and summer solstice, respectively. For the Creole goats under southern latitudes in South America, a shorter breeding season with concentration of sexual activity during the months of June and July has been reported (Santa Maria *et al.*, 1990). For feral goats at similar latitudes in Australia, spontaneous ovulation was shown beginning in April, with peak incidence occurring in June and no ovulation between September and February (Restall, 1992).

For Damascus goats reared under desert conditions in Egypt, oestrus occurrence is correlated with the decrease in day length. The percentage of goats manifesting oestrus is high from July to October, while the percentage of kidding is high in January, February and March (Shalaby *et al.*, 2000).

It has often been assumed that the goat is similar to the sheep with regard to reproduction, but differences between the two species exist, and extrapolation from one species to the other could be misleading. Seasonal variation in reproductive activity in Australian goats was investigated and showed that does did not begin spontaneous ovulation until April, with peak incidence occurring in June (90%), and no ovulations were recorded between September and February (Restall, 1992). In contrast, the sheep began to ovulate in February with a peak incidence in March (95%), and the proportion ovulating remained high until July and then declined with no ovulations observed between August and December.

In Tunisia, the breeding season of local goats occurred from September to March (Lassoued and Rekik, 2005), during which 80% of the nanny goats exhibited oestrus at least once and 53% of all oestrous cycles were accompanied by ovulations. The proportion of ovulations increased gradually from September to reach 100% during December, and then declined to reach a minimum of 14% during March. This season preceded a period of sexual inactivity (March–August) that was distinctly longer than for local breeds of sheep in the same environment (Khaldi, 1984). The trends for the local Tunisian goat are similar to what has been reported for the Bedouin goat native to the Algerian Sahara, where seasonal anoestrus extends from the end of winter to the end of summer (Charallah *et al.*, 2000). Some examples related to sexual activity in several goat breeds in tropical, subtropical and temperate zones are reported (Table 5.2).

Unlike most goat breeds, Boer goats are partially seasonal breeders. Anoestrus does not occur and Boer does will cycle virtually all year round if favourable rearing conditions are provided (Greyling, 1990). This is similar to indigenous goats in Zimbabwe, which exhibit ovarian activity throughout the year (Llewelyn *et al.*, 1993).

At locations close to the equator and in the tropics, local breeds of goats either are non-seasonal breeders or exhibit only a weak seasonality of reproduction (Chemineau, 1986). Under these conditions, seasonal reproductive patterns relate more to variations in rainfall, available nutrition and temperature than to day length (Devendra and Burns, 1983). In the Mediterranean and tropical areas, the majority of goats are maintained in extensive or semi-extensive systems and may be mated more than once a year and subject to seasonal variations in food availability. It is often thought that nutrition is responsible for the seasonal reproductive pattern; in particular, insufficient nutrition is often responsible for prolonged anoestrous and anovulatory periods, a reduction in fertility and prolificacy. In the dual-purpose goats of north Senegal, most goats (65%) kid once a year, but some (22%) fail to reproduce, while others (13%) reproduced three times in 2 years (reviewed by Walkden-Brown and Bocquier, 2000).

The social environment, in particular the presence of bucks, is an important factor that affects the length of the breeding season. In Australian goats, exposure to males extended the period of ovulatory activity both before and after the period of spontaneous ovulation. In the absence of males, does only ovulated between April and August, but exposure to bucks, either intermittently or continuously, extended the ovulatory period from March to September (Restall, 1992). Other results support the suggestion that continuous exposure to the male prolonged, to some extent, the ovulatory period and also an earlier initiation of the ovulatory activity in Creole goats than in females kept isolated from male stimuli (Rivera *et al.*, 2003). This may suggest that the annual ovulatory rhythm in does reflects a changing sensitivity to exteroceptive stimuli that initiate reproductive activity.

The effects of season on reproduction in sheep and goats are mediated by similar mechanisms. The change in the duration of the night-time melatonin pattern alters the hypothalamic responsiveness to oestradiol, and this leads to changes in the frequency of LH pulses. A low pulse frequency switches off the reproductive system and a high frequency switches it on. During the sexual season, the hypothalamus shows little response to oestrogen and the LH pulse frequency is high, whereas, during anoestrus, the hypothalamus is very responsive to oestrogen and the LH pulse frequency is low.

**Table 5.2.** Examples of seasonality of sexual activity of some goat breeds in tropical, subtropical and temperate zones.

| Cyclicity | Country | Breed | Sexual season | Reference |
|---|---|---|---|---|
| Seasonal | Tunisia | Local Maure | September–March | Lassoued and Rekik (2005) |
| Seasonal | Algeria | Bedouine | End summer–end winter | Charallah *et al.* (2000) |
| Seasonal | Argentina | Creole | March–October | Rivera *et al.* (2003) |
| Non-seasonal | Guadeloupe | Creole | All year | Chemineau (1986) |
| Non-seasonal | Morocco | D'Man | All year | Derquaoui and El Khaledi (1994) |

At the onset of the breeding season, the secretion of gonadotropins, particularly the frequency of LH pulses, increases to more than three pulses every 6 h in mid-September (Chemineau *et al.*, 1988). These seasonal changes in pulse frequency are mediated by melatonin-induced differences in responsiveness to gonadal feedback (Chemineau *et al.*, 1988). This gradual enhancement of gonadotropic activity stimulates ovarian folliculogenesis and therefore may be responsible for the gradual onset of cyclic ovarian activity.

### 5.2.3 The ovarian cycle and related endocrine events

During the natural breeding season, female goats are seasonally polyoestrous with a spontaneous ovulation. When the does are not pregnant, they will display a succession of oestrous cycles lasting 21 days on average, which can be very irregular as we will discuss later. At each cycle, the doe can ovulate producing one or more oocytes at approximately 30–36 h after the beginning of oestrus (Gonzalez-Stagnaro *et al.*, 1984). Following ovulation, the Graafian follicle transforms into the corpus luteum (the luteinization process), which is active in secreting progesterone during the luteal phase, which lasts 16 days. If the female does not conceive, the corpus luteum regresses (luteolysis) and a new follicular phase starts.

Several hormonal changes are associated with the main events of the oestrous cycle. These hormonal changes are mainly the result of interactions between the ovaries, the hypothalamic–pituitary axis and the uterus. Throughout the luteal phase, the frequency of LH pulses is negatively correlated ($r = -0.97$) with the level of circulating plasma progesterone (Sutherland, 1987). These concentrations exert a very potent negative feedback on LH secretion during the luteal phase, hence keeping the amplitude of LH pulses to less than 1 ng/ml. The maximum plasma progesterone concentrations are attained around day 10 (day 0 is ovulation time) and remain high until day 15 (Simões *et al.*, 2006).

Starting on days 16–17 of the oestrous cycle, the prostaglandin F2α (PGF2α) released from the non-pregnant uterus, most likely influenced by ovarian oxytocin (Homeida, 1986), will induce luteolysis. The rapid decline of plasma progesterone concentrations is associated with an acceleration of the frequency of LH pulses and an increase in their amplitude (Mori and Kano, 1984). This is the major event triggering the start of the ovulatory follicular phase during which there is stimulation of the growth of follicles with a mean diameter >1 mm (Akusu *et al.*, 1986). These growing follicles will increase their steroidogenic activity and hence start secreting oestradiol 17β at high concentrations >10 pg/ml (Rubianes and Menchaca, 2003). The sustained increase in oestradiol concentration will in turn induce a positive feedback action on the pituitary gland, culminating in the pre-ovulatory LH surge. This surge usually lasts between 8 and 10 h; its maximum is reached 3 h after the peak of oestradiol and 10–15 h after the onset of oestrus (Chemineau and Delgadillo, 1994). Ovulation usually occurs 20 h after the pre-ovulatory LH surge with some variation, as will be discussed later.

These endocrine events are those that have been reported for natural oestrous occurrence or when oestrus has been induced following luteolysis with exogenous PGF2α. Nevertheless, it is worth mentioning that when oestrus is induced with equine chorionic gonadotropin (eCG), growth of the follicles is faster and some features of the endocrine changes are modified, such as a longer interval between the peak of oestradiol 17β and the pre-ovulatory surge of LH, which occurs earlier in relation to the onset of oestrus (5.6 versus 15.7 h; Chemineau and Delgadillo, 1994).

In parallel to the endocrine changes, the doe oestrous cycle is also characterized by major changes in follicular turnover. Useful information on this topic can be found in recent reviews (Rubianes and Menchaca, 2003; Simões *et al.*, 2006) and we shall report here the main characteristics of follicular dynamics in the doe.

The use of ultrasonography has facilitated a rapid increase in our knowledge of animal reproductive physiology and its control. The efficiency of transrectal ultrasonography to study ovarian function in goats provides a means for repeated, direct, non-invasive monitoring and measuring of follicles >2 mm, regardless of their depth within the ovary. The results of daily ultrasonographic studies indicate that the inter-ovulatory cycle of goats is characterized by a wavelike pattern of follicular development, as has been reported for other ruminant species. A follicle wave involves the emergence of a group of small antral follicles from which commonly one or two follicles are selected to grow to >5 mm in diameter. According to different authors, the number of follicular waves ranges between two and five waves per cycle, but the predominant pattern for goats that developed an inter-ovulatory cycle of normal length (19–22 days) is of four waves (Fig. 5.1). The emergence of waves one, two, three and four (the ovulatory wave) occurs on days 0, 5–6, 10–11 and around day 15 post-ovulation, respectively. However, some breed differences have been reported and the duration of the first inter-wave interval in Serrana goats (5.6 ± 0.3 days; Simões *et al.*, 2006) was between the values observed in Boer goats (4.0 ± 1.4 days; Schwarz and Wierzchos, 2000) and in Saanen goats (7.3 ± 0.9 days; de Castro *et al.*, 1999).

Some of the more frequently observed characteristics of the follicular waves are as follows: (i) the diameter of the largest follicle of a wave differs between waves; commonly the largest follicles of waves two and three attain smaller maximum diameters than both the largest follicle of wave one and the ovulatory follicle; (ii) two or more follicles per wave frequently attain 5 mm or more; (iii) the growth rate between the day of emergence (i.e. first day with a size of 3 mm) and the day of maximum diameter is around 1 mm/day; (iv) as the luteal phase progresses, follicular turnover increases and the inter-wave intervals are shorter than during the early luteal phase; (v) during the mid- to late luteal phase, the follicles that do not grow beyond 4 mm often are not a part of the wave phenomenon and it is suggested that they represent a dynamic underlying pool; (vi) most of the ovulatory follicles are the largest follicles on the day of luteolysis; (vii) in most double-ovulatory goats, the ovulatory follicles emerge as part of the same follicular wave but in a few cases also as a part of different waves; and (viii) double ovulations occur on the same day in most cycles.

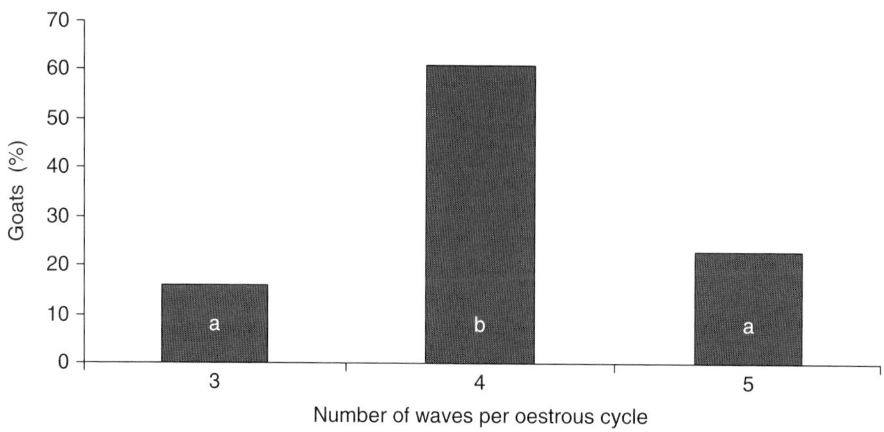


**Fig. 5.1.** Distribution of the number of follicular waves per oestrous cycle in goats. Different letters within columns (a versus b) show that the difference was significant ($P < 0.001$) (Simões *et al.*, 2006).

There is good evidence showing that the pattern of fluctuations of serum follicle-stimulating hormone (FSH) concentrations is tightly associated with the emergence of most follicular waves in sheep (Evans *et al.*, 2000). An increase in FSH concentrations commonly precedes the emergence of the wave and this is followed by a decrease, which is negatively correlated with the oestradiol produced by the largest follicle of the wave. Reports regarding the relationship between FSH and follicular dynamics are scarce in goats, and several reports related to FSH fluctuations along the oestrous cycle of the goat are inconclusive. However, the positive effects of an exogenous FSH treatment on follicle recruitment are well established in goats (Rubianes and Menchaca, 2003).

Goats are reported to have more irregular oestrous cycles than ewes. The period of the season (beginning or end of the sexual season) or post-partum are the major factors related to irregular oestrous cycles. However, buck introduction to anoestrous females or the administration of exogenous hormones can also contribute to the appearance of irregular oestrous cycles. Irregularity of the oestrous cycle refers to the length of the cycle, silent ovulation or oestrus without ovulation (Lassoued and Rekik, 2005). When studying the seasonality of oestrus and ovulation in Creole goats of Argentina (Rivera *et al.*, 2003), it was found that the mean length of the oestrous cycle was 21.7 ± 0.6 days with >90% of the cycles between 19 and 22 days. Frequencies of short (<16 days) and long (>27 days) oestrous cycles were, respectively, <5% and around 15% of all observed cycles. Such data are similar to those reported for the Boer goat (Greyling, 1988) in which the frequencies of short and long cycles have been recorded as being 16.6 and 10.2% and also for the Barbari goat breed (Prasad and Bhattacharyya, 1979) for which the oestrous cycle was categorized into short, medium and long cycles, with the frequency of each category being 19.7, 68.8 and 11.5%. Oestrous cycles were found to be significantly shorter during periods of the year with moderate climatic conditions, compared with extreme cold/dry and hot/wet periods. For the local Maure goat, it was reported that, on average, a goat has a mean of 1.2 normal cycles during the whole breeding season with an average length of 21.1 ± 1.5 days (Lassoued and Rekik, 2005). All other cycles were irregular, delimited either by silent ovulations or by oestruses without ovulation. The mean length of the oestrous cycle for different breeds of meat goats in many countries are summarized in Table 5.3.

### 5.2.4 Oestrous expression and occurrence of ovulation

Oestrous expression in goats is hormone dependent and is triggered by the sustained increase of oestradiol following regression of the corpus luteum at the end of the luteal

**Table 5.3.** Mean length (days) of the oestrous cycle for some breeds of meat-producing goats.

| Breed | Country | Length of oestrous cycle (days) | Specific information | Reference |
|---|---|---|---|---|
| Boer | South Africa | 20.7 ± 0.7 | – | Greyling (2000) |
| Creole | Argentina | 21.7 ± 0.7 | Regular cycles only | Rivera *et al.* (2003) |
| Maure | Tunisia | 21.1 ± 1.5 | Regular cycles only | Lassoued and Rekik (2005) |
| Dwarf | Pakistan | 19.7 ± 1.5 | Post-partum period | Khanum *et al.* (2007) |
| Damascus | New Zealand | 21.2 ± 1.5 | – | Zarkawi and Soukouti (2001) |
| Serrana | Portugal | 20.6 ± 1 | Induced oestrus with PGF2α | Simões *et al.* (2006) |
| Matou | China | 19.7 ± 1.5 | – | Moaeen-ud-Din *et al.* (2008) |

PGF2α, Prostaglandin F2α.

phase and the rapid fall in progesterone plasma concentrations. On the day of oestrus, progesterone plasma concentrations are very low, i.e. <0.5 ng/ml. The duration of the oestrous period in goats appears to be variable in length, but on average it lasts 36 h, with a variation of between 22 and 60 h (Riera, 1982). The mean duration of the natural oestrous period in the mature Boer goat is 37.4 ± 8.6 h, with a variation of 24–56 h among individuals (Greyling, 2000). No significant difference was recorded between multiparous, biparous and primiparous does (38.2, 34.0 and 38.6 h, respectively). When in oestrus, the doe is more expressive than other females of mammalian domestic species (Fabre-Nys, 2000). During the first phase of oestrus, behaviour known as 'proceptivity' or pro-oestrus occurs, when the female seeks partners for stimulation through intermittent approaches. The female then moves on to the second phase of 'receptivity' or oestrus, when she is in heat, expressing the following signs: seeking out bucks more frequently, wagging of the tail, mounting behaviour, bleating, clear mucous discharge from the vulva, urine emission, and reddening and swelling of the vulva. These signs will end with the female accepting mounting. When in oestrus, the female goat can express a homosexual behaviour by mounting other goats in heat. The intensity of the oestrous signs may vary; this often happens with young does for which these behaviours are not so obvious.

The time of ovulation in the goat is reported as occurring towards the end of the oestrous period (van der Westhuizen *et al.*, 1985). Ovulation is a complex process, initiated by the surge of gonadotropins, and is characterized by resumption of meiosis, culminating in rupture of the follicular surface, release of a mature fertilizable (secondary oocyte) ovum and restructuring of the follicular wall. To decide the precise time of mating and also to make effective use of modern reproductive tools such as artificial insemination and embryo transfer technology for improving goat production, information on ovulation profiles is essential. In the Boer goat doe, the time of ovulation is recorded as occurring 36.8 h after the onset of oestrus (86.7% of does ovulating by 38 h after the onset of oestrus), with the mean time interval between the LH peak and ovulation being 24.7 h (Greyling, 2000). The timing of the LH peak relative to the onset of oestrus was recorded as being between 4 and 20 h after the onset of oestrus. These data on the Boer goat are similar to what have been reported for the Jakhrana goat for which ovulation occurred 28 h after the onset of oestrus (towards the end of the oestrous period) and reached its peak at 36 h post-oestrus (Goel and Agrawal, 2003) and also for local Egyptian goats for which ovulation occurred at about 27 h after the onset of oestrus (Salama, 1972).

Timings of oestrus and ovulation are hastened when the does are hormonally treated. Knowledge of time, rate and synchrony of ovulation after treatment with intravaginal devices is important to establish a suitable schedule in fixed-time artificial insemination programmes. For the Tunisian local goat, it has been found that treatment of the does with eCG advances oestrus by approximately 10 h (N. Lassoued and M. Rekik, 2010, unpublished results). This is similar to other results (Ritar *et al.*, 1989) showing that ovulation had occurred by the final laparoscopy (65–75 h) in all females ($n$ = 22) injected with 200 or 400 international units (IU) eCG at 0 or –48 h in relation to withdrawal of the progestogen treatment, whereas only seven of the does not injected with eCG (control) had ovulated by this time.

### 5.2.5 Ovulation rate and reproductive wastage

For meat production breeds of goats, ovulation rate has a major impact on their litter size and hence on their reproductive effectiveness. A high ovulation rate is an important characteristic, referring to the number of ova liberated and eventually the number of kids born per doe kidding. Physiologically, the number of ova shed by a goat at each ovarian cycle is the interaction between three levels of regulatory mechanisms: the concentrations of circulating gonadotropins (mainly FSH), the response of the follicles to these hormones and the internal regulations at the ovarian level, whether paracrine (between follicles) or autocrine (within a follicle) (Driancourt *et al.*, 1990). Ovulation rate is influenced by the stage of breeding season, genotype, parity and nutrition.

When studying seasonal variations of reproductive traits of the local Maure goat in central Tunisia, an important effect of season on ovulation rate was reported (Lassoued and Rekik, 2005). Ovulation rate dropped from a mean value of 1.51 at the start of the breeding season in September–October to <1.25 at the end of the breeding season towards January and February. These findings are consistent with other published data (Restall, 1992) reporting a decreasing gradient of ovulation rate between the beginning and the end of the breeding season for Australian goats, with a trend for the ovulation rate to be higher for does kept in the presence of males. When ovulation rate shows seasonal variation, maximum prolificacy in the flocks can be obtained by presenting does to bucks at the start of the breeding season. For breeds that are not seasonal, as in the case of the Creole goat in Guadeloupe (Chemineau, 1986) or the D'Man in Morocco (Derquaoui and El Khaledi, 1994), seasonal variations of ovulation rate are less important. However, under such conditions, this trait can show variations associated with other environmental factors such as nutrition (Delgadillo *et al.*, 1997) or health status.

Breed is also a major variation factor for ovulation rate. The Boer doe with a mean ovulation rate of 1.72 is considered one of the more prolific breeds in the world (Greyling, 2000). In terms of litter size, the percentage of singletons, twins, triplets and quadruplets born in the Boer goat are quoted as being 24.5, 59.2, 15.3 and 1%, respectively (Campbell, 1994). Because of the close relationship between ovulation rate and litter size, we would expect most of Boer goats to shed twin ovulations. The ovulation rate of the Boer goats is higher than that quoted for Malawian goats (1.68; Campbell, 1994) or Jakhrana goats native to semi-arid India (1.33; Goel and Agrawal, 2003). Other breeds can have much higher ovulation rates, such as the Black Bengal goat with a mean rate of 4 (Rao and Bhattacharrya, 1980). For the prolific Matou breed in China, it is anticipated that ovulation rate, although not measured, would be high on the basis of litter sizes recorded in the flocks: mean litter size was reported to be 2.09 (Moaeen-ud-Din *et al.*, 2008), with the percentages of singletons, twins, triplets and quadruplets being 27.4, 45.4, 16.3 and 10.9%, respectively. Lower frequencies of does gave birth to quintuplets or sextuplets. However, the survival rate of kids was 90.8% and was negatively associated with litter size (Fig. 5.2).

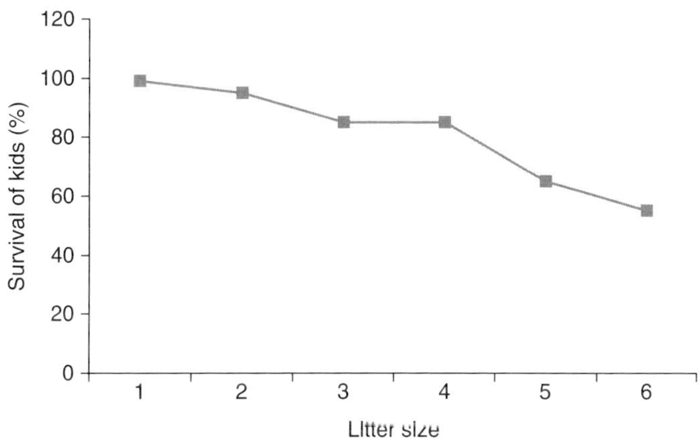


**Fig. 5.2.** Association of survival rate at puberty of kids with litter size in the Matou breed (redrawn from Moaeen-ud-Din *et al.*, 2008).

Nutrition is known to play a fundamental role in controlling several reproductive events, including hormone production, gametogenesis, fertilization and early embryonic development in farm animals. Like sheep, but in contrast to cattle, most goats have the potential for multiple ovulations, but this ability may be impaired by inadequate nutrition. Long-term effects (3 weeks) of overfeeding on ovulation rate are mediated through improved body condition (dynamic component of nutrition on ovulation rate), whereas short-term effects are achieved through the provision of nutrients that modify the hormonal environment, with no alteration of body condition (static component). The appearance and continuation of oestrous activity of goats are less dependent on nutrition than ovulation rate. In British Saanen and Toggenburg thin goats, severe energy deprivation during the 19 days before a synchronized oestrus did not affect the proportion of goats coming into oestrus but did decrease the ovulation rate, and timing of ovulation was delayed (Mani *et al.*, 1992). Ovulation rate increased asymptotically in small prolific Indonesian goats that gained body weight at 30 g/day before mating (Henniawati and Fletcher, 1986). Such an effect could be mediated through an increase in blood glucose, insulin and leptin, leading to increased folliculogenesis and ovulation rate (Viñoles, 2003). No negative effect of overconditioning on reproduction has been documented in goats, as it has been in sheep (West *et al.*, 1991). Further examples on the interaction between nutritional inputs, ovulation rate and litter size will be discussed in section 5.4.

Similarly to ewes, ovulation rate for does varies with age, although the effect can be confounded by parity. If we assume that the effect of age on ovulation rate is similar to the effect on litter size, maximum ovulation rate is reached when does are 5–7 years old. Investigations into the genitalia of pregnant Teddy goats in Pakistan (Anwar and Ahmad, 1999) showed that ovulation rate increased from 1.6 to 2.3 for young does with no permanent teeth and old does with six or eight permanent teeth, respectively. In goats with no permanent teeth, 57.6% of the does had multiple ovulations compared with 96.3% for goats with six or more permanent teeth. Interestingly, embryo losses increased from 7.7 to 16.1%, but it was not possible to conclude whether the trend in embryo loss was correlated with age or ovulation rate.

In mammals, reproductive losses can occur from mating time up to the end of pregnancy. These losses can be summarized as being due to fertilization failure, early embryonic losses and fetal losses (abortion). In goats, many of the data related to fertilization failure emanate from studies on superovulation and embryo recovery. Fertilization failure in goats has been summarized as being the result of a poor synchronization of oestrus and ovulation, especially following fixed-time artificial insemination, and also of abnormal maturation of the oocytes following superovulation (Lehloenya, 2008).

In sheep, most early embryo losses occurred during the period immediately preceding day 18 after mating during either blastulation or extension of the embryonic membranes (Quinlivan *et al.*, 1966). In goats, it was also concluded that limited conceptus loss is expected after the first 3 weeks of pregnancy (Anwar and Ahmad, 1999). Some of the factors involved in early embryonic mortality are: body condition (does that are too fat or too thin at mating time); dietary changes within the first few weeks after breeding negatively influencing functioning of the corpora lutea; extreme hot weather within the first month after breeding; genetic factors of the sire, dam or embryo leading to congenital defects and early conceptus loss; maternal hormone imbalances; age of the dam (embryo losses are greater in young does and goats over 6 years); excessive oestrogen in the diet (ingestion of plants containing phyto-oestrogens); deficient uterine environment (insufficient uterine space for the number of embryos or too few placental attachment sites or caruncles); and immunologic incompatibility (blood types and serum antigens).

During later stages of pregnancy and in the absence of the incidence of abortive diseases, it was shown that a severe reduction in the energy allowances well below the requirements of goats through the third month of pregnancy was associated with increased embryo loss (Mani *et al.*, 1992). In late gestation, an energy deficit, particularly for does carrying multiple fetuses, will result in pregnancy toxaemia (ketosis) and fetal death. Signs of ketosis are anorexia (off feed), lethargy and pain. Caught early, ketosis can be treated with propylene glycol given orally.

### 5.2.6 Gestation and uterine involution

For a pregnancy to be established, communication between conceptus and mother (maternal recognition) and implantation of the conceptus on the uterine wall must occur. The establishment of intimate contact between the embryo and the mother follows a succession of common critical steps whose chronology and timing may vary considerably from species to species. In goats, implantation begins around days 14–17 of pregnancy (Bazer *et al.*, 1997) when the embryo starts placentation. The placentome serves as immunological barrier membrane and mediates gas and micronutrient exchange (Wani, 1996). The placenta secretes steroid hormones (oestradiol and progesterone) and placental lactogen (cPL). The latter plays a major role in maintaining pregnancy and in growth of the mammary gland (Hayden *et al.*, 1979).

The length of pregnancy is called the gestation period. For goats, the gestation period is about 149 days with a usual range of 145–153 days. The length of gestation is variable according to breed and the individual. The mean gestation period for the Boer goat is recorded as being 148.2 ± 3.7 days (Greyling, 1988). In Black Bengal does, the length of gestation is 144 days (Jainudeen *et al.*, 2000). Gestation may be shorter in twin-bearing does and in extreme weather. However, for Boer goats, there was no significant difference in the gestation length between does bearing singletons or triplets, and the season of mating had no significant effect on gestation length (Greyling, 1988). The influence of nutrition on fetal development during certain months of pregnancy does tend to shorten or lengthen the gestation period, but the variation due to this factor was only 1.5 days (Riera, 1982).

The endocrine balance during pregnancy in goats is complex and involves several hormones such as progesterone, oestrone sulfate, cPL and pregnancy-associated glycoproteins (PAGs). Progesterone is required to maintain pregnancy (Meites *et al.*, 1951). Early during pregnancy, progesterone facilitates the implantation process and thereafter maintains pregnancy by ensuring closure of the cervix. During gestation, there is little evidence of any dramatic increase in serum progesterone concentration above the levels attained during the luteal phase of the oestrous cycle, suggesting the corpus luteum as the only source of progesterone (Wango *et al.*, 1991). A positive relationship between multiple pregnancies and maternal serum progesterone levels was not observed (Thorburn and Schneider, 1972).

cPL is somewhat related to prolactin and growth hormone and is secreted by the placenta. The hormone becomes detectable in the maternal circulation from day 44 of pregnancy (Currie *et al.*, 1990). The hormone is important for the development and activity of the mammary gland, and its secretion increases between weeks 10 and 16 of pregnancy, coinciding with the development of the epithelial tissue of the mammary gland (Hayden *et al.*, 1979).

PAGs are of feto-placental origin and are detected in the peripheral circulation of pregnant animals. PAGs are immunosuppressants and are believed to play a role in preventing the immune system of the pregnant female from attacking the embryo (Shaw and Morton, 1980; Bose *et al.*, 1989). These proteins are secreted by trophoblastic cells and are found in the peripheral circulation of pregnant dairy and beef cows (Zoli *et al.*, 1992), goats (Benitez-Ortiz, 1992) and in Churra, Merino and Assaf ewes (Ranilla *et al.*, 1994, 1997). During pregnancy of Moxto and Caninde goats, PAG was detected on day 24 after artificial insemination

(Sousa *et al.*, 1999). The profiles for PAG were not different between the breeds throughout pregnancy. However, a significant effect of the stage of pregnancy and the number of fetuses on PAG concentrations was found (Fig 5.3).

Morphological changes or their delay in the post-partum uterus and ovaries of farm animals exerts limitations on the reproductive performance of females following parturition. After distension and distortion of uterine tissues during pregnancy and the heightened glandular development for support of the conceptus, the uterus must undergo contractions and loss of weight, together with extensive regeneration of its epithelial layers during the process of uterine involution (Hunter, 1981). The interval from parturition to a subsequent pregnancy is a factor of major economic importance and hence the involution of the post-partum uterus must be seen as one of the important limitations in achieving the goal of optimal reproduction efficiency. The time required for uterine involution in goats is not very precise. In the Boer goat, macroscopic changes of the post-partum uterus show a rapid decline in weight and volume from parturition to approximately day 12 post-partum. This is demonstrated by the fact that, by day 12 post-partum, the uterus weight is 15% of its

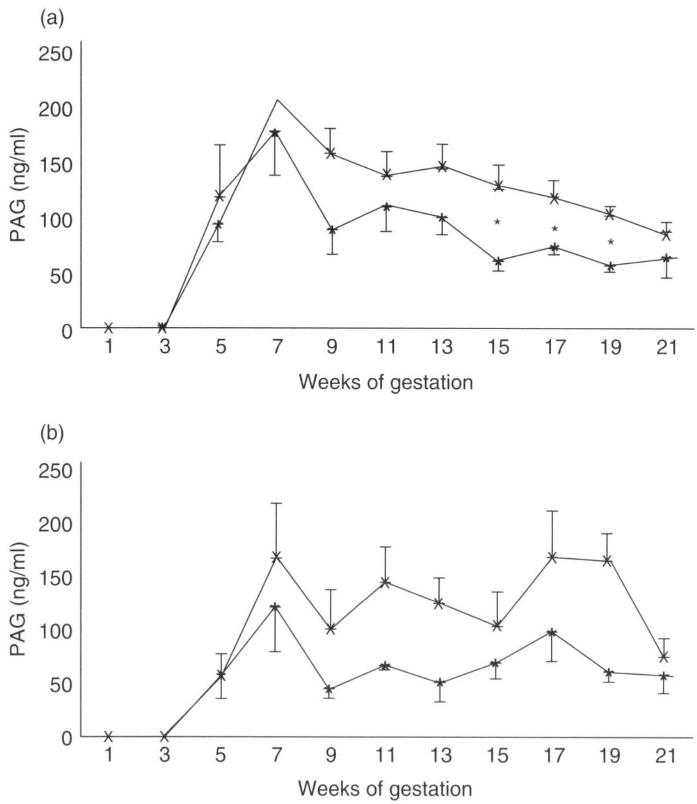


**Fig. 5.3.** Pregnancy-associated glycoprotein (PAG) profiles during pregnancy in Moxoto (a) and Caninde (b) goats with either single (–) or multiple (×) pregnancies. Significant differences ($P < 0.05$) in PAG concentrations between prolificacy levels are indicated by asterisks (*) (Sousa *et al.*, 1999).

weight at parturition. By day 20, it is 8% of that at parturition and only 27% more than the uterus weight of maiden Boer goats. Decreases in uterine horn length and diameter were less pronounced and the external diameters were back to normal by day 28 post-partum. According to these observations, it would seem that the involution process of the Boer goat uterus is macroscopically complete by approximately 28 days post-partum (Greyling and van Niekerk, 1991).

Pseudopregnancy occurs when the corpus luteum persists in the absence of a viable conceptus in the uterus. During pseudopregnancy, aseptic fluid accumulates in the uterus (visible on ultrasonography), and this pathological uterine condition is known as hydrometra (Pieterse and Taverne, 1986). Progesterone secreted by the corpus luteum is also high and the lifespan of the corpus luteum is longer in comparison with a cyclic corpus luteum and can prolong the duration of a gestation period. From field studies based on the use of ultrasonography, it appears that the incidence of hydrometra in herds of dairy goats varies between 3 and 21%, although hydrometra is rare in young goats (Mialot *et al.*, 1991; Hesselink, 1993a; Leboeuf *et al.*, 1994). There are no specific reports on the incidence of hydrometra in meat goats. Treatment of pseudopregnant goats with a luteolytic dose of PGF2α will cause discharge of the uterine fluid (Pieterse and Taverne, 1986). It has been demonstrated under field conditions that a second treatment with PGF2α given 12 days after the induced discharge significantly improves the reproductive performance of the goats when they are mated during the oestrus induced by the second injection (Hesselink, 1993b).

### 5.2.7 Post-partum anoestrous period

Early resumption of ovarian and oestrous activities during the post-partum period is a necessary requirement for successful rebreeding in domestic animals. Such a requirement is more critical for meat-producing livestock breeds than for dairy breeds, which are usually managed to reach the target of one reproductive cycle per year, i.e. one lactation per year. In goats, an early restart of oestrous activity during the post-partum period is important to obtain a suitable kidding interval. Physiologically, the endocrine mechanisms underlying post-partum anoestrus in goats are similar to those described for sheep and cattle and are illustrated by an increased oestradiol feedback at the hypothalamo-pituitary axis leading to inadequate pulsatility of LH to promote follicular growth at the ovarian level.

When comparing sheep and goats, the latter are reported to have an earlier resumption of post-partum ovarian activity (Mbayahaga *et al.*, 1998). However, various abnormalities characterize the post-partum ovarian activity in goats, namely a higher incidence of atypical cycles with short and long luteal phases. Regular ovarian cycles are usually observed in sheep. The first post-partum oestrus precedes the first ovarian activity in goats (4 days for local Burundian goats) but occurs later in sheep (36 days) (Mbayahaga *et al.*, 1998). These differences could be related to the requirement for progesterone priming between the two species. In fact, the central nervous system of the ewes needs to be primed with progesterone before they can become responsive to oestrogen, but goats do not require this progesterone priming for oestrous behaviour.

For goat breeds, a strong correlation was observed between kidding interval and the latitude of the area where the breeds originate from (Fig. 5.4; Delgadillo *et al.*, 1997). For breeds thriving near the equator, the kidding interval is approximately 250 days, which means that the post-partum anoestrous period is relatively short. In contrast, when the latitude increases (such as areas at latitude 30°), the kidding interval becomes longer, around 1 year. This relationship between kidding interval and latitude can be modified by several environmental factors such as seasonal anoestrus, breed, nutrition and body condition (a statement that is true for all reproductive traits), number of

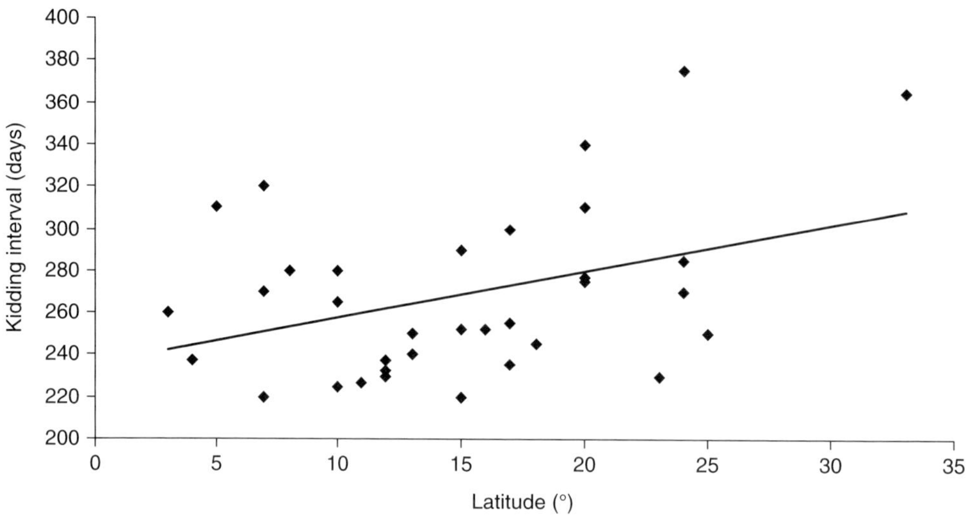


**Fig. 5.4.** Relationship between the length of the kidding interval and the latitude of the geographical area where the breed of goat originates from (each point represents a breed) (redrawn from Delgadillo *et al.*, 1997).

suckled kids and parity. Another factor that has a role to play in resumption of ovarian activity is the presence of the male. The practical applications of the buck effect will be discussed later.

One explanation for a delayed resumption of reproductive activity following kidding is an interaction between seasonal anoestrus and post-partum anoestrus. For local goats of North Mexico, 50% of females resumed reproductive activity 3.5 months after parturition in May, while the same percentage was reached at 5.5 months if parturition occurred in January (Delgadillo *et al.*, 1997). Intervals for resumption of reproductive activity after kidding are therefore shorter when the kidding season coincides with the natural breeding season, and this can be the basis for management practices to accelerate kiddings in meat-producing breeds.

Breed of goat is another factor that alters length of the post-partum anoestrus. In Brazil, local Caninde and mixed breeds of goats resumed post-partum reproductive activity on average between 46 and 52 days after kidding (Freitas *et al.*, 2004); this interval was much shorter than for Anglo-Nubian and Saanen goats raised in the same environment. The uniqueness of these two latter breeds, rendering them more sensitive to post-partum endocrine interactions, may be one of the reasons for the large interval between parturition and the occurrence of first oestrus. Dwarf goats are also known to have a shorter period of post-partum anoestrus than other meat breeds (e.g. Black Bengal, West African Dwarf or Malaysian Katjang). For Dwarf goats in Pakistan, a mean post-partum anoestrous period of only 28 days (range 15–59 days) was reported (Khanum *et al.*, 2007).

In several mammalian species, the presence of the young strongly inhibits the resumption of post-partum sexual activity, and the sensory stimulation caused by suckling is an important component in this process, although its mechanisms of action may vary between species. In domestic ruminants, there is also some effect of the young on post-partum anoestrus duration. For example, in sheep, there is a positive correlation between nursing activity and post-partum anoestrus duration (Fletcher, 1971), and ewes separated early from their young and submitted to milking return to heat sooner than the mothers that keep nursing, an effect partly due to an inhibitory action of prolactin. The effect of suckling stimulus on post-partum anoestrus in goats remains controversial. Twin-bearing goats tend to have a longer post-partum anoestrus than single-bearing does (Mbayahaga *et al.*, 1998), but the numbers of animals were not sufficient to demonstrate this statistically. Similarly, the mean duration of the post-partum anoestrous period in the Boer goat is quoted as being 55.5 ± 24.9 days with this period being 53.2 ± 14.3 days for does bearing singletons, 58.5 ± 30.0 days for does with twins and 61.7 ± 30.7 days for does bearing triplets (no significant differences) (Greyling, 2000). The duration of the suckling period did not have an effect on the length of the post-partum period. In North Mexican goats, weaning of the kids at 2, 30 or 90 days of age did not influence the time of resumption of post-partum ovarian activity (Delgadillo *et al.*, 1994).

Parity seems also to be a source of variation of the length of post-partum anoestrus. It was shown that goats in their first and second lactation had a longer post-partum anoestrus when compared with goats in later lactations (Freitas *et al.*, 2004). It is possible that this difference is due to a better ability for the return to oestrous activity after kidding in higher parity goats. Such a fact may be related to the faster uterine involution and/or to the return of the responsiveness to gonadotropin-releasing hormone (GnRH) post partum.

## 5.3 The Buck Reproductive Physiology

### 5.3.1 Neuroendocrine control and seasonality of reproduction

In bucks, spermatogenesis (all the processes that converge to the production of spermatozoa) are under the control of LH and FSH. These two pituitary hormones participate in the differentiation and multiplication of germinal cells and also in the synthesis and secretion of testosterone by Leydig cells. Testosterone in turn maintains spermatogenesis, triggers male sexual behaviour and exerts feedback control on gonadotropins. LH is secreted in pulses controlled by GnRH from the hypothalamus. Between pulses, basal secretion is recorded. It has been shown that the frequency of the pulses and their amplitude are key factors determining the response of the gonads to LH (Delgadillo and Chemineau, 1992).

In seasonal breeds, variations in the gonadotropic activity are responsible for the low reproductive activity of the bucks during spring when the photoperiod is increasing and for the intense activity during autumn and winter under a decreasing photoperiod. In the northern hemisphere, the basal plasma concentrations, pulse frequency and amplitude of LH are low from January to May (Chemineau and Delgadillo, 1994). The amplitude of the pulses then shows sustainable increases from June to January. In September, the pulse frequency increases while the amplitude declines because of the inverse relationship between the two features. It is also likely that the reduction in the LH pulse amplitude results from the negative feedback action of testosterone, which reaches its highest concentrations in autumn. After the peaks reached by LH and testosterone concentrations in autumn, these two hormones then show a gradual decrease until January when a new annual cycle begins. The close relationship between changes in gonadotropin plasma levels, the weight of the testes (as an indicator of spermatogenesis) and sexual behaviour confirms that seasonal changes in reproduction of the buck are controlled by neuroendocrine activity. Seasonal changes in reproductive activity of bucks have also been reported in areas where changes in photoperiod are less important than in temperate countries. This is the case for the low libido and semen quality of Zaraibi bucks in Egypt during the spring season, which has been attributed to the low level of circulating plasma testosterone and a reduction in the thickness of the seminiferous tubules due to the lower number of spermatic layers (Barkawi *et al.*, 2006). Seasonal changes in reproductive activity of bucks are also illustrated by a number of other observations where the month of collection significantly affected the mass motility, progressive motility, percentage of abnormal sperm and volume of ejaculate. Additionally, different trends in seasonality of ejaculate volume have been reported for Murciano-Granadina (at latitude 37°) and Verata (at latitude 40°) bucks (Roca *et al.*, 1992; Perez and Mateos, 1996). In these studies, higher semen volumes were recorded in summer and autumn (June–November). As has already been discussed for the doe, changes in photoperiod are the primary cue for the seasonal changes in neuroendocrine activity. Photoperiod would act through the secretion of melatonin from the pineal gland, hence modulating the sensitivity of the hypothalamic–pituitary axis to feedback by steroids.

Differences between breeds with regard to seasonal variation of reproduction can also have a genetic component. Under the same environment, photoperiod did not have the same effect on semen characteristics of the Verata and Malaguetia bucks. The Verata breed was more affected by photoperiod, with higher semen production and better semen quality during the decreasing photoperiod (Perez and Mateos, 1996).

### 5.3.2 Other factors affecting reproduction in bucks

Nutrition is one of the factors of utmost importance in the expression of the reproductive characteristics. Nutrition can have long-, medium- and short-term effects on the reproductive function of small ruminants. To alter reproduction, nutrition can act on the hypothalamic–pituitary axis through changes in gonadotropin secretion or directly on the gonads, interfering with the processes of gamete production. It is well established that the effects of nutrition are mediated through a number of metabolites and metabolic hormones, such as insulin, leptin, insulin-like growth factors, glucose and amino acids. In most tropical and subtropical climates, undernutrition poses serious limitations on animal production and animals may lose up to 40% of their body weight because of the poor quality of the pastures in the dry season (Clariget *et al.*, 1998). The effect of seasonal weight loss on several reproductive parameters has been described extensively for several goat breeds such as Damascus (Al-Ghalban *et al.*, 2004) and Zairabi (Barkawi *et al.*, 2006) goats. For Boer bucks in arid areas of South Africa, the feeding of non-supplemented winter veld hay significantly reduced the reproductive performance in the Boer goat buck (de Waal and Combrinck, 2000). From this study, it was possible to conclude that the feeding of only winter veld hay had a significant detrimental effect on testicular development and semen quality in the young Boer goat buck. Such interactions between nutrition and reproduction have also been well described for Cashmere bucks receiving a lucerne-based diet, which had a frequency of seven LH pulses in 7 h in comparison with only two pulses in their counterparts fed a tropical grass with a low protein content (Walkden-Brown, 1991).

Age is also considered an important factor affecting the reproductive efficiency of bucks. Mature bucks of the Damascus breed were heavier, with a greater scrotal circumference, total number of sperm per ejaculate and sperm concentration when compared with yearling bucks (Al-Ghalban *et al.*, 2004). This is in line with previous reports (Osinowo *et al.*, 1988), showing that mature rams generally have higher ejaculate volumes, sperm concentrations and total number of sperm per ejaculate than younger rams. The percentage of abnormal sperm was lower in mature bucks compared with yearlings (Al-Ghalban *et al.*, 2004), although other workers (Osinowo *et al.*, 1988) found no significant differences in this trait between yearlings and mature rams.

Although the male goat has an exceptionally high libido, particularly during the breeding season, infertility is reported only in polled bucks, which are, in reality, intersex (hermaphroditism and pseudohermaphroditism). The presence of horns (as an indication of masculinity) has a significant effect on sperm concentration, total sperm per ejaculate and viable sperm concentration for all breeds. This statement is in agreement with other results showing that fertility improved when using horned rather than polled Damascus bucks during the breeding season (Hasan and Shaker, 1990). In addition, the number of live kids per litter was greater in the case of horned bucks. Polledness was also associated with an increasing incidence of intersexuality in goat kids (Hancock and Louca, 1975). Because of the genetic link between the locus for the presence of horns and the fertility characteristics of bucks (Devendra and Burns, 1983), it is suggested that replacement sires should preferably be horned, and all bucks from two polled parents should be excluded from the breeding flock.

### 5.3.3 Sexual behaviour in bucks

In the buck, the level of sexual activity fluctuates during the year in relation to the concentration of testosterone. However, such dependence is less accentuated than in other species and can be modulated by the social environment, such as the presence of other males and regular exposure to receptive females. Sexual behaviour of the buck has been reviewed extensively (Fabre-Nys, 2000) and we report here the major elements of this review. The first step of the buck sexual behaviour is illustrated, as in the ram, by the adoption of a posture where the head is stretched to the front with the ears in a down position. It is then followed by sniffing of the anogenital area of the female, and the buck urinates and displays the characteristic 'flehmen' reaction. If the doe shows signs of receptivity, then the buck moves on to a courtship behaviour, rotating his head in the direction of the female, emitting brief sounds at a low frequency and striking the doe with his foreleg.

The second sequence of steps starts by copulation attempts, ending with the following sequence of events: erection, mounting, intromission and ejaculation. In the buck, ejaculation follows the first intromission and usually lasts a few seconds. Following ejaculation, the buck usually turns to feeding if available.

One major factor explaining variation of sexual behaviour in the buck is the level of testosterone. In temperate zones, sexual behaviour is usually preceded by an increase in the concentration of testosterone approximately 6 weeks earlier (Ahmad and Noakes, 1995). The two parameters remain high during autumn and winter, after which testosterone concentrations start to decline. This is followed by a diminution of sexual behaviour a few weeks afterwards.

Other factors are also implicated in the control of sexual behaviour of the buck. The effect of age and previous experience are difficult to dissociate in the young buck. Only 40% of young bucks show an interest in females at the age of 12 weeks (Ahmad and Noakes, 1996). At 21 weeks of age, all bucks will mount the females and collection of semen becomes possible at the age of 24 weeks.

If bucks are raised exclusively in groups of males, they can develop homosexual behaviour at the adult age and a sexual inhibition in the presence of receptive females (Price and Smith, 1984). In the adult, sexual behaviour shows little improvement with age. Nevertheless, regular stimulation of bucks with receptive females maintains a high level of sexual activity, even outside the breeding season (Khaldi, 1984). The same author also reports that males kept near other males expressing sexual behaviour will have a shorter period of inactivation between two ejaculations and a higher frequency of ejaculations than males kept alone.

### 5.3.4 Photoperiodic treatment of bucks and semen biotechnology

Bucks kept under photoperiodic treatments, initially developed for rams, would suppress the problem of seasonality of reproduction, particularly sperm production. Accelerated alternation of treatments of increasing photoperiod (1–2 months of long days:16 h light:8 h dark) and decreasing photoperiod (1–2 months of short days: 16 h dark:8 h light) has been shown to abolish seasonal variations of testicular weight and sperm production (Chemineau *et al.*, 1999). Over a period of 3 years, all reproductive traits were improved for treated in comparison with untreated bucks maintained under a natural photoperiod (Delgadillo *et al.*, 1993). For photoperiodically treated bucks collected twice a week, sperm yield was improved by 61% and the total number of doses for artificial insemination over a 2-year period was increased by 62%. However, the fecundity of the bucks tested on artificially inseminated does was not improved by the treatment.

The increase of sperm output by photoperiod-treated bucks can be explained by improvement of the biological efficiency of spermatogenesis. Spermatogonia and spermatogenic divisions in treated bucks reach maximum yields similar to those usually observed during the breeding season (Delgadillo *et al.*, 1995).

A specific problem during conservation of buck sperm in artificial insemination centres is the deleterious effect of the seminal plasma on the viability of spermatozoa after freeze-thawing when milk- or egg yolk-based diluents are used. There is an enzyme in the seminal plasma that reduces survival of the spermatozoa *in vitro* (Corteel, 1975; Memon *et al.*, 1985). The protein that is incriminated in these deleterious effects is found in the fraction of the seminal plasma that is produced by the bulbo-urethral glands. It is a glycoprotein of 55–60 kDa (named BUSgp60) and causes a reduction in mobile spermatozoa, rupture of the acrosome and death of spermatozoa when the semen is diluted in skimmed milk (Pellicer-Rubio *et al.*, 2008). So far, a washing step during which the seminal plasma is removed prior to sperm conservation is necessary before cryopreservation of the bucks' semen. More recently, commercial extenders with no biological components have been developed to improve sanitary safety in semen processing (Gil *et al.*, 2003). Techniques of sperm-cell cryopreservation and fresh semen production require further studies in order to increase the efficiency of semen production in artificial insemination centres.

### 5.3.5 Managing bucks for reproduction

Bucks must be in excellent condition at mating, otherwise mating performance and subsequent kidding performance could be depressed. Bucks must be capable of serving at least ten does each day when introduced during the second oestrous cycle of the breeding season. Spermatogenesis is susceptible to outside influences such as elevated temperature, the season of the year and nutrition, and breeding males need to be evaluated for reproductive soundness 3–4 weeks prior to the mating season (Memon *et al.*, 2007), while other actions are required several months before the breeding season:

- Sufficient feed.
- Fresh, clean water.
- Adequate shade during summer and in the time leading up to mating. If this is not done, semen production will be reduced and sperm numbers may be inadequate for fertilization of more than a few does.
- Regularly trimming of bucks' feet to make sure they are kept in good order. Bucks place most of their body weight on their hind legs and feet during mounting. If they are in pain from bad feet, they will refuse to mate.

Examination (palpation) of the external genitalia (scrotum and scrotal content, sheath and penis) for signs of infections and other abnormalities is also an important step of the breeding soundness examination (Memon *et al.*, 2007). The testes should be firm and spongy and free of lumps. A male with large symmetrical testes, free of defects, is likely to produce semen of good quality. There are currently no age and breed standards for scrotal circumference in meat-type breeds, and there is a need for guidelines to be developed. Alternatively, examination involves the collection and evaluation of an ejaculate. In trained bucks, this is achieved using an artificial vagina, but in most instances an electroejaculator or battery-operated ejaculator is used (Memon *et al.*, 1986). The ejaculate scored is immediately assessed for sperm motility using a light microscope and a pre-warmed slide. It is also possible to assess libido by allowing the buck access to an oestrous doe as a teaser. Animals deficient in any part of the examination should be considered questionable and retested after several weeks.

For young bucks, it is important to ensure that both testicles have descended. The breeding potential and value of a buck are considerably reduced if it has only one descended testicle. This abnormality (cryptorchidism) can be passed to the next generation (Mickelsen and Memon, 2007). Young, inexperienced bucks need extra management to ensure that they get on with mating. Young bucks should be confined in yards or small secure paddocks with mature does in oestrus. When the young bucks show interest in mating and staying with the flock, they can be put into larger paddocks used for mating.

The best method of checking if bucks are working is to use mating crayons. Bucks are fitted with a harness that holds a coloured crayon over the sternum. When a buck mates, a coloured mark is left on the rump of the doe. It is essential that the harness is correctly fitted. Records can be kept of when does are mated for rational management of kidding. By changing the colours of the crayons every 21 days, it is easy to determine when does have conceived.

## 5.4 Manipulating Reproduction for Increased Meat Production

### 5.4.1 Induction of puberty

Inducing early puberty in young female goats is one way for intensive meat production. Initiation of early reproductive life has been studied extensively in sheep, but little information is available in goats. We summarize below the techniques that have proved to be promising for early induction of puberty in goats.

Social interactions can play an important role in modifying expression of some reproductive traits in both sexes. There is evidence that the presence of the buck may modify the age of puberty in the goat (Greyling, 1996). In an early study (Amoah and Bryant, 1984), it was suggested that contact with the male goat has an effect on the timing of puberty and is associated with rapid and highly synchronous attainment of puberty in the majority of kids. Similarly, on the basis of the oestrous response, it was shown that the permanent, rather than the intermittent, presence of bucks had a marked beneficial effect on the number of young kids exhibiting oestrus (Greyling, 1996).

Exogenously administered melatonin from continuous slow-release implants has been shown to advance the onset of the breeding season in sheep and goats by mimicking the stimulatory effect of short days. It is well known that melatonin is implicated in the sequence of events leading to the onset of puberty in small ruminants, but there are very few reports on the use of exogenous melatonin to advance the onset of puberty. Melatonin implants administered during the last month of spring in autumn-born female Damascus goats advanced their breeding season by about 11–12 weeks when joined with young males also implanted with melatonin (Papachristoforou *et al.*, 2007). In the anticipation that such improvement would not affect later performances with regard to the earlier breeding of Damascus female kids, it was found to have positive effects on their lifetime performance (Mavrogenis and Constantinou, 1983).

Insufficient dietary energy/protein intake before puberty retards growth and delays puberty in livestock. Energy restriction prevents or slows the maturation process at the hypothalamus–pituitary level, and ovarian steroidogenesis may be compromised in energy-/protein-restricted animals. In a study designed to examine the effects of dietary supplementation with maize and cottonseed cake given at different ratios on the age and weight at first oestrus in nulliparous Savannah Brown young does (Fasanya *et al.*, 1992), it was demonstrated that supplemented animals were younger and heavier at the first pubertal oestrus. Dietary supplementation allowed animals to undergo fairly rapid growth, hence attaining a desirable size and weight at an early age.

### 5.4.2 Out-of-season breeding

For seasonal breeds, out-of-season breeding and resumption of post-partum anoestrus are necessary to intensify the rhythms of kidding and to increase meat output per doe per year. In this section, we will address the techniques that are used most in diverse production systems in order to induce breeding in anoestrous does. Two main techniques will be discussed: the use of melatonin and the buck effect. Other pharmacological means to induce breeding in anoestrous does such as the association between progestogens and exogenous gonadotropins (such as eCG) will be reported in the section dedicated to the discussion on oestrous induction and synchronization.

#### *Use of exogenous melatonin*

An alternative pharmacological means of modifying the seasonal breeding patterns is through manipulation of the melatonin signal. Treatments with exogenous melatonin in sheep and goats have increased the duration of melatonin elevation in spring and summer, which caused an increase in gonadotropin secretion and an early onset of the breeding season, as well as decreased prolactin secretion (Lincoln and Clarke, 1997).

Exogenous melatonin is administered continually to supplement endogenous release and thus mimic the 'short days' associated with the onset of breeding season in autumn, while animals' eyes perceive long days of spring and summer (Chemineau *et al.*, 1988). Giving (orally or by intramuscular injection) a large amount of melatonin during the early afternoon in June to animals artificially exposed to long days induces ovulation in goats (Chemineau *et al.*, 1986). The mechanism that can be put forward to explain the action of melatonin is that it provides a short-day signal (O'Callaghan *et al.*, 1989). Short-day treatment can therefore be replaced by melatonin treatments so that the maintenance of constantly high levels of melatonin mimics a short-day effect.

Melatonin can be supplied either as an orally active compound, by injection or as a subcutaneous implant for about 3 months (several commercial products are available in many countries), all of which have been shown to be similarly effective to increase the pregnancy rate in Spanish does (Wuliji *et al.*, 2003). A prerequisite for the advancement of the breeding season through melatonin treatment is for animals to have experienced a sufficient period of long days. Two months of photoperiod treatment, followed by melatonin treatment (daily injection or drenching or subcutaneous implants) allows cycles with ovulation to be maintained for a few months. In the seasonal breeds of sheep and goats, originating from northern Europe, melatonin alone cannot be used too early in the season and is only able to advance the natural breeding season by about 1.5 months (English *et al.*, 1986). Conversely, in Mediterranean breeds of sheep and goats, which are normally mated in the spring, melatonin can be used alone without previous light treatment.

The results indicate that melatonin treatment, in comparison with untreated animals, improved the kidding rate (71 versus 36%) and the percentage of kids born (180 versus 160%) (Wuliji *et al.*, 2003). Furthermore, the ability of melatonin treatment to interrupt seasonal or post-partum anoestrus implies that an accelerated out-of-season breeding system with goats, scheduling kidding twice, in both the autumn and spring, is feasible. Such a system should increase the total annual meat-goat production. However, it must be remembered that the type of treatment used and likely rate of success will depend greatly on the extent of breed susceptibility to photoperiod, time of the treatment application in relation to onset of the breeding season year (depth of anoestrus is greatest in late spring/early summer), physiological state, body-condition score and nutritional status.

### The buck effect

The existence of a two-peak abnormal distribution of lambing and kidding 5 months after the reintroduction of males in sheep and goat flocks was described very early in the literature. In the 19th century, this technique was presented as 'being able to fertilize all adult ewes of the flock in the shortest time possible' (Girard, 1813). In Angora goats, careful description of the distribution of lambing and kidding induced by the voluntary reintroduction of males was given (Shelton, 1960). The existence of two peaks of kidding, clearly separated by some days, suggested that the underlying physiological mechanisms were probably not so simple.

They suggested that this reintroduction probably provoked induction of synchronous ovulations and oestrous behaviour, being able to induce such synchronizations of parturitions. This phenomenon has attracted attention during recent years and the reasons for this distribution have been investigated.

Many authors have separated the short-, medium- and long-term responses to the male effect to allow a rationalization in the understanding of female responses (reviews in goats by Chemineau, 1987 and Walkden-Brown *et al.*, 1999).

Immediately after introduction of males (day 0), LH pulse frequency increases and remains elevated if the male remains present among females in the flock. The gonadotropin stimulation of the ovarian follicles provokes an increase in plasma oestradiol 17β, which centrally triggers the onset of the pre-ovulatory surge of LH, around 20 h after day 0, and females ovulate before day 3 after the introduction of males. After the first male-induced ovulation, in one group of females, the corpora lutea develop and secrete progesterone during normal duration, leading to a second ovulation around day 23 in goats. The second group of females experience a very early luteolysis, after only 1.5 days (i.e. days 4–5). After this short cycle of highly constant duration (5–6 days), these females reovulate a second time, around day 29. If does continue to cycle, subsequent ovulations generally occur at the normal interval of approximately 21 days and are accompanied by fertile oestrus. The induced ovulations are associated with oestrus in a variable proportion of goats: in 68% of Creole goats (Chemineau, 1983) and in 35% of native goats in Tunisia (Lassoued *et al.*, 1995).

The first ovulation is usually of low fertility and the second ovulation 5 days later is accompanied by a fertile oestrus with a luteal phase of normal length. The ovulation rate may be enhanced at this second ovulation (Chemineau, 1987; Lassoued *et al.*, 1995).

Injecting adequate doses of progesterone (i.e. 20 mg per doe) at exactly the same time as the introduction of males provokes a delay in the induced ovulations and prevents short cycles in 85% of females (Lassoued *et al.*, 1995). Similarly, progestogen pre-treatment through vaginal sponges is efficient to control short ovarian cycles in 96% of Creole meat goats (Chemineau, 1985).

A global explanation of the underlying physiological mechanisms controlling these short cycles points to the uterus being involved in the early regression of the corpus luteum, induced by the bucks through the secretion of PGF2α (Chemineau *et al.*, 2000). However, it appears that the quality of the follicles induced to ovulate are poor because of the unsustained long-term gonadotropin activity during anoestrus.

In goats, the reproductive condition in the buck seems to be the limiting factor determining the response of anoestrous does to the male effect (Flores *et al.*, 2000). A seasonal decline in the intensity of the inductive stimulus from the male may also be involved. Treating bucks with long days and melatonin increased their teasing capacity to induce sexual activity in females during anoestrus, which indicated that the absence of response to teasing at this time of the year is not due to female unresponsiveness but to insufficient stimulation from the male.

Recent published results (Luna-Orozco *et al.*, 2008) demonstrated that parity of female goats does not influence their oestrous and ovulatory responses to the male effect. The same results indicated that, regardless of parity, female goats respond to male introduction if they are stimulated by males that were previously exposed to artificial long days to increase their sexual behaviour.

A model of factors influencing the male effect was developed and presented as one component of a complex cycle of social interaction between sexes, markedly influenced by environmental, social and physiological factors. Work on Australian Cashmere goats indicated that improving buck nutrition, selecting bucks for maximum libido and exposing bucks to oestrous does prior to or during joining are all likely to enhance the ovulatory response to bucks by advancing and synchronizing it (Walkden-Brown *et al.*, 1993b).

Exposure to buck fleece alone may induce an ovulatory response in seasonally anovulatory does. However, this response is attenuated in comparison with that induced by bucks, with fewer does ovulating, and fewer ovulating does going on to reovulate. The response is not enhanced by the addition of buck urine. The intensity and duration of exposure to buck stimuli appeared to influence not only the proportion of does ovulating but also the timing and persistence of the ovulatory response (Walkden-Brown *et al.*, 1993a). Nevertheless, there are reports of the successful induction of ovulation in seasonally anovulatory does following short-term intermittent exposure to male fleece odours.

Suppression of the sense of smell by irrigation of nasal mucosa with zinc sulfate solution markedly modified the ovulatory and oestrous responses of female goats to the introduction of males. The proportions of responding females were reduced by ~50% but were not completely suppressed, probably because the females detected the males using other senses (Chemineau *et al.*, 1986). Overall, the data suggest that the male effect in these goats is not a simple reflex response to olfactory cues but rather a complex response involving the integration of a range of exteroceptive stimuli from the buck.

In addition to the 'male effect' in which sexually active males induce pulses of GnRH, female–female effects have been demonstrated in goats (Restall *et al.*, 1995). It was clearly established that oestrous females can induce ovulation in anovulatory Australian Cashmere goats. The oestrous state is essential for the phenomenon, as the presence of non-oestrous females had no effect. The nature of the induced ovulatory response is similar to that following exposure to males. Ewe–ewe interactions play no part in timing seasonal transitions in reproductive activity.

The proportion of initial ovulation with oestrus and of initial ovulation followed by a short luteal phase was shown to vary with 'depth of anoestrus' as reflected by the percentage of anovular goats in the group at the time of stimulation (Chemineau, 1983). In less seasonal breeds, the depth of anoestrus as defined by the percentage of anovulatory females before introduction of males is deep when >50% of Creole goats are in anoestrous and shallow when <50% of the goats are anoestrus (Chemineau, 1983). The mean interval (days) between teasing and ovulation seems to be related to the proportion of non-cycling females at the time of introduction of males. In the case of Creole goats, it averaged 3.3 days when the proportion was >50% and only 1.8 days when it was <50% (Chemineau, 1983).

### 5.4.3 Hormonal manipulation of the oestrous cycle

This section will be dedicated exclusively to synchronization protocols leading up to a grouping of oestrus and ovulation over a short time (days) using exogenous hormones. Other means for synchronization of oestrus over longer intervals (weeks) include melatonin implants or the buck effect, which are discussed above.

Oestrus synchronization in livestock focuses on the manipulation of either the luteal or the follicular phase of the oestrous cycle. In does and ewes, the opportunity for control of the cycle is greater during the luteal phase, which is of longer duration and is more responsive to manipulation. Strategies can be employed either to extend the lifespan of the corpus luteum by administration of exogenous progesterone or to shorten this phase through premature regression of the existing corpus luteum using PGF2α.

Oestrus synchronization allows for concentration of breeding time and parturition at suitable seasons to take advantage of the following:

- The use of reproductive biotechnologies for the dissemination of genetic improvement (e.g. artificial insemination, embryo transfer). With oestrus synchronization, producers are able to use complementary techniques more efficiently for reproductive management, including artificial insemination and embryo transfer, so that genetic material is more easily obtained or transferred domestically and internationally. Good examples can be given for European dairy goat breeds where synchronization of oestrus has been used extensively (up to 10% of the French goat population) to develop fixed-time artificial insemination with improved bucks while obtaining reasonably high pregnancy rates.
- Oestrus induction of out-of-season breeding.
- Improvement of flock management, taking advantage of niche markets, feed supplies, labour and rising price trends.

Since the initial introduction of techniques based on the use of progestogens to synchronize the reproductive cycle and eCG to stimulate ovarian follicular growth in ruminants, reproductive technologies applied to goats have not advanced greatly. In goats, the treatments of choice are still those based on administering intravaginal progestogens for 11 days, followed by an injection of eCG and PGF2α, or their analogues, 2 days before withdrawal of the progestogen-releasing device (Lopez-Sebastian *et al.*, 2007). According to different authors, mean fertility results obtained using this protocol combined with the use of semen refrigerated at 5°C and cervical artificial insemination performed 43–46 h after progestogen treatment can be variable. Under the best experimental conditions, mean fertility rates approaching 60% have been achieved (Leboeuf *et al.*, 1998). One major advantage of the treatment is its potential use irrespective of seasonal effects (breeding or anoestrus season). For example, when inducing oestrus during the non-breeding season, an 18-day sponge treatment in combination with 150 or 200 IU of eCG at sponge withdrawal allowed 100% of indigenous Damascus does to be mated, with a 65.8% conception rate, a 64.1% kidding rate and a 192.2% kid crop compared with buck use alone (no animals expressed oestrus or were mated; Zarkawi *et al.*, 1999).

The most commonly used synchronization protocols include methyl acetoxy progesterone (MAP; 60 mg) or fluorogestone acetate (FGA; 45 mg) vaginal sponges. A controlled internal drug-releasing device (CIDR) in the form of a silicone intravaginal progesterone insert is also available for use in goats. Comparison of the use of MAP, FGA and CIDR vaginal inserts in Boer and South African indigenous goats during the breeding season showed no influence of progestogen treatment on oestrous response, but time to the onset of oestrus was advanced by 3–5 h in the CIDR group compared with the FGA and MAP groups (Montlomelo *et al.*, 2002). These results are confirmed by those in Nubian goats, indicating that the use of CIDR, MAP and FGA treatments plus PGF2α following progestogen withdrawal is equally efficient in synchronizing oestrus, with similar fertility between treatments (Romano, 2004). The need for an intravaginal hormone-releasing device gives rise to problems such as vaginitis (Lopez-Sebastian *et al.*, 2007), which has adverse effects on fertility as a result of impaired sperm transport and viability. Moreover, the use of intravaginal progestogens has negative implications for the legal limits established for progestogen residues in milk and meat.

Besides these shortcomings related to the use of intravaginal devices to deliver progestogens, the development of anti-eCG antibodies in repeatedly treated does notably reduces fertility, particularly in artificially inseminated females (Chemineau *et al.*, 1999). Repeated use of eCG delays the timing of oestrus as a result of a delayed pre-ovulatory LH surge with a major drawback on fertility (Fig. 5.5). These limitations have prompted the development of alternative methods of synchronizing oestrus in female goats.

Recently, a new concept of synchronizing oestrus was developed, based on inducing ovulation by the male effect in progesterone-treated goats (injection of a single dose in oil) and provoking early corpus luteum lysis using cloprostenol (a prostaglandin analogue) during the non-breeding season 9 days after exposure to bucks (Lopez-Sebastian *et al.*, 2007). The proposed method was an adequate alternative for oestrus synchronization prior to artificial insemination during the non-breeding season in the absence of previous oestrus detection in goats. The outcome of the protocol provided higher fertility rates than those observed in response to the classic method based on the use of progestogens and eCG.

Another convenient method to administer progestogens without having the adverse effects of intravaginal devices is

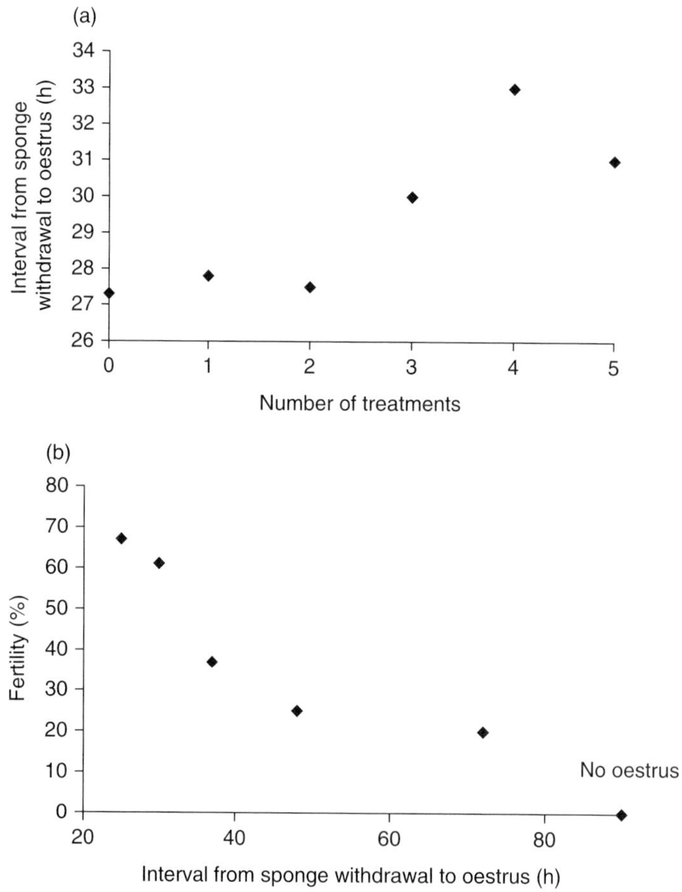


**Fig. 5.5.** Relationship between the interval of withdrawal of sponges to oestrus and the number of equine chorionic gonadotropin treatments (a) and between the same interval and fertility of the does (b) (redrawn from Chemineau *et al.*, 1999).

oral dosing in feeds. The use of melengestrol acetate has been documented but not as many studies have determined its potential value in goats.

During the breeding season, early studies with a few animals have shown that PGF2α and its analogues (as with cattle and sheep) can be used for synchronization in cycling females. In Black Bengal goats ($n$ = 6 or 8 per treatment), a comparison was made between 15 mg PGF2α injections given intramuscularly 11 days apart with treatments that included progesterone injections for 16 days with or without eCG (Ishwar and Pandey, 1992). Overall, PGF2α-treated does showed maximum fertility compared with other treatments, with oestrus and ovulation in four out of six and kidding (per doe mated) in three out of four treated does. In Boer goats, it was shown that two injections of cloprostenol at 62.5, 125 or 250 μg, administered 14 days apart, were effective in synchronizing does during the breeding season (Greyling and van Niekerk, 1986). Although the highest dose given (250 μg) apparently increased the percentage of does in oestrus (100%, compared with 87.5% for 125 μg and 93.8% for 62.5 μg), overall fertility appeared to decrease with increasing dose, with seemingly lower conception rates and numbers born per doe bred. In addition, Dwarf goats under different environmental/nutritive conditions responded positively to two injections of a PGF2α analogue 10 days apart (Khanum *et al.*, 2006). The treatment resulted in regression of the corpus luteum within 48–56 h, and in 19/20 animals oestrus signs appeared 56–72 h after the second injection.

Another breakthrough in the development of oestrus synchronization in goats during the breeding season is the method known as 'Ovsynch', which is derived from cows and generates control of the follicular waves as well as of the lifespan of the corpus luteum (Holtz *et al.*, 2008). The method includes a GnRH–PGF2α–GnRH treatment sequence. When the females are cycling with a large follicle present at the time of injection, the first GnRH injection triggers ovulation and formation of an accessory corpus luteum. At the same time, it reprogrammes follicle development by initiating the emergence of a new follicular wave. To get full control of ovarian function, the Ovsynch treatment includes a prostaglandin injection 7 days after the first GnRH treatment to time luteal regression (natural and accessory corpus luteum) in all females. Ovulation is timed by the second GnRH administration. Fertility and prolificacy following the Ovsynch synchronization scheme compared favourably with the 'classical' sponge impregnated with progestogens and eCG treatment, provided the animals are cycling (i.e. in season).

### 5.4.4 Improvement of litter size

Like other litter-bearing animals, litter size has a major impact on the reproductive efficiency of goats. This trait shows large variations between breeds and production environments, which make changes in management a primary cause to alter litter size in meat-producing goats. It is important to note that the scientific literature related to improvements of litter size in sheep is much more important than for goats. This can be explained by the higher natural prolificacy of goat breeds in comparison with sheep in most parts of the world and also by the presence of goats in more marginal production systems, making improvement of litter size a risky step in the management of the flock.

With regard to breed differences, we shall give only a few examples of the relationship between litter size and breed. There is a growing body of literature in different environments which indicates that Nubian goats present a higher frequency of multiple births than goats of European origin (Mellado *et al.*, 2006). When studying the reproductive performance and pre-weaning growth in the West African Dwarf goat with respect to variations in coat colour, it was reported that black or chocolate-brown goats had the highest prolificacy of 210 ± 24, while white goats had the lowest figure of 162 ± 31 (Ebozoje and Ikeobi, 1998). Goats with other colours had intermediate levels of prolificacy. The authors concluded that coat colour plays an important role in the adaptation and survival of the West African Dwarf goat breed. Selection for breed identification mark on the basis of coat colour would probably favour the black goats based on the results of their performance. Although it is not immediately known why performance increased with an increase in coat pigmentation intensity, it could be associated with the suppressive action of the non-pigmented (white) gene. The top dominant allele at the agouti locus (*Awh*), which produces white coat colour, depressed ewe fertility in Icelandic sheep by about 0.15 lambs per ewe mated (Adalsteinsson, 1975). A similar gene action could have been responsible for the differences in performance observed in the West African Dwarf goat. The production environment is also reported to affect the litter size of goats. Goats mated in the autumn were almost half as likely to present multiple births as does mated in the hottest part of the year (summer) (Mellado *et al.*, 2006). Other studies in hot environments (reported by Mellado *et al.*, 2006) also report significantly larger litters in goats mated in the spring or summer compared with cooler seasons of the year. However, no clear explanation of these results can be put forward at this stage. For Korean native goats (Song *et al.*, 2006), mean litter sizes were 1.69 ± 0.03 and 1.78 ± 0.16 at birth and 1.31 ± 0.03 (77.5%) and 1.52 ± 0.17 (85.4%) at weaning (which could passively be seen as an indication of mothering ability) for range and intensively managed groups, respectively. For Creole goats in Guadeloupe, a significant correlation was found between prolificacy and rainfall 1 month before conception (Chemineau and Xandé, 1982). Prolificacy depends on the feeding conditions during mating, which is a consequence of the relationship between natural feed availability, body condition and ovulation rate of the females. Indeed, ovulation rate was higher ($P < 0.01$) in does with a better body condition (1.9 ± 0.1) than in those with a worse body condition (1.6 ± 0.1) (de Santiago-Miramontes *et al.*, 2008). A higher ovulation rate in ewes with a better body condition is well documented for a wide range of breeds and under various production systems.

For local goats browsing in the harsh conditions of central Tunisia, condensed tannins in *Acacia cyanophylla* Lindl. had a detrimental effect on ovulation rate (Lassoued *et al.*, 2006). Deactivation of these condensed tannins by polyethylene glycol increased the ovulation rate to 1.76 ± 0.60 in comparison with 1.25 ± 0.45 for untreated goats. The probable increased availability of proteins in the polyethylene glycol-receiving goats could explain their higher ovulation rate. The prospects for nutritional manipulation of litter size through alteration of ovulation rate and embryo survival, similar to what is currently established in sheep, should be investigated further for different genotypes under different production systems.

The effect of the social environment of the doe at mating on later litter size remains equivocal. There was no beneficial effect of buck stimulation prior to the breeding period in terms of kidding rate or litter size, either in goats exposed to bucks immediately before the breeding period (litter size of 1.30) or goats teased by bucks 15 days prior to the breeding season (litter size of 1.40) (Mellado *et al.*, 1994).

Litter size can also be manipulated by pharmacological means. Increases in litter size have been achieved through the immunization of does to steroids. Steroid immunization has become commercially available in many countries, for example, using Fecundin®, which immunizes females to androstenedione. Immunization (active or passive) is achieved by two subcutaneous injections (2 ml each) administered initially 2–3 weeks apart and in single annual boosters thereafter. A period of 3 weeks is suggested between the booster immunization and the time of optimum ovulation. Due to the long-term effects and the relative ease of application of the product, steroid immunization can be used for the improvement of ovulation rate and subsequently litter size in more extensively managed flocks. The animal response in terms of ovulation rate and litter size varies with breed and location, but improvements of ovulation rate (+1) and litter size (+0.5) have been achieved in does. Immunization against androstenedione increases ovulation rate and does not affect embryo losses. Therefore, prolificacy is always increased and the improvement is higher in naturally lowly prolific breeds. For the hardy Greek dairy breed, litter size at birth was 1.25 ± 0.43 in comparison with 1.63 ± 0.64 for untreated and immunized goats, respectively (Driancourt *et al.*, 1990). Most treated goats tend to have twin pregnancies.

When using hormone therapies to increase litter size, the association between sponges impregnated with a synthetic progestogen followed by the injection of eCG can also yield increases in litter size. Such treatment is classic in sheep and is perhaps the most widely used for both synchronization of oestrus and increases in litter size. There are limited published data on the use of eCG in goats to increase litter size. Our unpublished results (N. Lassoued and M. Rekik, 2010, Fig. 5.6) on Tunisian local breeds show that, at doses of 200 and 300 IU of eCG, there is an improvement in conception rate at the induced oestrus. However, prolificacy was similar to the does not receiving eCG after withdrawal of the sponges. A substantial increase in prolificacy is obtained only when the eCG dose is increased to 400 IU. However, at this dose, fertility at the induced oestrus is very low because of a high incidence of reproductive wastage (unfertilized ova and early embryo losses).

A number of other pharmacological treatments to manipulate litter size in goats are under investigation and development under research conditions. However, it is not clear to what extent these approaches will be biologically and/or economically feasible. Among the concepts under investigation are: (i) immunization against inhibin, which selectively suppresses FSH but not LH; and (ii) the use of GnRH in conjunction with progestogen-based superovulation treatments. Both techniques have the potential to be used in superovulation treatments as part of embryo transfer programmes. However, their use to increase litter size under farming conditions is precluded.

### 5.4.5 Reproductive biotechnologies for improved meat production

The application of assisted reproductive technologies (ART) in livestock production allows animals of high genetic merit to produce more offspring than would be possible by natural breeding. ART present producers of breeding stock with unique opportunities

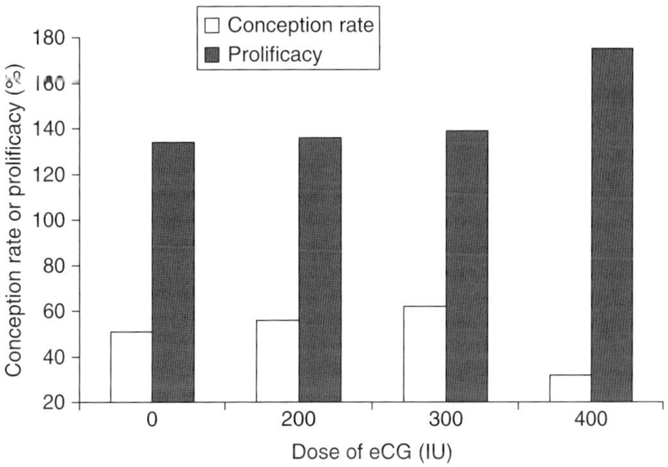


**Fig. 5.6.** The effect of increasing doses of equine chorionic gonadotropin on conception rate and prolificacy at the induced oestrus of local Tunisian female goats (N. Lassoued and M. Rekik, 2010, unpublished results).

to move their germplasm around with relative ease. The costs involved in ART are most likely prohibitive for producers of goats that are marketed for meat. In the fledgling meat-goat industry, the recent introduction of the Boer goat in several countries is an excellent example of the need to apply ART for the dissemination of stock. As other superior meat-producing germplasm is identified, the application of artificial insemination and embryo transfer is likely to rise in the area of meat-goat production.

### *Artificial insemination*

Artificial insemination can contribute to the optimization of the selection schemes of the main seasonal meat breeds and is considered a powerful tool to control kidding dates in order to adapt production to market demands, especially during the non-breeding season in the goat. In addition, proper selection of young bucks and appropriate healthcare of males, as well as utilizing artificial insemination, may control some infectious diseases. However, at present, genetics is the main justification for using artificial insemination (and frozen semen) because of its formidable ability to produce many offspring per male in multiple environments over time. This ability is necessary to create and diffuse genetic gains and facilitates the application of recent molecular genetics techniques in selection programmes (Leboeuf *et al.*, 1998).

Three methods of semen preservation (fresh, refrigerated and frozen/thawed) and two techniques of insemination (cervical and intrauterine) are used worldwide in goats (Evans and Maxwell, 1987; Chemineau and Cognié, 1991; Amoah and Gelaye, 1997; Leboeuf *et al.*, 2000).

Does can be inseminated with fresh and extended, chilled semen stored for up to 48 h. Fresh semen is the preferred method when the bucks are collected on the farm, especially during the breeding season when semen production and quality are at their peak. The use of refrigerated semen is a common strategy in circumstances where a particular male is shared by a group of farmers located within a relatively small area. In such cases, the semen is stored at ~4°C and can be used for up to 24 h after collection. However, for most practical purposes, semen originates frozen from outside the farm, having a long-term preservation allowing it to be marketed over a wide area and used throughout the year, and conserving the genetic material in case the animal dies.

Transcervical insemination involves deposition of semen in the body of the uterus. The conception rate using this method ranges from 50 to 70% depending on the season of insemination. The conception rate is low during spring and summer, due to lower sperm motility, than in the autumn and winter (Table 5.4; Tuli and Holtz, 1995). However, photoperiod treatment of bucks (Delgadillo *et al.*, 1995), which enables quality sperm collection all year round, may alleviate such seasonal variation in sperm quality. In their review of the effectiveness of artificial insemination

**Table 5.4.** Pre- and post-freezing percentage of progressive motility and percentage of live spermatozoa in Boer goat semen at different seasons of the year (from Tuli and Holtz, 1995).

| | | Progressive spermatozoa motility (%) | | | Live spermatozoa (%) | | |
|---|---|---|---|---|---|---|---|
| Season | No. doses | Pre-freezing ± SEM | Post-freezing ± SEM | Freezing loss (%) | Pre-freezing ± SEM | Post-freezing ± SEM | Freezing loss (%) |
| Spring | 35 | 62b ± 2 | 29b ± 2 | 54 | 61b ± 2 | 32b ± 3 | 47 |
| Summer | 45 | 65b ± 1 | 35b ± 2 | 46 | 71c ± 2 | 40c ± 3 | 44 |
| Autumn | 60 | 71d ± 2 | 39c ± 2 | 45 | 75cd ± 3 | 44c ± 2 | 41 |
| Winter | 37 | 73d ± 2 | 45d ± 3 | 38 | 76d ± 3 | 49d ± 3 | 36 |

Within columns, different letters (b, c, d) indicate significant differences ($P < 0.05$).

in dairy breeds in France (Leboeuf *et al.*, 1998), the authors enumerated some of the factors affecting variation in fertility after artificial insemination:

- Ovarian response to hormonal treatment: 4.3% of the females had a low progesterone concentration, i.e. a probable lack of ovulation following the synchronization treatment.
- Ovulation time after hormonal treatment: variability of the ovulation time seems to be a limiting factor for the efficiency of hormonal treatment plus artificial insemination.
- Repeated hormonal treatments and antibodies against eCG: this was discussed in an earlier section.
- Parity of inseminated females: artificial insemination of nulliparous does often yields lower fertility rates than the adult goats. These low fertility rates could be caused by factors such as weaning age, growth rate, body condition and age at first artificial insemination.

Laparoscopic insemination involves the use of a laparoscope and depositing fresh or frozen-thawed semen directly into the uterine horns. Laparoscopic insemination procedures are described for sheep and goats (Ritar and Ball, 1991) and a >80% conception rate has been reported. However, the technique requires a skilled operator. Due to the high cost of the procedure, laparoscopic insemination has a very limited use in meat goats.

Semen is usually collected from bucks trained to serve an artificial vagina. Once a collection schedule is initiated, bucks can be collected two to three times daily on alternate days. Semen is immediately evaluated for quality and the concentration determined. The semen is then extended in a medium containing egg yolk, sugars and buffer to provide an insemination dose of 20 million (frozen, laparoscopic intrauterine) to 300 million (fresh, cervical) spermatozoa, depending on the intended insemination technique. The success of the actual insemination depends to a large degree on appropriate timing in relation to oestrus and ovulation. Does must be observed closely for the onset of oestrus, or can be synchronized (see oestrus synchronization discussed above) and should be inseminated 12–18 h after the onset of oestrus.

### *Embryo transfer*

The first successful cryopreservation of goat embryos was reported by Bilton and Moore (1976). Since then, large numbers of goats have been successfully produced from frozen-thawed embryos (Baril *et al.*, 1989). The technique of multiple ovulations and embryo transfer (MOET) is often referred to as a method of producing more offspring from a genetically valuable female than would be possible by natural breeding. However, MOET has not yet become a widespread tool for genetic improvement for a variety of reasons including its cost, technical demands and variable and unpredictable efficiency (Baril *et al.*, 1989; Cognié *et al.*, 2003). The main factors contributing to the unpredictability of this technique are the variability of the superovulatory response, poor fertilization associated with high ovulatory responses and early regression of the corpus luteum (Cognié *et al.*, 2003). An average of six to eight transferable embryos per donor can be produced in a successful goat MOET programme (Baril *et al.*, 1989; Cognié *et al.*, 2003). It is common for the number of transferable embryos to range from 0 to 30 per donor, with up to 30% of the donors failing to produce any transferable embryos due to fertilization failure and early regression of corpus luteum (ERCL) (Pintado *et al.*, 1998). The causes of ERCL are not fully understood, but its occurrence has been associated with inadequate nutrition (Jabbour *et al.*, 1991) and the use of eCG in superovulatory regimes (Pintado *et al.*, 1998).

Techniques used for oestrus synchronization of donor and recipient and for superovulation of the donor with gonadotropins (FSH and eCG) are similar to those described above (see oestrus synchronization discussed above). Insemination of the donor does should occur either naturally or through cervical rather than intrauterine artificial insemination, to avoid additional manipulation of the uterus and oviducts. For the collection, the uterus of the donor is flushed 3–5 days after mating. Particularly in the case of repeated collections, this may cause adhesions interfering with subsequent collections. Recent embryo collection techniques using laparoscopy have been developed with successful outcome in goats (76% pregnancy). Following collection, the flushing medium is examined to identify fertilized (cleaved) embryos, determine the recovery of embryos (based on the number of corpora lutea) and evaluate embryo quality. Only high-grade embryos should be used for freezing and storage, whereas embryos of less quality may be used for fresh transfer. Embryos should be transferred into the uterine horn of the same side containing an ovary with a corpus luteum. Following a sufficient period of rest, donor does can be repeatedly collected.

It has been reported that blastocysts can be cryopreserved better than morulae (Puls-Kleingeld *et al.*, 1992). Goat embryos are successfully traded internationally using ethylene glycol, which is a better cryoprotectant than glycerol (35 versus 22% kids born from embryos thawed, respectively; Le Gal *et al.*, 1993). Vitrification of goat embryos involving embryo exposure to high concentrations of cryoprotectants followed by direct immersion in liquid nitrogen has been reported to be successful (Yuswiati and Holtz, 1990).

Superovulation is an important part of a MOET programme and has the potential to increase the reproductive performance of selected donors of goat breeds in high demand. Superovulation accomplished by gonadotropins (primarily FSH and eCG), used at higher (pharmacological) doses to elicit a superovulatory response, is commonly used in embryo transfer programmes. eCG is more easily administered than FSH, usually as a single injection of up to 1500–2000 IU, but the superovulatory response to eCG can be quite variable and is usually lower than in an FSH-induced superovulation. Currently, the major factor leading to a variable ovulation rate and embryo output seems to be the follicular status of the donor at the onset of superovulatory treatment (Gonzalez-Bulnes *et al.*, 2004). Several strategies have been suggested for increasing the number of small recruitable ovarian follicles at the time of FSH treatment, while avoiding the presence of large (dominant) follicles. Among these concepts are: (i) immunization against inhibin, which selectively suppresses FSH but not LH: (ii) the use of GnRH agonist/antagonists and the administration of FSH shortly after an induced oestrus/ovulation; and (iii) the use of FSH plus GnRH treatment. Immunization of goats against inhibin has proved to be a practicable means of producing embryos for transfer purposes (Wang *et al.*, 2009). Pre-treatment with GnRH antagonist for 10 days prior to superovulation resulted in an increased number of small follicles at the time of FSH administration and an increased number of ovulations (Cognié *et al.*, 2003).

### *Pregnancy diagnosis*

Early and accurate diagnosis of pregnancy is important for effective livestock management. While not of immediate concern in extensive goat operations that utilize extended natural mating, the early determination of pregnancy can be a useful management tool under more intensive production conditions, or when artificial insemination and embryo transfer are employed. Inability to detect early pregnancy can result in economic losses in milk and kid production due to longer kidding intervals. Lack of knowledge of techniques to differentiate pregnant from non-pregnant animals may result in heavy reproduction and production losses in the form of abortions, stillbirths and the production of weak kids. It also results in uneconomical feeding of non-pregnant animals. Pregnancy diagnosis will identify the females requiring repeat breeding or insemination and/or will allow the separation of pregnant and open females for differential management. When fetal numbers can be determined as part of the pregnancy diagnosis, different feeding regimes can be applied to single- and litter-bearing females. To be most useful to the producer, pregnant animals need to be identified as early as possible in gestation. A variety of approaches have been explored for the early detection of pregnancy and possibly fetal numbers. The techniques have either focused on the detection of physical changes resulting from pregnancy (fluid accumulation and presence of a detectable fetus) through palpation and ultrasound (Wani and Sahni, 1980; Buckrell, 1988; Goel and Agrawal, 1989) or been concerned with the identification of maternal and fetal physiological signals (progesterone, oestrone sulfate and pregnancy proteins) associated with pregnancy (Restall *et al.*, 1990; Ishwar, 1995). Only a few methods seem reliable and applicable under field conditions.

NON-RETURN TO OESTRUS The length of the oestrous cycle ranges from 19 to 24 days (average 21 days) in goats. Non-return to oestrus after breeding is considered a sign of pregnancy. During the breeding season, goats return to oestrus within 7–23 days if there is a fertilization failure. The sign of non-return to oestrus due to pregnancy is not physically different from seasonal anoestrus at the end of the breeding season and out-of-breeding season. Non-return to oestrus is an unreliable method when does are synchronized and bred during the non-breeding season. In addition, pathological conditions of the uterus or ovaries may cause anoestrus in non-pregnant does (Ishwar, 1995).

Therefore, pregnancy diagnosis based on non-return to oestrus is not reliable in goats that exhibit seasonality in oestrous behaviour. The use of vasectomized bucks with raddle harnesses with the does after mating to detect the return to service appeared to be unreliable in detecting pregnancy in Thai goats, as 36.5% of pregnant does came into oestrus, as evidenced by raddle marks on the rump, during early pregnancy, and this can lead to serious errors (Restall *et al.*, 1990). Non-return determination is a method recommended for traditional goat owners only, who do not have other facilities for pregnancy diagnosis.

ABDOMINAL PALPATION In the late stages of pregnancy, does can be examined by abdominal palpation. This technique becomes easier and more reliable as pregnancy advances. The gravid uterus or fetus can sometimes be palpated through the relaxed abdominal wall by placing a hand on either side of the abdomen and squeezing or lifting upwards (Ishwar, 1995). If the doe is pregnant, the fetus is felt to drop on to the palpating hand (Arther *et al.*, 1982). Withholding feed and water for at least 12 h before examination increases the ease of the examination (Ishwar, 1995). It is easier in thin does than in fat animals. Pregnancy can be diagnosed to an acceptable accuracy (up to 70%) after 80 days of gestation by the abdominal palpation method. As the method is simple and does not involve any equipment, it can be used by goat owners to screen their flock.

ULTRASOUND TECHNIQUES There has been increasing interest in the use of ultrasound techniques for pregnancy diagnosis in goats by various workers (Aswad *et al.*, 1976; Wani and Sahni, 1980; Shelton, 1982; Buckrell, 1988; Goel and Agrawal, 1990; Haibel, 1990). Pregnancy may be detected with all three types of ultrasonographics available: amplitude depth (A-mode), Doppler and real-time B-mode ultrasonics. Each can be used under field conditions. The accuracy of diagnosis, timing of examination, fetal numbers and age and fetal viability vary considerably among these techniques. One of the most important features of ultrasound, when used for tissue examination, is its safety to the operator and patient. Today, the most used method is the real-time B-mode ultrasonics.

Real-time, B-mode ultrasonic scanning appears to offer an accurate, rapid, safe and practical means of diagnosing pregnancy and determining fetal numbers. The technique allows visualization of the fetus in the uterus, fetal numbers and fetal viability. Two types of scanners are available on the market and both are utilized for pregnancy diagnosis: (i) linear array scanners; and (ii) sector scanners. There are a number of abdominal/rectal probes. A mid-range abdominal (5 MHz) or rectal (7 MHz) probe is ideal for goats (Goel and Agrawal, 1992). Sector scanners provide a much wider angle of view and the entire uterus can be visualized from either side of the animal. They have some advantages over linear scanners: they require less skin surface contact, have a reduced scanning time and have superior resolution. In goats, transrectal ultrasonography with a 5 MHz probe allows visualization of the embryo vesicle starting from day 19 of pregnancy (Martinez *et al.*, 1998). Nevertheless, it is only from day 25 onward that precision and sensitivity are >90%.

Scanning Thai native goat does by real-time ultrasonic imaging between 55 and 65 days post-coitus was 100% accurate in detecting pregnancy using an abdominal probe (Restall *et al.*, 1990). It is therefore apparent that day 45–50 is the ideal time to make a pregnancy diagnosis by transabdominal scanning with a high degree of accuracy (Buckrell, 1988). Accuracy of counting numbers of fetuses with real-time ultrasonography is an advantage over other ultrasound techniques. The optimal time for counting fetal numbers is between 45 and 90 days of gestation (Haibel, 1990). After 90–100 days of gestation, fetuses become too large to be consistently differentiated from each other.

In addition to pregnancy diagnosis, ultrasound scanners may be helpful for early diagnosis of fetal malformation, factors influencing fetal growth and diagnosis of diseases of the reproductive tract (Buckrell, 1988). Fetal age in ewes and does can also be determined by the use of real-time ultrasonics at 40–100 days of gestation by measuring the width of the fetal skull (Reichle and Haibel, 1991). This technique would be helpful in predicting parturition date when the actual date of breeding is not known.

The main limiting factor preventing greater use of this technique in developing countries is its high cost. In addition, training and considerable experience are required to obtain better accuracy in terms of image interpretation. With a decrease in equipment cost and an increase in practical experience, it may be possible in the future to significantly increase the use of portable ultrasound scanners in goats.

HORMONAL ASSAYS Progesterone levels can be obtained using a radioimmunoassay (RIA) or enzyme-linked immunosorbent assay (ELISA). Assays can be performed on serum, plasma or milk samples. The concentration of plasma progesterone can be determined on days 19–23 post-breeding in does with high accuracy (Gonzalez *et al.*, 2004). Progesterone concentration in plasma as well as in milk 21–22 days after artificial insemination in the goats was estimated (Thibier *et al.*, 1982). The level of plasma progesterone was 7.64 ± 4.17 and 0.86 ± 0.73 ng/ml in pregnant and non-pregnant goats, respectively. Its accuracy in early pregnancy diagnosis was 100 and 87.5%, for non-pregnant and pregnant does, respectively.

A progesterone test in does is a good test for non-pregnancy as it allows early identification of open does with 80–100% accuracy, but it is only a fair test for pregnancy.

Oestrone sulfate is produced by the placenta in sheep and goats. Oestrone sulfate can be detected in the plasma of does from around 40–50 days post-breeding. Around day 60 of pregnancy, the average concentration is 0.6 ng/ml in non-pregnant goats and 6.1 ng/ml in those that are pregnant (Refstal *et al.*, 1991). A positive test indicates a viable fetus.

In the pregnant doe, PAG concentrations are detectable from day 17–18, reaching concentrations of 3–5 ng/ml on days 21–22. Pregnancies can be detected by day 24 (Humblot *et al.*, 1995) or day 25 (Folch *et al.*, 1993). PAG determination was highly accurate on days 24 and 26 (99 and 100%, respectively) (Gonzalez-Bulnes *et al.*, 2004).

The PAG milk-test provides an accurate pregnancy diagnosis from day 32 after breeding and, in combination with good management practices, this test would be suitable under farm conditions to confirm pregnancies tentatively identified by non-return to oestrus on day 21 after breeding (Gonzalez *et al.*, 2001). A summary of the effectiveness and practical applications of different techniques for pregnancy diagnosis is presented in Table 5.5.

**Table 5.5.** Comparisons of techniques available for pregnancy diagnosis in the doe.

| Technique | Sensitivity range(days) | Fetal numbers | Accuracy (%) | Practical application |
|---|---|---|---|---|
| Vasectomized harnessed male | >20 | No | 65–90 | High |
| Abdominal palpation | 60–115 | No | 60–90 | Moderate |
| Progesterone assay | 18–22 | No | 90–95 | Moderate |
| Oestrone assay | >60 | No | 90–95 | Low |
| Real-time ultrasound | 40–100 | Yes | 90–95 | High |
| Doppler ultrasound | 60–90 | No | 85–90 | Moderate |
| Radiography | >50 | Yes | 90–95 | Low |

### 5.4.6 Genetic improvement of reproductive traits

For animals kept primarily for meat production, reproductive rate is the single most important factor contributing to the efficiency of production. The economic importance of reproductive traits should not be abandoned while selecting for performance traits associated with improving meat production. This is because reproductive rate directly influences the income and profitability for commercial production of meat from goats. Reproductive traits of interest in meat goat enterprises are conception rate, kidding rate and ability to breed out of season. In addition, several studies have demonstrated that twins and triplets produce more total weight of kid per doe per year. Therefore, prolificacy, defined as the number of kids born per doe, is another important reproduction trait. Genetic improvement of these traits can be undertaken by selection within purebred strains or by crossbreeding. Prospects for improvement of reproductive traits in several meat breeds of goats by selection are available (Shrestha and Fahmy, 2007). The heritability estimate for reproductive traits, although negligible in magnitude, is positive (Table 5.6). However, for some traits such as age at first kidding, heritability estimates are high and this can yield important genetic progress.

**Table 5.6.** Estimates of heritabilities and repeatabilities for some reproductive traits of meat-producing breeds of goat (from Shrestha and Fahmy, 2007).

| Trait/Breed | Heritability | Repeatability |
|---|---|---|
| Age at first kidding | | |
| Alpine x Beetal | 0.56 ± 0.08 | |
| Beetal | 0.48 ± 0.09 | |
| Litter size | | |
| Beetal | 0.15 | |
| Beetal, Black Bengal | 0.09 ± 0.25 | |
| Alpine × African common | 0.02 | |
| Multiple births | | |
| Egyptian Baladi | 0.25 | 0.29 |
| Beetal | 0.15 | 0.22 |
| Black Bengal | 0.09 | 0.15 |
| Bengal | 0.17 ± 0.20 | |

From a practical point of view, it is useful to quantify reproductive performance using the number of weaned animals (and their weight) relative to the number of reproductive females. In this case, litter size is considered to be one of the most important components of reproductive performance, due to its high correlation with the number of animals at weaning. Unfortunately, litter size is a sex-limited characteristic with a low heritability value, which acts as a limiting factor for the improvement of reproductive performance. A good example in this respect is given for African common goats in Rwanda for which heritability and repeatability estimates are 0.025 and 0.061, respectively, for the number of kids born, indicating that the possibility of selection to improve these reproductive traits would take a long time (Mourad, 1994, 1996). However, the genetic correlation between the number of kids born and the number of kids alive was positive, which means that selection for one character may tend to improve the other.

Most of the breeding programmes in goats ignore the frequency of year-round kidding, an important characteristic in many goat breeds, particularly goats from tropical regions that are non-seasonal breeders and kid all year round. Therefore, incorporating this trait of non-seasonality into a meat-goat enterprise would be advantageous. An alternative would be to include reproductive traits and a pre-weaning growth rate that can be combined into an index to give a measure of productivity of the doe. An example of such an index is presented by Gipson (2008) as:

Productivity index
= conception rate × litter size
× survivability to weaning
× 365/kidding interval
× (birth weight + pre-weaning growth rate × age at weaning)

As for most livestock species, improvement of reproductive traits in goats has also been attempted through crossbreeding. In Egypt, reproduction, growth and carcass traits of kids from Alpine and Rove breeds and Alpine × Rove crosses were evaluated (Anous and Mourad, 1993). The study revealed that the Alpine breed was more fecund than the Rove breed but was similar in carcass traits and weighed less. The estimates of heterosis of 24% for prolificacy and 4% for fertility were important, but not for the rate of abortion. Prolificacy is the trait that is most improved by crossbreeding a local with a prolific breed, while no significant superiority in fertility of crossbred over purebred goats is reported (Anous and Mourad, 1993).

In Tunisia, genetic improvement of the local goat was undertaken through crossbreeding with imported breeds, namely the Alpine, Boer and Saanen, in the north of the country. The impact of such crossbreeding on the reproductive traits and productivity of the does is reported in Table 5.7 (Rekik *et al.*, 2005). It was concluded that, under the pastoral production systems of northern Tunisia, there is no benefit of crossbreeding with imported breeds; the absence of a clear effect of crossbreeding on the improvement of productivity could be explained by problems of adaptation of the crossbred animals to the prevailing difficult conditions.

The specific three-breed cross has demonstrated increased productivity in pig and sheep species and should also have potential in meat goats (Shrestha and Fahmy, 2007). In order to produce a specific three-breed-cross offspring, does of the fecund-type dam breeds that excel in rate of reproduction, milk production, mothering ability and survival are first mated to bucks of an alternative breed to produce fecund-type two-breed-cross offspring. The crossbred offspring selected as the female parent are mated to bucks of a third breed, usually a meat-type terminal sire breed recognized for superior growth performance and meat quality. The crossbred offspring is expected to exhibit maximum individual and maternal heterosis, resulting in rapid growth and increased uniformity in the marketing of meat and meat products to the consumer.

**Table 5.7.** Compared reproductive and productivity performances of indigenous goat and its crosses with imported breeds (from Rekik *et al.*, 2005).

| Trait | Local goat | $F_1$ Alpine | $F_1$ Boer | $F_1$ Saanen |
|---|---|---|---|---|
| Fertility rate (%) | 94.1 | 95.6 | 98.0 | 89.3 |
| Prolificacy rate (%) | 148.8 | 144.5 | 144.5 | 109.1 |
| Pre-weaning mortality rate (%) | 5.58 | 23.1 | 0 | 12.69 |
| Kids weaned per goat (%) | 134 | 115 | 141 | 85 |
| Kilograms weaned per goat | 16.9 | 16.6 | 18.7 | 12.1 |

## References


Adalsteinsson, S. (1975) Depressed fertility in Icelandic sheep caused by a single colour gene. *Annales de Génétique et de Sélection Animale* 7, 445–447.

Ahmad, N. and Noakes, D.E. (1995) Seasonal variations in testis size, libido and plasma testosterone concentrations in British goats. *Animal Science* 61, 553–559.

Ahmad, N. and Noakes, D.E. (1996) Sexual maturity in British breeds of goat kids. *British Veterinary Journal* 152, 93–103.

Akusu, M.O., Osuagwuh, A.I.A., Akpokodje, J.U. and Egbunike, G.N. (1986) Ovarian activities of the West African goat (*Capra hircus*) during estrus. *Journal of Reproduction and Fertility* 78, 459–462.

Al-Ghalban, A.M., Tabbaa, J. and Kridli, R.T. (2004) Factors affecting semen characteristics and scrotal circumference in Damascus bucks. *Small Ruminant Research* 53, 141–149.

Amoah, E.A. and Bryant, M.J. (1984) A note on the effect of contact with male goats on occurrence of puberty in female goat kids. *Animal Production* 38, 141–144.

Amoah, E.A., and Gelaye, S. (1997) Biotechnological advances in goat reproduction. *Journal of Animal Science* 75, 578–585.

Anous, M.R. and Mourad, M.M. (1993) Crossbreeding effects on reproductive traits of does and growth and carcass traits of kids. *Small Ruminant Research* 12, 141–149.

Anwar, M. and Ahmad, K.M. (1999) Ovulation rate, number of fetuses and embryo loss in Teddy goats of Pakistan. *Small Ruminant Research* 31, 281–283.

Arther, G.H., Noakes, D.E. and Pearson, H. (1982) *Veterinary Reproduction and Obstetrics (Theriogenology)*, 5th edn. The English Language Book Society and Baillière Tindall, London, UK.

Aswad, A., Abdou, M.S.S., Al-Bayaty, F. and El-Sawaf, S.A. (1976) The validity of the 'ultrasonic method' for pregnancy diagnosis in ewes and goats. *Zentralblatt für Veterinärmedizin* 23, 467–474.

Attwood, B.M. (2007) *A Guide to the Grazing Requirements of Fibre and Meat Goats.* Agriculture Notes AG0998, Department of Primary Industries, Victoria, Australia.

Baril, G., Casamitjana, P., Perrin, J. and Vallet, J.C. (1989) Embryo production, freezing and transfer in Angora, Alpine and Saanen goats. *Zuchthygiene* 24, 101–115.

Barkawi, A.H., Elsayed, E.H., Ashour, G. and Shehata, E. (2006) Seasonal changes in semen characteristics, hormonal profiles and testicular activity in Zaraibi goats. *Small Ruminant Research* 66, 209–213.

Bazer, F.W., Spencer, T.E. and Ott, T.L. (1997) Interferon tau: a novel pregnancy recognition signal. *American Journal of Reproductive Immunology* 37, 412–420.

Benitez-Ortiz, W. (1992) Diagnostic de pregnancy et étude de la mortalité embryonnaire chez les ruminants par dosage de la pregnancy associated glycoprotein. PhD thesis, Institut de Médecine Tropicale Prince Leopold, Antwerp, Belgium.

Bhattacharrya, B., Sanwall, P.C., Pandle, J.D. and Varshney, V.P. (1984) Plasma levels of sex hormones in female kids approaching puberty. *Animal Breeding Abstracts* 52, 758.

Bilton, R.J. and Moore, N.W. (1976) In vitro culture, storage and transfer of goat embryos. *Australian Journal of Biological Sciences* 29, 125–129.

Bose, R., Heng, H., Sabbadini, E., McCoshen, J., MaHadovan, M. and Fleetham, J. (1989) Purified human early pregnancy factor from pre-implantation embryo possesses immuno-suppressive properties. *American Journal of Obstetrics and Gynecology* 160, 954–960.

Buckrell, B.C. (1988) Applications of ultrasonography in reproduction in sheep and goats. *Theriogenology* 29, 71–84.

Campbell, Q.P. (1994) Information regarding the improved Boer goat. *Boer Goat News* 11, 49–50.

Charallah, S., Khammar, F., Amirat, Z. and Lakhdari, Y. (2000) Evaluation de l'activité sexuelle male et femelle, caractérisations zootechnique et nutritionnelle chez la chèvre bédouine. In: *Proceedings of the 7th International Conference on Goats*, France, Vol. I, 15–21 May, p. 460.

Chemineau, P. (1983) Effect on estrus and ovulation of exposing creole goats to the male at three times of the year. *Journal of Reproduction and Fertility* 67, 65–72.

Chemineau, P. (1985) Effects of a progestogen on buck-induced short ovarian cycles in the creole meat goat. *Animal Reproduction Science* 9, 87–94.

Chemineau, P. (1986) Influence de la saison sur l'activité sexuelle du Cabri Créole mâle et femelle. PhD thesis, Université des Sciences Techniques de Languedoc, Montpellier, France.

Chemineau P. (1987) Possibilities for using bucks to stimulate ovarian and oestrous cycles in anovulatory goats – a review. *Livestock Production Science* 17, 135–147.

Chemineau, P. and Cognié, Y. (1991) *Training Manual on Artificial Insemination in Sheep and Goats.* FAO publications, Rome, Italy.

Chemineau, P. and Delgadillo, J.A. (1994) Neuroendocrinologie de la reproduction chez les caprins. *INRA Productions Animales* 7, 315–326.

Chemineau, P. and Xandé, A. (1982) Reproductive efficiency of Creole meat goats permanently kept with males. Relationship to a tropical environment. *Tropical Animal Production* 7, 98–104.

Chemineau, P., Normant, E., Ravault, J.P. and Thimonier, J. (1986) Induction and persistence of pituitary and ovarian activity in the out-of-season lactating dairy goat after a treatment combining a skeleton photoperiod, melatonin and the male effect. *Journal of Reproduction and Fertility* 78, 497–504.

Chemineau, P., Martin, G.B., Saumande, J. and Normant, E. (1988) Seasonal and hormonal control of pulsatile LH secretion in the dairy goat (*Capra hircus*). *Journal of Reproduction and Fertility* 88, 91–98.

Chemineau, P., Malpaux, B., Delgadillo, J.A., Guerin, Y., Ravault, J.P., Thimonier, J. and Pelletier, J. (1992) Control of sheep and goat reproduction: use of light and melatonin. *Animal Reproduction Science* 30, 157–184.

Chemineau, P., Baril, G., Leboeuf, B., Maurel, M.C., Roy, F., Pellicer-Rubio, M., Malpaux, B. and Cognie, Y. (1999) Implications des progrès récents en physiologie de la reproduction pour la conduite de la reproduction dans l'espèce caprine. *INRA Productions Animales* 12, 135–146.

Chemineau, P., Pellicer-Rubio, M.T., Lassoued, N., Khaldi, G. and Monniaux, D. (2006) Male-induced short oestrous and ovarian cycles in sheep and goats: a working hypothesis. *Reproduction Nutrition and Development* 46, 417–429.

Clariget, R.P., Forsberg, M. and Rodriguez-Martinez, H. (1998) Seasonal variation in live weight, testes size, testosterone, LH secretion, melatonin and thyroxine in Merino and Corriedale rams in a subtropical climate. *Acta Veterinaria Scandinavae* 39, 35–47.

Cognié, Y., Baril, G., Poulin, N. and Mermillod, P. (2003) Current status of embryo technologies in sheep and goat. *Theriogenology* 59, 171–188.

Corteel, J. (1975) Effect of washing on deep frozen goat semen preservation. *Annales de Biologie Animale Biochimie Biophysique* 15, 525–528.

Currie, W.B., Card, C.E., Michel, F.J. and Ignotz, G. (1990) Purification, partial characterization, and development of a specific radioimmunoassay for goat placental lactogen. *Journal of Reproduction and Fertility* 90, 25–36.

de Castro, T., Rubianes, E., Menchaca, A. and Rivero, A. (1999) Ovarian dynamics, serum estradiol and progesterone concentrations during the interovulatory interval in goats. *Theriogenology* 52, 399–411.

Delgadillo, J.A. and Chemineau, P. (1992) Abolition of the seasonal release of luteinizing hormone and testosterone in Alpine male goats (*Capra hircus*) by short photoperiodic cycles. *Journal of Reproduction and Fertility* 94, 45–55.

Delgadillo, J.A. and Malpaux, B. (1996) Seasonal variations in testicular weight in Alpine and Nubian male goats in subtropical conditions (Northern Mexico). In: *Proceedings of the Sixth International Conference on Goats.* International Academic Publishers, Beijing, China, pp. 785–793.

Delgadillo, J.A., Leboeuf, B. and Chemineau, P. (1993) Maintenance of sperm production in bucks during a third year of short photoperiodic cycles. *Reproduction Nutrition Development* 33, 609–617.

Delgadillo, J.A., Flores, J.A., Luna, M.C., Duarte, G., Carrilo, E., Hoyos, G. and Nava, P. (1994) El anestro postparto de las cabras de la comarca laguna que paren en enero no es modificado por el momento en que se realiza el destete. In: *IX Reunion Nacional de Caprinocultura*, La Paz, Mexico, pp. 157–160.

Delgadillo, J.A., Hochereau-de Reviers, M.T., Daveau, A. and Chemineau, P. (1995) Effect of short photoperiodic cycles on male genital tract and testicular parameters in male goats (*Capra hircus*). *Reproduction Nutrition Development* 35, 549–558.

Delgadillo, J.A., Malpaux, B. and Chemineau, P. (1997) La reproduction des caprins dans les zones tropicales et subtropicales. *INRA Productions Animales* 10, 33–41.

Derquaoui, L. and El Khaledi, O. (1994) Evaluation de l'activité sexuelle pendant la saison de baisse de fertilité chez la chèvre de race D'Man. In: *Proceedings of the 2nd Conference of the African Small Ruminant Research Network*, Arusha, Tanzania, 7–11 December, 1992, Addis Ababa, Ethiopia, pp. 49–51.

Devendra, C. and Burns, M. (1983) Reproductive performance. In: Devendra, C. and Burns, M. (eds) *Goat Production in the Tropics.* Commonwealth Agricultural Bureaux, Farnham Royal, UK, pp. 74–89.

de Santiago-Miramontes, M.A., Malpaux, B. and Delgadillo, J.A. (2008) Body condition is associated with a shorter breeding season and reduced ovulation rate in subtropical goats. *Animal Reproduction Science* 114, 175–182.

de Waal, H.O. and Combrinck, W.J. (2000) The development of the Dorper, its nutrition and a perspective of the grazing ruminant on veld. *Small Ruminant Research* 36, 103–117.

Driancourt, M.A., Philipon, P., Terqui, M., Molenat, G., Mirman, B., Louault, C., Avdi, M., Folch, J. and Cognie, Y. (1990) Possibilités de l'immunisation contre les stéroïdes pour améliorer les performances ovulatoires et la taille de la portée des ovins et caprins. *INRA Productions Animales* 3, 31–37.

Drymundsson, O.R. (1983) The influence of environmental factors on the attainment of puberty in ewe lambs. In: Haresign, W. (ed.) *Sheep Production*. Butterworth, London, UK, pp. 393–408.

Ebozoje, M.O. and Ikeobi, C.O.N. (1998) Colour variation and reproduction in the West African dwarf (WAD) goats. *Small Ruminant Research* 27, 125–130.

English, J., Poulton, A.L., Arendt, J. and Symons, A.M. (1986) A comparison of the efficiency of melatonin treatments in advancing estrus in ewes. *Journal of Reproduction and Fertility* 77, 321–327.

Evans, A.C., Duffy, P., Hynes, N. and Boland, M.P. (2000) Waves of follicle development during the estrous cycle in sheep. *Theriogenology* 53, 699–715.

Evans, G. and Maxwell, W.M.C. (1987) Frozen storage of semen. In: *Salamon's Artificial Insemination of Sheep and Goats*. Butterworths, Sydney, Australia, pp. 122–141.

Fabre-Nys, C. (2000) Le comportement sexuel des caprins: contrôle hormonal et facteurs sociaux. *INRA Productions Animales* 13, 11–23.

Fasanya, O.O.A., Molokwu, E.C.I., Eduvie, L.O. and Dim, N.I. (1992) Dietary supplementation in the Savanna brown goat. 1. Effect on attainment of puberty in the doe. *Animal Reproduction Science* 29, 157–166.

Fletcher, I.C. (1971) Relationships between frequency of suckling, lamb growth and post-partum oestrous behaviour in ewes. *Animal Behaviour* 19, 108–111.

Flores, J.A., Véliz, F.G., Pérez-Villanueva, J.A., Martínez de la Escalera, G., Chemineau, P., Poindron, P., Malpaux, B. and Delgadillo, J.A. (2000) Male reproductive condition is the limiting factor of efficiency in the male effect during seasonal anestrus in female goats. *Biology of Reproduction* 62, 1409–1414.

Folch, J., Benitez, W., Alabart, J.L. and Beckers, J.F. (1993) Determination de la concentracion plasmatica de PAG (pregnancy-associated glycoprotein) en cabras blanca celtibericay su utilizacion como diagnostico de gestation. *ITEA* 12, 364.

Foster, D.L. and Ryan, K.D. (1981) Endocrine mechanisms governing transition into adulthood in female sheep. *Journal of Reproduction and Fertility Supplement* 30, 75–90.

Freitas, V.J.F., Rondina, D., Nogueira, D.M. and Simplicio, A.A. (2004) Post-partum anestrus in Anglo-Nubian and Saanen goats raised in semi-arid area of North-eastern Brazil. *Livestock Production Science* 90, 219–226.

Gil, J., Rodriguez-Irazoqui, M., Lundeheim, N., Soderquist, L. and Rodriguez-Martinez, H. (2003) Fertility of ram semen frozen in Bioexcell® and used for cervical artificial insemination. *Theriogenology* 59, 1157–1170.

Gipson, T.A. (2008) Meat goat breeds and breeding plans <http://www.GoatWorld.com>.

Girard, L. (1813) Moyens employés avec succès, par M. Morel de Vindé, Membre de la société d'Agriculture de Seine et Oise, pour obtenir, dans le temps le plus court possible, la fécondation du plus grand nombre des brebis portières d'un troupeau. In: *Ephémérides de la Société d'Agriculture du Département de l'Indre pour l'An 1813*, 5 September, Chateauroux, Département de l'Indre, France, Cahier VII, pp. 66–68

Goel, A.K. and Agrawal, K.P. (1989) Pregnancy diagnosis in sheep. *Indian Journal of Animal Science* 59, 974–976.

Goel, A.K. and Agrawal, K.P. (1990) Pregnancy diagnosis in goats. *Indian Veterinary Medical Journal* 14, 77–78.

Goel, A.K. and Agrawal, K.P. (1992) A review of pregnancy diagnosis techniques in sheep and goats. *Small Ruminant Research* 9, 255–264.

Goel, A.K. and Agrawal, K.P. (2003) Ovulation in Jakhrana goats native to tropical climates. *Small Ruminant Research* 50, 209–212.

Gonzalez, F., Sulon, J., Calero, P., Batista, M., Gracia, A. and Beckers, J.F. (2001) Pregnancy-associated glycoproteins (PAG) detection in milk samples for pregnancy diagnosis in dairy goats. *Theriogenology* 56, 671–676.

Gonzalez, F., Cabreraa, F., Batistaa, M., Rodriguez, N., Alamoa, D., Sulon, J., Beckers, J.F. and Gracia, A. (2004) A comparison of diagnosis of pregnancy in the goat via transrectal ultrasound scanning, progesterone, and pregnancy-associated glycoprotein assays. *Theriogenology* 62, 1108–1115.

Gonzalez-Bulnes, A., Santiago-Moreno, J., Garcia-Garcia, R.M., Souza, C.J.H., Lopez-Sebastian, A. and McNeilly, A.S. (2004) Effect of GnRH antagonists treatment on gonadotrophin secretion, follicular development and inhibin A secretion in goats. *Theriogenology* 61, 977–985.

Gonzalez-Stagnaro, C., Pelletier, J., Cognié, Y., Locatelli, A., Baril, G. and Corteel, J.M. (1984) Descarga preovulatoria de LH y momento de oculacion en cabras lecheras durante el celo natural o inducido por via hormonal. In: *Proceedings of the 10th International Congress on Animal Reproduction and Artificial Insemination,* University of Illinois at Urbana-Champaign, IL, USA, 10–14 June, Vol. 2, p. 10 (abstract).

Greyling, J.P.C. (1988) Reproductive physiology in the Boer goat doe. PhD thesis, University of Stellenbosch, Republic of South Africa.

Greyling, J.P.C. (1990) Sexual activity of the Boer goat doe. *Boer Goat News* 9, 51–53.

Greyling, J.P.C. (1996) The induction of puberty in female Boer goat kids. In: Lebbie, S.H.B. and Kagwini, E. (eds) *Small Ruminant Research and Development in Africa. Proceedings of the Third Biennial Conference of the African Small Ruminant Research Network*, UICC, Kampala, Uganda, 5–9 December, 1994. International Livestock Research Institute, Nairobi, Kenya.

Greyling, J.P.C. (2000) Reproduction traits in the Boer goat doe. *Small Ruminant Research* 36, 171–177.

Greyling, J.P.C. and van Niekerk, C.H. (1986) Synchronization of estrus in the Boer goat doe: dose effect of prostaglandin in the double injection regime. *South African Journal of Animal Science* 16, 146–150.

Greyling, J.P.C. and van Niekerk, C.H. (1991) Macroscopic uterine involution in the post-partum Boer goat. *Small Ruminant Research* 4, 277–283.

Haibel, G.K. (1990) Use of ultrasonography in reproductive management of sheep and goat herds. In: Smith, M.C. (ed.) *Advances in Sheep and Goat Medicine.* The Veterinary Clinics of North America Food Animal Practice, Vol. 6, no. 3. W.B. Saunders, Philadelphia, PA, USA, pp. 597–613.

Hancock, J. and Louca, A. (1975) Polledness and intersexuality in the Damascus breed of goat. *Animal Production* 21, 227–231.

Hasan, N.A. and Shaker, B. (1990) *Goat Resources in Arab States in Syrian Arab Republic.* Publication of the Arab Center for the Study of Arid Zones and Dry Lands (ACSAD), Syrian Arab Republic.

Hayden, T.J., Thomas, C.R. and Forsyth, I.A. (1979) Effect of number of young born (litter size) on milk yield of goats: role for placental lactogen. *Journal of Dairy Science* 62, 53–63.

Henniawati, I. and Fletcher, C. (1986) Reproduction in Indonesian sheep and goats at two levels of nutrition. *Animal Reproduction Science* 12, 77–84.

Hesselink, J.W. (1993a) Incidence of hydrometra in dairy goats. *Veterinary Record* 132, 110–112.

Hesselink J.W. (1993b) Hydrometra in dairy goats: reproductive performance after treatment with prostaglandins. *Veterinary Record* 133, 186–187.

Holtz, W., Sohnrey, B., Gerland, M. and Driancourt, M.A. (2008) Ovsynch synchronization and fixed-time insemination in goats. *Theriogenology* 69, 785–792.

Homeida, A.M. (1986) Role of oxytocin during the oestrous cycle of ruminants with particular reference to the goat. *Animal Breeding Abstracts* 54, 263–268.

Humblot, P., Brice, G., Chemineau, P. and Broqua, B. (1995) Mortalité embryonnaire chez la chèvre laitière après synchronisation des chaleurs et insémination artificielle à contre saison. *Rencontres Recherches Ruminants* 2, 387–390.

Hunter, R.H.F. (1981) *Physiology and Technology of Reproduction in Female Domestic Animals.* Academic Press, London/New York.

Ishwar, A.K. (1995) Pregnancy diagnosis in sheep and goats: a review. *Small Ruminant Research* 17, 37–44.

Ishwar, A.K. and Pandey, J.N. (1992) Estrus synchronization and fertility in Black Bengal goats following administration of progesterone and gonadotropins. *Research in Veterinary Science* 52, 141–146.

Jabbour, H.N., Ryan, J.P., Evans, G. and Maxwell, W.M.C. (1991) Effect of season, GnRH administration and lupin supplementation on the ovarian and endocrine response of Merino ewes treated with PMSG and FSH-P to induce superovulation. *Reproduction, Fertility and Development* 3, 699–707.

Jainudeen, M.R., Wahid, H. and Hafez, E.S.E. (2000) Sheep and goats. In: Hafez, E.S.E. and Hafez, B. (eds) *Reproduction in Farm Animals*, 7th revised edition. Blackwell Publishers, UK, pp. 172–181.

Karsch, F.J., Bittman, E.L., Foster, D.L., Goodman, R.L., Legan, S.J. and Robinson, J.E. (1984) Neuroendocrine basis of seasonal reproduction. *Recent Progress in Hormone Research* 40, 185–225.

Kassem, R. (2005) Small ruminant breeds of Syria. In: Iniguez, L. (ed.) *Characterisation of Small Ruminant Breeds in West Asia and North Africa*, Vol. 2. International Centre for Agricultural Research in the Dry Areas (ICARDA), Aleppo, Syria, pp. 183–237.

Khaldi, G. (1984) Variations saisonnières de l'activité ovarienne, du comportement d'estrus et de la durée de l'anestrus post-partum des femelles ovines de race Barbarine: influence du niveau alimentaire et de la présence du mâle. PhD thesis, Université des Sciences et Techniques du Languedoc, Académie de Montpellier, France.

Khanum, S.A., Hussain, M. and Kausar, R. (2006) Manipulation of estrous cycle in Dwarf goat (*Capra hircus*) using estrumate under different management conditions. *Animal Reproduction Science* 92, 97–106.

Khanum, S.A., Hussain, M. and Kausar, R. (2007) Assessment of reproductive parameters in female Dwarf goat (*Capra hircus*) on the basis of progesterone profiles. *Animal Reproduction Science* 102, 267–275.

Lassoued, N. and Rekik, M. (2005) Variations saisonnières de l'estrus et de l'ovulation chez la chèvre locale Maure en Tunisie. *Revue d'Elevage et de Médecine Vétérinaire des Pays Tropicaux* 58, 69–73.

Lassoued, N., Khaldi, G., Cognié, Y., Chemineau, P. and Thimonier, J. (1995) Effet de la progestérone sur le taux d'ovulation et la durée du cycle ovarien induits par effet mâle chez la brebis Barbarine et la chèvre locale Tunisienne. *Reproduction Nutrition Development* 35, 415–426.

Lassoued, N., Rekik, M., Ben Salem, H. and Dargouth, M.A. (2006) Reproductive and productivity of goats grazing *Acacia cyanophylla* Lindl., with and without daily polyethylene glycol supplementation. *Livestock Science* 105, 129–136.

Leboeuf, B., Renaud, G., de Fontaubert, Y., Broqua, B. and Chemineau, P. (1994) Echographie et pseudo-gestation chez la chèvre. In: *Proceedings of the 7th International Meeting on Animal Reproduction*, Murcia, Spain, pp. 251–255.

Leboeuf, B., Manfredi, E., Boue, P., Piacere, A., Brice, G. and Baril, G. (1998) Artificial insemination of dairy goats in France. *Livestock Production Science* 55, 193–203.

Leboeuf, B., Restall, B. and Salamon, S. (2000) Production and storage of goat semen for artificial insemination. *Animal Reproduction Science* 62, 113–141.

Le Gal, F., Baril, G., Vallet, J.C. and Leboeuf, B. (1993) In vivo and in vitro survival of goat embryos after freezing with ethylene glycol or glycerol. *Theriogenology* 40, 771–777.

Legan, S.J. and Karsch, F.J. (1983) Importance of retinal photoreceptors to the photoperiodic control of seasonal breeding in the ewe. *Biology of Reproduction* 29, 316–325.

Lehloenya, K.C. (2008) Multiple ovulation and embryo transfer in goats. PhD thesis, University of the Free State, Bloemfontein, Republic of South Africa.

Lincoln, G.A. and Clarke, I.J. (1997) Refractoriness to a static melatonin signal develops in the pituitary gland for the control of prolactin secretion in the ram. *Biology of Reproduction* 57, 460–467.

Llewelyn, C.A., Ogaa, J.S. and Obwolo, M.J. (1993) Plasma progesterone profiles and variation in cyclic ovarian activity throughout the year in indigenous goats in Zimbabwe. *Animal Reproduction Science* 30, 301–311.

Lopez-Sebastian, A., Gonzalez-Bulnes, A., Carrizosa, J.A., Urrutia, B., Diaz-Delfa, C., Santiago-Moreno, J. and Gomez-Brunet, A. (2007) New estrus synchronization and artificial insemination protocol for goats based on male exposure, progesterone and cloprostenol during the non-breeding season. *Theriogenology* 68, 1081–1087.

Luna-Orozco, J.R., Fernández, I.G., Gelez, H. and Delgadillo, J.A. (2008) Parity of female goats does not influence their estrous and ovulatory responses to the male effect. *Animal Reproduction Science* 106, 352–360.

Mani, A.U., McKelvey, W.A.C. and Watson, E.D. (1992) The effects of low level of feeding on response to synchronization of estrus, ovulation rate and embryo loss in goats. *Theriogenology* 38, 1013–1022.

Martinez, M.F., Bosch, P. and Bosch, R.A. (1998) Determination of early pregnancy and embryonic growth in goats by transrectal ultrasound scanning. *Theriogenology* 49, 1555–1565.

Mavrogenis, A.P. (2005) Small ruminant breeds of Cyprus. In: Iniguez, L. (ed.) *Characterisation of Small Ruminant Breeds in West Asia and North Africa*, Vol. 1. International Centre for Agricultural Research in the Dry Areas (ICARDA), Aleppo, Syria, pp. 417–458.

Mavrogenis, A.P. and Constantinou, A. (1983) *Performance of Damascus Goats Bred as Yearlings or as Two-year Olds.* Technical Bulletin No. 45. Agricultural Research Institute, Nicosia, Cyprus.

Mbayahaga, J., Mandiki, S.N.M., Bister, J.L. and Paquay, R. (1998) Body weight, oestrous and ovarian activity in local Burundian ewes and goats after parturition in the dry season. *Animal Reproduction Science* 51, 289–300.

Meites, J., Webster, H.D., Young, W., Thorp, F.J.R. and Hatch, R.N. (1951) Effects of corpora lutea removal and replacement with progesterone on pregnancy in goats. *Journal of Animal Science* 10, 411–416.

Mellado, M., Vera, A. and Loera, H. (1994) Reproductive performance of crossbred goats in good or poor body condition exposed to bucks before breeding. *Small Ruminant Research* 14, 45–48.

Mellado, M., Valdez, R., Garcia, J.E., Lopez, R. and Rodriguez, A. (2006) Factors affecting the reproductive performance of goats under intensive conditions in a hot arid environment. *Small Ruminant Research* 63, 110–118.

Memon, M.A., Bretzlaff, K.N. and Ott, R.S. (1985) Effect of washing on motility and acrosome morphology of frozen-thawed goat spermatozoa. *American Journal of Veterinary Research* 46, 473–475.
Memon, M.A., Bretzlaff, K.N. and Ott, R.S. (1986) Comparison of semen collection techniques in goats. *Theriogenology* 26, 823–827.
Memon, M.A., Mickelsen, W.D. and Goyal, H.O. (2007) Examination of the male reproductive tract and evaluation of potential breeding. In: Youngquist, R.S. and Threlfall, W.R. (eds) *Current Therapy in Large Animal Theriogenology*. W.B. Saunders, Philadelphia, PA, USA, pp. 515–518.
Mialot, J.P., Saboureau, L., Gueraud, J.M., Prengere, E., Parizot, D., Pirot, G., Duquesnel, R., Petat, M. and Chemineau, P. (1991) La pseudogestation chez la chèvre: observations préliminaires. *Recueil de Médecine Vétérinaire Spécial Reproduction Ruminants* 1, 383–390.
Mickelsen, W.D. and Memon, M.A. (2007) Infertility and diseases of the buck. In: Youngquist, R.S. and Threlfall, W.R. (eds) *Current Therapy in Large Animal Theriogenology*. W.B. Saunders, Philadelphia, PA, USA, pp. 519–523.
Moaeen-ud-Din, M., Yang, L.G., Chen, S.L., Zhang, Z.R., Xiao, J.Z., Wen, Q.Y. and Dai, M. (2008) Reproductive performance of Matou goat under subtropical monsoonal climate of Central China. *Tropical Animal Health and Production* 40, 17–23.
Molina, S., Fernández, J.L., Fernández, M. and Martin, G.O. (1997) Frecuencia y distribución mensual de pariciones en majadas de caprinos criollos. *Revista Argentina de Producción Animal* 17 (Suppl. 1), 271–272.
Montlomelo, K.C., Greyling, J.P.C. and Schwalbach, L.M.J. (2002) Synchronisation of estrus in goats: the use of different progestagen treatments. *Small Ruminant Research* 45, 45–49.
Mori, Y. and Kano, Y. (1984) Changes in plasma concentrations of LH, progesterone and oestradiol in relation to the occurrence of luteolysis, estrus and time of ovulation in the Shiba goat (*Capra hircus*). *Journal of Reproduction and Fertility* 72, 223–230.
Mourad, M. (1994) Estimation of genetic and phenotypic parameters of some reproductive traits of African Common goats in Rwanda. *Small Ruminant Research* 15, 67–71.
Mourad, M. (1996) Estimation of repeatability of litter size of common African goats and crosses with alpine in Rwanda. *Small Ruminant Research* 19, 263–266.
O'Callaghan, D., Roche, F.J., Boland, M.P. and Karsch, F.J. (1989) Does a melatonin implant mimic a short day photoperiodic effect in ewes? *Journal of Animal Science* 67, Abstract 879.
Osinowo, O.A., Amed, M.S. and Ekpe, G.A. (1988) Semen quality and sperm output of Yankasa rams at different ages. *Theriogenology* 29, 381–386.
Papachristoforou, C., Koumas, A. and Photiou, C. (2007) Initiation of the breeding season in ewe lambs and goat kids with melatonin implants. *Small Ruminant Research* 73, 122–126.
Pellicer-Rubio, M.T., Leboeuf, B., Bernelas, D., Forgerit, Y., Pougnard, J.L., Bonné, J.L., Senty, E., Breton, S., Brun, F. and Chemineau, P. (2008) High fertility using artificial insemination during deep anestrus after induction and synchronisation of ovulatory activity by the 'male effect' in lactating goats subjected to treatment with artificial long days and progestagens. *Animal Reproduction Science* 109, 172–188.
Perez, B. and Mateos, E. (1996) Effect of photoperiod on semen production and quality in bucks of Verata and Malagueña breeds. *Small Ruminant Research* 22, 163–168.
Pieterse, M.C. and Taverne, M.A.M. (1986) Hydrometra in goats: diagnosis with real-time ultrasound and treatment with prostaglandins or oxytocin. *Theriogenology* 26, 813–821.
Pintado, B., Gutierrez-Adan, A. and Perez Llano, B. (1998) Superovulatory response of murciana goats to treatments based on PMSG/anti-PMSG or combined FSH/PMSG administration. *Theriogenology* 50, 357–364.
Prasad, S.P. and Bhattacharyya, N.K. (1979) Estrus cycle and behaviour in different seasons in Barbari nannies. *Indian Journal of Animal Science* 49, 1058–1062.
Price, E.O. and Smith, V.M. (1984) The relationship of male–male mounting to mate choice and sexual performance in male dairy goats. *Applied Animal Behaviour Science* 13, 71–82.
Puls-Kleingeld, M., Nowshari, M.A. and Holtz, W. (1992) Cryopreservation of goat embryos by the one-step or three-step equilibration procedure. In: Lokeshwar, R.R. (ed.) *Recent Advances in Goat Production*. Nutan Printers, New Delhi, India, pp. 1388–1391.
Quinlivan, T.D., Martin, C.A., Taylor, W.B. and Cairney, I.M. (1966) Estimates of pre-and perinatal mortality in the New Zeland Romney Marsh ewe. *Journal of Reproduction and Fertility* 11, 379–390.
Ranilla, M.J., Sulon, J., Carro, M.D., Mantecon, A.R. and Beckers, J.F. (1994) Plasmatic profiles of pregnancy-associated glycoprotein and progesterone levels during gestation in Churra and Merino sheep. *Theriogenology* 42, 537–545.

Ranilla, M.J., Sulon, J., Mantecon, A.R., Beckers, J.F. and Carro, M.D. (1997) Plasma pregnancy-associated glycoprotein and progesterone concentrations in pregnant Assaf ewes carrying single and twin lambs. *Small Ruminant Research* 24, 125–131.

Rao, V.H. and Bhattacharrya, N.K. (1980) Ovulation in Black Bengal nanny goats. *Journal of Reproduction and Fertility* 58, 67–69.

Refstal, K.R., Marteniuk, J.V., Williams, C.S.F. and Nachreiner, R.F. (1991) Concentrations of estrone sulfate in peripheral serum of pregnant goats: relationships with gestation length, fetal number and the occurrence of fetal death in utero. *Theriogenology* 36, 449–461.

Reichle, J.K. and Haibel, G.K. (1991) Ultrasonic biparietal diameter of second trimester Pygmy goat fetuses. *Theriogenology* 35, 689–694.

Rekik, M., Aloulou, R. and Ben Hamouda, M. (2005) Small ruminant breeds of Tunisia. In: Iniguez, L. (ed.) *Characterisation of Small Ruminant Breeds in West Asia and North Africa*, Vol. 2. International Centre for Agricultural Research in the Dry Areas (ICARDA), Aleppo, Syria, pp. 91–140.

Restall, B.J. (1992) Seasonal variation in reproductive activity in Australian goats. *Animal Reproduction Science* 27, 305–318.

Restall, B.J., Milton, J.T.B., Klongyutti, P. and Kochapakdee, S. (1990) Pregnancy diagnosis in Thai native goats. *Theriogenology* 34, 313–317.

Restall, B.J., Restall, H. and Walkden-Brown, S.W. (1995) The induction of ovulation in anovulatory goats by oestrous females. *Animal Reproduction Science* 40, 299–303.

Riera, S. (1982) Reproductive efficiency and management in goats. In: *Proceedings of the Third International Conference on Goat Production and Disease*, Tuscon, Arizona, USA, 10–15 January 1982. Dairy Goat Publishing Co., AZ, USA, pp. 162–174.

Ritar, A.J. and Ball, P.D. (1991) Fertility of young Cashmere goats after laparoscopic insemination. *Journal of Agricultural Science* 117, 271–273.

Ritar, A.J., Salamon, S., Ball, P.D. and O'May, P.J. (1989) Ovulation and fertility in goats after intravaginal device PMSG treatment. *Small Ruminant Research* 2, 323–331.

Rivera, G.M., Alanis, G.A., Chaves, M.A., Ferrero, S.B. and Morello, H.H. (2003) Seasonality of estrus and ovulation in Creole goats of Argentina. *Small Ruminant Research* 48, 109–117.

Roca, J., Martinez, E., Vazquez, J.M. and Coy, P. (1992) Characteristics and seasonal variations in the semen of Murciano-Granadina goats in the Mediterranean area. *Animal Reproduction Science* 29, 255–262.

Romano, J.E. (2004) Synchronization of estrus using CIDR, FGA or MAP intravaginal pessaries during the breeding season in Nubian goats. *Small Ruminant Research* 55, 15–19.

Rubianes, E. and Menchaca, A. (2003) The pattern and manipulation of ovarian follicular growth in goats. *Animal Reproduction Science* 78, 271–287.

Salama, A. (1972) Ovarian changes in goats during estrus. *Indian Journal of Animal Science* 42, 436–438.

Santa Maria, A., Cox, J., Muñoz, E., Rodriguez, R. and Caldera, L. (1990) Estudio del ciclo sexual, estacionalidad reproductiva y control del estro en la cabra Criolla en Chile, Liverstock reproduction in Latin America. In: *Proceedings of the Final Research Co-ordination Meeting*, Bogotá, 19–23 September 1988. International Atomic Energy Agency, Vienna, Austria, pp. 363–385.

Schwarz, T. and Wierzchos, E. (2000) Relation between FSH and ovarian follicular dynamics in goats during the estrous cycle. *Theriogenology* 53, 381.

Shalaby, A.S., Shalawy, S.M., Saleh, N.H. and Medan, M.S. (2000) Reproductive performance of Damascus goats in semi-arid areas in Egypt. In: *Proceedings of the 7th International Conference on Goats*, France, 15–21 May, Vol. 1, pp. 424–425.

Shaw, F.D. and Morton, H. (1980) The immunological approach to pregnancy diagnosis: a review. *Veterinary Record* 106, 268–270.

Shelton, M. (1960) Influence of the presence of a male goat on the initiation of estrous cycling and ovulation of Angora does. *Journal of Animal Science* 19, 368–375.

Shelton, M. (1982) Methods of pregnancy diagnosis in Angora and dry meat-type goats. In: *Proceedings of the 3rd International Conference on Goat Production and Disease*, Tucson, Arizona, USA, 10–15 January, p. 496 (abstract).

Shrestha, J.N.B. and Fahmy, M.H. (2007) Breeding goats for meat production: 3. Selection and breeding strategies. *Small Ruminant Research* 67, 113–125.

Simões, J., Almeida, J.C., Valentim, R., Baril, G., Azevedo, J., Fontes, P. and Mascarenhas, R. (2006) Follicular dynamics in Serrana goats. *Animal Reproduction Science* 95, 16–26.

Smith, M.C. (1980) Caprine reproduction. In: Morrow, D.A. (ed.) *Current Therapy in Theriogenology*. W.B. Saunders, Philadelphia, PA, USA, pp. 975–977.

Song, H.B., Jo, I.H. and Sol, H.S. (2006) Reproductive performance of Korean native goats under natural and intensive conditions. *Small Ruminant Research* 65, 284–287.

Sousa N.M., Garbayo, J.M., Figueiredo, J.R., Sulon, J., Gonçalves, P.B.D. and Beckers, J.F. (1999) Pregnancy-associated glycoprotein and progesterone profiles during pregnancy and postpartum in native goats from the north-east of Brazil. *Small Ruminant Research* 32, 137–147.

Sutherland, S.R.D. (1987) Progesterone concentration and pulsatile LH secretion during normal oestrous cycles in Angora-cross does. In: *Proceedings of the 4th AAAP Animal Science Congress*, Hamilton, New Zealand, p. 246.

Thibier, M., Jenguyot, N. and Montigny, G.D. (1982) Accuracy of early pregnancy diagnosis in goats based on plasma and milk progesterone concentrations. *International Goat and Sheep Research* 2, 1–6.

Thorburn, G.D. and Schneider, W. (1972) The progesterone concentration in the plasma of the goat during the estrus cycle and pregnancy. *Journal of Endocrinology* 52, 23–36.

Tuli, R.K. and Holtz, W. (1995) Effect of season on the freezability of Boer goat semen in the northern temperate zone. *Theriogenology* 43, 1359–1363.

van der Westhuizen, J.M., Wentzel, D. and Grobler, M.C. (1985) *Angora Goats and Mohair in South Africa.* NKB Printers, Port Elizabeth, Republic of South Africa, pp, 138–140.

Viñoles, C. (2003) Effect of nutrition on follicle development and ovulation rate in the ewe. PhD thesis, University of Uppsala, Sweden.

Walkden-Brown, S.W. (1991) Environmental and social influences on reproduction in Australian cashmere goats. PhD thesis, University of Queensland, Australia.

Walkden-Brown, S.W. and Bocquier, F. (2000) Nutritional regulation of reproduction in goats. In: *Proceedings of the 7th International Conference on Goats*, Tours, France, pp. 389–395.

Walkden-Brown, S.W., Restall, B.J. and Henniawati (1993a) The male effect in the Australian cashmere goat. 2. Role of olfactory cues from the male. *Animal Reproduction Science* 32, 55–67.

Walkden-Brown, S.W., Restall, B.J. and Henniawati (1993b) The male effect in the Australian cashmere goat. 3. Enhancement with buck nutrition and use of oestrous females. *Animal Reproduction Science* 32, 69–84.

Walkden-Brown, S.W., Martin, G.B. and Restall, B.J. (1999) Role of male–female interactions in regulating reproduction in sheep and goats. *Journal of Reproduction and Fertility Supplement* 54, 243–257.

Wang, X.L., El-Gayar, M., Knight, P.G. and Holtz, W. (2009) The long-term effect of active immunization against inhibin in goats. *Theriogenology* 71, 318–322.

Wango, E.O., Heap, R.B. and Wooding, F.B.P. (1991) Progesterone and 5b-pregnanediol production by isolated fetal placental binucleate cells from sheep and goats. *Journal of Endocrinology* 129, 283–289.

Wani, G.M. (1996) *Embryo Biotechnology in Sheep and Goats.* Valley Book House Publisher, University Road, Hazratbal Srinagar, Kashmir.

Wani, G.M. and Sahni, K.L. (1980) An ultrasonic technique for detection of pregnancy in sheep and goats. *Veterinary Research Journal* 2, 35–73.

West, K.S., Meyer, H.H. and Nawaz, M. (1991) Effects of differential ewe condition at mating and early postmating nutrition on embryo survival. *Journal of Animal Science* 69, 3931–3938.

Wuliji, T., Litherland, A., Goetsch, A.L., Sahlu, T., Puchala, R., Dawsonc, L.J. and Gipson, T. (2003) Evaluation of melatonin and bromocryptine administration in Spanish goats: I. Effects on the out of season breeding performance in spring, kidding rate and fleece weight of does. *Small Ruminant Research* 49, 31–40.

Yuswiati, E. and Holtz, W. (1990) Work in progress: successful transfer of vitrified goat embryos. *Theriogenology* 34, 629–632.

Zarkawi, M. and Soukouti, A. (2001) Serum progesterone levels using radioimmunoassay during oestrous cycle of indigenous Damascus does. *New Zealand Journal of Agricultural Research* 44, 165–169.

Zarkawi, M., Al-Merestani, M.R. and Wardeh, M.F. (1999) Induction of synchronized oestrus in indigenous Damascus goats outside the breeding season. *Small Ruminant Research* 33, 193–197.

Zoli, A.P., Guilbault, L.A., Delahaut, P., Benitez-Ortiz, W. and Beckers, J.F. (1992) Radioimmunoassay of a bovine pregnancy-associated glycoprotein in serum: its application for pregnancy diagnosis. *Biology of Reproduction* 46, 83–92.

# 6 Nutrition of the Meat Goat

J.R. Kawas[1], O. Mahgoub[2] and C.D. Lu[3]
[1]*Cuerpo Académico de Producción Animal y Recursos Naturales, Facultad de Agronomía, Universidad Autónoma de Nuevo León, Nuevo León, Mexico;* [2]*Department of Animal and Veterinary Sciences, College of Agriculture and Marine Sciences, Sultan Qaboos University, Oman;* [3]*University of Hawaii, Hilo, Hawaii, USA*

## 6.1 Abstract

This chapter aims to provide a summary of some applications of basic nutrition to the meat goat. Nutrient requirements of goats and nutritional limitations associated with forages and other feed resources are discussed. Some aspects of the feeding habits, feed composition, nutrient requirements, feed formulation and feeding of goats will be discussed. Present knowledge of nutrient requirements and nutritional management based on information available for the meat goat of native and specialized breeds in the literature is summarized. Emphasis is also placed on the use of feed supplements to optimize the production potential of goats in semi-arid range environments. Where applicable, comparisons with other relevant species are made.

## 6.2 Introduction

Nutrition involves the consumption, digestion, absorption and metabolism of nutrients required for tissue repair and health, reproduction, growth and lactation of the meat goat. Browsing constitutes the main diet of free-ranging goats. As intermediate opportunistic domestic feeders, goats possess morphological, physiological and behaviour characteristics that allow them to select and have a greater availability of the native vegetation than other ruminants (Malechek and Provenza, 1983). Among these are the mobility of their superior lip, their ability to stand in a biped posture and their capability for travelling long distances (Askins and Turner, 1972).

Forage fibre and other feed components are fermented in their rumen, generating energy precursors, synthesizing microbial protein for post-ruminal absorption as amino acids, and synthesizing B-complex vitamins and vitamin K (Haenlein and Caccese, 1992). Water conservation and nitrogen recycling are important mechanisms that allow goats to withstand better the adverse conditions of semi-arid regions of the world (Silanikove, 2000).

Many environmental and nutritional factors influence goat production in rangelands. Reduced forage availability and quality lead to low consumption of feed nutrients, which are further exacerbated by the effect that plant secondary metabolites (PSMs) in browse and parasitism have on nutrient utilization. Frequently, available feed resources fail to satisfy the maintenance requirements of small ruminants (Ben Salem and Nefzaoui, 2003). Supplementation is required to mitigate both nutrient deficiencies and the effect of PSM toxicity.

Supplementary feeds include leguminous fodders, which can serve as good protein and energy sources, cacti, which require very little water, and energy sources such as molasses, cereal grains and by-products and oilseed meals. Mineral and protein supplementation of goats consuming high-fibre/low-protein forages generally improves intake and performance. Nutrient-specific supplements should take into consideration the protein content and mineral profiles of forages in each region. These should be formulated to be cost-effective and supply those nutrients required to optimize growth and reproduction of range goats (Kawas *et al.*, 2010).

## 6.3 Gastrointestinal Tract

The goat gastrointestinal tract has evolved on the basis of adaptation to feed resources. The goat is a ruminant animal with a digestive tract consisting of mouth, oesophagus, a four-compartment stomach, small intestine (duodenum, jejunum and ileum), and large intestine (caecum and colon). Accessory organs such as the salivary glands, liver and pancreas contribute to digestion. Ruminant comes from the word 'rumen', which is the first major compartment of the four-compartment stomach of goats where microbial fermentation takes place. Some anatomical and physiological aspects of the digestive tract of the goat have been discussed by several authors (Haenlein and Caccese, 1992; Silanikove, 2000; Lu *et al.*, 2005; Hart, 2008) and will be summarized in this section.

The primary prehensile structures of the goat are the mouth, lips, teeth and tongue. Like other ruminant animals, goats lack upper incisor and canine teeth. They depend on the hard and rigid dental pad, lower incisor teeth, lips and tongue to bite and take feed into their mouths. The lips are important prehensile organs. This characteristic enables the greater selectivity of goats when grazing under range conditions. Feed components are broken down into smaller particles by the mechanical action of the teeth. The tongue helps in mixing the feed in the mouth and swallowing it. Five pairs of salivary glands (parotid, submaxillaries, sublingual, inferior molar and buccal glands) secrete a mixture of serous and mucous fluids, released by the stimulation of chewing activities including eating and rumination (Haenlein and Caccese, 1992). The saliva secreted aids in cud chewing and swallowing, and is needed for buffering and maintenance of rumen pH within an optimal range (pH 6.2–6.8) for microbial growth (Lu *et al.*, 2005).

The insalivated feed then enters the oesophagus, which is a tube-like duct that carries food from the mouth into the stomach at the reticulum. The feed bolus is forced down the oesophagus as a result of voluntary muscular contractions in its proximal portion, while smooth muscles in the remainder of the oesophagus cause peristaltic contractions that transport the bolus further to the reticulum. During resting, the ball-like bolus of fibrous and coarse feed, called the cud, is regurgitated from the reticulo-rumen into the mouth through the oesophagus. The rumination process includes regurgitation of the cud, which is thoroughly chewed and swallowed again. Goats may spend more than 7 h a day ruminating, depending on the fibre content of the feed (Lu *et al.*, 2005). As chewing or mastication is comprised of both consumption and rumination, excess fibre in the diet may increase the latter, affecting the amount of feed dry matter (DM) the goat can eat. Chewing aids in particle size reduction, allowing a greater surface area for microbial digestion and for the smaller particles to flow out of the rumen. Lateral movements of the jaw allow effective grinding of forage and other feed particles (Haenlein and Caccese, 1992).

The four-compartment stomach is comprised of the reticulum, rumen, omasum and abomasum, the latter being the true stomach (Fig. 6.1) where enzymatic digestion occurs. Each of these compartments has a specific function in the digestive process. When the goat kid is born, the rumen, reticulum and omasum are underdeveloped,

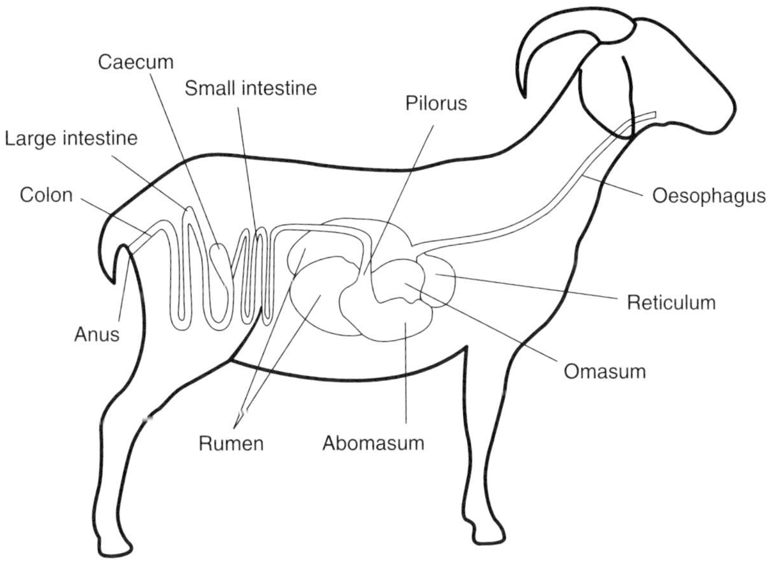


**Fig. 6.1.** Diagram of the digestive tract of the meat goat.

and the abomasum or gastric stomach is the largest of the four compartments. The abomasum functions similarly to the stomach of non-ruminant animals. The utilization of nutrients and absorption of antibodies (immunoglobulins) from colostrum that bypasses the rumen and reticulum through the oesophageal groove and into the abomasum is high during the first days of life. The four stomach compartments reach their adult size after 2 months of age. In the adult goat, the rumen is the largest of the four stomach compartments.

Feed first undergoes microbial digestion in the rumen and reticulum, and these compartments are collectively referred to as the reticulo-rumen. The reticulum is separated from the rumen by the reticulo-ruminal fold. The capacity of the goat reticulum is only 1–2 l, whereas the rumen contains a considerable amount of fluid (12–25 l; >20 l in the adult goat), which depends on the type of feed consumed (Silanikove, 2000). Small finger-like projections called papillae that cover the rumen surface increase the absorptive surface area.

As goat kids grow, they continuously increase their consumption of solid feed, allowing the rumen to develop and microorganisms to thrive. Microbial fermentation accomplished by millions of microorganisms such as bacteria, protozoa and fungi produce enzymes that break down forage fibre and other feed components, generating energy precursors know as volatile fatty acids (VFAs), and synthesizing B-complex vitamins and vitamin K and providing protein from microbial origin for post-ruminal digestion and absorption as amino acids by the host animal.

Carbohydrates are fermented in the reticulo-rumen producing VFAs, primarily acetic, propionic and butyric acids, which are absorbed through the rumen wall providing most of the energy requirements. Fermentation by rumen microorganisms also generates gases, primarily methane and carbon dioxide, which are lost through eructation. With the rapid consumption of large amounts of lush legumes or grain, goats may be unable to eliminate these accumulated gases, producing a condition known as bloat, which may be lethal.

After forage has been digested and broken down into small particles in the reticulo-rumen, it passes through the omasum where water absorption occurs, and then enters the abomasum, the true stomach, where hydrochloric acid and digestive enzymes are secreted. Here, the digestion process continues before passing into the small intestine. The capacity of the abomasum is approximately 4 l in the adult goat. The mucosa of the abomasum contains parietal cells that secrete hydrochloric acid, and peptic or chief cells that secrete the enzyme pepsin, and in the young goat the enzyme rennin is responsible for the formation of milk curdles and digestion of casein, the major protein in milk. The acidic condition of the abomasum is maintained at approximately pH 3. Lipase in gastric secretions degrades fat into glycerol and fatty acids.

Feed passes from the abomasum into the first segment of the small intestine, the duodenum, through the pyloric sphincter. The middle and later segments of the small intestine are the jejunum and ileum, respectively. In the small intestine, enzymatic digestion continues and most of the absorption occurs here. While digestion in the abomasum occurs in an acid environment of pH 3, duodenal digestion takes place in an alkaline medium of up to pH 8. Secretions from the pancreas, liver and duodenum help alkalinize the partially digested aqueous feed flowing from the stomach into the duodenum, known as chyme. Enzymes secreted by the pancreas are important in the small intestinal digestion of carbohydrates, proteins and fats. Bile produced by the liver, stored by the gall bladder and secreted into the small intestine, helps emulsify fat, allowing lipase enzymes to break down fats. Peristaltic contractions move the chyme through the small intestine helping to drive the flow of blood and lymph in the intestinal wall.

The large intestine is comprised of the caecum, colon and rectum. Its major function is the removal of water from the intestinal contents to concentrate the feed residues before excretion, forming faecal pellets at the end of the colon. Maximum water removal is attained by reducing the rate of passage of the intestinal contents, which appears to be an important water-conservation mechanism in goats (Silanikove, 2000). The capacity of the large intestine of goats ranges from 5 to 6 l.

## 6.4 Feeding Habits of Goats

Research on feeding behaviour and diet selection of meat goats in rangelands explains why goats select certain plants and avoid others, as well as their dietary preferences in different seasons and curious diet selection profile based on the nutrient composition of plants.

Some morphological, physiological and behaviour characteristics of goats are inherent in the strategy used to utilize the available feed resources, and these are manifested in their feeding behaviour, which we refer to as 'selectivity' (Malechek and Provenza, 1983). Among these are the mobile superior lip and their ability to stand in a biped posture. The biped posture greatly maximizes the available forage in a specific area in which the goat can browse. Another characteristic of goats is that they can travel greater distances than cattle or sheep, exposing them to a greater quantity of forage and variety of forage species (Askins and Turner, 1972).

Goats are known as opportunistic herbivores due to their versatile feeding habits and ease of adaptation to vegetation changes compared with cattle or sheep (Lu and Coleman, 1984). In contrast to other ruminants fed in confinement, goats on grazing have the liberty to select their diet from a complex variety of native plant species available in the range. The selection of some plant species in relation to others depends on previous experience, and on the nutritional and physiological state of the animal. Additionally, the diet selected in a particular situation depends on diverse interactions between soil, plants and the environment (Malechek and Provenza, 1983).

In semi-arid climates, meat goats are exposed to the changing vegetation diversity that occurs with the extreme climate conditions of these regions. With changes in diet availability, diet preferences are related to the fibre and nutrient composition of plants. Thus, goats consume mixed diets and forages based on season and opportunities. These animals prefer diets based on the concentrate components of plant material and will select shrubs and browse (Silanikove, 2000).

Ruminants have been classified according to their morphological feeding type into three groups: the concentrate selectors, the intermediate type and the grass/roughage eaters (Hofman, 1989). As intermediate opportunistic domestic feeders, goats tend to select more nutrient-dense diets such as forbs and shrub foliage, whereas cattle and sheep may consume grass and roughage (Lu, 1988).

One theory suggests that the superior digestion capacity of goats, in comparison with other ruminant species, is due to the presence of large salivary glands, the large absorptive area of their rumen epithelium and the ability to rapidly increase foregut volume when high-fibre feeds are consumed (Silanikove, 2000). However, this advantage appears to be true only for certain breeds or strains having particular anatomical dimensions and is not a characteristic of the entire goat population (Huston, 1978).

The second theory suggests that the ability for selective feeding of cattle, sheep and goats is inversely related to their ability to retain and digest fibre in the rumen. The lower ability to digest fibrous carbohydrates has been attributed to their adaptation for selective discrimination of forages for quality. Thus, high-fibre diets may significantly affect their energy intake, as organic matter digestibility and forage intake may be lower (Huston *et al.*, 1986).

Although forage quality decreases during the dry season, diet composition may depend on availability and the botanical profile of the rangeland grazed by goats. The goat diet in the range is variable and consists of shrubs, herbs and grasses. Feeding habits have been defined as follows (Askins and Turner, 1972):

1. *Grazing* is the habit of feeding on vegetation in its natural state at surface level, including leaves of deciduous trees and the stems of some shrubs. Goats graze grass and herbs when they are more abundant and prefer browsing when woody vegetation predominates. In north-east Brazil, low dietary crude protein (CP) and neutral detergent fibre (NDF) contents of the diet of goats during the wet season suggest that the goats' diet may have been composed largely of grasses (Kawas *et al.*, 1999).
2. *Browsing* is the habit of feeding on intact foliage, buds, flowers, stems and woody trees. Browsing constitutes the main diet of the range goat in the semi-arid regions of north-east Brazil (Schacht *et al.*, 1992), north-east Mexico (Ramirez *et al.*, 1990) and the state of Texas in the USA (Askins and Turner, 1972). With this feeding habit, goats can select and have a greater availability of the native vegetation than other ruminants (Malechek and Provenza, 1983). The CP content of diet samples of goats grazing in a dense stand of caatinga woodland in north-east Brazil was high at the end of the wet season (12%) in comparison with the dry season (8.4%). Botanical analyses indicated that >75% of the diet during the dry season was composed of dead herbaceous and fallen tree leaves (Schacht *et al.*, 1992).

Generally, the influence of humans on the grazing behaviour of goats is limited to decisions about the season and location of grazing, carrying capacity and herd composition (Malechek and Provenza, 1983). However, the diet selected in a particular situation is ultimately dependent on seasonal variations and geography. In addition, palatability is determined by particular characteristics of each plant, and these have a wide influence on acceptability or rejection of forage available to goats.

## 6.5 Feed Nutrients

Feedstuffs commonly consumed by meat goats are mostly of plant origin. Forages contain varying quantities of water, and the DM fraction is composed of organic and inorganic components. Organic matter includes nutrients such as carbohydrates, proteins, fats and vitamins, whereas minerals are inorganic components. Both organic and inorganic entities may provide nutrients for body maintenance, growth, reproduction, pregnancy and lactation.

DM may be partitioned based on the nutritive value of its components into cell contents and cell wall components (Fig. 6.2) as follows (Moore and Hartfield, 1994):

1. Cell contents:

- Organic acids
- Mono- and oligosaccharides (sugars)
- Non-structural polysaccharides (starches)
- Protein and other nitrogen-containing compounds
- Lipids, fatty acids and other components soluble in ether
- Soluble minerals

2. Cell wall:

- Structural polysaccharides (partially nutritive matter)
  - Cellulose
  - Hemicellulose
  - Pectic compounds
- Non-nutritive matter
  - Lignin
  - Non-nutritive ether extracts (wax, terpenes, etc.)
  - Insoluble minerals (silica)

### 6.5.1 Water

Water is an essential nutrient of high consumption and may be the most critical of all nutrients. Life is dependent more on the availability of water than on other nutrients. Reduced water availability can restrict feed intake and feed efficiency, and adversely affects growth, reproduction and milk production. Water is responsible for the structure and function of cells and serves as a solvent for digestion, absorption, transport and metabolism of nutrients. Water is needed as a medium for metabolic reactions and for excretion of metabolic residues.

Water intake can be affected by factors related to the animal, the environment, feed composition and water quality.

#### *Animal factors*

Water intake depends on the body size, age, physical activity and health status of goats, as well as environmental factors. Goats drink 2 litres of water per kg of DM consumed in the humid cool winter season, with the ratio increasing to 6:1 with heat stress in the hot summer season (Denek *et al.*, 2006). With greater physical activity, water consumption is also increased. Younger goats require more water than older animals. However, more water is required for milk production, as goat's milk is >87% water. Desert goats are among the most efficient of domestic animals in their use of water, withstanding prolonged periods of water deprivation and grazing far away from watering sites (Silanikove, 2000). Some mechanisms that make goats less susceptible to high temperature stress than other species of domestic livestock include the ability to maintain water economy together with the capacity to endure severe dehydration and undergo rapid rehydration. The role of the rumen as a water reservoir is more pronounced in desert mammals, particularly desert goats. The capacity of the desert goat to secrete

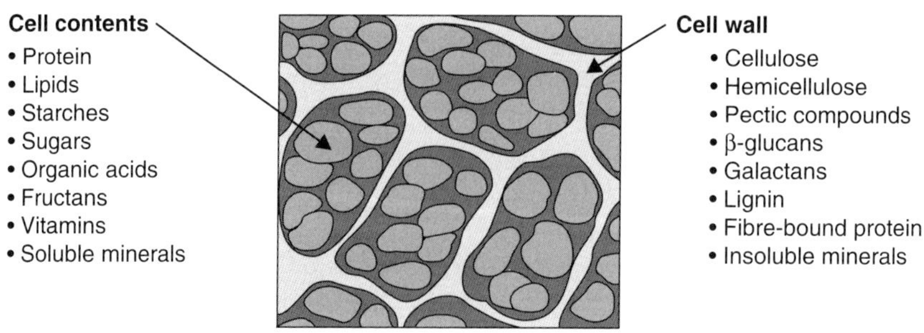


**Fig. 6.2.** Plant cell composition.

large amounts of saliva allows them to achieve efficient retention of water following rehydration. In addition to a reduced need for body water evaporation for maintaining comfort in hot climates, goats can conserve body water by decreased losses in urine and faeces (Silanikove, 2000; Merck, 2008).

*Physical activity*

In comparison with confined goats, activity under grazing conditions is more demanding for water. Goats need to walk more to obtain the feed they need, and for this additional activity more water is required.

*Environmental factors*

Water is also needed for body temperature regulation. Ambient temperature and humidity are factors affecting water intake. During the summer, especially on hot humid days, water evaporation from the organism through sweat may cause greater water consumption than normal. During cold winter days, freezing of water may reduce its availability for intake.

*Feed quality and composition*

Protein levels in the ration affect water consumption. During protein metabolism, urea is synthesized in the liver and excreted through the kidney. With more protein in the diet, which may include the use of feed-grade urea, water intake is greater as more urine is excreted to eliminate the greater quantity of urea being produced in the liver (Quinisa and Boomker, 1998). An excess of salt in the ration will also cause an increase in water consumption.

### 6.5.2 Fibre and non-fibre carbohydrates

Carbohydrates comprise the greatest proportion of the goat's diet. Forages primarily contain fibre carbohydrates, whereas grains contain mostly non-fibre polysaccharides. The carbohydrate fractions may be determined analytically using chemical and enzymatic methods, which should represent nutritionally relevant fractions. On the other hand, the non-structural carbohydrates are comprised of carbohydrates from the cell contents including monosaccharides, oligosaccharides (low-molecular-weight carbohydrates), fructans and starch.

### 6.5.3 Fibre fraction

In nutrition, the term fibre has been defined as 'polymeric compounds from plants and other feedstuffs that are not digested by the action of mammalian digestive enzymes' (Moore and Hartfield, 1994). This definition of fibre in the diet has been accepted by both ruminant and non-ruminant nutritionists. The mammal digestive tract does not secrete enzymes that are able to hydrolyse the β1-4 linkages that predominate in plant cell wall polysaccharides, and thus depend on microorganisms in the ruminant gastrointestinal tract to ferment these structural carbohydrates, producing volatile fatty acids that are further absorbed. Ruminants are among the most specialized herbivores using this symbiotic relationship to utilize plant cell walls as a source of nutrients (Moore and Hartfield, 1994; van Soest, 1994).

Plant cells can be divided into two fractions based on their nutritional availability (Fig. 6.3). One of these fractions is soluble in a neutral detergent solution (plant cell contents), while the other is an insoluble fraction that represents the plant cell wall, also known as neutral detergent fibre (NDF). In forages commonly fed to livestock, fibre refers to the plant cell wall. Another chemical entity that represents the less digestible fibre is acid detergent fibre, which is obtained from the dry residue that results from the further solubilization of hemicellulose with an acid detergent solution (Moore and Hartfield, 1994).

As the biological functions of the plant cell wall have resulted in a chemical structure of variable and often low digestibility, it appears obvious that the

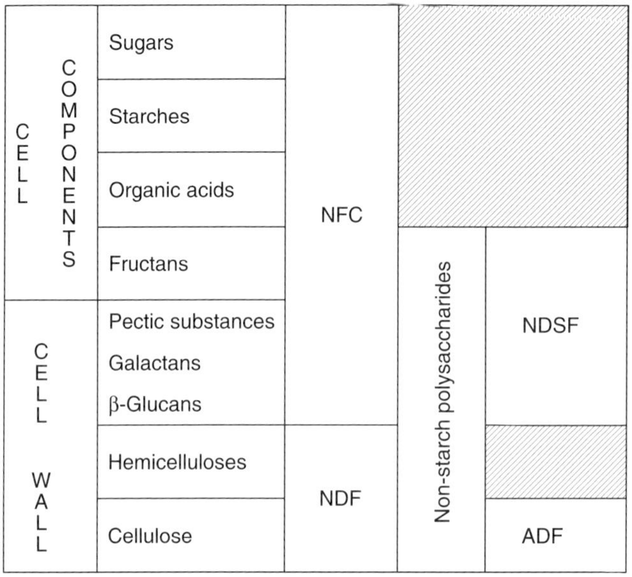


**Fig. 6.3.** Chemical determination of carbohydrate fractions. NDF, Neutral detergent fibre; NDSF, neutral detergent-soluble fibre; NFC, non-NDF carbohydrates; ADF, acid detergent fibre (adapted from Hall, 2003).

cell wall did not evolve to serve as a feed for ruminants. In the rumen, cell walls physically occupy volume, affecting feed intake and animal performance. Analysis of the amount of fibre or cell wall present in feedstuffs of meat goats is of major importance as their diet often contains large amounts of forage.

### 6.5.4 Non-structural carbohydrates

As the determination of non-structural carbohydrates is a time-consuming analytical procedure, for practical purposes, analogously, the amount of non-fibre carbohydrates is commonly calculated by subtracting from the DM content, the NDF, CP, ether extract and ash fractions determined in the laboratory. This fraction encompasses organic acids, mono- and disaccharides, oligosaccharides, fructans, starch, pectic substances, (1→3)(1→4)-β-glucans, and other carbohydrates exclusive of the hemicelluloses and cellulose found in NDF (Hall, 2003).

### 6.5.5 Protein

Proteins are nitrogen-containing molecules, found in the body tissue and milk of the meat goat. Proteins contain one or more amino acid chains, and their quality relates to their amino acid profile. Proteins are essential components of blood, muscle, skin and bone cells. Enzymes, hormones and antibodies are also protein molecules, and are involved in digestion, metabolism, reproduction and immunity. Amino acids from protein in the diet are required for body maintenance, growth, gestation and lactation of meat animals.

Protein from feed is partially degraded in the rumen, with $NH_3$ being available for utilization by rumen microorganisms and synthesis of microbial protein (Fig. 6.4). Urea is commonly used as an economic source of non-protein nitrogen for inclusion in ruminant feeds. Nitrogen from urea is utilized for microbial protein synthesis, for which energy is also required. Excess $NH_3$ in the rumen is absorbed and transported through the portal vein to the liver, where

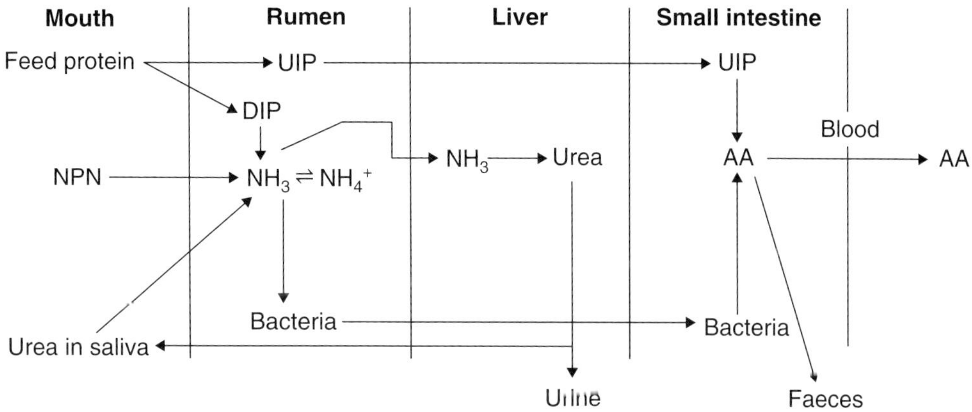


**Fig. 6.4.** Nitrogen utilization in the ruminant. AA, Amino acids; DIP, degradable intake protein; NPN, non-protein nitrogen; UIP, undegradable intake protein (adapted from NRC, 1985).

urea is synthesized and may be further recycled in saliva or excreted in urine. The carbon skeletons from the protein molecule will provide energy for the goat. As lush forages and silages contain a high concentration of non-protein nitrogen, the use of urea should be limited in these diets or supplements for meat goats, whereas urea will be more useful when goats are consuming straw or other low-quality forage.

The degradable protein fraction is known as degradable intake protein, whereas the intact protein fraction that resists rumen degradation is known as undegradable intake protein or bypass protein. Microbial protein and feed proteins that escape degradation in the rumen are further broken down by the action of digestive enzymes in the abomasum and small intestine, absorbed in the small intestine and used to replace body losses and for the synthesis of new body tissue. Unlike meat goats, a high percentage of bypass protein is required by high-yield-producing dairy goats. Meat goats generally do not require much bypass protein, except when grazing or browsing lush forage that contains a high proportion of degradable intake protein.

Improvements in the nutritive value of low-quality forage diets for goats often depend on increasing the supply of rumen nitrogen, thereby increasing fibre digestibility (Ben Salem and Smith, 2008). DM intake and digestibility are reduced if dietary protein content is <7%, further causing an energy deficiency. Studies on the effect of energy or protein supplementation have demonstrated a response in feed intake of beef cattle (NRC, 1987) and goats (Negesse *et al.*, 2001) consuming low-level CP diets based on low-quality roughages. Supplements that provide non-protein nitrogen and true protein may increase forage consumption as the minimum nitrogen requirements for rumen microorganisms are satisfied. Because an important proportion of the protein associated with the undegradable fibre fraction may not be utilized by rumen microbes, setting a minimum CP level for normal rumen function may be especially critical with high-fibre/low-protein roughages.

As the CP content of forage increases, the magnitude of the response in production with additional protein supplementation may be in response to changes in forage intake rather than digestibility or to a greater metabolic efficiency in nutrient utilization, which includes the effects of intake of degradable or undegradable protein (NRC, 1987). As forages from temperate areas have a high level of degradable protein, undegradable intake protein supplementation can improve the performance of ruminants under grazing conditions. After satisfying the degradable protein needs of the rumen, additional feeding of undegradable protein may improve performance without affecting intake (Kawas *et al.*, 1997). With range goats browsing shrub vegetation, protein supplementation may improve performance as consumption of high levels of condensed tannins may reduce protein degradability due to binding of food proteins and inactivation of digestive tract enzymes (Kumar and Vaithyanathan, 1990).

### 6.5.6 Fats

Whereas carbohydrates are the main source of energy for ruminants, fats contain more energy per unit weight. Fats, also known as lipids, are composed of triglycerides, which are esters of glycerol and three fatty acids. Fats can be of animal or plant origin. Plants contain very small quantities of fat, this being the reason why the meat-goat diet is low in fat. In the laboratory, the fat content of feed is determined as that fraction that is soluble in ether. Other ether-soluble plant compounds such as waxes or terpenes can be found in the diet of browsing or grazing goats. However, these compounds do not supply energy to the goat.

The fat content of the range goat is generally <2%. The inclusion of fat in range supplements is not common. Although consumption of small quantities of fat can improve energy intake and goat performance, too much fat in the diet may reduce diet fibre digestibility. Animal and vegetable fat sources can be processed to be inert in the rumen, so that fibre digestion is not affected. These soaps (long-chain fatty acid calcium salts) are commonly called 'bypass fats'.

In confinement, fat may be added to the diet of goats to increase their energy content. This is not a common practice with meat goats, unlike with high-level-producing dairy goats. Feeding of fat could alter the milk fatty acid composition in lactating dairy goats (Lu, 1993). With dairy goats, a high dietary fat level may depress fibre fermentation, the rumen concentration of acetic acid and milk fat synthesis. Feeding meat goats high levels of crushed whole soybeans (0–33% of the whole diet) significantly reduced fibre digestibility, which appeared to be related to the fat content of the diet. Nitrogen retention also increased as the level of whole soybeans in the diet increased in response to a greater nitrogen intake (Kadzere and Jingura, 1993). With low-level supplementation of range goats, overfeeding of fat is not common, and the fat level may represent a greater proportion of the supplement in contrast to its inclusion in a complete feed offered in confinement (Kawas *et al.*, 2010).

### 6.5.7 Minerals

The ash contents of herbage and other feeds represent the inorganic matter fraction, or minerals. Minerals are inorganic nutrients and are subdivided into two groups, macrominerals and trace minerals. The macrominerals are those required in percentage quantities in the diet, whereas microminerals or trace minerals are those required at levels of parts per million (ppm). Macrominerals include calcium, phosphorus, sodium, chloride, potassium, magnesium and sulfur. The microminerals required in the diet of ruminants are iron, manganese, zinc, copper, iodine, selenium and cobalt.

The diverse functions of minerals in the body include the structural components of bones and teeth (calcium and phosphorus) and haemoglobin (iron), electrolytes involved in water balance (sodium and potassium) and components of enzymes involved in the regulation of metabolism (iodine is a structural component of the hormones thyroxine and tri-iodothyronine (NRC, 2007).

Under rangeland grazing systems, minerals consumed by goats depend almost exclusively on forage intake. Mineral concentrations of forages are dependent on the interactions between factors such as soil, forage species, stage of maturity, climate and season. The abundant vegetation available for goats during the rainy season may provide energy and protein that allow maximum growth. Although an abundant source of feed is available for range goats, unbalanced mineral concentrations in forages may be a cause of low production and reproductive problems (McDowell *et al.*, 1983).

Specific mineral supplementation of small ruminants is possible by mapping or studying the mineral concentrations in soil, forage, drinking water and animal tissue in a specific region. Supplementation of major minerals and trace minerals should be considered to satisfy the requirements for growth and reproduction (NRC, 2007). Other elements required, although not needing to be supplemented, are molybdenum and fluorine. By knowing the mineral profiles in forages consumed by goats of different regions, specific mineral supplements can be designed to satisfy requirements and optimize productivity and health (Kawas *et al.*, 2010).

### 6.5.8 Vitamins

Vitamins are organic compounds that are required in small quantities to maintain body functions, participating as cofactors in many metabolic processes. A deficiency of a vitamin will slow or block the metabolic path in which it is involved, resulting in the appearance of clinical symptoms. These symptoms become worse if the vitamin is not supplemented in the diet (NRC, 2007).

Vitamins can be grouped into water-soluble and fat-soluble. Among the water-soluble vitamins are the so-called B-complex vitamins such as thiamine ($B_1$), riboflavin ($B_2$), niacin ($B_3$), pantothenic acid ($B_5$), pyridoxine ($B_6$), biotin ($B_7$), folic acid ($B_9$) and cyanocobalamin ($B_{12}$). Another water-soluble vitamin, vitamin C, is synthesized by body tissue in quantities that meet the animal's needs. Fat-soluble vitamins include vitamins A, D, E and K. Vitamin K is the only fat-soluble vitamin synthesized in the rumen and is required for blood clotting (NRC, 2007).

All the water-soluble vitamins are synthesized by rumen microorganisms. However, a deficiency of two B vitamins may occur in meat goats:

- *Vitamin $B_{12}$.* A cobalt deficiency in the diet may result in not enough vitamin $B_{12}$ being synthesized by microbes in the rumen to satisfy needs; and
- *Thiamine (vitamin $B_1$).* A thiamine deficiency may cause polioencephalomalacia, a neurological disorder of ruminants characterized by necrosis of the cerebral cortex. The disease may appear in some cases of rumen dysfunction. Some predisposing factors include high-concentrate feeding, combined with a small particle size of forage and other feed ingredients and high sulfur intake. High sulfur content in water and feedstuffs such as molasses and distillers' dried grains in the diet may trigger the disease. Clinical signs include blindness, ataxia (incoordination) and recumbency with seizures. Treatment requires immediate injection of large quantities of thiamine (Gould, 1998).

Deficiency symptoms of vitamin A include xerophthalmia (failure to produce tears) and diarrhoea. Vitamin A is required for normal vision and the prevention of reproductive and respiratory problems. The precursor of vitamin A is β-carotene found in green forages. Of the fat-soluble vitamins, vitamin A is of importance for goats consuming hay or forage that have been exposed to and deteriorated by rain and unfavourable climate, especially during a prolonged drought period in semi-arid regions. If the dry season is not very long, vitamin A supplementation may not be needed as the liver stores may last up to 4 months in adult goats (Frier *et al.*, 1974). Vitamin A requirements are generally expressed as international units (IU). One IU per kg is equivalent to 1 United

States pharmacological unit (USP) per kg. Synthetic forms of the vitamin are available commercially. One milligram of β-carotene provides 400 IU of vitamin A.

Vitamin D is required for absorption and metabolism of calcium and phosphorus for bone growth and development. Vitamin D deficiency results in rickets, a disease characterized by lameness, weak bones and bowed and crooked legs. Goats exposed to sunlight can meet their vitamin D requirements by the action of ultraviolet light, which penetrates the skin. Meat goats continuously exposed to the sun would not normally require vitamin D supplementation. The liver is the main site for storage of vitamin D in the body. Vitamin $D_2$ is synthesized during haymaking of sun-cured forages.

Both vitamin E and selenium function as antioxidants, the requirement of either one being partially met by the other; this is the reason why vitamin E is important in areas in which selenium deficiency has been detected. A marginal deficiency of vitamin E can depress the immune system and cause reproductive failure. In the goat kid, a vitamin E deficiency may cause white muscle disease. Supplementation of vitamin E will prevent this disease. Vitamin E is commonly supplemented in relatively high quantities in dairy cow and dairy goat diets to stimulate the immune system, reducing the severity and duration of subclinical mastitis. High levels of vitamin E may be found in green grass and green sun-cured hay. One milligram of α-tocopherol is equivalent to 1 IU of vitamin E.

### 6.5.9 Energy

Energy is obtained from the oxidation of organic nutrients in the body and is required for metabolic reactions. Carbohydrates are the main source of energy, but energy may also be obtained from protein and fat. The latter is the most energy-dense nutrient. The carbon skeletons of proteins that remain after deamination of amino acids to liberate ammonia in the rumen are utilized in energy metabolism.

Energy is needed for biochemical reactions and for normal daily activities of the goat. When goats move to select and consume their diet, energy is being utilized. Energy is also needed for chewing activities, digestion, absorption and metabolism of feed nutrients. During both cold winters and warm summers, energy is required for the maintenance of body temperature.

Energy is first measured as gross energy by the combustion of feed samples using an adiabatic bomb calorimeter. Oxidation of feed organic components generates water and carbon dioxide, with the release of heat. Ash, the mineral fraction of feeds, will not burn and therefore does not provide energy. Digestible energy is a term that represents the gross energy consumed minus the gross energy lost in the faeces.

Metabolizable energy (ME) is the energy fraction left after deducting energy losses in faeces, urine and gases produced during rumen fermentation. Another term, net energy, is the fraction of gross energy consumed by the goat that is utilized for body maintenance and meat production, after considering all body losses including heat generated by body cell metabolism and rumen fermentation, which is known as heat increment.

Carbohydrates, especially structural fibre components, supply most of the energy consumed by range goats. Fibre carbohydrates from browse or grazed forage are fermented in the rumen and the end products of rumen fermentation are volatile fatty acids, primarily acetic, propionic and butyric acids, which are used as energy precursors (Lu *et al.*, 2005). In confinement, grain represents an important source of energy when included in the ration.

Energy is the major limiting factor in animal production under grazing conditions prevailing in arid and semi-arid regions. This restriction may increase several fold with scarce forage and water conditions. Thus, herbage energy availability is a key factor in adaptation to the environment and the behaviour and feeding strategies of goats

(Lachica and Aguilera, 2008). Accurate estimates of overall energy expenditure depend on the additional energy expenditure and heat production due to grazing activity.

Free-ranging goats may lose weight during the dry season as forage availability and quality limit the energy supply for body maintenance. During the wet season, the energy consumed by goats will generally be enough for maintenance, and additional energy may be used to recover body condition. These changes in body condition represent changes in energy balance. In confinement, with high-level grain diets, excess energy intake may be stored as fat.

The nutrient demand of the animal is one of the conditions that influences those factors used to predict the activity energy cost of grazing, and therefore forage intake (Fierro and Bryant, 1990). Forage availability can influence both grazing time and the nutritive value of ingested forage (Seman *et al.*, 1991; Krysl and Hess, 1993; Herselman *et al.*, 1999). As forage availability decreases, bite size declines, which results in at least partial compensatory changes in grazing time and rate of biting (Davies and Southey, 2001). Decreased forage quality also increases time spent chewing (Sahlu *et al.*, 1989). Factors likely to be responsible for increased energy expenditure and the grazing activity energy cost with increasing stocking rate are decreased forage mass, which elicits increased grazing time, and the number of steps or distance travelled (Animut *et al.*, 2005).

## 6.6 Nutritional Constraints

In semi-arid regions of the world, drastic climate changes cause feeding constraints that may affect the performance of range goats (Lu, 1989a; Silanikove, 2000; Ben Salem and Nefzaoui, 2003; Alexandre and Mandonnet, 2005; Ben Salem and Smith, 2008). These constraints may include low forage availability (underfeeding), low forage quality (low concentrations of nutrients) and PSMs in browse, which affect nutrient utilization and parasitism (Kawas *et al.*, 2010). Depression of feed intake and reduction in production are commonly observed in heat-stressed goats when environmental temperature exceeds 25–30ºC (Lu, 1989a).

### 6.6.1 Forage availability and quality

Fibre digestion and utilization in goats has been thoroughly reviewed (Lu *et al.*, 2005). Under range conditions, the potential for goat meat production is limited by many environmental and nutritional factors. Seasonal fluctuations in forage availability and quality are one of the main causes of nutritional stress that limit animal production in semi-arid regions of the world. During the dry season, inadequate foliage intake occurring as a result of reduced availability in the range, the low protein content and an increase in fibre components and lignification can reduce the intake of nutrients that are required by goats for maintenance and production (Kawas and Huston, 1990; Ben Salem and Nefzaoui, 2003; Alexandre and Mandonnet, 2005; Lachica and Aguilera, 2005). Therefore, the physical and chemical characteristics of high-fibre feeds may not provide enough nutrients for cost-efficient goat production (Silanikove, 2000; Alexandre and Mandonnet, 2005).

#### *Forage availability*

Frequently, available feed resources fail to satisfy the maintenance requirements of small ruminants. Forage availability may be as limiting as forage quality to goat performance. In north-eastern Brazil, during the dry season, the main component of the goat diet is the bed of dry leaves from deciduous trees, which become scarce at the end of the dry season. As a result of the reduced availability of feed, seasonal variations in body weight of >50% can be observed within a period of 5–6 months (Figueiredo *et al.*, 1980).

#### *Forage quality*

The two most important factors affecting forage quality and utilization are forage species and maturity. The cell wall content of leguminous fodders is less than that of grasses, and most forages decrease in quality with advancing maturity. The cell wall of legumes has a faster rate of digestion than that of grasses, which explains why a higher intake is observed for legumes (Kawas *et al.*, 1989). When legumes and grasses of equal digestibility are fed to animals, the voluntary intake of legumes exceeds that of grasses. This has led to the association of cell wall type with feed intake (Thornton and Minson, 1973). With forage maturity, the non-structural carbohydrates decrease and plant cell wall content increases and becomes more complex due to an increase in the presence of polymeric compounds such as lignin and silica. Due to the greater stem:leaf ratio, mature forages contain more fibre and less soluble carbohydrate and protein (Kawas, 1983; Kawas *et al.*, 1990).

Forage quality is also affected by other factors including cutting date and climate. The nutritive value of grasses and legumes generally decreases as temperature increases. Compared with temperate forages, tropical forages have increased annual DM yield. However, increased yield is usually associated with decreased forage quality and subsequent feeding value (Arthington and Brown, 2005). Although irrigation is the primary means of altering climatic constraints in agriculture, physical and economic considerations often limit its use in semi-arid regions where tropical forages are grown (Sotomayor-Rios and Pitman, 2001).

Cacti are commonly used as an emergency feed supplement in semi-arid regions of countries of the American (Brazil, Chile and Mexico) and African (Morocco, South Africa and Tunisia) continents (Ben Salem and Smith, 2008). Some species of cactus are high in fibre (up to 58% NDF on DM basis) and ash (up to 25%) contents, which may reduce energy intake, and a significant amount of the protein, determined as acid detergent insoluble nitrogen, may be associated with the less degradable fraction of the plant cell walls (Dávila-Gutiérrez, 1996). Ben Salem and Nefzaoui (2003) reported the composition of spineless cactus (*Opuntia ficus indica f. inermis*) having a CP content of <5%. This level of CP may provide less nitrogen for rumen bacteria than the 7% CP needed in the diet for normal rumen function (NRC, 1987). The ash content was reported as 23.8%, of which 5.2% was calcium and only 0.1% was phosphorus (Ben Salem and Nefzaoui, 2003).

### 6.6.2 Plant secondary metabolites

The nutrient availability and palatability of certain plant species appear to be affected by anti-nutritional compounds, also known as plant secondary metabolites (PSMs), such as tannins, phenolic acids and alkaloids (Melechek and Provenza, 1983; Pfister, 1983; Lu, 1992), which are found at higher concentrations in the diet of range goats and deer than in that of other ruminants (Silanikove, 2000; Nantoumé *et al.*, 2001). Tannins are heterogeneous polyphenolic compounds varying in molecular weight, with the ability to form complexes with proteins and other nutrients due to the binding properties of their hydroxyl and carboxyl groups (Silanikove, 2000; Makkar, 2003). Tannins are usually classified either as hydrolysable tannins or condensed tannins; the latter are also known as proanthocyanidins, based on their molecular structure (Min and Hart, 2003):

- *Hydrolysable tannins* contain a carbohydrate (generally D-glucose) as a central core. The hydroxyl groups of these carbohydrates are esterified with phenolic groups, such as ellagic acid or gallic acid. Hydrolysable tannins can be further metabolized to compounds such as pyrogallol by some rumen bacteria involved in these degradative pathways, which are potentially toxic to ruminants (Min and Hart, 2003). However, it appears that goats can adapt to the consumption of large amounts of tannins without suffering any ill effects (Silanikove *et al.*, 1996).
- *Condensed tannins* are the most common type of tannin in forage legumes, trees and shrubs (Barry and McNabb, 1999).

Structurally, condensed tannins are complexes of oligomers and polymers of flavonoid units linked by carbon–carbon bonds. Condensed tannins exist as oligomers of flavan-3-ols (catechin) or flavan-3,4-diols (epicatechin), and those occurring in temperate forages have a relative molecular mass of 2000–4000 comprising 10–12 oligomers of condensed tannins (Foo *et al.*, 1986). Together, these differences can produce an infinite variety of chemical structures, which in turn affect the physical and biological properties of the condensed tannins. Condensed tannins accumulate in the vacuoles of cells in various tissues of many forage species. The different structures and range of molecular weights for forage condensed tannins have been reported (Min and Hart, 2003).

Tannins can induce detrimental effects when consumed by herbivores, such as reduced protein availability, lower palatability, gut irritation and systemic toxicity (Kumar and Vaithyanathan, 1990; Makkar, 2003):

- Condensed tannins of the plant quebracho were shown to cause epithelial degeneration and ulceration of the gastrointestinal tract and reduced DM, fibre and nitogen digestibility (Dawson *et al.*, 1999).
- Guajillo (*Acacia berlandieri Benth.*), a shrub species consumed by goats and whitetail deer and high in nitrogen content, is widely distributed in southern Texas and northern Mexico. It is particularly valuable when grasses are dormant or during periods of extended drought (Nantoumé *et al.*, 2001). Guajillo can cause ataxia of the hind limbs of goats (Price and Hardy, 1953), which appears to be related to the presence of phenolic compounds and alkaloids of which *N*-methyl-β-phenethylamine is found in high concentrations (Clement *et al.*, 1997). This compound has profound physiological effects on the hypothalamic–adrenal–gonadal axis (Forbes *et al.*, 1994; Vera-Avila *et al.*, 1996). With increasing levels of guajillo in the diet of goats, palatability, nutrient digestibility and both energy and nitrogen balance decreased (Nantoumé *et al.*, 2001). An increased demand for glucose to detoxify the PSM at high intake levels of guajillo (>50%) suggests that supplements are needed to improve digestibility and availability of nutrients, and provide energy and other nutrients to mitigate the toxic effects of the PSM ingested.

Secondary plant compounds at a concentration below the level of toxicity can be leveraged for disease prevention, control and treatment, and may fill the vacuum from the absence of the use of chemicals in organic goat production. In combination with other naturally occurring materials, they have the potential to improve nutrient digestion and utilization in goats. Goats are versatile in harvesting plant materials and are able to survive under adverse foraging conditions. The tolerance of goats towards the bitterness of PSMs can have an anthelmintic role and make goats more suitable for high-forage organic production systems than other ruminant species (Lu, 2011).

### 6.6.3 Internal parasites

Gastrointestinal parasites are a major health problem for livestock production in tropical and subtropical environments (Sykes, 1994). Economic losses are primarily due to lower performance rates. The incidence of parasite infestation may be high for grazing goats because most parasites are found close to the ground, whereas browsing goats are usually parasite free. A reduction in parasite exposure, detection of clinical signs of parasitism and proper treatment of infected goats are recommended management practices (Mobley *et al.*, 2007).

Control and treatment of gastrointestinal parasite infections are usually accomplished by administering anthelmintic drugs. Anthelmintics are chemical dewormers used to treat infections of parasitic worms (helminths), whereas coccidia are normally treated with sulfa drugs. As worm parasites may have acquired resistance over decades of use of chemicals, treatment with these chemical dewormers is becoming less effective for parasite control of infected goats, sheep and cattle. Therefore, alternative treatments have been evaluated to reduce the dependence on these chemical compounds (Waller, 1994).

Nutrition plays a major role in how well animals are able to withstand the detrimental effects of internal parasites. In fact, signs of parasitism are usually used as a symptom of poor nutrition (Wallace *et al.*, 1998). Research has shown that improved nutrition increases the animal's ability to resist and defend itself against infection. Some alternative non-drug methods that appear promising for farm production systems, especially those focused on organic goat production (Lu *et al.*, 2010), are nutritionally related:

1. *Protein supplementation.* Increased protein intake appears to improve the resistance of the host to gastrointestinal parasites, apparently mediated by an enhanced host immunity, which may be especially important with selection for immunity to gastrointestinal parasites (Min and Hart, 2003). Field studies have shown that supplementation with urea molasses blocks can result in increased live weight of lambs at weaning, increased reproduction rates in ewes and a reduction in faecal egg output of parasites in grazing sheep. However, pen studies with young goats have shown that urea supplements alone gave no production benefits but, when accompanied by 100 g/day of cottonseed meal, beneficial responses were observed (Knox and Steel, 1996).
2. *Copper administration.* Copper oxide wire particle boluses also reduce gastrointestinal parasite load (Burke *et al.*, 2005, 2007) and were shown to help control *Haemonchus contortus* or barber pole worm in sheep and goats, a highly prolific nematode with a short life cycle. This parasite causes a reduction in production and death losses when severe outbreaks occur during warm humid months because of its voracious appetite for blood in the abomasum in which adult worms thrive and produce eggs (Knox, 2002; Burke *et al.*, 2004).
3. *Condensed tannins.* Alternative parasite management strategies using forages containing condensed tannins have been suggested (Niezen *et al.*, 1995; Barry *et al.*, 2001; Min *et al.*, 2002). Condensed tannins have been reported to reduce gastrointestinal parasite loads in goats by reducing worm fertility, eliminating adult worms and retarding the establishment of incoming larvae (Min and Hart, 2003; Waller and Thamsborg, 2004; Knox *et al.*, 2006; Waller, 2006). These compounds can be found in leguminous plants and browse. It seems possible that the consumption of forage condensed tannins may reduce gastrointestinal parasite numbers and improve animal performance through direct and indirect mechanisms. Direct effects of condensed tannins may be mediated through condensed tannin–parasite interactions, thereby affecting the physiological functioning of the parasites. Condensed tannins extracted from various forages can markedly decrease the viability of the larval stages of several nematodes. Condensed tannins also may react directly by interfering with parasite egg hatching and the development of infective-stage larvae. Indirectly, condensed tannins can improve protein nutrition by binding to plant proteins in the rumen and preventing microbial degradation, thereby increasing amino acid flow to the duodenum. The condensed tannins bind proteins and other molecules at the near-neutral pH of the rumen, dissociating in the acidic pH of the abomasum, allowing them to be digested (Min and Hart, 2003). However, condensed tannins from some plants appear to be effective only against parasites that affect the small intestine and not against those found in the abomasums, such as the barber pole worm.

## 6.7 Nutrient Requirements of the Goat

Meat, milk and fibre production are physiological processes that require large quantities of readily available nutrients. A schematic representation of feed nutrients, intake, digestion and utilization in the meat goat is presented in Fig. 6.5. As the genetic capability of goat breeds is improved, there is a greater need for improvement of management, nutrition, health and reproduction. The last report of the US National Research Council was published in 2007 (NRC, 2007). However, more research has since been published on requirements and the response to different feeding regimes, which can be used to further complement feeding guidelines for growth and reproduction of meat goats.

Nutrient concentrations in the diet of goats are affected by feed intake. One of the most limiting factors in terms of establishing the requirements and feed guidelines for goats is the lack of precision for predicting DM intake. Feed intake has been reported for lactating (Lu, 1989b; Lu *et al.*, 1991) and growing (Lu and Potchoiba, 1990) goats. A National Research Council (NRC, 1987) report presented a revision of research conducted to predict the feed intake of some domestic animals but excluded goats. More recently, DM intake prediction equations for growing and mature meat goats were reported (Sahlu *et al.*, 2004) and used to calculate DM intakes needed to supply the required energy concentrations of 1.91, 2.39 and 2.87 Mcal ME/kg, taking into consideration goat body weights (NRC, 2007). A summary of the nutrient requirements of goats reported by the NRC (2007), considering a range of body weights, is presented in Table 6.1.

### 6.7.1 Energy

Energy is the nutrient component required in greater quantities in the diet of meat goats. Energy requirements depend on factors such as physiological state (maintenance, growth, gestation and lactation), level of production, activity, ambient temperature and nutrition (fibre level, energy density and additives such as sodium monensin).

Information in the literature concerning energy requirements of goats in open range is scarce. The energy expenditure of goats in semi-intensive and intensive production systems can represent a significant proportion of their total energy requirements. The average requirements for growth derived from regression analysis were 4.6 kcal ME/g of average daily gain of growing goats of 4–8 months of age (Lu and Potchoiba, 1990). The average energy requirement based on 144 lactating dairy goats during the first 16–20 weeks of lactation was 119 kcal ME/kg of 4% fat-corrected milk (Lu, 1989b). The daily activities of grazing goats such as time spent eating and

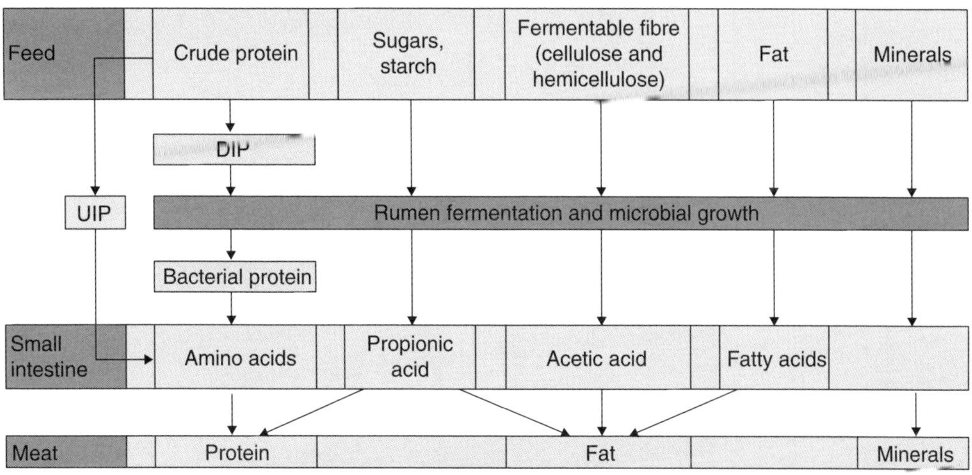


**Fig. 6.5.** Feed nutrients, intake, digestion and utilization in the meat goat. DIP, Degradable intake protein; UIP, undegradable intake protein.

**Table 6.1.** Nutrient requirements of meat indigenous breeds and Boer goats of various production stages (adapted from NRC, 2007).[a]

| Stage | ME[b] (Mcal/kg) | CP[c] (%) | MP[c] (%) | DIP[c] (%) | Ca[d] (%) | P[d] (%) | Vitamim A[e] (RE/kg) | Vitamin E[e] (IU/kg) |
|---|---|---|---|---|---|---|---|---|
| Starter (<15 kg BW; 200 g average daily gain/day) | | | | | | | | |
| Indigenous breeds | | | | | | | | |
| Doelings and male castrates | 3.25 | 22.4 | 15.3 | 8.1 | 1.21 | 0.52 | 2451 | 245 |
| Intact males | 3.21 | 20.5 | 14.1 | 8.0 | 1.11 | 0.50 | 2252 | 225 |
| Grower (15–30 kg BW; 250 g average daily gain/day) | | | | | | | | |
| Boer | | | | | | | | |
| Doelings and male castrates | 3.28 | 26.5 | 18.2 | 8.2 | 1.11 | 0.50 | 2252 | 225 |
| Intact males | 3.25 | 24.7 | 17.0 | 8.2 | 1.05 | 0.46 | 2101 | 210 |
| Indigenous breeds | | | | | | | | |
| Doelings and male castrates | 2.73 | 18.0 | 12.4 | 6.9 | 0.93 | 0.43 | 2679 | 268 |
| Intact males | 3.18 | 19.4 | 13.4 | 8.0 | 0.98 | 0.44 | 2885 | 288 |
| Boer | | | | | | | | |
| Doelings and male castrates | 3.24 | 25.1 | 17.2 | 8.1 | 0.99 | 0.45 | 2922 | 292 |
| Intact males | 3.21 | 23.1 | 15.9 | 8.0 | 0.93 | 0.43 | 2695 | 269 |
| Finisher (>30 kg BW; 300 g average daily gain/day) | | | | | | | | |
| Indigenous breeds | | | | | | | | |
| Doelings and male castrates | 2.40 | 15.4 | 10.6 | 6.0 | 0.77 | 0.37 | 2823 | 282 |
| Intact males | 2.60 | 15.4 | 10.6 | 6.5 | 0.77 | 0.37 | 2823 | 282 |
| Boer | | | | | | | | |
| Doelings and male castrates | 2.61 | 19.6 | 13.4 | 6.5 | 0.78 | 0.37 | 2846 | 285 |
| Intact males | 2.87 | 19.9 | 13.7 | 7.2 | 0.79 | 0.38 | 2893 | 289 |
| Mature does | | | | | | | | |
| Maintenance | 1.91 | 7.1 | 4.8 | 4.8 | 0.17–0.24 | 0.14–0.16 | 1256–1835 | 212–310 |
| Breeding | 1.91 | 7.1 | 4.8 | 4.8 | 0.17–0.24 | 0.14–0.16 | 1142–1662 | 193–281 |
| Gestation | | | | | | | | |
| Early | | | | | | | | |
| One kid | 1.91 | 8.4–9.1 | 5.8–6.3 | 4.8 | 0.27–0.53 | 0.18–0.27 | 981–1503 | 166–254 |
| Two kids | 1.91 | 9.0–9.7 | 6.2–6.7 | 4.8 | 0.33–0.69 | 0.21–0.33 | 872–1359 | 147–229 |
| Three or more kids | 1.91 | 9.1–9.6 | 6.3–6.6 | 4.8 | 0.38–0.67 | 0.23–0.33 | 924–1279 | 156–216 |

| | | | | | | | | |
|---|---|---|---|---|---|---|---|---|
| Late | | | | | | | | |
| One kid | 1.91–2.39 | 9.1–11.5 | 6.3–7.9 | 4.8–6.0 | 0.23–0.47 | 0.17–0.25 | 1213–1557 | 149–192 |
| Two kids | 1.91–2.87 | 9.4–13.6 | 6.6–9.4 | 4.8–7.2 | 0.26–0.65 | 0.13–0.32 | 1167–1280 | 144–158 |
| Three or more kids | 2.39–2.87 | 11.1 -13.4 | 7.6–9.3 | 6.0–7.1 | 0.33–0.62 | 0.20–0.31 | 1219–1437 | 150–177 |
| Lactation | | | | | | | | |
| Early | | | | | | | | |
| One kid | 1.91 | 5.9 | 4.1 | 4.8 | 0.29–0.56 | 0.21–0.33 | 1244–2040 | 130–214 |
| Two kids | 1.91–2.39 | 5.7 | 3.9 | 4.8–5.9 | 0.39–0.90 | 0.25–0.50 | 1216–1660 | 127–174 |
| Three or more kids | 1.91–2.39 | 5.5 | 3.8 | 4.8–5.9 | 0.43–0.89 | 0.28–0.49 | 1216–1450 | 127–152 |
| Mid | | | | | | | | |
| One kid | 1.91 | 6.2 | 4.3 | 4.8 | 0.31–0.62 | 0.22–0.36 | 1446–2304 | 151–241 |
| Two kids | 1.91 | 5.8 | 4.0 | 4.8 | 0.41–0.89 | 0.26–0.49 | 1139–1957 | 101–205 |
| Three or more kids | 1.91 | 5.6 | 3.9 | 4.8 | 0.50–0.89 | 0.3[illegible]–0.49 | 1216–1764 | 127–185 |
| Late | | | | | | | | |
| One kid | 1.91 | 6.5 | 4.5 | 4.8 | 0.34–0.73 | 0.23–0.40 | 1726–2660 | 181–278 |
| Two kids | 1.91 | 6.3 | 4.3 | 4.8 | 0.48–1.10 | 0.30–0.59 | 1529–2408 | 160–252 |
| Three or more kids | 1.91 | 6.2 | 4.2 | 4.8 | 0.60–1.12 | 0.36–0.61 | 1605–2261 | 168–237 |
| Mature bucks | | | | | | | | |
| Maintenance | 1.91 | 6.6 | 4.5 | 4.8 | 0.17 | 0.15 | 1377–1812 | 232–306 |
| Pre-breeding | 1.91 | 6.6 | 4.5 | 4.8 | 0.17 | 0.14 | 1806–2386 | 222–294 |

[a]Nutrient requirement for growing kids and adult does and bucks (greater requirements are for younger does with lower body weight).
[b]Energy requirements, expressed as metabolizable energy (ME).
[c]Protein requirements expressed as crude protein (CP), metabolizable protein (MP) and degradable intake protein (DIP). The CP requirement differs with the proportion of undegradable intake protein because of the minimal DIP required.
[d]Calcium (Ca) and phosphorus (P) are presented as minimum requirements that should be considered in diet formulation.
[e]Vitamins A, which is often deficient in the diet of goats, is expressed as retinol equivalents (RE = 1.0 g of all-*trans* retinol, 5.0 g of all-*trans* β-carotene and 7.6 g of other carotenoids), whereas vitamin E is expressed as international units (IU). Inclusion of vitamin E is expensive and is generally not recommended for range supplements.

distance travelled are different from those of confined animals (Lachica and Aguilera, 2003). These activities result in greater energy expenditure than in confinement, which can limit the energy available for maintenance and production.

Huston (1978) suggested that goats can travel greater distances than sheep or cattle. For goats, previous recommendations (NRC, 1981) suggested an addition of 25% of the suggested ME requirement for maintenance with light activity, 50% with semi-arid rangeland and slightly hilly conditions, and 75% with sparsely vegetated rangeland or mountainous transhumance pasture. Because of the close relationship between grazing time and energy expenditure (Osuji, 1974), predictions based primarily on time spent grazing and walking, herbage digestibility, distance travelled and terrain ruggedness or topography have been suggested (Sahlu et al., 2004).

Presently, the tabulated requirements for goats (NRC, 2007) consider the necessary muscular activities depending on the availability of feed and water, distance travelled and land characteristics to calculate the ME requirements under grazing conditions. When goats satisfy their energy requirements, DM intake starts to diminish; thus, energy requirement has a relationship with the requirements for other nutrients.

An inadequate energy intake will cause delayed puberty, suppressed oestrus and ovulation in females and suppressed libido and spermatozoa production in males. Excessive energy intake will be reflected in low conception and abortion rates, dystocia, a retained placenta in females and reduced libido in males (Bearden and Fuquay, 1992; NRC, 2007).

### 6.7.2 Protein

Meat goats should receive diets with enough protein to meet their requirements for optimum performance. Protein requirement is expressed as g/day or percentage of DM. Average requirements for growth derived from regression analysis were 0.26 g CP/g of average daily gain for growing goats of 4–8 months of age (Lu and Potchoiba, 1990). Average protein requirement based on 144 lactating dairy goats during the first 16–20 weeks of lactation was 84–100 g CP/kg of 4% fat-corrected milk (Lu, 1989b). As the degradable intake protein requirements of goats have not been well studied, the NRC (2007), considering several recent studies with goats (Prieto *et al.*, 2000; Soto-Navarro *et al.*, 2003, 2004), estimated a recommended level of 9% of total digestible nutrient intake. A deficient protein intake may cause suppressed oestrus, low conception rates, fetal resorption, premature parturition and weak offspring, whereas an excessive protein intake may cause a low conception rate (Bearden and Fuquay, 1992; NRC, 2007).

### 6.7.3 Minerals

Drinking water in some semi-arid regions may contain high concentrations of total dissolved minerals, of which sodium, sulfur, iron, manganese and arsenic are of main concern. Calcium and magnesium are other minerals that should be considered in drinking water for diet formulation, especially in confinement. The deleterious effects of high concentrations of minerals in water include reduced growth and reproduction levels. Goats can tolerate a high concentration of sodium in water but will tend to reduce DM intake and consequently will lose weight.

Mineral concentrations required in the diet vary with growth performance, reproductive status (maintenance, breeding, gestation or lactation) and breed (i.e. indigenous breeds versus Boer goats). Some important facts with respect to excessive or deficient mineral intake (Bearden and Fuquay, 1992; NRC, 2007) are listed below:

- *Calcium and phosphorous* deficiency will cause anoestrus and irregular oestrus. Prolific does require more calcium and phosphorus in their diet.
- *Potassium* deficiency will reduce feed intake and daily weight gain. A marginal potassium deficiency is common.

Dietary potassium concentration is a good indicator of deficiency.

- *Sulfur.* A marginal sulfur deficiency reduces feed intake and digestibility and the synthesis of microbial protein in the rumen. Sulfur is needed for synthesis of the sulfur amino acids cysteine and methionine. When urea is used as the main source of nitrogen, sulfur supplementation is needed. On the other hand, an excess of sulfur in drinking water or feed may cause the copper requirement to increase, and may be a predisposing factor for the incidence of polioencephalomalacia, as discussed previously. Sulfur metabolism and requirements in dairy, meat and fibre-producing goats have been reported (Qi *et al.*, 1993a,b, 1994a,b,c).
- *Iron.* The iron requirement for growing goats is 95 mg/kg DM, and for pregnant and lactating does is 35 mg/kg DM.
- *Manganese* deficiency will cause a reduction in reproductive performance (reduced and irregular oestrus), reduced conception and abortions.
- *Selenium* deficiency will cause white muscle disease in kids and a retained placenta in does.
- *Copper.* Normal plasma copper concentrations in the goat may vary from 0.9 to 1.39 mg/l (Galbraith *et al.*, 1997; Solaiman *et al.*, 2001). It is important to consider that the toxic level of copper for sheep may not be the same for goats, as the copper requirements of goats may actually be higher (Solaiman *et al.*, 2001). The copper requirements for lactating does (15 mg/kg DM), mature goats and bucks (20 mg/kg DM) and growing goats (25 mg/kg DM) (with both forage and supplement) have been recommended taking into consideration normal molybdenum (1–2 mg/kg DM) and sulfur (0.15–0.25%) intakes (NRC, 2007). These values are much higher than the recommended level of 10 mg/kg DM for beef cattle for all physiological stages (NRC, 2000). Furthermore, high levels of copper supplementation may improve goat performance. Copper supplementation of 100 mg/day improved gain efficiency, altered the serum lipid profile, decreased carcass fat depth over the 12th rib and improved carcass boneless closely trimmed retail cuts in goat kids fed a high-concentrate diet, without producing signs of copper toxicity (Solaiman *et al.*, 2006). The maximum copper tolerance level in the diet of goats has been set at 40 mg/kg DM (NRC, 2007). A copper deficiency will cause depressed reproduction, an impaired immune system and impaired ovarian function (Bearden and Fuquay, 1992).
- *Iodine.* The iodine requirement for growing and mature non-lactating does is 0.5%, whereas that of lactating does is 0.8%.
- *Zinc* deficiency will cause reduced spermatogenesis in bucks.
- *Cobalt.* The cobalt requirement is 0.11 mg/kg DM (NRC, 2007).

### 6.7.4 Vitamins

Under normal conditions, B-complex vitamins are synthesized in sufficient quantities in the rumen to satisfy requirements, so supplementation is generally not required (NRC, 2007). Vitamin K is also normally synthesized in adequate quantities by bacteria in the digestive tract and goats do not generally need to be supplemented.

Fat-soluble vitamins, specifically vitamins A and E, must be consumed by goats. Goats grazing good-quality green forages will usually consume enough of these vitamins. Supplementation of these vitamins depends on the physiological status and production potential of goats. A deficiency of vitamin A will cause impaired spermatogenesis, anoestrus, low conception rates, abortion, weak offspring and a retained placenta (Bearden and Fuquay, 1992; NRC, 2007).

## 6.8 Nutrient Composition of Feeds

The availability and utilization of feed resources in North America and Europe have been reviewed (Lu and Rubino, 1992). In traditional livestock production systems, feed is the single largest expense. An international feed nomenclature (Table 6.2) has been established (NRC, 2000). Feeds can be classified in two large groups, namely forages or concentrates. Forage feeds include grasses, herbaceous legumes and tree legumes consumed by grazing livestock. In confinement, forages can be fed as freshly cut fodder, or offered as hay or silage. Crop residues from cereal crops and hulls from some oilseeds are also classified as forages.

Whereas forages are high in fibre content, concentrate feeds are more energy dense and less fibrous. Energy in the concentrate feeds comes primarily from starch, sugars, other non-structural carbohydrates, and oil. Concentrates include cereal grains (maize, grain sorghum, oats and barley), grain milling, brewing and distillation milling by-products (wheat middlings, brewers' dried grains, maize gluten feed, distillers' dried grains, etc.), oilseed meals (soybean meal, cottonseed meal, etc.), molasses, vegetable and animal fat, vitamin and mineral sources, and feed additives. The mineral content of selected feedstuffs is described in Table 6.3.

### 6.8.1 Feed additives

Few feed additives have been approved as supplements for goats, among them decoquinate and monensin, both recommended for the prevention of coccidiosis in young animals (Animal Health Institute, 2002). Although decoquinate is only used to control coccidiosis, monensin can be included in feeds and supplements to improve the growth rates of goats on high-roughage diets (Huston *et al.*, 1990; NRC, 2007; Flythe and Andries, 2009).

Monensin, an ionophore, modifies the movement of ions across the membranes of rumen microbes, altering fermentation patterns by reducing methane and acetate production and increasing propionate production (Bergen and Bates, 1984; Schelling, 1984). An increase in nitrogen digestibility due to monensin was reported with goats fed a low-protein diet (Beede *et al.*, 1978). Monensin appears to inhibit ammonia

**Table 6.2.** International feed nomenclature (adapted from NRC, 2000).

| Class | Description |
|---|---|
| Pasture, range plants and forages fed fresh | Forage feeds either not cut, or cut and fed fresh |
| Silages | This class includes ensiled forages such as sorghum, maize, grass and legumes. It does not include grain or its by-products, fruits or tubercles |
| Dry forages and roughages | All forages and roughages cut and cured, and fibrous by-products with >18% CF or >35% NDF (legume and grass hay, stover and straw) |
| High protein sources | Products with >20% CP (dry basis) of animal or plant origin such as meat meal, oilseed meals, urea, maize gluten and other by-product meals |
| High energy sources | Products with <20% CP and <18% FC or 35% NDF, on a dry basis, such as cereal grains and their by-products, molasses, tubercles and fats |
| Mineral supplements | Bone meal and other feeds that contain a mineral as the main component |
| Vitamin supplements | Synthetic supplements of the water-soluble and fat-soluble vitamins |
| Non-nutritive additives | Antibiotics, dewormers (febendazole), flavours, ionophores (sodium monensin, etc.), polyethylene glycol, etc. |

CP, Crude protein; CF, crude fibre; FC, fibre carbohydrate; NDF, neutral detergent fibre.

**Table 6.3.** Mineral content of selected feedstuffs (adapted from NRC, 2007).

| Nutrient | Concentrations/characteristics |
|---|---|
| Calcium and phosphorus | Leguminous fodders have high calcium contents, whereas grasses are generally low. Browse and grasses in semi-arid regions with calcareous soils may have very high calcium levels, and commonly low phosphorus levels. With maturity, the phosphorus content of forages is reduced. Cereal grains, oilseed meals and animal by-products have low levels of calcium and moderate to high levels of phosphorus. Excess calcium in the diet may affect the metabolism of some mineral elements such as phosphorus, magnesium and zinc. A high calcium intake may be tolerated by ruminant animals, especially if goats consume diets with enough phosphorus. Calcium carbonate is commonly used as a source of calcium, whereas inorganic sources of phosphorus include calcium phosphates, monosodium phosphate, ammonium phosphates and rock phosphate |
| Sodium | Animal by-product feeds generally contain more sodium than feeds of plant origin. Sodium is commonly deficient in the diet of range goats. Salt is generally supplemented to provide sodium |
| Potassium | Good-quality forages have a high potassium content (1–4%). Cereal grains are low in potassium, whereas oilseed meals are good sources of potassium. Potassium chloride is the most common inorganic source of potassium |
| Magnesium | Concentrations vary with forage species, magnesium concentration in soil, stage of growth, season of the year and ambient temperature. Leguminous fodders contain more magnesium than grasses. Cereal grains contain 0.11–0.17% magnesium, whereas oilseed meals contain twice as much magnesium. Inorganic sources of magnesium include magnesium oxide and magnesium sulfate |
| Sulfur | When urea is used as the main source of nitrogen, sulfur supplementation is needed. Mature forages are sulfur deficient, although most confinement diets are adequate. Sulfur levels in water of >500 ppm can be deleterious to meat production, especially if feed ingredients are high in sulfur. The maximum tolerable level in the diet of ruminants is 0.4%. Elemental sulfur and ammonium sulfate are common sources of sulfur |
| Manganese | The content in forage depends on forage species, soil pH and soil drainage. Cereal grains contain 5–40 mg/kg, oilseed meals contain 30–50 mg/kg and animal protein products have 5–15 mg/kg. Manganese sulfate and manganese oxide are manganese sources included in supplements and feed |
| Zinc | The content in forages depends on plant species and maturity and the zinc concentration in soil. Legumes have higher zinc concentrations than grasses. Cereal grains contain 20–30 mg/kg, whereas oilseed protein sources have 50–70 mg/kg. The most common sources of zinc are zinc sulfate and zinc oxide |
| Copper | Legumes normally contain more copper than grasses. Cereal grains contain 4–8 mg/kg, whereas oilseed meals contain 15–30 mg/kg. Milk contains low levels of copper. Copper sulfate is commonly used as a copper source |
| Iodine | The iodine content in feed ingredients depends on soil concentration. Calcium and sodium iodide and ethylenediamine dihydroiodide (EDDI) are used as iodine sources |
| Cobalt | Soils can be deficient in cobalt, and soil pH affects its concentration in forages. Cobalt absorption is reduced with a soil pH close to 7.0. Legumes contain more cobalt than grasses. Cobalt sulfate and cobalt carbonate are cobalt sources |
| Selenium | The selenium content of forages depends on forage species and selenium concentrations in soils. Some plants such as *Astragalus* and *Stanleya* naturally accumulate selenium (up to 3,000 mg/kg). Sodium selenite is used as a selenium source |

production in the goat rumen, which may reduce its excretion in the urine and increase the flow of amino acids into the small intestine (Flythe and Andries, 2009). Monensin greatly reduced supplement intake in sheep, with a limited effect in goats (Huston *et al.*, 1990). Monensin may even have environmental benefits as it is reported to reduce methane production by 25%. It does not present a hazard to human health by increasing resistant food-borne bacteria, as ionophores are not used in human therapy due to their narrow therapeutic index (Tedeschi *et al.*, 2003). More research is needed to evaluate the effectiveness of monensin as a feed additive in supplements of range goats to improve growth.

Polyethylene glycol (PEG) is another additive that has been evaluated more recently (Silanikove *et al.*, 1994; Decandia *et al.*, 2000; Gilboa *et al.*, 2000; Ben Salem and Nefzaoui, 2003) to reduce the anti-nutritional effects of condensed tannins. Makkar (2003) presented a thorough review of the use of tannin-binding agents such as PEG that may alleviate the negative dietary effects of condensed tannins. Dose level may vary depending on the amount of condensed tannins in the goat's diet (Decandia *et al.*, 2000; Makkar, 2003).

The results reported by Villalba and Provenza (2002) implied that PEG may be used to modify foraging distribution by goats browsing tannin-rich fodder, as PEG appeared to be an effective substance to attract goats to underutilized feeding sites in plant communities dominated by species with high concentrations of tannins. However, before a large-scale implementation of this technology, more research is needed to determine the recommended dose for specific situations and its profitability (Decandia *et al.*, 2000).

## 6.9 Strategic Nutrient Supplementation

The need for supplementation depends on the presence or lack of adequate forage nutrients in forage and can help increase the grazing capacity under pasture and range conditions (Lusby, 1990; Arthington and Brown, 2005). Supplementation can also provide a vehicle for carrying non-nutritive additives, antimicrobials and other compounds for the prevention or treatment of potential health problems such as parasitism, and to facilitate management (Lusby, 1990).

### 6.9.1 Nutritional profile of supplements

Supplements should be formulated considering the animal's requirements in terms of its physiological status, the effect of the environment on the animal and characteristics of the available feed (Alexandre and Mandonnet, 2005; Lachica and Aguilera, 2005). To improve the digestion and intake of grazed forage, a supplementation strategy should not supply nutrients in excess of the goat's requirements. When grazing ruminants are supplemented, changes in forage consumption occur as a result of the changes in digestion and passage of feed through the digestive tract that are associated with the additional nutrient intake provided by the supplement (Bowman and Sowell, 1997; Caton and Dhuyvetter, 1997). Nutrient supplementation should facilitate the synchronization of nitrogen and carbohydrate degradability in the rumen and increase the supply of nutrients for absorption in the small intestine (Wu *et al.*, 2005).

Strategic nutrient supplementation of goats should consider dietary factors such as the level of protein needed to satisfy the nitrogen requirements of microbes in the rumen, the quantity of undegradable intake protein, the effects of supplementation on ruminal pH, the quantity and frequency of supplementation, the type of carbohydrates that comprise the supplement (sugars, starch or rapidly fermented fibre) and the physical form of the supplement, as well as existing interactions among them (Paterson *et al.*, 1994). However, few studies have considered all nutrients when formulating the supplement evaluated. Only a few of the studies reviewed by Ben Salem and Nefzaoui (2003) reported using a mineral/vitamin pre-mix in their hand-crafted feed blocks.

### 6.9.2 Ingredient characteristics

The particle size of forages and grain affects chewing activities and consequently normal rumen function. The reduction in forage intake appears to be related to the amount, form (whole grain versus processed) and source (sugars and starch versus rapidly digestible fibre) of the energy supplement (Galloway *et al.*, 1993).

Maintaining an appropriate forage particle size and including whole or cracked grain helps maintain normal rumen function in goats fed high-concentrate diets or supplements. Some advantages of on-farm supplementation of whole grain include simplicity and lower costs by not grinding the whole grain, fewer hand-labour expenses and prevention of the negative effect of the readily available starch, especially at high intake levels (Kawas *et al.*, 2010).

### 6.9.3 Effect of supplementation on forage intake

Three scenarios with respect to forage DM intake can be expected with specific nutrient supplementation of grazing goats (NRC, 2007). Complementary, substitution and additive responses on forage DM intake may be observed depending on the type and quantity of supplement offered:

**1.** *Complementary response.* Overcoming mineral, vitamin or rumen degradable nitrogen deficiencies may increase forage DM intake; protein and minerals are utilized by bacteria needed for optimum rumen fermentation (Wu *et al.*, 2005; Kawas *et al.*, 2010). Protein supplementation may include true protein sources such as oilseed meals or non-protein nitrogen sources such as urea. The latter is commonly used as a source of nitrogen in ruminant feeds. When urea is supplemented, a readily available energy source is needed for its utilization by rumen microbes (van Soest, 1994). For goats grazing native vegetation in north-eastern Brazil during both wet and dry seasons, the growth response was not different when supplemented with either urea or molasses alone. During the wet season, when goats consumed a low-fibre/high-protein diet (44.9% NDF and 12% CP), rates of weight gain were not improved with a molasses/urea supplement. As forage quality declined during the dry season, diet NDF significantly increased to 52.1% and CP decreased to 8.4%, and the liquid molasses/urea supplement allowed weight gains to continue throughout the dry season (Schacht *et al.*, 1992). As grain alone does not provide enough nitrogen for maximum microbial fermentation, additional nitrogen supplementation is needed to improve fibre digestion and roughage DM intake (Abebe *et al.*, 2004).

**2.** *Substitution response.* Consumption of high-energy/low-protein supplements may decrease forage intake. The rate of substitution is defined as the unit of change in forage intake per unit increment in supplement intake (Paterson *et al.*, 1994; Caton and Dhuyvetter, 1997). As the digestibility of forage increases, the rate of substitution of forage is also incremented. Energy supplementation should not be excessive, as it can take the form of substitution when grazing forage nutrients are removed from animal diets in exchange for supplement (Caton and Dhuyvetter, 1997), so that most of the energy consumed by range goats should be provided through fibre fermentation of consumed forage (Lu *et al.*, 2005). As the starch level increased in the diet of grazing cattle, the negative effects on forage intake were accentuated, possibly as rumen fermentation is negatively affected or in response to a protein and/or energy imbalance in the diet (Paterson *et al.*, 1994). A substitution effect, in which forage intake was reduced, was also observed in a study with range does supplemented with increasing levels of whole sorghum grain (Kawas *et al.*, 1999).

**3.** *Additive response.* Supplementation may not affect forage intake. A decrease in forage intake by ruminants cannot always be observed with starch-based supplements. Small quantities of a grain-based supplement (0.2–0.3% of body weight) did not affect forage DM intake of beef cattle, whereas greater levels depressed forage intake (Vanzant *et al.*, 1990; Pordomingo *et al.*, 1991).

### 6.9.4 Hand-crafted supplement blocks

The production of hand-crafted blocks is economical and environmentally friendly, as no electric energy or heat is required to manufacture them. Compared with liquid or mash supplements, hand-crafted blocks hold several advantages including simplicity of transport and management, a more homogeneous intake among animals, a reduced need for salt as an intake regulator and less risk in the use of urea as a source of non-protein nitrogen (Ben Salem and Nefzaoui, 2003; Kawas *et al.*, 2010). Flexibility in the formulation of supplements is required so that specific nutrient requirements are satisfied according to the physiological status of goats (Kawas *et al.*, 2010).

Multi-nutrient blocks based on molasses and urea can be hand-crafted by goat producers (Ben Salem and Nefzaoui, 2003; Kawas, 2008). This type of block consists of about 35% molasses, 4–6% urea, 10–12% salt, 8–10% lime, cereal by-products (wheat middlings, maize bran or rice bran), oilseed meals (cottonseed meal, soybean meal) and a mixture of a source of phosphorus, trace minerals and feed additives. A fibrous by-product such as soybean hulls will aid in compaction. The lime will provide more hardness than cement, which has frequently been used to produce blocks in developing countries, and lime is generally much cheaper (Kawas, 2008). A vitamin A supplement should be added to the protein/mineral mixture during the dry or winter seasons, when only low-quality forage is available. Ben Salem and Nefzaoui (2003) have listed the formulas and chemical composition of molasses-containing and molasses-free feed blocks from several studies reviewed from the literature.

With respect to ingredient characteristics, a high level of urea (15–20%) may reduce block intake, as urea is unpalatable. Excessive urea intake is more critical, as it can cause toxicity (Ben Salem and Nefzaoui, 2003; Ben Salem and Smith, 2008; Kawas, 2008). As the CP level of molasses/urea blocks increased from 17 to 24%, the coefficient of variation (CV) of individual block intake of grazing sheep decreased from 132 to 82% (Kendall *et al.*, 1983).

The distribution and quantity of blocks in the paddocks affect their consumption. If block intake is high, they should be placed far from the water troughs, whereas, if it is low, they should be located closer to the water source. If a higher intake is desired, more blocks should be offered (Kawas, 2008).

Kawas (2008) recommended two types of multi-nutrient lick blocks that can be manufactured for wet and dry seasons, respectively, with the following characteristics:

- *Mineral–protein blocks*. This is a low-intake block (~0.2–0.3% of body weight) with 30% CP or more (Fig. 6.6). The purpose of offering this block is to supplement protein, macrominerals (calcium, phosphorus and magnesium), trace minerals (iron, manganese, zinc, copper, iodine, cobalt and selenium), and vitamin A during the dry season. This product is practical to use in situations in which it is difficult to provide and handle supplements, as many animals can be supplemented with only one block. This supplement can be offered during dry or wet seasons as a reproductive management tool for goats. It is important to consider that, during the wet season, forages contain more protein and energy, so it is important to emphasize this in mineral supplementation so that these nutrients can be better utilized. During the wet season, plenty of vitamin A is available in forages, so no vitamin A supplementation is required. Protein supplementation produces a greater response on DM intake with forages of lower protein content compared with forage with a

**Fig. 6.6.** Hand-crafted mineral–protein blocks for small ruminants (Kawas *et al.*, 2010).

relatively high protein content. Sudana and Leng (1986) reported that lambs supplemented with a molasses/urea block consumed 26% more wheat straw compared with a non-supplemented diet. Block intake was approximately 90 g/day.

- *Protein–energy blocks.* These blocks contain minerals, but they contain more energy and less protein (20–25% CP) than a mineral–protein block. Intake of this block should be greater (~0.4–0.5% of body weight). Intake can vary depending on various factors. This supplement is recommended for the dry season, when availability and quality of forages limit energy intake by goats (Kawas, 2008).

### 6.9.5 Reducing intake variation

Supplements should be designed to reduce the variation in intake between individual animals (Bowman and Sowell, 1997), especially those that contain high grain content (Caton and Dhuyvetter, 1997). The effectiveness of supplementation programmes is affected by the ability to reduce intake variation by individual animals and to meet target supplementation consumption. If grazing ruminants consume less than the target amount of supplement, then the formulated nutrient intake is not received, whereas, if consumption is more than the target amount, supplementation costs are increased and there can be potential negative impacts on forage intake and digestibility (Bowman and Sowell, 1997).

Two common ways of restricting supplement intake (self-limiting or self-regulation) include the addition of salt and/or offering supplements as a block. The type of supplement is important to minimize wastage and to maximize its impact on productivity (Bowman and Sowell, 1997). On-farm manufactured multi-nutrient supplements can be offered as mash or blocks (Ben Salem and Nefzaoui, 2003; Ben Salem and Smith, 2008; Kawas, 2008).

Adequate block intake is important so that expected results with supplementation are obtained. If block intake is less than expected, utilization of low-quality forages will not be maximized and performance (growth, reproduction and milk production) will be lower than expected (Bowman and Sowell, 1997; Ben Salem and Nefzaoui, 2003). Block intake is generally greater with small ruminants, on a body weight basis, in comparison with beef cattle. Initially, block intake may be low, until animals become accustomed to it (Birbe *et al.*, 2006).

Factors that affect block intake include hardness, season of the year (wet or dry) and the availability and quality of forages. Temperature and relative humidity affect block hardness. Other factors that have an influence on hardness and/or intake are the technique used for manufacturing, the temperature and relative humidity, the type and level of reactant used for compaction, the particle size of ingredients, ingredient characteristics (i.e. protein, fat or fibre content), the level of compaction, the time and type of storage, flavour and odour. If the block is too soft, intake may be greater than desired. In contrast, if it is too hard, intake is less than required. A soil penetrometer may be used to measure block hardness (Birbe *et al.*, 2006).

A harder block will restrict its intake, especially during the dry season, so blocks should be left to harden before being supplemented (Ben Salem and Nefzaoui, 2003). Zhu *et al.* (1991) observed that, as the hardness index of blocks increased, supplement intake decreased linearly, with a concomitant increase in the CV from 29% for soft blocks to 58% for hard blocks. Kendall *et al.* (1983) also observed a greater individual intake CV of molasses/urea block supplements offered to sheep, with an individual CV of 29% for soft blocks and 50% for hard blocks.

During the dry or winter seasons, forages have lower quality and, consequently, block intake generally increases. In contrast, with good-quality forages, block intake may be low or negligible. During the wet season, supplements may not improve fibre digestion as green forages available during this season are relatively high in nitrogen and low in fibre (Ben Salem and Nefzaoui, 2003).

Blocks have the advantage of self-limiting supplement intake, which is especially important when non-nutritive additives such as monensin or PEG are supplemented. The advantages of incorporating PEG in lick blocks have been discussed (Ben Salem and Nefzaoui, 2003; Makkar, 2003). Under extensive range conditions, supplementing goats with PEG can be labour intensive and complicated to implement (Villalba and Provenza, 2002; Ben Salem and Nefzaoui, 2003). Range blocks that contain PEG may have the advantage of self-regulation of PEG intake, which may aid in reducing the cost of PEG treatment (Villalba and Provenza, 2002).

### 6.9.6 Preventing urea toxicity

Urea can be toxic if consumption is excessive. Urea in feeds and supplements should be used with some precautions (Ben Salem and Nefzaoui, 2003; Ben Salem and Smith, 2008; Kawas, 2008), including the following:

- Adding salt in mash supplements or imparting enough hardness to blocks to restrict intake.
- Allowing for adaptation of rumen microorganisms for utilization of urea by gradually increasing the amount of supplement offered.
- Preventing the supplement from getting wet as urea is very soluble and toxic.
- Including an energy source such as molasses or grain so that nitrogen from the urea is better utilized by the rumen bacteria.
- Introducing blocks gradually to adapt for rumen utilization of urea, and limiting the amount or the time blocks are offered.

## 6.10 Effect of Nutrition on Reproduction

The nutritional status of the goat herd is the most important factor influencing reproduction (Webb *et al.*, 2004). Several reviews have discussed various aspects of nutrition on fertility in ruminants (Hurley and Doane, 1989; Beam and Butler, 1999; Butler, 2000; Armstrong *et al.*, 2003; Lucy, 2003; Webb *et al.*, 2004). Nutrition can easily be controlled by the producer by either increasing or reducing nutrient consumption (Webb *et al.*, 2004).

Nutritional status has been correlated with embryo survival and is a major factor controlling efficiency in assisted reproduction technologies. Nutrition has an effect on ovarian activity by acting at various levels within the hypothalamus–pituitary–ovarian axis. This can occur without significant variation in circulating gonadotropin concentrations but appears to be influenced by changes in other circulating concentrations of metabolic hormones, including insulin, insulin-like growth factor 1 (IGF-1), growth hormone and leptin. Nutrition can also affect the expression of mRNA encoding components of the ovarian IGF system to regulate the sensitivity/response of follicles towards gonadotropins. The detailed physiological mechanisms through which nutrition exerts many of these effects remain to be determined (Webb *et al.*, 2004).

Undernutrition may cause abortion as a response to low blood glucose concentration, which can be almost totally prevented by adequate nutrition, especially by supplementing with energy (Wentzel, 1982; Morand-Fehr, 1987). Increasing levels of whole sorghum grain (0, 0.6, 1.2 and 1.8%) were offered as an energy supplement to pre-pubertal female Moxoto does grazing native vegetation in north-east Brazil (Kawas *et al.*, 1992). First oestrus was recorded significantly earlier (75–78 days) for the supplemented groups (0.6–1.8% of body weight) than for the unsupplemented control group (104 days). Fertility, prolificacy, fecundity and gestation length were greater for does offered whole sorghum grain compared with does that were not supplemented.

Nutritional management for reproducing goats under organic production systems has been reviewed recently (Lu, 2011). Rapid fetal growth associated with the physical fill of forage limits feed intake during late pregnancy and therefore presents a challenge for nutritional balance in goats under high-forage organic production systems. Understanding regulation of feed intake and fibre digestion and utilization can lead to nutritional balance, minimizing metabolic disorders associated with pregnancy, parturition and lactation.

The body condition of does strongly affects the time at which puberty starts, the conception rate at first oestrus in young does, the length of the post-partum interval and the health and vigour of newborn kids. Body condition or changes in body condition before and during the breeding season affect reproductive performance in terms of services per conception, lambing and kidding intervals, and the percentages of open does. Does should be in good body condition at kidding and should maintain a good body condition during the breeding season. Increasing the level of energy offered to does should continue throughout the breeding season and for approximately 30–40 days after removing the bucks, for adequate implantation of the fetuses in the uterus (Luginbuhl *et al.*, 2009).

Flushing is a reproductive management tool in which a supplement is offered to does prior to breeding to increase body weight, ovulation rate and litter size. Body condition is used to determine whether flushing will be of benefit to breeding does. Does in extremely good body condition (body-condition score = 7) will tend not to respond to flushing, whereas does that are in relatively poor condition (body-condition score = 4) as a result of poor feed quality and supply, being highly parasitized and with late kidding of twins or triplets will respond favourably to flushing by improving their body condition. Flushing can be accomplished by moving breeding does to a lush, nutritious pasture or offering a high-energy supplement 3–4 weeks prior to the introduction of the bucks (Luginbuhl *et al.*, 2009).

## References

Abebe, G., Merkel, R.C., Animut, G., Sahlu, T. and Goetsch, A.L. (2004) Effects of ammoniation of wheat straw and supplementation with soybean meal or broiler litter on feed intake and digestion in yearling Spanish goat wethers. *Small Ruminant Research* 51, 37–46.

Alexandre, G. and Mandonnet, N. (2005) Goat meat production in harsh environments. *Small Ruminant Research* 60, 53–66.

Animal Health Institute (2002) *Feed Additive Compendium.* Miller Publishing Company, Minnetonka, Minnesota.

Animut, G., Goetsch, A.L., Aiken, G.E., Puchala, R., Detweiler, G., Krehbiel, C.R., Merkel, R.C., Sahlu, T., Dawson, L.J., Johnson, Z.B. and Gipson, T.A. (2005) Grazing behavior and energy expenditure by sheep and goats co-grazing grass/forb pastures at three stocking rates. *Small Ruminant Research* 59, 191–201.

Armstrong, D.G., Gong, J.G. and Webb, R. (2003) Interactions between nutrition and ovarian activity in cattle: physiological, cellular and molecular mechanisms. *Reproduction Suppl* 61, 403–414.

Arthington, J.D. and Brown, W.F. (2005) Estimation of feeding value of four tropical forage species at two stages of maturity. *Journal of Animal Science* 83, 1726–1731.

Askins, G.D. and Turner, E.E. (1972) A behavioral study of Angora goats on West Texas Range. *Journal of Range Management* 25, 82–87.

Barry, T.N. and McNabb, W.C. (1999) The implications of condensed tannins on the nutritive value of temperate forages fed to ruminants. *British Journal of Nutrition* 81, 263–272.

Barry, T.N., McNeill, D.M. and McNabb, W.C. (2001) Plant secondary compounds: their impact on nutritive value and upon animal production. In: *Proceedings of the International Grassland Conference*, São Paulo, Brazil, pp. 445–452.

Beam, S.W. and Butler, W.R. (1999) Effects of energy balance on follicular development and first ovulation in postpartum dairy cows. *Journal of Reproduction Fertility* (Suppl. 4) 411–424.

Bearden, H.J. and Fuquay, J.W. (1992) Nutritional management. In: *Applied Animal Reproduction.* Prentice Hall, New Jersey, pp. 283–292.

Beede, D.K., Trabue, P.J., Schelling, G.T., Mitchell, G.E. Jr and Tucker, R.E. (1978) Nitrogen balance and energy digestibility in growing goats fed a low protein diet with or without monensin. *Journal of Animal Science* 47 (Suppl. 1), 114.

Ben Salem, H. and Nefzaoui, A. (2003) Feed blocks as alternative supplements for sheep and goats. *Small Ruminant Research* 49, 275–288.

Ben Salem, H. and Smith, T. (2008) Feeding strategies to increase small ruminant production in dry environments. *Small Ruminant Research* 77, 174–194.

Bergen, W.C. and Bates, D.B. (1984) Lonophores: their effects on production efficiency and mode of action. *Journal of Animal Science* 58, 1465–1483.

Birbe, B., Herrera, P., Colmenares, O. and Martínez, N. (2006) *El consumo como variable en el uso de bloques multinutricionales. X Seminario de Pastos y Forrajes.* Universidad Nacional Experimental Simón Rodríguez, Estación Experimental La Iguana, Valle de la Pascua, Venezuela.

Bowman, J.G.P. and Sowell, B.F. (1997) Delivery method and supplement consumption by grazing ruminants: a review. *Journal of Animal Science* 75, 543–550.

Burke, J.M., Miller, J.E., Olcott, D.D., Olcott, B.M. and Terrill, T.H. (2004) Effect of copper oxide wire particles dosage and feed supplement level on *Haemonchus contortus. Veterinary Parasitology* 123, 235–243.

Burke, J.M., Miller, J.E., Larsen, M. and Terrill, T.H. (2005) Interaction between copper oxide wire particles and *Duddingtonia flagrans* in lambs. *Veterinary Parasitology* 134, 141–146.

Burke, J.M., Terrill, T.H., Kallu, R.R., Miller, J.E. and Mosjidis, J. (2007) Use of copper oxide wire particles to control gastrointestinal nematodes in goats. *Journal of Animal Science* 85, 2753–2761.

Butler, W.R. (2000) Nutritional interactions with reproductive performance in dairy cattle. *Animal Production Science* 60–61, 449–457.

Caton, J.S. and Dhuyvetter, D.V. (1997) Influence of energy supplementation on grazing ruminants: requirements and responses. *Journal of Animal Science* 75, 533–542.

Clement, B.A., Gulf, C.M. and Forbes, T.D.A. (1997) Toxic amines and alkaloids from *Acacia berlandieri. Phytochemistry* 46, 249–254.

Davies, H.L. and Southey, I.N. (2001) Effects of grazing management and stocking rate on pasture production, ewe liveweight, ewe fertility and lamb growth on subterranean clover-based pasture in Western Australia. *Australian Journal of Experimental Agriculture* 41, 161–168.

Dávila-Gutiérrez, X.D. (1996) Celular de cactaceas consumidas por la tortuga (Xerobates berlandien). BSc thesis, Universidad Autónoma de Nuevo León. Monterrey, NL, Mexico.

Dawson, J.M., Buttery, P.J., Jenkins, D., Wood, C.D. and Gill, M. (1999) Effects of dietary quebracho tannin on nutrient utililization and tissue metabolism in sheep and rats. *Journal of Science Food and Agriculture* 79, 1423–1430.

Decandia, M., Sitzia, M., Cabiddu, A., Kababya, D. and Molle, G. (2000) The use of polyethylene glycol to reduce the anti-nutritional effects of tannins in goats fed woody species. *Small Ruminant Research* 38, 157–164.

Denek, N., Can, A., Tufenk, S., Yazgan, K., Ipek, H. and Iriadam, M. (2006) The effect of heat load on nutrient utilization and blood parameters of Awassi ram lambs fed different types and levels of forages. *Small Ruminant Research* 63, 156–161.

Fierro, L.C. and Bryant, F.C. (1990) Grazing activities and bioenergetics of sheep on native range in southern Peru. *Small Ruminant Research* 3, 135–146.

Figueiredo, E.A., Simplicio, A., Ribora, G.S., Melo, E.L. and Olivera, E.R. (1980) *Comportamento ao longo do ano em cabras criolas, em sistema tradicional de manejo.* Embrapa, Comunicado Técnico 4, 3.

Flythe, M.C. and Andries, K. (2009) The effects of monensin on amino acid catabolizing bacteria isolated from the Boer goat rumen. *Small Ruminant Research* 81, 178–181.

Foo, L.Y., Jones, W.T., Porter, L.J. and Williams, V.N. (1986) Proanthrocyanidin polymers of fodder legumes. *Phytochemistry* 21, 933–935.

Forbes, T.D.A., Carpenter, B.B., Tolleson, D.R. and Randel, R.D. (1994) Effects of *N*-methyl-β-phenethylamine on GnRH stimulated luteinizing hormone release and plasma catecholamine concentrations in wethers. *Journal of Animal Science* 72, 464–469.

Frier, H.I., Gorgacz, E.J., Hall, R.C., Gallina, A.M., Roussen, J.E., Eaton, H.D. and Nielsen, S.W. (1974) Formation and absorption of cerebrospinal fluid in adult goats with hypo and hypervitaminosis A. *American Journal of Veterinary Research* 35, 45–55.

Galbraith, H.W., Chigwada, W., Scaife, J.R. and Humphries, W.R. (1997) The effect of dietary molybdenum supplementation on tissue copper concentrations, mohair fibre and carcass characteristics of growing Angora goats. *Animal Feed Science and Technology* 67, 83–90.

Galloway, D.L., Goetsch, A.L., Forster, L.A. Jr, Brake, A.C. and Johnson, Z.B. (1993) Digestion, feed intake, and live weight gain by cattle consuming Bermudagrass and supplemented with different grains. *Journal of Animal Science* 71, 1288–1297.

Gilboa, N., Perevolotsky, A., Landau, S., Nitsan, Z. and Silanikove, N. (2000) Increasing productivity in goats grazing Mediterranean woodland and scrubland by supplementation of polyethylene glycol. *Small Ruminant Research* 38, 183–190.

Gould, D.H. (1998) Polioencephalomalacia. *Journal of Animal Science* 76, 309–314.

Haenlein, G.F.W. and Caccese, R. (1992) Digestion. In: *Extension Goat Handbook.* University of Delaware, Newark, Delaware.

Hall, M.B. (2003) Challenges with nonfiber carbohydrate methods. *Journal of Animal Science* 81, 3226–3232.

Hart, S. (2008) Meat goat nutrition. In: *Proceedings of the 23rd Annual Goat Field Day*, Langston University, Langston, Oklahoma, pp. 58–83.

Herselman, M.J., Hart, S.P., Sahlu, T., Coleman, S.W. and Goetsch, A.L. (1999) Heat energy for growing goats and sheep grazing different pastures in the summer. *Journal of Animal Science* 77, 1258–1265.

Hofman, R.R. (1989) Evolutionary steps of ecophysiological adaptation and diversification of ruminants: a comparative view of their digestive system. *Oecologia* 78, 443–457.

Hurley, W.L. and Doane, R.M. (1989) Recent developments in the role of vitamins and minerals in reproduction. *Journal of Dairy Science* 72, 784–804.

Huston, J.E. (1978) Forage utilization and nutrient requirements of the goat. *Journal of Dairy Science* 61, 988–993.

Huston, J.E., Rector, B.S., Ellis, W.C. and Allen, M.L. (1986) Dynamics of digestion in cattle, sheep, goats, and deer. *Journal of Animal Science* 62, 208–215.

Huston, J.E., Engdahl, B.S. and Calhoun, M.C. (1990) Effects of supplemental feed with or without ionophores on lambs and Angora kid goats on rangeland. *Journal of Animal Science* 68, 3980–3986.

Kadzere, C.T. and Jingura, R. (1993) Digestibility and nitrogen balance in goats given different levels of crushed whole soybeans. *Small Ruminant Research* 10, 175–180.

Kawas, J.R. (1983) The significance of fiber level on the nutritive value of alfalfa hay-based diets for ruminants. PhD thesis, University of Wisconsin-Madison, Madison, Wisconsin.

Kawas, J.R. (2008) Producción y Utilización de Bloques Multinutrientes como Complemento de Forrajes de Baja Calidad para Caprinos y Ovinos: La Experiencia en Regiones Semiáridas. *Tecnología e Ciência Agropecuaria (Brazil)* 2, 63–69.
Kawas, J.R. and Huston, J.E. (1990) Nutrient requirements of hair sheep in tropical and subtropical regions. In: Shelton, M. and Figueiredo, E.A.P. (eds) *Hair Sheep Production in Tropical and Subtropical Regions*, Chapter 4. Small Ruminant Collaborative Research Support Program, US-AID.
Kawas, J.R., Danelón, J.L., Craig, J.L. and Jorgensen, N.A. (1989) Efecto del origen del inoculo, especie y estado de madurez de los forrajes sobre la digestión *in vitro* de la pared celular. *Revista Argentina de Producción Animal* 9, 15–21.
Kawas, J.R., Jorgensen, N.A. and Lu, C.D. (1990) Influence of alfalfa maturity on feed intake and site of nutrient digestion in sheep. *Journal of Animal Science* 68, 4376–4386.
Kawas, J.R., Foote, W.C. and Simplicio, A. (1992) Nutritional aspects of female reproduction. In: *Proceedings of the Fifth International Conference on Goats*, New Delhi, India, Vol. 2, pp. 342–354.
Kawas, J.R., Osmin, O., Hernández, J., Leal, R., Garza, F. and Danelón, J.L. (1997) Performance of grazing bull calves supplemented with increasing levels of ruminally undegradable protein. In: *XVIII International Grassland Congress*, Winnipeg, Manitoba and Saskatoon, Saskatchewan, Canada.
Kawas, J.R., Schacht, W.H., Shelton, J.M., Olivares, E. and Lu, C.D. (1999) Effects of grain supplementation on the intake and digestibility of range diets consumed by goats. *Small Ruminant Research* 34, 49–56.
Kawas, J.R., Andrade-Montemayor, H. and Lu, C.D. (2010) Strategic nutrient supplementation of free-ranging goats. *Small Ruminant Research* 89, 234–243.
Kendall, P.T., Ducker, M.J. and Hemingway, R.G. (1983) Individual intake variation in ewes given feedblock or trough supplements indoors or at winter grazing. *Animal Production* 36, 7–19.
Knox, M.R. (2002) Effectiveness of copper oxide wire particles for *Haemonchus contortus* control in sheep. *Australian Veterinary Journal* 80, 224–227.
Knox, M. and Steel, J. (1996) Nutritional enhancement of parasite control in small ruminant production systems in developing countries of south-east Asia and the Pacific. *International Journal of Parasitology* 26, 963–970.
Knox, M.R., Torres-Acosta, J.F.J. and Aguilar-Caballero, A.J. (2006) Exploiting the effect of dietary supplementation of small ruminants on resilience and resistance against gastrointestinal nematodes. *Veterinary Parasitology* 139, 385–393.
Krysl, L.J. and Hess, B.W. (1993) Influence of supplementation on behavior of grazing cattle. *Journal of Animal Science* 71, 2546–2555.
Kumar, R. and Vaithyanathan, S. (1990) Occurrence, nutritional significance and effect on animal productivity of tannins in tree leaves. *Animal Feed Science and Technology* 30, 21–38.
Lachica, M. and Aguilera, J.E. (2003) Estimation of energy needs in the free-ranging goat with reference to the assessment of its energy expenditure by the $^{13}$C-bicarbonate method. *Small Ruminant Research* 49, 303–318.
Lachica, M. and Aguilera, J.E. (2005) Energy needs of the free-ranging goat. *Small Ruminant* Research 60, 111–125.
Lachica, M. and Aguilera, J.E. (2008) Methods to estimate the energy expenditure of goats: from the lab to the field. *Small Ruminant Research* 79, 179–182.
Lu, C.D. (1988) Grazing behavior and diet selection of goats. *Small Ruminant Research* 1, 205–216.
Lu, C.D. (1989a) Effect of heat stress on goat production. *Small Ruminant Research* 2, 151–162.
Lu, C.D. (1989b) Energy and protein nutrition in lactating dairy goats. In: *Proceedings of the 24th Pacific Northwest Animal Nutrition Conference*, Boise, Idaho, pp. 133–142.
Lu, C.D. (1992) Effect of antiquality substances on utilization of leaf protein by animals. *World Review of Animal Production* 26, 29–35.
Lu, C.D. (1993) Implication of feeding isoenergetic diets containing animal fat on milk composition of Alpine does during early lactation. *Journal of Dairy Science* 76, 1137–1147.
Lu, C.D. (2011) Nutritionally related strategies for organic goat production. *Small Ruminant Research* 98, 73–82.
Lu, C.D. and Coleman, L.J. (1984) Grazing behavior and diet selection of goats. In: *Proceedings of the 1st Regional Meat Goat Conference.* Division of Agricultural Science, Florida A&M University, Tallahassee, Florida, pp. 56–71.
Lu, C.D. and Potchoiba, M.J. (1990) Feed intake and weight gain of growing goats fed diets of various energy and protein levels. *Journal of Animal Science* 68, 1751–1759.

Lu, C.D. and Rubino, R. (1992) Recent advancements on availability and utilization of feed resources for goats in North America and Europe. In: *Proceedings of the 5th International Conference on Goats*, Vol. 2, pp. 105–120.

Lu, C.D., Akinsoyinu, A.O. and Qi, K. (1991) Predicting dry matter intake of lactating goats. In: *Proceedings of the 7th National Conference on Goat Production*, Monterrey, Mexico, pp. 258–289.

Lu, C.D., Kawas, J.R. and Mahgoub, O.G. (2005) Fibre digestion and utilization in goats. *Small Ruminant Research* 60, 45–52.

Lu, C.D., Gangyi, X. and Kawas, J.R. (2010) Organic goat production, processing and marketing: opportunities, challenges and outlook. *Small Ruminant Research* 89, 102–109.

Lucy, M.C. (2003) Mechanisms linking nutrition and reproduction in postpartum cows. *Reproduction Suppl* 61, 415–417.

Luginbuhl, J.M., Poore, M.H., Mueller, J.P. and Green, J.T. (2009) *Breeding and Kidding Management in the Goat Herd*. University of Kentucky College of Agriculture, Lexington, Kentucky.

Lusby, K.S. (1990) Supplementation of cattle on rangeland. In: *Proceedings of the 16th Southern Pasture and Forage Crop Improvement Conference*, Overton, Texas, pp. 64–71.

Makkar, H.P.S. (2003) Effects and fate of tannins in ruminant animals, adaptation to tannins, and strategies to overcome detrimental effects of feeding tannin-rich feeds. *Small Ruminant Research* 49, 241–256.

Malechek, J.C. and Provenza, F.D. (1983) Feeding behaviour and nutrition of goats in rangelands. *World Animal Review* 47, 38–48.

McDowell, L.R., Conrad, J.H., Ellis, G.L. and Loosli, J.K. (1983) *Minerals for Grazing Ruminants in Tropical Regions*. Department of Animal Science and Center for Tropical Agriculture, University of Florida, Gainsville, USA, and US Agency for International Development.

Merck (2008) *Merck Veterinary Manual*. Merck Company, Whitehouse Station, New Jersey.

Min, B.R. and Hart, S.P. (2003) Tannins for suppression of intestinal parasites. *Journal of Animal Science* 81, 102–109.

Min, B.R., Attwood, G.T., Reilly, K., Sun, W., Peters, J.S., Barry, T.N. and McNabb, W.C. (2002) *Lotus corniculatus* condensed tannins decrease *in vivo* populations of proteolytic bacteria and affect nitrogen metabolism in the rumen of sheep. *Canadian Journal of Microbiology* 48, 911–921.

Mobley, R., Kahan, T., Okpebholo, F., Nurse, G., Beaudoin, J., Lyttle-N'guessan, C. and Peterson, T. (2007) *Practical Management of Internal Parasites in Goats*. Cooperative Extension Program, College of Engineering Sciences, Technology and Agriculture, Florida A&M University. Tallahassee, FL, USA.

Moore, K.J. and Hartfield, R.D. (1994) Carbohydrate and forage quality. In: Fahey, G.C. Jr, Collins, M.C., Mertens, D.R. and Moser, L.E. (eds) *Forage Quality Evaluation, and Utilization*. ASA-CSSA-SSSA, Madison, Wisconsin.

Morand-Fehr, P. (1987) Management programs for the prevention of kid losses. In: *Proceedings of the III International Conference on Goats*, 8–13 March. International Goat Association, Brazilia, Brazil, p. 571.

Nantoumé, H., Forbes, T.D.A., Hensarling, C.M. and Sieckenius, S.S. (2001) Nutritive value and palatability of guajillo (*Acacia berlandieri*) as a component of goat diets. *Small Ruminant Research* 40, 139–148.

Negesse, T., Rodehutscord, M. and Pfeffer, E. (2001) The effect of dietary crude protein level on intake, growth, and protein retention and utilization of growing male Saanen kids. *Small Ruminant Research* 39, 243–251.

Niezen, J.H., Waghorn, T.S., Charleston, W.A. and Waghorn, G.C. (1995) Growth and gastrointestinal parasitism in lambs grazing one of seven herbages and dosed with larvae for six weeks. *Journal of Agricultural Science* 125, 281–288.

NRC (1981) *Nutrient Requirements of Goats: Angora, Dairy and Meat Goats in Temperate and Tropical Countries*. National Research Council, National Academy Press, Washington, DC, USA.

NRC (1985) *Ruminant Nitrogen Usage*. National Academy Press, Washington, DC, USA.

NRC (1987) *Predicting Feed Intake of Food-producing Animals*. National Research Council, National Academy Press. Washington, DC, USA.

NRC (2000) *Nutrient Requirements of Beef Cattle*, 7th revised edn. National Research Council, National Academy Press. Washington, DC, USA.

NRC (2007) *Nutrient Requirements of Small Ruminants: Sheep, Goats, Cervids and New World Camelids*. National Research Council, National Academy Press. Washington, DC, USA.

Osuji, P.O. (1974) The physiology of eating and the energy expenditure of the ruminant at pasture. *Journal of Range Management* 27, 437–443.

Paterson, J.A., Belyea, R.L., Bowman, J.P., Kerley, M.S. and Williams, J.E. (1994) The impact of forage quality and supplementation regimen on ruminant animal intake and performance. In: Fahey, G.C. Jr, Collins, M.C., Mertens, D.R. and Moser, L.E. (eds) *Forage Quality Evaluation, and Utilization.* ASA-CSSA-SSSA, Madison, Wisconsin.

Pfister, J.A. (1983) Nutrition and feeding behavior of goats and sheep grazing deciduous shrub-woodland in northeastern Brazil. PhD thesis, Utah State University, Logan, Utah.

Pordomingo, A.J., Wallace, J.D., Freeman, A.S. and Galyean, M.L. (1991) Supplemental corn grain for steers grazing native rangeland during summer. *Journal of Animal Science* 69, 1678–1687.

Price, D.A. and Hardy, W.T. (1953) Guajillo poisoning of sheep. *Journal of the Animal Veterinary Medical Association* 122, 223–225.

Prieto, I., Goetsch, A.L., Banskalieva, V., Cameron, M., Puchala, R., Sahlu, T., Dawson, L.J. and Coleman, S.W. (2000) Effects of dietary protein concentration on postweaning growth of Boer crossbred and Spanish goat wethers. *Small Ruminant Research* 78, 2275–2281.

Qi, K., Lu, C.D. and Owens, F.N. (1993a) Sulfate supplementation of Angora goats: sulfur metabolism and interactions with zinc, copper and molybdenum. *Small Ruminant Research* 11, 209–225.

Qi, K., Lu, C.D. and Owens, F.N. (1993b) Sulfate supplementation of growing goats: effects on performance, acid–base balance, and nutrient digestibilities. *Journal of Animal Science* 71, 1579–1587.

Qi, K., Owens, F.N. and Lu, C.D. (1994a) Effects of sulfur deficiency on performance of fiber-producing sheep and goats: a review. *Small Ruminant Research* 14, 115–126.

Qi, K., Lu, C.D. and Owens, F.N. (1994b) Effects of sulfate supplementation on performance, acid–base balance, and nutrient metabolism in Alpine kids. *Small Ruminant Research* 15, 9–18.

Qi, K., Lu, C.D., Owens, F.N. and Lupton, C.J. (1994c) Effects of sulfate supplementation on performance, acid–base balance, and nutrient metabolism in Angora kids. *Small Ruminant Research* 15, 19–30.

Quinisa, M.M. and Boomker, E.A. (1998) Feed selection and water intake of indigenous goat wethers under stall-feeding conditions. *South African Journal of Animal Science* 28, 173–178.

Ramirez, R.G., Rodriguez, A., Del Valle, L.A. and Gonzalez, J. (1990) Nutrient content and intake of forage grazed by range goats in northeastern Mexico. *Small Ruminant Research* 3, 435–448.

Sahlu, T., Jung, H.G. and Morris, J.G. (1989) Influence of grazing pressure on energy cost of grazing by sheep on smooth Bromegrass. *Journal of Animal Science* 67, 2098–2105.

Sahlu, T., Goetsch, A.L., Luo, J., Nsahlai, I.V., Moore, J.E., Galean, M.L., Owens, F.N., Ferrell, C.L. and Johnson, Z.B. (2004) Nutrient requirements of goats: developed equations, other considerations and future research to improve them. *Small Ruminant Research* 53, 191–219.

Schacht, W.H., Kawas, J.R. and Malechek, J.C. (1992) Effects of supplemental urea and molasses on dry season weight gains of goats in semiarid tropical woodland, Brazil. *Small Ruminant Research* 7, 235–244.

Schelling, G.T. (1984) Monensin mode of action in the rumen. *Journal of Animal Science* 58, 1518–1527.

Seman, D.H., Frere, M.H., Stuedemann, J.A. and Wilkinson, S.R. (1991) Simulating the influence of stocking rate, sward height and density on steer productivity and grazing behavior. *Agriculture Systems* 37, 165–181.

Silanikove, N. (2000) The physiological basis of adaptation in goats to harsh environments. *Small Ruminant Research* 35, 181–193.

Silanikove, N., Nitsan, Z. and Perevolotsky, A. (1994) Effect of daily supplementation of polyethylene glycol on intake and digestion of tannin-containing leaves (*Ceratonia siliqua*) by sheep. *Journal of Agriculture and Food Chemistry* 42, 2844–2847.

Silanikove, N., Gilboa, N., Perevolotsky, A. and Nitsan, Z. (1996) Goats fed tannin-containing leaves do not exhibit toxic syndromes. *Small Ruminant Research* 21, 195–201.

Solaiman, S.G., Maloney, M.A., Qureshi, M.A., Davies, G. and D'Andrea, G. (2001) Effects of high copper supplements on performance, health, plasma copper and enzymes in goats. *Small Ruminant Research* 41, 127–129.

Solaiman, S.G., Shoemaker, C.E., Jones, W.R. and Kerth, C.R. (2006) Effects of high levels of supplemental copper on the serum lipid profile, carcass traits, and carcass composition of goat kids. *Journal of Animal Science* 84, 171–177.

Sotomayor-Rios, A. and Pitman, W.D. (2001) *Tropical Forage Plants: Development and Use.* CRC Press, Boca Raton, Florida.

Soto-Navarro, S.A., Goetsch, A.L., Sahlu, T., Puchala, R. and Dawson, L.J. (2003) Effects of ruminally degraded nitrogen source and level in a high concentrate diet on site of digestion in yearling Boer x Spanish wether goats. *Small Ruminant Research* 50, 117–128.

Soto-Navarro, S.A., Goetsch, A.L., Sahlu, T. and Puchala, R. (2004) Effects of level and source of supplemental protein in a concentrate-based diet on growth performance Boer × Spanish wether goats. *Small Ruminant Research* 51, 101–106.

Sudana, I.B. and Leng, R.A. (1986) Effects of supplementing a wheat straw diet with urea or urea-molasses blocks and/or cottonseed meal on intake and live weight change of lambs. *Animal Feed Science and Technology* 16, 25–35.

Sykes, A.R. (1994) Parasitism and production in farm animals. *Animal Production* 59, 155–172.

Tedeschi, L.O., Fox, D.G. and Tylutki, T.P. (2003) Potential environmental benefits of ionophores in ruminant diets. *Journal of Environmental Quality* 32, 1591–1602.

Thornton, R.F. and Minson, D.J. (1973) The relationship between apparent retention time in the rumen, voluntary intake and apparent digestibility of legume and grass diets in sheep. *Australian Journal of Agricultural Research* 24, 889–906.

Van Soest, P.J. (1994) *Nutritional Ecology of the Ruminant*, 2nd edn. Cornell University Press, Ithaca/London.

Vanzant, E.S., Cochran, R.C., Jacques, K.A., Beharka, A.A., Delcurto, T. and Avery, T.B. (1990) Influence of level of supplementation and type of grain in supplements on intake and utilization of harvested, early-growing-season, bluestem-range forage by beef steers. *Journal of Animal Science* 68, 1457–1468.

Vera-Avila, H.R., Forbes, T.D.A. and Randel, R.D. (1996) Plant phenolic amines: potential effects on sympathoadrenal medullary, hypothalamic–pituitary–adrenal, and hypothalamic–pituitary–gonadal function in ruminants. *Domestic Animal Endocrinology* 13, 285–296.

Villalba, J.J. and Provenza, F.D. (2002) Glycol influences selection of foraging location by sheep consuming quebracho tannin. *Journal of Animal Science* 80, 1846–1851.

Wallace, D.S., Bairden, K., Duncan, J.L., Eckersall, P.D., Fishwick, G., Gill, M., Holmes, P.H., McKellar, Q.A., Murray, M., Parkins, J.J. and Stear, J. (1998) The influence of dietary supplementation with urea on resilience and resistance to infection with *Haemonchus contortus*. *Parasitology* 116, 67–72.

Waller, P.J. (1994) The development of anthelmintic resistance in ruminant livestock. *Acta Tropica* 56, 233–243.

Waller, P.J. (1997) Anthelmintic resistance. *Veterinary Parasitology* 72, 391–412.

Waller, P.J. (2006) Sustainable nematode parasite control strategies for ruminant livestock by grazing management and biological control. *Animal Feed Science Technology* 126, 277–289.

Waller, P.J. and Thamsborg, S.M. (2004) Nematode control in green ruminant production systems. *Trends in Parasitology* 20, 493–497.

Webb, R., Garnsworthy, P.C., Gong, J.G. and Armstrong, D.G. (2004) Control of follicular growth: local interactions and nutritional influences. *Journal of Animal Science* 82, 63–74.

Wentzel, D. (1982) Non-infectious abortion in Angora goats. In: *Proceedings of the 3rd International Conference on Goat Production and Disease*. Dairy Goat Journal Publication Company, Scottsdale, AZ, USA, p. 155.

Wu, Y.M., Hu, W.L.L. and Liu, J.X. (2005) Effects of supplementary urea-minerals lick block on the kinetics of fibre digestion, nutrient digestibility and nitrogen utilization of low quality roughages. *Plant and Animal Sciences and Biotechnology* 6, 793–797.

Zhu, X., Deyoe, C.W., Behnke, K.C. and Seib, P.A. (1991) Poured feed blocks using distillery by-products as supplements for ruminants. *Journal of the Science of Food and Agriculture* 54, 535–547.

# 7 Growth, Development and Growth Manipulation in Goats

E.C. Webb[1], N.H. Casey[1] and L. Simela[2]
[1]*Department of Animal and Wildlife Sciences, University of Pretoria, South Africa;*
[2]*New Emerging Red Meat Producers Organization, Lynnwood, South Africa*

## 7.1 Abstract

A thorough understanding of growth and development in goats and the factors that affect these processes is important, because it affects the efficiency of production and it has a direct bearing on product quality. Although the use of growth promotants or molecules that change the efficiency of growth are not generally employed in goats, other more natural pathways exist to manipulate growth and development. These approaches do not detract from product quality and provide the opportunity to produce meat, milk, fibre, leather and related products that meet the demands of consumers for more environmentally friendly and natural production systems.

Growth and development in goats occur in a similar way compared with other land-living mammals, and these processes are directly dependent on influences in the external and internal environment.

There is great significance in the way in which the growth and developmental processes have evolved in goats to ensure the propagation of this unique species. There are obvious advantages in ensuring a minimum size at birth in terms of survival value, while the size of the birth canal limits excessive size at birth.

The skeletal size of goats is rather variable due to the effects of natural selection and adaptation of the various breeds or types of goat to different ecological regions in the world. Goats are subdivided into three categories based on frame size, namely, dwarf breeds, small breeds and large breeds. Embryological development includes organogenesis, skeletal growth and development, myogenesis and adipose tissue synthesis and deposition. The growth and development of fatty tissue in goats occur at a lower rate and to a lower extent compared with sheep. Accumulation of abdominal fat occurs at an earlier stage compared with subcutaneous fat in all genders. Growth is a chronological process and its inevitable consequence is an enlargement in size and ageing. Ageing is associated with an increasing inability to sustain the functional integrity of cells, organs and systems.

Animal growth is the consequence of complex interactions between the genetic potential of the animal and several hormones, the nutrient supply and the environment. This complexity is beneficial in terms of manipulating growth and development because it provides the possibility for many different ways of manipulating growth. The most significant means of manipulating growth and development in goats include genetic selection, castration and intensive fattening. Selection pressure on different production characteristics has resulted in the development of different types of goat for different purposes.

The effects of nutrition on the growth and development of goats are well documented. These effects are sometimes described as 'production system' effects and include 'intensive' feeding and conditioning, or 'extensive' grazing. Certain goat genotypes respond better to feeding or feed supplementation, and this ability differs mainly between the dairy and meat types of goat. Conditioning of goats has a marked effect on body composition. Consumer preferences for goat carcasses differ among countries and this makes it difficult to implement universally acceptable grading or classification systems, as well as standard carcass cuts. In addition, the variation in subcutaneous fat thickness between goats of different ages, sex and types is extremely limited, which makes it difficult to use fat thickness for purposes of carcass classification or grading.

## 7.2 Introduction

Domesticated goats are of the class Mammalia, order Artiodactyla, family Bovidae, subfamily Caprina, genus *Capra* and species *hircus*. This genetic lineage has equipped goats with interesting morphological and physiological attributes that make them adaptable to virtually all climatic zones. The morphology of goats ranges from small breeds characteristic of the tropical regions to large breeds found in the Himalayan regions. Their coat ranges from short-haired, which is more suitable for hot and humid regions, to long-haired, suitable for colder regions. The Angora goat, which originates from the mountainous regions of Central Asia, has long, fine, silvery-white hair, which makes it suitable for both hot and cold regions but not humid regions. Human intervention has had its influence on both the morphology and the physiology of goats. Morphological characteristics became embedded through selection for specific traits and identities. This yielded the range of coat colours that can be associated with human population groups or tribes. Selection for milk as opposed to carcass characteristics gave the lean, wedge-shaped doe as opposed to the square, stocky-shaped meat goat such as the Boer goat. A discussion on the growth and development of goats needs to take this range into account, although, despite the variety of breeds, as they are of the same species, the inherent traits expressed in their growth and development remain similar.

## 7.3 Growth and Development in Goats

A thorough understanding of growth and development in goats and the factors that affect these processes is important, because it affects the efficiency of production and it has a direct bearing on product quality. The rate and efficiency of growth and the subsequent effects on product quality can be managed in livestock, and this is a deliberate, active process that reaches from conception to consumption (Casey and Webb, 2010). Although the use of growth promotants or molecules that change the efficiency of growth is not generally employed in goats, other more natural pathways exist to manipulate growth and development. These approaches do not detract from product quality and provide the opportunity to produce meat, milk, fibre, leather and related products that meet the demands of consumers for more environmentally friendly and natural production systems.

### 7.3.1 Principles of growth and development

Growth and development in goats occur in a similar way compared with other land-living mammals. Growth genes determine growth (size) and development (conformation), but the extent to which these genes are expressed is determined principally by single and interactive multiple factors of the external and internal environments.

Growth can be illustrated graphically by means of a sigmoidal or S-shaped curve

(Fig. 7.1), which represents the continuous changes in body conformation and composition. These changes include gradual increases in bone, muscle and fat in that order (Webb and Casey, 2005). An allometric relationship exists between different tissues in the body, which means that the growth of tissues occurs in a specific order. It is therefore possible to estimate the weight of a tissue if the weight of an associated tissue is known.

Growth can be described by means of a mathematical polynomial function of a suitable degree. The quantitative aspects of growth are best described by means of changes in size, mass, height or length, while qualitative changes are described by means of changes in conformation, such as secondary sexual development. Goats express distinct sexual dimorphism.

A distinction can be made between potential and actual growth. Potential growth refers to the expected growth based on production records (of lineage, i.e. parents, and siblings) and entered into a predictive mathematical model. Actual growth is what is being attained or has been attained, and the measured parameters can be entered into a descriptive mathematical model. The measurement of actual growth can include factors that influence growth. The data of a number of these models can then be combined to develop a predictive model that accommodates a number of determinant variables. An example of an actual growth model could be:

Post-weaning live mass gain
= weaning mass × age
× feed intake × season
× (any contributing factor that can be measured)

### 7.3.2 Rate of growth and development

An increase in cell numbers (hyperplasia) occurs rapidly after fertilization. While hyperplasia continues, significant conformational changes occur that result in the formation of the morula, blastula and gastrula stages typically observed during embryological development in mammals. During these early stages of embryological development, rapid growth and development of tissues and organs occur, which is known as organogenesis. This is followed by the development of specialized systems to provide structural support, facilitate digestion, absorption and utilization of feedstuffs in a structured way, facilitate locomotion and ultimately ensure the propagation of the species through carefully adapted neural and reproductive systems and processes. Most structural development in animals occurs before birth. In fact, if development is plotted against time, the general trend is reflected as a negative logarithmic curve. In

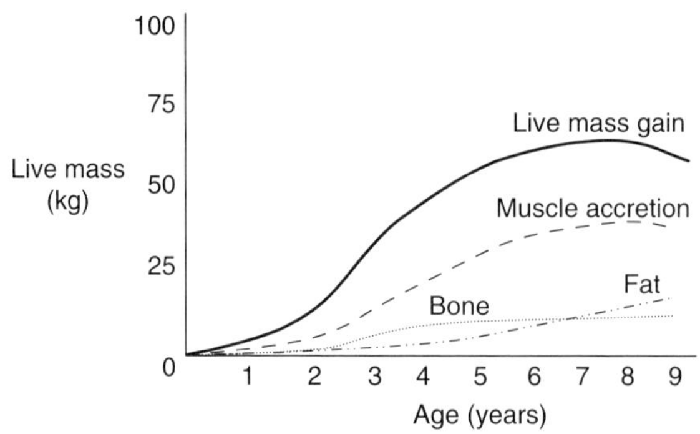


**Fig. 7.1.** Graphical representation of growth in Boer goats, illustrating the typical sigmoidal growth response over time. Average birth mass is 3.5 kg, and kids are weaned at about 29 kg and slaughtered at 43 kg live mass.

contrast, if growth is expressed in quantitative terms (mass, height, length, width, volume) on the *y*-axis versus time on the *x*-axis, as outlined in Fig. 7.1, the general trend is reflected as a sigmoidal curve.

In terms of quantitative growth characteristics, goats experience accelerated growth (weight gain) shortly after birth, followed by a self-accelerating phase more or less until weaning. This latter phase represents the phase during which goats tend to gain most weight (or change in dimensions) relative to time compared with any other phase in the growth process.

### 7.3.3 Growth, development and survival

Growth and development of goats has evolved over time. For example, kids are relatively small at birth but they grow and develop quickly to ensure the survival of this unique species under extensive conditions. The average birth mass of goats varies between 2.5 and 3.5 kg, while the average litter size varies between 1.0 and 1.7. Twins tend to have a lower birth mass. There are obvious advantages in ensuring a minimum size at birth in terms of survival value, while the size of the birth canal limits excessive size within the breed (Webb and Casey, 2010). Does usually kid-down easily and are not prone to dystocia, although care should be taken – as with all livestock – not to use large, heavy males, especially of a large breed, for breeding with young or small females. The temptation remains for producers to attempt to improve the growth and carcass yield by breeding females of small local breeds with exotic large breeds. The risk is the loss of the kid at birth and permanent damage to the doe.

## 7.4 Tissue Growth and Development

### 7.4.1 Growth and development of the skeletal system

The brain and spinal cord are the first anatomical developments that occur after implantation of the zygote on the endometrium (mucosal lining of the uterus) and placentation (attachment of the embryo to the uterus). The skeletal system grows and develops rapidly to support the growing cell mass. Skeletal growth continues until well after puberty, when the rate of growth in length of long bones starts to decrease, while the diameter of bones continues to increase until maturity. Castration prolongs skeletal growth, particularly growth of long bones, while an increase in bone diameter is retarded compared with their intact male counterparts. Castration has long been the sought-after procedure to manipulate growth and restrict development of masculine traits in male goats.

The skeletal size of goats is rather variable due to the effects of natural selection and adaptation of the various breeds or types of goat to different ecological regions in the world (Webb, 2007). Goats are subdivided into three categories based on frame size, namely, dwarf breeds, small breeds and large breeds. Goats with a small frame size vary between 15 and 30 kg at about 15 months of age, while large-framed breeds may weigh up to 55 kg live weight at a similar age. Dwarf goats are small and seldom weigh more than 25 kg at 15–24 months of age.

### 7.4.2 Growth and development of muscle

Although muscle fibre characteristics and muscle biochemistry differ between species, the growth and development of muscle tissue occur in a similar way to that in other livestock species. Muscle precursor cells originate from the mesodermal progenitor cells during early embryonic development. The dorsolateral and ventrolateral edges of the dermomyotome portion of the somite proliferate and form multinucleated myofibrils, which are also known as primary fibres or slow fibres, while secondary myoblasts form secondary fibres or fast fibres (Kokta *et al.*, 2004). During the post-natal stage, muscle development is characterized by the proliferation of satellite cells present between the sarcolemma and basal lamina of myofibrils (Fig. 7.2).

Although the origin of satellite cells during post-natal development is uncertain, it appears that intrinsic growth factors are secreted more or less from puberty onwards, which stimulates myogenic cells to re-enter the cell cycle and form precursor cells that differentiate and fuse with myofibrils and form myotubes (Fig. 7.3). It has been shown that these processes are particularly evident after exercise or injury.

Myogenesis per se is influenced by insulin-like growth factor (IGF-1 and IGF-2), insulin-like growth factor-binding protein (IGFBP) and myostatin. IGF and IGFBP promote muscle cell proliferation and differentiation, while myostatin inhibits muscle hyperplasia and thus the number of myofibrils. An autocrine effect has been proposed in muscle cell proliferation because IGF-1 and the number of IGF-1 receptors increase during cellular proliferation and differentiation, and IGF-2 is also

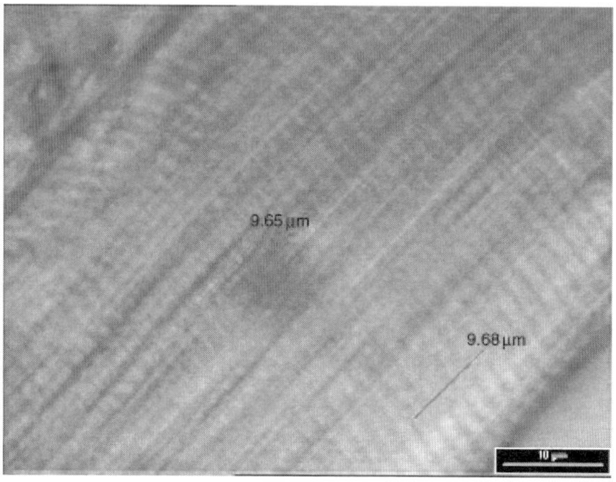


**Fig. 7.2.** Histological section through the longissimus dorsi muscle of a Boer goat, illustrating longitudinal striations and sarcomere lengths as viewed under a visual image analyser (100× magnification). The lines indicate the distance between five sarcomeres (from Simela, 2005).

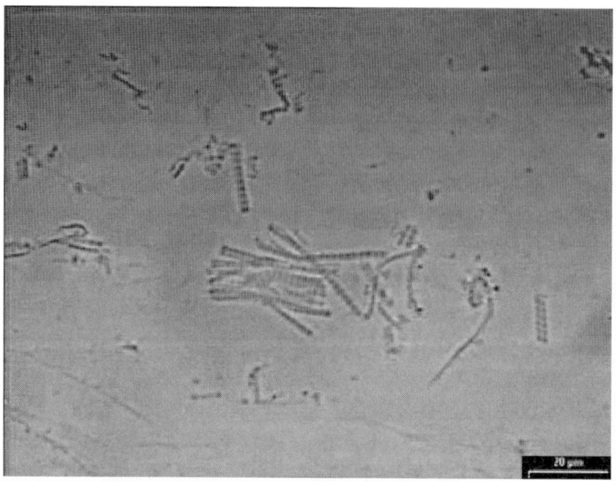

**Fig. 7.3.** Histological section through the longissimus dorsi muscle of a Boer goat illustrating myofibrillar fragments as viewed under a visual image analyser at a magnification of 40× (from Simela, 2005).

involved. Of the six isoforms of IGFBPs that are known, it appears that IGFBP2, -4, -5 and -6 are expressed in skeletal muscle tissue.

An important contribution to our current knowledge of the characteristics of muscle tissue in goats is associated with the sensitivity of goats to pre-slaughter stress, which significantly affects the conversion of muscle to meat (Webb *et al.*, 2005). Goats are prone to peri-mortem stress, as evidenced by the high ultimate pH ($pH_u$) values of carcasses (Simela *et al.*, 2004). Muscle glycogen concentrations in goat carcasses tend to be low (about 33 μmol/g), which results in low muscle glucose concentrations and inadequate conversion of glucose to lactic acid under anaerobic conditions. The latter process is crucial for the normal decrease in muscle pH observed post mortem in most bovine and ovine species. This condition is partially due to the relatively high proportion of red muscle fibres in goat muscles (Fig. 7.4).

### 7.4.3 Growth and development of fat

The growth and development of fatty tissue in goats occur at a lower rate and to a lower extent compared with sheep, although differences are also noted between sexes and in different anatomical locations with ageing. Subcutaneous and intermuscular fat accumulation is limited in males, while more accumulation occurs in females and castrates, but fat also accumulates with ageing as in other species. Accumulation of abdominal fat occurs at an earlier stage compared with subcutaneous fat in all genders. Louveau and Gondret (2004) showed that the accumulation of fat is due to both proliferation and differentiation of adipose precursor cells in the stromal–vascular fraction of adipose tissue, and subsequent hypertrophy of mature adipocytes. Triacylglycerols represent the main lipid fraction in adipose tissue of animals, and fluctuations in this fraction of the adipose depots are due to hydrolysis of lipoproteins and lipoprotein lipase, uptake of free fatty acids, *de novo* synthesis of fatty acids from carbon precursors (lipogenesis) and hydrolysis of triacylglycerols (lipolysis) (Webb and Casey, 2005; Webb and O'Neill, 2008).

### 7.4.4 Effects of ageing on growth and development

Growth is a chronological process and its inevitable consequence is an increase in

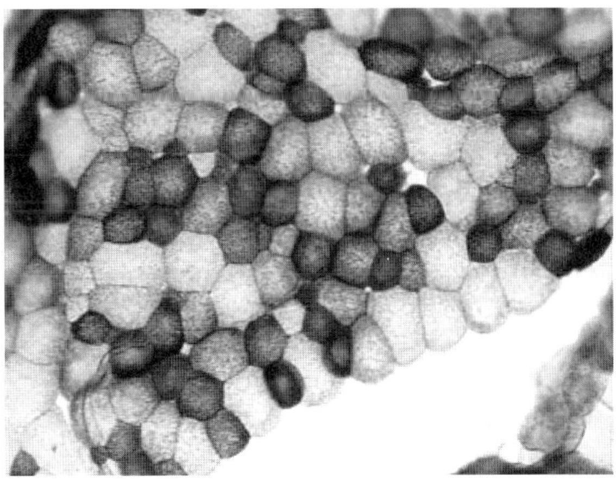

**Fig. 7.4.** Histological section indicating muscle fibre types in the m. longissimus thoracis of the Boer goat (10× magnification). Dark areas are red muscle fibres, grey areas are intermediate fibres and white areas are white muscle fibres (from Simela, 2005).

size and ageing. Ageing is associated with an increasing inability to sustain the functional integrity of cells, organs and systems (Webb and Casey, 2005). This is confirmed by the observations of Chen (2004) that ageing is associated with a noticeable decrease in certain pituitary hormones, namely growth hormone (GH), prolactin, thyrotropin, luteinizing hormone and follicle-stimulating hormone. In contrast, low concentrations of IGF-1 appear to increase the lifespan of certain organisms (Tatar *et al.*, 2003). Neuropeptide Y (NPY) also plays an important role in the regulation of gonadotropin-releasing hormone (GnRH) secretion in the hypothalamus. According to Chen (2004), ageing is associated with an increase in NPY, which appears to be a compensatory mechanism to counteract the age-related downregulation of the GnRH receptor mRNA. One way in which age-related changes in tissues can be retarded somewhat is by the restriction of metabolizable energy intake. Restricting dietary energy intake increased the lifespan of rats by downregulating the expression of about 19 genes (Chen, 2004), but the effect of short-term restriction appears to be more significant compared with long-term restriction in most tissues. This may explain why indigenous rural goats appear to have a greater longevity than intensively managed dairy-type goats.

## 7.5 Factors that Influence Growth and Development

Animal growth is the consequence of complex interactions between the genetic potential of the animal and several hormones, the nutrient supply and the environment. This complexity is beneficial in terms of manipulating growth and development because it provides the possibility for many different ways of manipulating growth (Webb and Casey, 2005). The emphasis of most livestock production systems is to improve feed conversion efficiency and thus maximize profit (Casey and Webb, 2010). Unfortunately, hormonal growth promoters were banned by the European Union in the 1980s, so virtually no research was done on the possibilities and effects of these molecules in goats. Even the new-generation products such as β-agonists and recombinant GH derivatives have not been researched in goats to any extent.

The most significant means of manipulating growth and development in goats includes genetic selection, castration and intensive fattening. The latter has been employed by farmers for decades to produce big and strong castrates that yield sought-after carcasses with acceptable meat-quality attributes. The differences between genders and the effects of age, pre-slaughter conditioning and reducing the effects of stress have also been listed as ways of improving growth and carcass characteristics in goats.

### 7.5.1 Effect of genetics and selection on growth and development

Genetics certainly plays a major role in any production programme for goat meat. One of the first objectives should be to select a breed that can adapt to the climatic and topographic environment, or that is suitable for a particular production system (Casey and Webb, 2010). If the emphasis is on meat production, then the breed must exhibit characteristics that are typical of a meat-producing animal. As most goats are kept on extensive rangelands, it is vital that the breed is robust, highly fertile and resistant or tolerant to endemic diseases.

Selection pressure on different production characteristics has resulted in the development of different types of goat for different purposes. The varieties of goat genotypes that are available worldwide attest to the impact and significance of genetic selection. Genetic selection has significant effects on the growth and development of goats; for example, selection for growth and carcass characteristics in the Boer goat has resulted in a masculine, robust and fast-growing goat with excellent meat-quality characteristics. This breed has excellent body length and depth and is generally large-framed with exceptional conformation. In contrast, selection for mothering ability and milk yield has resulted in the development of milk goats with finer features, smaller dimensions and relatively poor body conformation compared with Boer goats (Fig. 7.5). The differences between meat and dairy goats are significant and indicate that selection has a major effect on growth and development. Recording of pre- and post-weaning growth under rangeland conditions and performance testing have shown its merit in terms of improving dairy and meat-quality traits in goats (Casey and Webb, 2010).

### 7.5.2 Effect of age and sex on growth and development of goats

The effects of age and sex on growth and development have been clearly documented in a number of species, including goats of different breeds and types. The most significant effects of age on growth and development are generally associated with increasing live mass with obvious changes in body composition and conformation (Webb *et al.*, 2005). Table 7.1 shows the significant changes in body composition and conformation in indigenous goats with increasing age, resulting in increases in carcass mass, body dimensions, carcass lean and fat content and dressing percentage. It is interesting to note that the dressing percentage remains virtually unchanged from 0-tooth to 8-tooth stages, which is explained by the fact that most fat accumulates in the viscera (omentum). The viscera are removed during the slaughter process and do not contribute to carcass mass. Carcass fat content varies between 6% for 0-tooth goats up to about 16% for mature (8-tooth) goats.

Gender also influences growth and development in goats. Bucks exhibit the fastest growth rates and yield the leanest carcasses compared with does or castrates (Table 7.2). No significant differences in dressing percentages are observed between bucks, does and castrates, despite significant differences in carcass mass, carcass dimensions and fat content.

### 7.5.3 Nutritional manipulation of growth in goats

The effects of nutrition on animal growth and development are well documented, and similar studies confirm the significance of feeding in goats of different breeds, genders and ages (Casey, 1982; Simela, 2000; Webb *et al.*, 2005). These effects are sometimes described as 'production system' effects, which include 'intensive' feeding and conditioning or 'extensive' grazing. Crude fibre appears to

(a) (b)

**Fig. 7.5.** (a) Boer goat. These are masculine, robust and fast-growing goats with excellent meat-quality characteristics. (b) Saanen milk goat with finer features, smaller dimensions and relatively poor body conformation but good mothering ability and high milk yield.

**Table 7.1.** Effect of age on live animal and carcass characteristics of South African indigenous goats (means ± SD) (from Simela, 2005).

| Characteristic | Age class | | | | *P* value |
|---|---|---|---|---|---|
| | 0 teeth | 2 teeth | 4–6 teeth | 8 teeth | |
| Slaughter weight (kg) | 27.8a ± 3.81 | 33.1b ± 5.66 | 36.6c ± 6.39 | 42.7d ± 3.92 | <0.0001 |
| Cold carcass weight (kg) | 11.8a ± 11.43 | 13.7b ± 3.18 | 15.2bc ± 3.10b | 16.9c ± 2.88 | <0.0001 |
| Dressing percentage | 42.1 ± 5.99 | 41.0 ± 3.36 | 41.4 ± 2.90 | 39.0 ± 4.34 | 0.0868 |
| Chest girth (cm) | 71.1a ± 3.44 | 75.5b ± 4.74 | 79.c ± 6.89 | 84.1d ± 2.39 | <0.0001 |
| Carcass length (cm) | 66.3a ± 3.73 | 68.6b ± 4.22 | 70.1b ± 3.07 | 75.1c ± 3.22 | <0.0001 |
| Buttock circumference (cm) | 48.6a ± 3.36 | 51.2b ± 3.76 | 54.2c ± 5.36 | 56.0c ± 3.10 | 0.0001 |
| M. longissimus thoracis area (cm$^2$) | 11.1 ± 3.94 | 11.1 ± 3.35 | 12.5 ± 2.30 | 12.9 ± 3.07 | 0.3345 |
| Omentum fat (g) | 553a ± 382 | 554a ± 423 | 711a ± 229 | 1 197b ± 716 | <0.0001 |
| Kidney knob and channel fat (g) | 402ab ± 302 | 357a ± 275 | 533 ± 161b | 700c ± 445 | <0.0001 |

SD, Standard deviation.
Means within a row with different letters (a, b, c, d) differ significantly ($P < 0.05$).

**Table 7.2.** Effect of sex on live animal and carcass characteristics of South African indigenous goats (means ± SD) (from Simela, 2005).

| Characteristic | Sex | | | *P* value |
|---|---|---|---|---|
| | Castrates | Females | Intact males | |
| Slaughter weight (kg) | 36.03b ± 6.47 | 31.41a ± 5.87 | 37.66b ± 7.17 | <0.0001 |
| Cold carcass weight (kg) | 14.86b ± 3.55 | 12.86a ± 2.84 | 15.49b ± 3.88 | 0.0021 |
| Dressing percentage | 40.79 ± 3.66 | 40.99 ± 4.46 | 40.88 ± 5.50 | 0.9772 |
| Chest girth (cm) | 78.20b ± 5.99 | 74.75a ± 5.21 | 79.52b ± 6.82 | 0.0032 |
| Carcass length (cm) | 69.80a ± 3.77 | 67.48a ± 4.11 | 72.66b ± 4.51 | <0.0001 |
| Buttock circumference (cm) | 52.83 ± 4.51 | 51.59 ± 4.32 | 53.00 ± 4.58 | 0.4793 |
| M. longissimus thoracis area (cm$^2$) | 12.24 ± 3.05 | 11.07 ± 3.19 | 12.39 ± 4.57 | 0.2118 |
| Omentum fat (g) | 845 ± 376 | 739 ± 619 | 678 ± 438 | 0.2859 |
| Kidney knob and channel fat (g) | 547 ± 249 | 493 ± 368 | 455 ± 298 | 0.3458 |

SD, Standard deviation.
Means within a row with different letters (a, b) differ significantly ($P < 0.05$).

be a critical aspect in the normal growth and development of the digestive system (pre-gastric compartments like the rumen, reticulum and omasum, as well as the small intestine). Similar effects of dietary fibre were demonstrated on the intestinal development of monogastric species such as guinea pigs (Patten *et al.*, 2004).

Other factors that influence growth and development of goats are litter size, age of the dam, uterine environment and external environment. It is well known that supplementation of macro- and micronutrients improves growth and development, while the extent and duration of nutrient deficiencies, as well as the physiological status of goats, influence growth retardation and the subsequent ability to regain or even compensate for previous losses.

Certain goat genotypes respond better to feeding or feed supplementation, and this ability differs mainly between the dairy- and meat-type goats. Similar differences are evident in dairy and beef cattle. Beef cattle are generally classified as ‘accretion type’, while dairy cattle are classified as ‘secretion type' (Bellman *et al.*, 2004). A similar difference appears to exist in goats, which provides an interesting basis for the comparison of nutrient partitioning and hormonal regulation (Webb and Casey, 2005). The findings generally suggest that GH is positively correlated with and IGF-1 negatively correlated with genetic merit for milk yield, while both these hormones are positively correlated with growth rate. The implication is that dairy-type goats probably use fat as an easily available energy store, while goats selected for growth and carcass characteristics use protein as a long-lasting energy store.

Conditioning of goats has a marked effect on body composition. The increase in carcass fat content in pre-slaughter conditioned goats from 10 to ~19% is clearly indicated in Table 7.3, while moderate decreases in the proportion of lean and bone are demonstrated. Conditioning of goats has the potential to significantly improve the eating quality of goat meat without the adverse effects of unfavourable odours or flavour, which typically develop with ageing, particularly in male goats. In contrast to ageing and differences between genders, conditioning improves the dressing percentage of goat carcasses (Fig. 7.6).

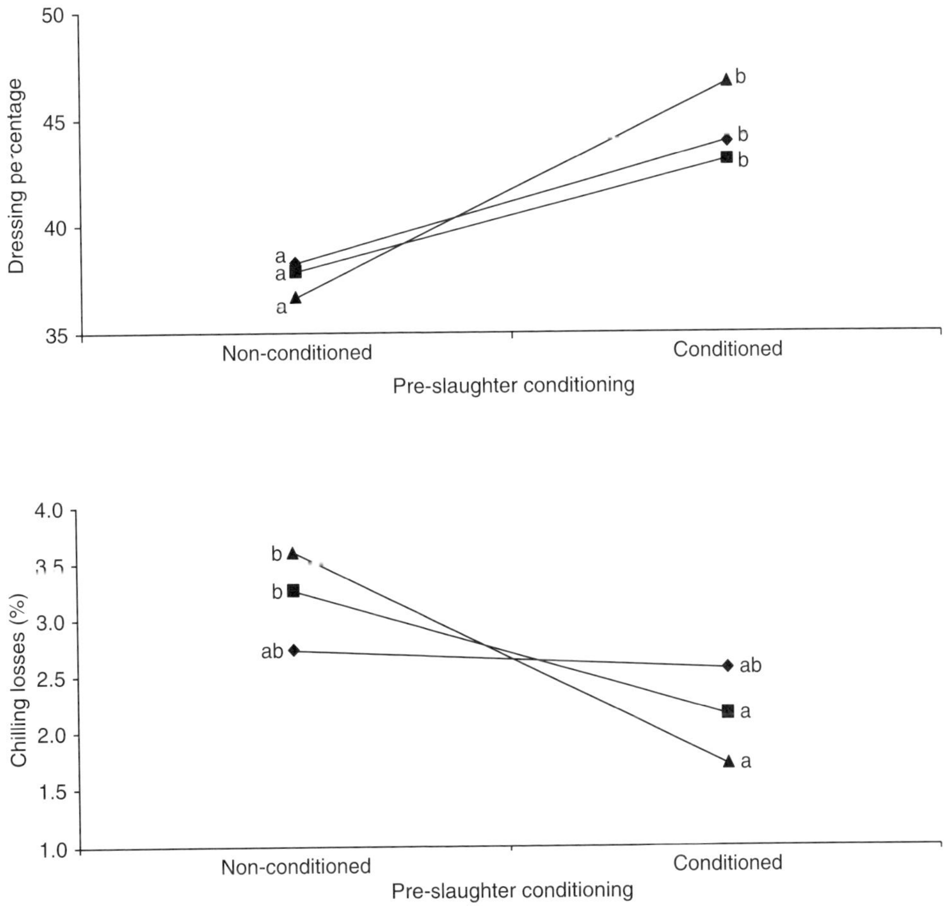


**Fig. 7.6.** Effect of pre-slaughter conditioning on the dressing percentage of South African indigenous goat carcasses. Differences (a and b) are indicated for bucks (▲), does (■) and castrates (♦) (from Simela, 2005).

**Table 7.3.** Effect of pre-slaughter conditioning on proportions of tissues (%) in joints of the right carcass halves of South African indigenous goats (means ± SD) (from Simela, 2005).

| Characteristic | Pre-slaughter conditioning | | *P* value |
|---|---|---|---|
| | Non-conditioned | Pre-slaughter conditioned | |
| Lean | 65.52 ± 2.88 | 59.88 ± 4.26 | <0.0001 |
| Bone | 23.46 ± 3.08 | 19.44 ± 1.74 | <0.0001 |
| Intermuscular fat | 6.76 ± 3.11 | 13.72 ± 3.77 | <0.0001 |
| Subcutaneous fat | 3.42 ± 1.74 | 6.00 ± 1.96 | <0.0001 |
| Total carcass fat | 10.18 ± 4.33 | 19.72 ± 4.74 | <0.0001 |

SD, Standard deviation.

## 7.6 Classification or Grading of Goat Carcasses

Consumer preferences for goat carcass differ among countries and this makes it difficult to implement universally acceptable grading or classification systems, as well as standard carcass cuts. Countries also differ in terms of their preferences for specific carcass cuts; for example, the cuts associated with the loin region or dorsal trunk and the hind limb are the most sought-after cuts in western countries (Casey *et al.*, 2003). Cuts from the hind limb are generally regarded as high-value cuts due to the low fat and high lean content (Casey, 1982; Simela, 2005). Although cuts from the dorsal trunk also have a low fat content, they are perceived to be bony.

In most countries, goat carcasses are graded or classified based on a system very similar to that employed for sheep carcasses. This creates problems in terms of fat code or fat scores, as goat carcasses are generally very lean, even in conditioned goats. Guidelines for goat carcass evaluation have been developed (Colomer-Rocher *et al.*, 1987), but there are still problems with conformation scoring in goats.

Many of the current problems relating to goat carcass grading are based on the unique fat accumulation in goats (Webb *et al.*, 2005). Fat accumulation in goats occurs approximately when they have reached their mature body mass (Owen *et al.*, 1978, 1983). However, body fat content tends to be rather variable, despite the normal variations due to age, sex, body weight and growth rate (Owen *et al.*, 1978; Kirton, 1988). Carcass fat occurs predominantly in the visceral depots rather than the typical subcutaneous depots (Devendra and Owen, 1983; Kirton, 1988), and this explains why goat carcasses are generally regarded as lean. As indicated earlier in this chapter, goat carcasses contain about 60% dissectible lean and 5–16% dissectible fat. Accumulation of fat in the subcutaneous fat depot is limited, even with ageing and conditioning (Webb *et al.*, 2005). In addition, the variation in subcutaneous fat thickness between goats of different ages, sex and types is extremely limited, which makes it difficult to use fat thickness for purposes of carcass classification or grading.

## 7.7 Conclusions

Growth and development in goats influence the efficiency of production with subsequent effects on product quality. The use of growth promotants or molecules that change the efficiency of growth are not popular in goat production, but other more natural pathways exist to manipulate growth and development that do not detract from product quality and meet consumer demands for more environmentally friendly and natural production systems. There is great significance in the way in which the growth and developmental processes have evolved in goats to ensure the propagation of this unique species. The skeletal size of goats is rather variable due to the effects of natural selection and adaptation of the various breeds or types of goat to different ecological regions in the world. Goats are subdivided into three categories based on frame size; namely, dwarf breeds, small breeds and large breeds. Fatty tissue accumulates in goats at a lower rate and to a lesser extent compared with sheep. Animal growth is the consequence of complex interactions between the genetic potential of the animal and several hormones, the nutrient supply and the environment. This complexity is beneficial in terms of manipulating growth and development because it provides the possibility for many different ways of manipulating growth. The most significant means of manipulating growth and development in goats include genetic selection, castration and intensive fattening.

## References


Bellman, O., Wegner, J., Rehfeldt, C., Teuscher, F., Scheider, F., Voigt, J., Derno, M., Sauerwein, H., Weingärtner, J. and Ender, K. (2004) Beef versus dairy cattle: a comparison of metabolically relevant hormones, enzymes, and metabolites. *Livestock Production Science* 89, 41–54.

Casey, N.H. (1982) Carcass and growth characteristics of four South African sheep breed and the Boer goat. PhD thesis, University of Pretoria, South Africa.

Casey, N.H. and Webb, E.C. (2010) Managing goat production for meat quality. *Small Ruminant Research* 89, 218–224.

Casey, N.H., Webb, E.C. and van Niekerk, W.A. (2003) Goat meat. In: Caballero, B., Trugo, L. and Finglass, P. (eds) *Encyclopedia of Food Sciences and Nutrition.* Academic Press, London, pp. 2937–2944.

Chen, H. (2004) Gene expression by the anterior pituitary gland: effects of ageing and caloric restriction. *Molecular and Cellular Endocrinology* 222, 21–31.

Colomer-Rocher, F., Morand-Fehr, P. and Kirton, A.H. (1987) Standard methods and procedures for goat carcass evaluation, jointing and tissue separation. *Livestock Production Science* 17, 149–159.

Devendra, C. and Owen, J.E. (1983) Quantitative and qualitative aspects of meat production from goats. *World Animal Review* 47, 19–29.

Kirton, H. (1988) Characteristics of goat meat, including carcass quality and methods of slaughter. In: *Goat Meat Production in Asia.* Proceedings of a workshop held in Tando Jam, Pakistan, 13–18 March 1988. IDRC, Ottawa, Canada, pp. 87–99

Kokta, T.A., Dodson, M.V., Gerler, A. and Hill, R.A. (2004) Intercellular signaling between adipose tissue and muscle tissue. *Domestic Animal Endocrinology* 27, 303–331.

Louveau, I. and Gondret, F. (2004) Regulation of development and metabolism of adipose tissue by growth hormone and the insulin-like growth factor system. *Domestic Animal Endocrinology* 27, 241–255.

Owen, J.E., Norman, G.A., Philbrooks, C.A. and Jones, N.S.D. (1978) Studies on the meat production characteristics of Botswana goats and sheep. Part III: Carcass tissue composition and distribution. *Meat Science* 2, 59–74.

Owen, J.E., Arias Cereceres, M.T., Garcia Macias, J.A. and Nunez Gonzalez, F.A. (1983) Studies on the Criolli goat of Northern Mexico. Part I. The effects of body weight on body components and carcass development. *Meat Science* 9, 191–204.

Patten, G.S., Bird, A.R., Topping, D.L. and Abeyawardena, M.Y. (2004) Effects of convenience rice congee on guinea pig whole animal and gut growth, caecal digesta SCFA and in vitro ileal contractility. *Asia Pacific Journal of Clinical Nutrition* 13, 92–100.

Simela, L. (2000) Demand and supply of chevon in urban markets of Zimbabwe. In: *Improvement of Market Orientated Small Ruminant Production Systems and Sustainable Land Use in Semi-arid Regions of Southern Africa.* Project TS3*-CT94-0312, Final Technical Report, pp. 72–85.

Simela, L. (2005) Meat quality characteristics of indigenous goats in Southern Africa. PhD thesis, Department of Animal and Wildlife Sciences, University of Pretoria, South Africa.

Simela, L., Webb, E.C. and Frylinck, L. (2004) Post-mortem metabolic status, pH and temperature of chevon from indigenous South African goats slaughtered under commercial conditions. *South African Journal of Animal Science* 24 (Suppl. 1), 204–207.

Tatar, M., Bartke, A. and Antebi, A. (2003) The endocrine regulation of ageing by insulin-like signals. *Science* 299, 1346–1351.

Webb, E.C. (2007) Food safety and quality: goat meat. In: *CABI Animal Health and Production Compendium.* CAB International, Wallingford, UK.

Webb, E.C. and Casey, N.H. (2005) Achievements of research in the field of growth and development. In: *WAAP Book of the Year 2005.* World Association for Animal Production, Wageningen Academic Publishers, pp. 85–90.

Webb, E.C. and Casey, N.H. (2010) Physiological limits to growth and the related effects on meat quality. *Livestock Science* 130, 33–40.

Webb, E.C. and O'Neill, H.A. (2008) The animal fat paradox and meat quality. *Meat Science* 80, 28–36.

Webb, E.C., Casey, N.H. and Simela, L. (2005) Goat meat quality. *Small Ruminant Research* 60, 153–166.

# 8 The Role of Objective and Subjective Evaluation in the Production and Marketing of Goats for Meat

B.A. McGregor

*Institute for Technology, Research & Innovation, Deakin University, Victoria, Australia*

## 8.1 Abstract

Objective and subjective evaluations of goats for meat production are related to important determinants of production and profitability. The most important attributes in assessment of goats for market are: live weight, body-condition score and age. As goats grow, their carcass and body organs increase in weight in proportion to the empty body weight. For farmers and field workers, the linear regression approach for estimating carcass weight by measuring live weight is the most suitable as it accounts for 88–97% of the variation in carcass, offal and boneless meat weight. Live-weight scales or heart girth tapes should be used, and the risks and errors associated with these methods are summarized. The proportion of a live goat that is the carcass, known as dressing percentage, increases from 35 to ~50% as goats grow. The usefulness and errors associated with dressing percentage in field estimation are discussed. A valuable subjective method for estimating the nutritional status of goats is the use of body-condition scoring, as it accounts for 60–67% of the variation in live-weight change, carcass weight and fat reserves of goats. A method for body-condition scoring and a similar fat scoring system are explained. Body-condition score is also associated with mortality risk and the reproductive performance of goats. The number of permanent incisors in the lower jaw of goats is a method of estimating the age of goats but is biased by differences in live weights of goats. The value and role of ultrasound scanning of goat carcasses is summarized. For the marketing of kid meat, no permanent incisors should have erupted. Other useful practices for the successful marketing of goat meat are discussed including: knowing market specifications and chemical withholding periods; animal health; prevention of bruising; identification of goats; size of consignments; timeliness; and provision of paperwork. A checklist is provided. The use of subjective and objective assessment techniques in evaluating goats for meat production will provide the best results. Where only subjective assessment techniques are available, they will provide satisfactory performance provided the skills have been learned and are applied.

## 8.2 Introduction

Objective and subjective evaluations of goats for meat production have important roles to play in developing and developed economies. These methods of assessment

 *Goat Meat Production and Quality* (eds O. Mahgoub, I.T. Kadim and E.C. Webb)

are related to important determinants of meat production and enterprise profitability. In many countries, low-cost subjective evaluation methods are the only methods available. The application of subjective evaluation methods in developing countries is a valuable tool that can greatly help goat farmers improve the health and survival, reproduction and meat production of their goats.

### 8.2.1 Purpose of evaluation in meat production and marketing

There are several important reasons to assess live goats during production and before they are sold for meat including (McGregor, 2007a):

- To select goats that closely match the specifications of a buyer.
- To decide whether the goats need additional feeding to maintain growth rate.
- To decide whether the goats need additional shelter or management inputs during pregnancy and adverse climatic conditions.
- To determine animal health treatments, such as the size of doses for drugs, e.g. anthelmintics.
- To estimate carcass attributes of goats before sale for meat.
- To avoid being penalized for failing to meet buyer specifications.
- To evaluate goats for selection and breeding.

### 8.2.2 Methods of live animal assessment

The three most important methods of assessing goats for market are:

- Live-weight measurement or estimation.
- Body-condition scoring.
- Estimation of the age of goats.

The use of ultrasound scanning of live goat carcasses is discussed later in the chapter. Other important descriptors of goats include sex and breed, which will not be discussed further in this chapter.

### 8.2.3 Preparing goats for meat marketing

In undertaking the assessment of live weight, body-condition score and age of goats for meat marketing, there are other attributes of goats and practices of farmers that are required for successful marketing. The last section of this chapter discusses ten important issues that farmers should be aware of for the successful marketing of goat meat.

## 8.3 Live-weight Measurement or Estimation

The most precise estimate of carcass production and offal yields from goats is made using empty body weight (ingesta-free live weight) (McGregor, 1985). However, farmers and field workers are unable to accurately determine empty body weight, and use either 24 h fasted live weight or live weight, with a subsequent loss of accuracy in prediction. Thus, for most goat farmers, the live weight of their animals is the most important aspect of a goat that determines meat yield.

### 8.3.1 The use of live weight in predicting carcass and organ yield

As goats grow, their carcass and other body organs increase in weight (Fig. 8.1). Fat reserves also increase as goats grow. These observations have been confirmed in numerous studies on a wide range of goat breeds as well as in all other farm animals (Tulloh, 1963; Gall, 1983; McGregor, 1985; Warmington and Kirton, 1990).

The relative growth coefficient (RGC) of body components relative to the fasted body can be estimated using the allometric growth equation:

$$\log y = \log a + b\log x$$

where $x$ is the fasted live weight at slaughter, $y$ is the fresh organ weight, $b$ represents the RGC of the organ relative to the fasted body and $a$ is a constant.

Allometric growth equations are usually more precise and account for more of

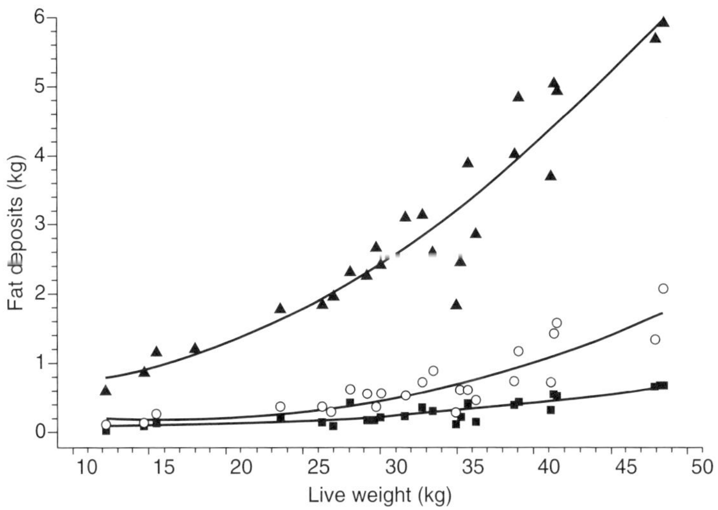


**Fig. 8.1.** The relationships of carcass fat (▲), omental fat (○) and perirenal fat (■) with the live weight of castrated male Angora goats (from McGregor, 1992).

the variation of dependent variables than linear-equation approaches. Allometric growth equations indicate that, during growth of goats from birth to maturity, fat reserves and the carcass develop at a slightly faster rate than the entire body, while bone develops more slowly than the entire body (Wilson, 1958; Owen *et al.*, 1977; McGregor, 1982, 1992). For example, with Saanen and Angora goats given similar grazing and slaughter management, the relative growth of carcass and various fat deposits was >1 while carcass protein was not >1 (Table 8.1). For Saanen goats, the perirenal and omental fat depots grew 2.6 times faster than the fasted body weight.

This chapter will focus on studies that used the linear-regression approach for estimating carcass weight as this approach is more suitable for application with farmers (McGregor, 1985; Warmington and Kirton, 1990). With linear regressions, the farmer estimates the carcass weight as follows:

$$\text{Carcass weight (kg)} = m \times \text{live weight (kg)} + c$$

where $m$ is the regression coefficient or slope and $c$ is the regression constant (which is often negative).

### 8.3.2 Prediction of saleable meat on goats

#### *Live weight and carcass weight*

Studies with goats indicate that, as goats grow, carcass weight increases by 0.43–0.54 kg for every 1 kg increase in live weight. Table 8.2 provides examples for different goat breeds grazed on pasture in a similar way in the same temperate environment and slaughtered using the same standardized procedures (Aus-Meat, 2001).

Using live weight to predict carcass weight accounted for 88–97% of the variation in the studies illustrated in Table 8.2. The residual standard deviation in these studies ranged from about 0.5 to 1.6 kg. The precision of prediction can be improved further when the subjectively assessed body-condition score is used with live weight in prediction equations. Body-condition scoring is discussed in the following section.

Nutrition does affect carcass development and carcass yields. Table 8.2 illustrates that the regression coefficients were lower for goats subjected to periods of maintenance of live weight or of low live weight gain when the proportion of the live animal

**Table 8.1.** The relative growth coefficient (RGC) of body components of grazing goats relative to fasted body weight for Saanen castrate goats (McGregor, 1982) and Angora castrate goats (McGregor, 1992).

| | Saanen | | Angora | |
|---|---|---|---|---|
| Component | RGC | 100 × $r^2$ | RGC | 100 × $r^2$ |
| Carcass[a] | 1.098 | 99 | 1.015 | 96 |
| Carcass fat[a] | 2.155 | 86 | 1.039 | 88 |
| Carcass protein | 0.930 | 98 | 1.003 | 96 |
| Perirenal fat | 2.665 | 83 | 1.068 | 72 |
| Omental fat | 2.620 | 86 | 1.074 | 76 |

[a]Excluding perirenal fat.

**Table 8.2.** Regression constants (± SEM) and correlation coefficients for linear relationships between carcass weight (kg) and live weight (kg) for Australian goats of different breeds, ages and sexes.

| Breed (age, years) | Sex | Regression coefficient | Constant | RSD | 100 × $r^2$ | Authority |
|---|---|---|---|---|---|---|
| Angora (0.4) | M | 0.488 ± 0.018 | –1.3 | 0.45 | 90 | McGregor (1996) |
| Angora (0.5–4) | C | 0.523 ± 0.021 | –2.3 | 1.00 | 96 | McGregor (1996) |
| **Cashmere (0.3)** | **M** | **0.450 ± 0.031** | **–0.9** | **0.46** | **88** | McGregor *et al.* (1988) |
| Cashmere (0.6) | M | 0.515 ± 0.025 | –1.7 | 0.94 | 88 | McGregor *et al.* (1988) |
| **Cashmere (2.5–4.5)** | **C** | **0.434 ± 0.014** | **–0.8** | **1.64** | **90** | McGregor (1990) |
| Saanen (0.3–3) | C | 0.504 ± 0.025 | –1.4 | | 97 | McGregor (1982) |

SEM, Standard error of the mean.
C, Castrated males; M, mixed-sex kids; RSD, residual standard deviation.
Data in bold relate to animals slaughtered after periods of nutrition that resulted in zero or low live weight gain.

that was carcass was less than for goats subject to periods of good nutrition.

### *Dressing percentage*

As goats gain live weight, the proportion of the body that is the carcass increases. This proportion, called the dressing percentage, is determined as: (carcass weight × 100)/ live weight. Dressing percentage is used as a rule of thumb for estimating carcass weight. For example, for goats at 10 kg live weight, the carcass may represent 35% of live weight, but, at 50 kg live weight, the carcass may represent 48% of live weight. Higher dressing percentage values are sometimes obtained. Dressing percentage values can suffer from two major deficiencies:

**1.** Standardized methods of determining dressing percentage are often not used and so published reports can be misleading. Descriptions of a standardized carcass have been developed and mandated by industry in goat-meat-exporting countries such as Australia (Colomer-Rocher *et al.*, 1987; Aus-Meat, 2001). Issues include the removal of pelvic channel fat and the position of removal of the head, feet and tail.
**2.** Unfortunately, dressing percentage can vary substantially depending on the management of the animal and its sex, whether it has been fasted, diet (which affects gut fill) and the amount of fleece. As many goat keepers and research workers obviously fail to recognize these influences, the most important influences are summarized in Table 8.3.

To enable goat farmers, market agents and researchers to accurately estimate the expected yields and therefore market value of carcasses from live goats, it is important that, in each environment, with a given

**Table 8.3.** Factors influencing the apparent dressing percentage of goats (adapted from McGregor, 1985).

| Dressing percentage change | Factor |
|---|---|
| Increased dressing percentage | Milk fed kids: have little rumen development and gut fill |
| | Concentrate feeding: reduces gut fill, increases fat deposits |
| | Fasting: reduces gut fill before weighing |
| | Age: older animals tend to be heavier |
| | Live weight: heavier animals usually have more muscle and fat |
| Reduction in dressing percentage | Weaning: increases gut fill and reduces fat reserves |
| | Hay and straw feeding: increases gut fill and live weight |
| | Lactation: usually reduces fat reserves |
| | Mating: for bucks, results in reduced appetite and weight loss |
| | Dry pastures: usually results in loss of live weight and fat reserves |
| | Heavy fleece: results in overestimates of true live weight |
| | Large horns and testicles: in bucks, these will reduce the relative carcass weight |

breed and management practice, dressing percentage be determined over a range of live weights.

*Boneless meat yield from carcasses*

For many manufactured goat meat products, such as mince required for sausages, goat meat is removed from the bones before processing. The yield of this boneless meat is a lower proportion of the live animal compared with the carcass yield, as the bone and some fat is discarded during the trimming process. There is a lack of knowledge about the actual boneless meat yield from entire goats.

Research on boneless meat yield from Angora goats in Texas (Smith *et al.*, 1972; Eggen *et al.*, 1973) indicated that culled 2–6-year-old does of mean live weight of 30.7 kg produced 12.8 kg carcasses that yielded 57% meat. With 9–14 kg Angora goat carcasses, the edible portion of meat rose from 56 to 63% as the degree of muscling and live weight increased. With heavy-weight 2.5–4.5-year-old Australian cashmere goats of 44–79 kg live weight, carcass production was 20.5–32.6 kg with a boneless meat yield of 64.2%. With lighter-weight 14 kg carcasses, the boneless meat yield declined to 61.1% (McGregor, 1990). Linear regression coefficients indicated that, for every 1 kg increase in live weight or hot carcass weight, boneless meat yield increased by 307 or 670 g, respectively (Fig. 8.2).

### 8.3.3 Live-weight measurement or estimation

Where possible, accurate livestock scales are recommended for weighing goats. Where this is not possible, particularly in the small-scale farming sector, the use of a heart girth tape or small scales for smaller animals are useful. However, these methods can have large errors. If small-scale farmers must estimate the body weight of their goats without any aids, then inaccuracies in decision making and husbandry will be the consequence.

*Livestock scales*

Live weight is best determined by weighing goats on scales designed for sheep and goats. Crates with specially fitted gates are best for this purpose. Modern electronic scales that have weigh bars can be used provided a suitable platform or crate is made to restrain the animals. Electronic scales are more expensive and need power either from a mains supply or from a car battery. There are important issues to carefully manage to

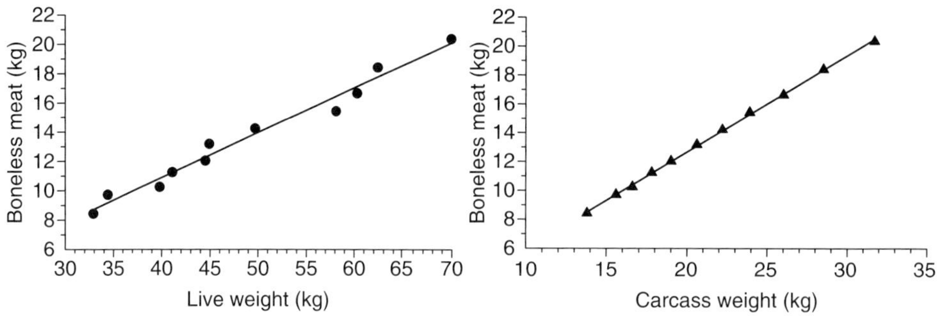


**Fig. 8.2.** Relationship between the live weight and the hot carcass weight of castrated male goats and the amount of boneless meat (McGregor, 1990).

ensure accurate operation of livestock scales (Table 8.4).

### *Heart girth tapes*

Heart girth tapes are calibrated to the live weight of a particular breed of goat. As there are differences in the physical frame size and time to maturity among breeds of goats (McGregor, 1985; Warmington and Kirton, 1990), heart girth tapes need to be calibrated for the breed and sex of animals to be evaluated and for pregnant and non-pregnant does (Pomroy *et al.*, 1987; Mohammed and Amin, 1997; Slippers *et al.*, 2000; Nsoso *et al.*, 2003; Mekasha *et al.*, 2008). This calibration needs to be done with the aid of those with livestock scales and the ability to complete linear regression analyses. In some countries, commercial or industry-sponsored heart girth tapes are available (e.g. USA, Australia).

The correct method for using a heart girth tape is as follows:

**1.** Use a non-elastic calibrated tape.
**2.** Have the goat standing squarely on all four legs.
**3.** Measure the heart girth around the chest, directly behind the forelegs and across the back.
**4.** Draw the tape in firmly, especially for fleece bearing goats, and read the value.

Examples of the relationship between girth measurements and live weight for some breeds of goats are shown in Table 8.5. Using girth tape measurements to estimate live weight has accounted for about 90% of the variation in live weight.

Care should be exercised in extrapolating from the data used to calibrate heart girth tapes to animals outside the ranges of those measured. As noted in numerous studies, the correlation between heart girth and live weight in adults is lower than for growing goats and for females in advanced pregnancy (Mohammed and Amin, 1997) and is lower when goats are losing live weight than when they are growing (Nsoso *et al.*, 2003).

The use of the upper 95% confidence limit when using the girth measurements has been recommended when determining dose rates for animal health use (Pomroy *et al.*, 1987). In practice, the upper 95% confidence limit required the addition of 11 kg to estimates of Saanen doe live weight and 6 kg for estimates of Angora and cashmere/feral doe live weight (Pomroy *et al.*, 1987). The addition for the upper 95% confidence limit potentially overestimates the live weight of small goats such as those <20 kg. As a consequence the use of girth measurements is not recommended for the estimation of body weight when using mineralized drenches or other drugs of narrow or uncertain safety limits (Pomroy *et al.*, 1987).

### *Errors in live-weight measurement*

There are three common errors related to live-weight measurement: (i) inaccurate operation of measuring equipment (discussed

**Table 8.4.** Risk factors that must be managed to operate livestock scales accurately.

| Risk factor | Operation guideline |
|---|---|
| Scales not calibrated accurately | Prior to, during and at the end of each operation, use check weights to ensure accurate operation. Recalibrate if necessary |
| Scales on uneven ground | Use scales only on level ground or a level floor |
| Scales do not move freely | Keep scales from touching hard objects during operation, e.g. away from fences, walls, sides of yards and handling equipment. Check scales frequently during operation. If possible, fix scales to level surface using bolts. |
| | During use, keep area underneath scales clean by removing any stones, sticks or other material from near the scales |
| Scales not tared properly | Ensure scales are tared to zero before use |
| | Check scale tare regularly during use, especially if scales are bumped, knocked or pushed by rough animals |
| Animal not weighed properly | Ensure all four feet are on the weighing platform |
| | Ensure no other goat has its foot on the weighing platform |

**Table 8.5.** Examples of regression constants (± SEM) and correlation coefficient for relationships between heart girth measurement (cm) and live weight (kg) for different breeds and sexes of goats from published sources.

| Breed | Comment | Regression | Constant | $100 \times r^2$ | Reference |
|---|---|---|---|---|---|
| Saanen | Doe | 1.42 | –74.8 | 88 | Pomroy *et al.* (1987) |
| Angora | Doe | 0.96 | –42.9 | 88 | Pomroy *et al.* (1987) |
| Sahel (Borno White) | Kids, multiple birth | 0.50 | –13.6 | 98 | Mohammed and Amin (1997) |
| Nguni | Doe | 1.08 | –47.7 | 94 | Slippers *et al.* (2000) |
| Nguni | Buck | 0.99 | –43.0 | 88 | Slippers *et al.* (2000) |
| Tswana | Wet season | 0.50 | –11.2 | 85 | Nsoso *et al.* (2003) |
| Tswana | Dry season | 2.97 | –181 | 74 | Nsoso *et al.* (2003) |

earlier for scales and tapes); (ii) fasting times; and (iii) variations in the time of day that weighing takes place.

Fasting refers to the amount of time that animals are deprived of food and water. The importance of fasting becomes apparent when it is realized that, for a grazing goat with a live weight of 32 kg, the gut contents (stomachs and intestines) may comprise nearly 25% of the live weight (McGregor, 1982, 1992). If goats are removed from feed and water for 24 h, it is normal that they commonly lose 1–2 kg of live weight or more in hot environments. It is important to standardize the method to be used for measuring the live weight of goats; for example, animals are taken straight from pasture and weighed without any fasting, or animals are taken from pasture and left in livestock yards for 4 h. Whichever procedure is chosen, it should be used routinely.

It is also important that, if comparisons are being made over time, the time of day when weighing takes place is standardized. This is important as animals usually have a grazing, resting and drinking routine. Goats can drink several litres of water in one session, so their live weight can increase by 2 kg or more. It is therefore important to choose a standard routine when weighing goats. Choose a time, preferably in the cool of the morning, and keep to this time of day for any future weighing of the goats.

## 8.4 Body-condition Scoring

Body-condition scoring is a subjective method to assess the relative nutritional status of animals. Body-condition scoring in goats has been shown to be related to goat live weight, milk production, carcass production, carcass fatness, reproductive performance and mortality. All of these production parameters are of commercial importance in goat meat production. Body-condition scoring is therefore an essential practical skill for farmers, extension agents, meat buyers and researchers in both developing and developed economies.

Body-condition scoring has been used on sheep in Australia since at least the 1940s and was first explained by McClymont and Lambourne (1958) and Jefferies (1961). Body-condition scoring has been applied with goats since at least 1982 (McGregor, 2010a). Body-condition scoring can be used to:

- Monitor the live-weight change of goats when no objective method is available.
- Monitor the nutritional state of goats: a decline in body-condition score is a good indication of a decline in nutrition.
- Assist in the selection of goats prior to slaughter.
- Assess the risk of goats to mortality in adverse weather (McGregor and Butler, 2008).

There are three methods of body-condition scoring: (i) body-condition scoring of the short ribs; (ii) fat scoring the long ribs; and (iii) palpating the sternum. Palpating the sternum is the preferred method for use with dairy goats (Aumont *et al.*, 1994; Morand-Fehr, 2005) and is not covered here. Details of this method are summarized by Smith and Sherman (2009).

This section discusses the methods of body-condition scoring, the relationships of body-condition scores to carcass attributes and the use of body-condition scoring to assist in selling goats for meat.

### 8.4.1 Body-condition scoring of short ribs

Body-condition scoring is the easiest method for farmers, meat buyers and researchers to use as it allows an easy 'hands-on' estimation of standing goats. Body-condition scores give a direct assessment of the amount of tissue present over one of the prime carcass sites.

Scientific studies have shown body-condition scoring to be reliable in predicting carcass weight when used with the live weight of goats. While goats may have less subcutaneous fat than sheep, it is easier to gain a more reliable estimate of the body condition and carcass yield of goats using body-condition scores than it is with sheep.

#### *How to undertake body-condition scoring*

**1.** The animal must be standing on all feet and 'relaxed', not tensed up or pushed into a corner. It is not possible to score if an animal is crouching under or jumping over other animals.
**2.** Use the 'balls' of the fingers and thumb rather than the tips.
**3.** Feel the body along the backbone, just behind the last long rib in the loin area. Feel for the prominence of the spine, its sharpness and the amount of flesh on each side of the spine (Table 8.6).
**4.** Now span the loin with the hand with fingers and thumb extended. Feel the ends of the spinal processes and press the fingers gently under the ends to assess the amount of flesh present (Table 8.6).
**5.** Finally, feel the eye muscle by feeling the thickness and coverage of flesh between the backbone and the spinal processes. Use the open flat palm of the hand and gently push against the eye muscle to feel its shape. Is it rounded, flat or depressed?
**6.** For animals with a dense fleece, the fleece should be parted to feel the skin more easily.

**Table 8.6.** What body-condition scores feel like and the cross-section appearance of the tissue reserves in the loin area of the short ribs on the carcass of a live goat (Jefferies, 1961; McGregor, 1983, 2005; Mitchell, 1983).

| Body-condition score | What the score means for meat production | Carcass cross-section in the loin area of the short ribs | What can be felt at each site | | |
|---|---|---|---|---|---|
| | | | Backbone | Spinal processes | Eye muscle |
| 1 | Very lean.<br>Poor meat yield. Should be fed more.<br>Further weight loss may result in death. | 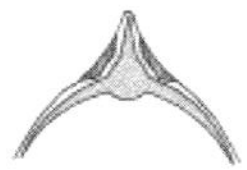 | Prominent and sharp. | Sharp ends. Fingers easily pass under ends. | Very thin and feels hollow. |
| 2 | Lean.<br>Moderate meat yield for adults.<br>Too low for prime kids. | 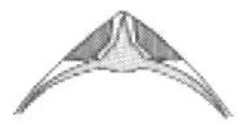 | Prominent but smooth. | Smooth and rounded. Fingers pass under ends. | Some tissue present. Feels flat. |
| 3 | Medium.<br>Ideal for prime kids.<br>May be too fat for adult goats where a slightly lower score is often preferred. |  | Smooth and round over the top but still elevated. | Smooth. Need pressure to feel ends. | Full coverage to end of spinal processes. Feels rounded. |
| 4 | Fat.<br>Too much feed has been used.<br>Fat has to be cut off meat when processed. |  | Only detected with pressure. | Cannot be felt. | Feels well rounded. |

Figure 8.3 shows a farmer condition scoring a goat in a livestock crate used for weighing. The goat is standing still on all its feet, not crouching or lying down. The farmer's hand is spanning the backbone while he feels the short spinal processes and the coverage of flesh using the balls of his fingers and thumb.

### *Reliability of body-condition scores*

The reliability of body-condition scoring improves with practice. It is recommended that scoring should be practised whenever goats are handled, yarded or fed. Body-condition scoring should be used at livestock shows and meat markets.

The original systems for body-condition scoring of sheep used six levels of body condition (0–5; McClymont and Lambourne, 1958; Jefferies, 1961), although level 0, indicating severe emaciation at the point of death following extended drought or disease, was not commonly used. Thus, most descriptions of the body-condition scoring system since this time have referred to only five levels of body condition, 1–5. Skilled assessors can assign body-condition scores that are intermediate between the main scores. Many Australian farmers assign one score between each main category providing an eight-step range, i.e. 1, 1.5, 2, 2.5, and so on. Research has been published where two scores were assigned between each of the main categories providing a 13-step range (McGregor, 1990, 1992, 2005, 2010a), for example ...1.7, 2.3, 2.7, 3, 3.3... The difference between these systems is not important. However, Australian experience indicates that the very high body-condition score of 5 for very fat sheep is not relevant to goats. This view is supported by the lower level of subcutaneous back fat deposits of goats compared with sheep (McGregor, 2005).

Within a year, a goat may experience an increase and a decrease in its body-condition score depending on nutrition and live-weight change (McGregor, 2010a). Within a mob of goats, it is usual to observe a range in body-condition scores (McGregor, 2005).

### *Body-condition score, live weight and carcass attributes*

BODY-CONDITION SCORE, NUTRITIONAL TREATMENT AND LIVE-WEIGHT CHANGE Changes in live weight associated with differences in nutrition are reflected in changes in body-condition score (McGregor, 1988, 2010a).

**Fig. 8.3.** A farmer weighing and body-condition scoring his goats.

Table 8.7 illustrates the typical response of body-condition score to long-term nutritional treatments that result in substantial changes in live weight. The data come from housed goats fed the same forage diet at different levels of energy provision (McGregor, 1988). Those goats fed to lose weight (below maintenance of live weight, 0.8 M) lost 4.9 kg and their body-condition score declined by 1 unit. The goats that gained live weight increased their body-condition score in proportion to the amount of live weight gain at a rate of approximately 6.5 kg per 1 unit score.

For grazing Angora goats, the change in live weight associated with a 1 unit change score in body condition is approximately 7.0 kg (McGregor, 1992) to 8.4 kg (McGregor, 2010a). The impacts of seasonal nutritional conditions and long-term stocking rate on the body-condition score of Angora goats are illustrated in Fig. 8.4. In small East African goats in Zimbabwe a change in condition score of 1 represented an average

**Table 8.7.** Live weight and body-condition score and their changes with time for individually housed goats fed forage diets at different levels of energy provision over a 5-month period (from McGregor, 1988).

| Nutrition treatment[a] | Live weight (kg) | | | Body-condition score | | |
|---|---|---|---|---|---|---|
| | 30/11/84 | 22/04/85 | Change | 30/11/84 | 22/04/85 | Change |
| 0.8 M | 28.4 | 23.5 | –4.9 | 2 | 1 | –1.0 |
| M | 28.4 | 27.9 | –0.5 | 2 | 2 | 0 |
| 1.25 M | 28.2 | 30.6 | +2.4 | 2 | 2.3 | +0.3 |
| 1.5 M | 28.5 | 33.3 | +4.8 | 2 | 2.7 | +0.7 |
| Ad libitum | 28.4 | 36.4 | +8.0 | 2 | 3.3 | +1.3 |

[a]Energy nutrition treatments are relative to the maintenance of live weight (M).

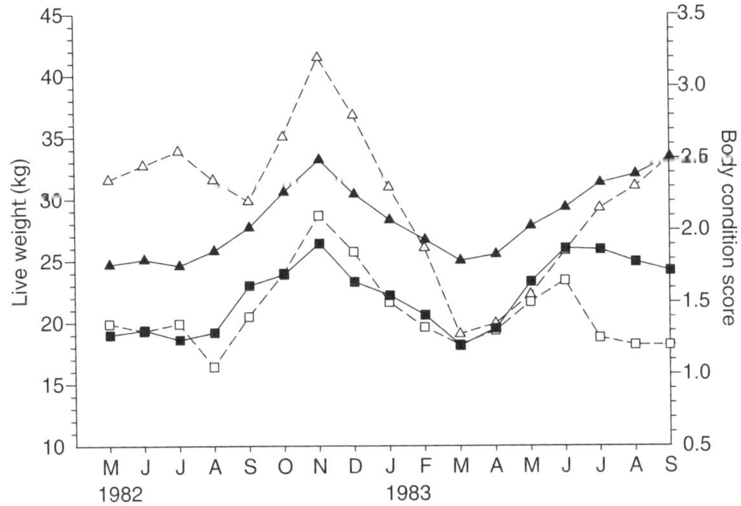


**Fig. 8.4.** Relationship between the body-condition score (Δ, □) and the fleece-free live weight (▲, ■) of Angora goats grazed on annual temperate pastures from May 1982 to September 1983 at stocking rates of 7.5 animals/ha (Δ, ▲) and 12.5 animals/ha (□, ■) (modified from McGregor, 2010a).

change of 12% in live weight (Honhold *et al.*, 1989). The association between body-condition score and live weight of goats can be quite high with regression correlation coefficients as high as 0.93 (McGregor, 2010a).

BODY-CONDITION SCORE AND CARCASS ATTRIBUTES Body-condition score of goats, when used in linear regressions, has been shown to account for 44–67% of the variation in a range of carcass attributes (Table 8.8). This indicates that, as a subjective method for assessing carcass attributes, body-condition scoring can be a useful and practical aid for farmers without livestock scales.

### 8.4.2 Fat scoring the long ribs

Fat scoring is used in livestock market reports in Australia as a method of estimating fat classes for sheep and for describing animals for sale. This system has been applied to specification of goat carcasses (Aus-Meat, 2001). Fat classes of carcasses are determined objectively in the meat works by measuring the tissue depth of the carcass at the grid reference (GR) site. The GR site is on the second last long rib (12th rib) at a site 110 mm from the midline (ridge of the spine) (Fig. 8.5).

The tissue depth at the GR site includes muscle and fat. The GR site is regarded as a good reference point as it provides a reliable indication of the meat and fat content of the carcass and is easy to measure. Examples of the relationship between GR tissue depth and other goat carcass attributes are available (McGregor, 1990, 1992, 1996). Fat scores and fat classes range from 1 to 5 and for goats are explained in Table 8.9. The fat class descriptions for sheep carcasses have different tissue depths at the GR site compared with those used for goat carcasses.

*How to fat score*

Fat scoring uses the sense of touch to estimate the fat class into which an animal will be assigned for sale and is carried out as follows:

**1.** The animal must be standing on all feet and 'relaxed', not tensed up or pushed into a corner. The side of the animal must be accessible. It is not possible to score

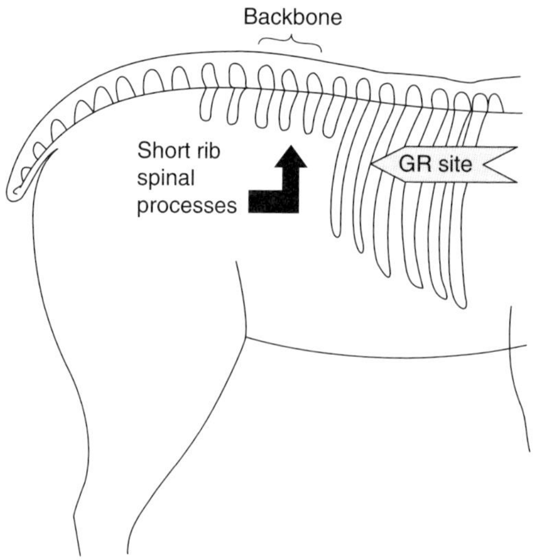


**Fig. 8.5.** The position of the GR site on the 12th long rib and the short ribs.

**Table 8.8.** Regression constants (± SEM) and correlation coefficients for linear relationships between carcass attributes (kg) and total body fat (kg) and body-condition score for Angora goats (modified from McGregor, 1992).

| Attribute | Regression coefficient | Constant | RSD | 100 × $r^2$ |
|---|---|---|---|---|
| Carcass weight | 4.13 ± 0.85 | 7.6 | 2.26 | 62 |
| Carcass fat | 1.56 ± 0.29 | 0.26 | 0.76 | 67 |
| Total body fat | 2.73 ± 0.51 | –0.26 | 1.36 | 66 |
| Subcutaneous back fat | 1.18 ± 0.34 | –0.83 | 0.91 | 44 |
| Carcass protein | 0.58 ± 0.14 | 1.62 | 0.37 | 55 |
| Fat-free carcass weight | 2.57 ± 0.62 | 7.30 | 1.65 | 53 |

SEM, Standard error of the mean; RSD, residual standard deviation.

**Table 8.9.** Relationship between goat fat classes and tissue depth at the GR site (from Aus-Meat, 2001).

| Fat class | Description | Tissue depth at GR site |
|---|---|---|
| 1 | Very lean | Up to 3 mm |
| 2 | Lean | 4–6 mm |
| 3 | Moderately lean | 7–9 mm |
| 4 | Moderately fat | 10–12 mm |
| 5 | Fat | Over 12 mm |

GR, Grid reference site at the 12th rib.

**Table 8.10.** What fat scores feel like on a live goat.

| Fat score | What is felt at the GR site |
|---|---|
| 1 | Fingers 'fall' between ribs<br>No tissue can be felt over ribs |
| 2 | Fingers fit between ribs<br>Slight amount of tissue over ribs |
| 3 | Fingers sit on ribs<br>Some tissue over ribs |
| 4 | Ribs can be felt<br>Lots of tissue present |
| 5 | Ribs only felt with pressure<br>Tissue very prominent and may be fluid |

GR, Grid reference site at the 12th rib.

properly if an animal is crouching or jumping over other animals.

**2.** Use the 'balls' of the fingers rather than the tips.

**3.** Feel the body over the 12th long rib where the GR measurement would be taken. Feel for the prominence of the rib and the amount of tissue over the ribs (Table 8.10).

**4.** The easier it is to feel the rib, the lower the fat score (Table 8.10).

**5.** For animals with a dense fleece, the fleece should be parted to feel the skin more easily.

No objective data relating the use of fat scores to either goat meat production or animal management issues have been found.

### 8.4.3 Other uses of body-condition scoring in goat meat production

Body-condition scoring has been shown to have important associations with other management issues of vital importance in goat meat production. In particular, body-condition scores are associated with the risk of mortality in adverse climatic conditions and from pregnancy toxaemia, and with reproductive performance.

*Mortality risk for goats*

Mortality in flocks of Angora goats grazing pastures and subjected to adverse climatic risks was most related to the body-condition score reached during the preceding 2 months (McGregor and Butler, 2008). For flocks of Angora goats, there was no mortality at a body-condition score ≥2.5 and mortality increased sharply at a mean body-condition score <2.0. For individual Angora goats, mortality increased as body-condition score declined, and stocking rate and grazing combinations were additive in effect on mortality. Grazing with sheep increased the mortality of Angora goats at higher stocking rates. Live weight loss was not related to mortality rates of goats once body-condition score had been accounted for. It was concluded that body-condition score and stocking rate were highly significant determinants of welfare risk in Angora goats. Analysis of individual goat mortality rate indicated that these results were applicable in many situations. Consequently, farmers and animal welfare assessors can confidently use body-condition scores to determine welfare risk in goats (McGregor and Butler, 2008).

Morand-Fehr *et al.* (1992) noted that the risk of pregnancy toxaemia to dairy goats was related more to a decline in body-condition score rather than to body-condition score per se.

#### *Reproductive performance*

Body-condition scores of <2.5 have been implicated with increased abortions and reduced kidding rates in Mexican native goats grazed under extensive conditions (Mellado *et al.*, 2004). Compared with all other does, the thinnest goats (body-condition score <1.5) were nine times more likely to abort. Body-condition score was not identified as a risk factor with regard to pregnancy in these goats.

## 8.5 Estimating the Age of Goats Using Dentition

The dentition of goats has commercial importance, particularly the age at eruption of the first pair of permanent incisors, as this affects the commercial value for meat production and the sale of animals for breeding purposes. For meat production, the eruption of permanent first incisors in small farm ruminants is used to signify a change in meat quality by altering the classification of lamb and kid carcasses. Thus, in many developed meat markets, it is essential to know the age of goats at sale.

For goats that are provided with ear tags in their year of birth, it is easy to determine their age. The systematic use of coloured ear tags, where the colour of the tags is different for each year of birth, allows easy identification of the age of goats. Goats of different birth years with different coloured ear tags can be easily separated by drafting in a race. However, if ear tags are not used, then the subjective assessment of the dentition of goats can be used to estimate the age of animals.

Goats have two successive dentitions, deciduous dentition ($n$ = 20) and permanent dentition ($n$ = 32). Upper incisors are absent and are replaced by a very thick connective tissue pad (palate). Permanent first incisors are easily distinguishable from the deciduous first incisors due to their relatively large size. The number of incisor and molar teeth that have erupted (broken through the gum surface) in the lower jaw of a goat is used to describe the age of a goat. Within a mob of goats of similar age, there will be a range in age for when individuals show the eruption of permanent incisors (Table 8.11).

**Table 8.11.** Estimation of age by dentition.

| Dentition (number of teeth erupted) | Age (months) |
|---|---|
| No permanent incisors | 0–15 |
| First pair of permanent mandibular molars | 3–5 |
| First pair of permanent incisors | 13–21 |
| Second pair of permanent incisors | 18–24 |
| Third pair of permanent incisors | 22–32 |
| Fourth pair of permanent incisors (full mouth) | 27 or more |

There are few scientific reports of eruption patterns of permanent incisors in goats (Wilson and Durkin, 1984; Matika *et al.*, 1992; Kwantes, 1994; McGregor and Butler, 2011), but text books (Gall, 1981; Pugh, 2002; Radostits *et al.*, 2007; Smith and Sherman, 2009) provide tables of eruption ages for goats. Photographs and X-rays of the lower jaw of different-aged goats are available to show the development of incisors and molars (Holst and Denny, 1980).

McGregor and Butler (2011) have shown that the time to reach similar development stages for first permanent incisor eruption was about 3 months longer for the lightest yearling goats compared with the heaviest yearling goats. Furthermore, where the eruption of permanent first incisors is used to estimate the age of goats, allowance needs to be made in estimates of the age of lighter goats compared with heavier goats within the same cohort, as each 1 kg decrease in live weight was associated with an increase in about 6 days in the time to reach each stage of permanent first incisor development, such as the loss of first deciduous incisors or the eruption of permanent first incisors. Thus, it should not be assumed that all lighter goats within a cohort are younger just because their permanent first incisors have not reached the same stage of development observed in heavier goats. Within this research flock, the differences in live weight of goats explained 3 months in the variation in eruption of permanent first incisors, which is about half of the reported variation in age at eruption shown in Table 8.11.

A practical application in goat meat marketing of the use of dentition is to describe goats as kids when they have no evidence of eruption of permanent incisors (Aus-Meat, 2001), even though these goats may be up to 15 months of age. For very young kids, the eruption of the first mandibular molar could be used for ageing (Holst and Denny, 1980). It is very common for goats to be described in livestock sales as two-tooth, four-tooth or full mouth for example, meaning that they are, respectively, 1 year old, 2 years old or adult, even though there could be a range in ages within such descriptions.

## 8.6 Ultrasound Scanning of Live Goats

Real-time ultrasound scanning (ultrasonography) is a non-invasive technique used in animal production to detect pregnancy status and live animal body and carcass attributes. Ultrasound scanning can be used as a method of indirect measurement of the eye muscle depth (Longissimus dorsi measured at the C site, 45 mm from the midline at the 12/13th rib), subcutaneous back fat depth and sternum fat deposits in goats using the same techniques that are used with sheep and pigs (Wood and Fisher, 1990; Stanford *et al.*, 1995; Hopkins *et al.*, 2007; Teixeira *et al.*, 2008). There has been much more intensive evaluation of the use of ultrasound scanning of the carcass attributes of sheep than of goats. However, the relevance and specific transfer of research findings with sheep to goat carcasses needs to be cautioned by the knowledge that fat distribution within goats differs significantly from that of sheep (Gall, 1981; McGregor, 1985) and goats have been subject to far less genetic selection for carcass traits than sheep.

Unfortunately, the costs of both equipment and hire of consultants to conduct ultrasound scanning are likely to result in these techniques being applicable only in larger commercial breeding flocks, during genetic selection programmes for carcass attributes and where carcass attributes are important in the classification of carcasses at meat works.

Eye muscle area has been shown to be positively related to hot carcass yield in Jamunapari goats (Amin *et al.*, 2000). Teixeira *et al.* (2008) reported that the best correlation for muscle depth in Spanish Celtiberica adult goats was found for ultrasound measurements taken between the third and fourth lumbar vertebrae. These estimates accounted for 70% of the variation in muscle depth. The lumbar vertebrae sites are the same as those used for body-condition scoring. The practical question is, therefore, to what extent does expensive ultrasound measurement provide better estimates of carcass yield, carcass composition and muscle attributes of goats than the easily applied technique of on-farm body-condition scoring? The on-farm measurements of live weight and body-condition score used together accounted for 58% of the total variation in eye muscle depth of Angora goats or 87% of that accounted for by the best model, which required carcass weight (McGregor, 2010b). It appears that ultrasound measurement of muscle depth does not account for all the variation in this attribute and that goat meat producers can achieve very similar results using other methods.

In centralized breeding schemes in Australia where ultrasound scanning has been used on farm to measure eye muscle depth in meat sheep (Hopkins *et al.*, 2007; LambPlan, 2008), improvements have been obtained in growth and carcass weight, and a significant medium-term return on investment has been obtained (Holst, 1999). While a centralized breeding scheme for goats has been available in Australia for some years (KidPlan, 2008), few breeders have invested in applying ultrasound measurements to evaluate their bucks. However, it has been shown that there are significant differences between the progeny of Angora bucks in eye muscle depth and subcutaneous back fat at 14 months of age with a range of 1.3 and 2.8 mm, respectively, between sire groups of progeny (Ferguson and McGregor, 2005). For Boer bucks, the range from the 1st percentile to the 100th percentile of measurements indicates differences of 3 mm in eye muscle depth and 2.2 mm in subcutaneous back fat depth (KidPlan, 2008).

However, the evidence that it is cost-effective to use ultrasound measurements must be questioned given the findings of two recent reports. Using ultrasound measurements of subcutaneous lumbar fat depth and eye muscle depth to predict commercial carcass yield of Angora goats added only an extra 2.4% to the 89.1% of variance accounted for by live weight, body-condition score and sire (McGregor, 2010b). For total muscle prediction, Teixeira *et al.* (2008) reported that using ultrasound measurement at the lumbar site only increased the precision of muscle prediction by 8% (to a total of 90% of variance accounted for) compared with using body weight alone. Thus, the available evidence suggests that, with goats, the use of body weight and body-condition scoring are adequate and cost-effective methods for goat meat producers to use to estimate meat yield and carcass attributes, and that the additional expense of using ultrasound measurements currently provides little extra benefit.

## 8.7 Preparing Goats for Meat Marketing

Commercial marketing of goats for meat involves identifying the market, correct husbandry and nutritional management, proper assessment of goats suitable for marketing and the correct preparation of goats prior to dispatch to the market. This section summarizes the correct preparation of meat goats prior to dispatch to the identified market (McGregor, 2007b).

Commercial market requirements can vary with seasons and between years so it is important that farmers intending to sell goats for meat contact potential buyers, agents or marketing networks in advance to ensure that they clearly understand the current market requirements.

When goats are being prepared for market, the farmer must time his/her work carefully to ensure that the buyer will accept delivery of the goats on time and according to specification. During the months prior to delivery, husbandry operations must be carefully planned to enable goats to arrive at the correct specifications and appearance.

Ideally, goats delivered for slaughter will:

- Meet the specification;
- Be outside any chemical withholding period;

- Be healthy;
- Be clean and dry;
- Have short fleeces;
- Have no bruises;
- Have clear identification;
- Be delivered in the agreed sized load;
- Be ready on time; and
- Be accompanied by the appropriate paper work.

Each of these points is discussed further below.

### 8.7.1 Meeting the specification

It is critical only to sell goats that closely match the specifications of the buyer. Usually, buyers will specify the age, live weight or carcass weight and condition score of the goats they wish to buy. The assessment procedures required for the marketing of goats have been discussed earlier in this chapter.

All goats that are being considered for sale should be inspected. Any goat that does not match the specifications should be rejected to avoid penalties for failing to meet the specifications. The main penalty will be not being paid for goats that are outside the specification. If the inspection of goats occurs well before marketing, a farmer can decide whether the goats that are currently unsuitable will benefit from additional feeding before sale. Live weight should be measured directly. The body condition should be monitored. Goats that do not have the correct body-condition score should not be sold. The age of sale goats should be determined from farm records or from dentition (teeth development).

### 8.7.2 Chemical withholding periods

In most developed markets, farmers must maintain and carefully check farm records to ensure that goats being sold will be outside the withholding periods for any chemical treatment that they may have received. It is common for goats to be treated with veterinary drugs such as vaccines, drenches to control internal parasites and chemicals to control lice. Each chemical treatment has an associated specified withholding period (Anon., 2009). Withholding periods are designed to ensure a reasonable time period between chemical treatment and slaughter, so that any chemical residues that may exist in the food are below the relevant maximum residue limit. Maximum residue limits apply to all food products sold in many countries and are legally binding. The withholding period is printed on chemical and drug labels.

Farmers selling goats destined for export need to be aware of any export slaughter interval (ESI) that may apply. For example, in Australia, the ESI reflects the differences between Australian and overseas maximum residue limits (Anon., 2009). The ESI may be longer than a chemical withholding period in order to satisfy lower overseas maximum residue limits.

It is the responsibility of farmers to ensure that withholding periods and ESIs are honoured. Products without goats on the label should not be used on goats for export meat production unless there is a permit for use issued by the National Registration Authority.

### 8.7.3 Animal health

Only healthy goats should be sent to market. It may be a breach of any Code of Welfare and Code of Transport that may apply to goats to send sick or injured animals to market. Animals with broken limbs, broken horns or other physical injuries should be removed from any mob of goats being sold and carefully treated. Such codes of practice apply for example in Australia (Anon., 2001, 2002).

Kids that are sold for meat should be weaned from their mothers just prior to transport. This means that farmers must be well organized so as not to delay the transport carrier. Kids do not have a large gut that can stay full of food. Prolonged periods of food deprivation will result in dark and dry carcasses that will be unsuitable for the high-value kid-meat markets.

### 8.7.4 Clean and dry animals

Goats contaminated with mud, weed seeds, dags or scours should be cleaned up. Wet and dry dags must be removed from the breech, tail and legs.

Where practical, goats should be loaded when they are dry. If it is raining and the yards are muddy, keep the stock under cover and if possible arrange to load the stock out of a shed.

Goats should have access to water up until the time of shedding or yarding. Feed and water should be withheld for 12 h prior to transport of adult goats. This will result in cleaner and safer transport and make unloading easier.

There is no advantage in having goats ready earlier than needed, as prolonged deprivation of feed and water results in a loss of body and carcass tissue weight.

Load goats only into a clean transport vehicle. Do not put straw or hay on to the floor of the vehicle. Such material will blow about and become lodged in the fleece of the goats.

If goats ready for sale have to be held for some time, place them in large holding yards or paddocks with ample feed, shade and water. Avoid using overhead hay racks as goats can become covered with seeds and litter.

Animals suffering from scouring should be removed from any consignment. Scouring animals foul themselves and other animals in the consignment and lead to higher rates of carcass contamination that will reduce the shelf life of goat meat products.

Animals given a chemical treatment to stop scouring or reduce internal parasitism cannot be sold until the withholding period has expired.

### 8.7.5 Short fleeces

Goats are best sold with short fleeces, ideally less than 3 cm long. Fleece-bearing Angora and cashmere goats should be shorn preferably 3 weeks prior to slaughter. A 3-week period will allow any cuts and bruising to heal. A short fleece will enable goats to be transported more efficiently. Angora goats should be sold no later than 10 weeks after shearing.

A short fleece will also reduce any contamination and make slaughter more efficient. Goats destined for the 'skin-on' carcass trade must have short fleeces, as it is difficult to remove long fleece during processing.

### 8.7.6 Absence of bruises

Bruising and dog bites result in downgrading, severe trimming or condemning of goat carcasses in the meat works. Bruising costs farmers and marketing agents hundreds of thousands of dollars each year. Any bruising caused by physical blows or pulling of the fleece will show on the carcass, possibly leading to trimming and downgrading.

To minimize bruising, goats should be handled quietly and carefully. Do not use electric prodders. Ensure that there are no projections in handling facilities such as yards and races. Keep handling to a minimum. Do not frighten the goats with dogs, loud noises or noisy machines. If dogs are used, they should be muzzled. Do not pull fleeces or the skin. Rough handling causes bruising.

When transporting, keep the pens small and not overcrowded. Goats tend to pack down and small pens avoid large pile-ups and suffocation of goats at the bottom. Put goats into groups of similar sex and size. The transport vehicle should drive and stop carefully. The vehicle should stop occasionally and the driver should check to ensure that the goats are comfortable. In Australia, transport drivers should be familiar with the Code of Practice for Welfare of Farm Animals during Transport. In Australia, livestock transport drivers are expected to have a quality assurance system in place such as TruckCare (2009). TruckCare has been developed for livestock transporters by the Australian Livestock Transporters Association. The programme is aimed at raising awareness and introducing a quality management system that can be audited by customers or by an externally qualified auditor and integrated with customers' or road transport quality assurance programmes. The Australian Livestock Transporters Association developed TruckCare in response to the need to improve animal welfare, the occupational health and safety of its members and to reduce biosecurity risks in the livestock industries.

### 8.7.7 Clear identification

Clear identification of each goat being sold supports the farmer being paid for his/her product. Discuss the identification of goats with the agent before goats are dispatched for market. Identification can be with ear tags, ear notches, leg tags, raddle or a coloured mark on the head or horns of the goats. Do not mark the body or fleece with coloured marks as this will downgrade the value of the skin.

In some countries, microchip ear tags are being used, which ensures national recognition for identified livestock. One benefit of microchip-identified livestock is that they cannot be lost in the system.

If a farmer is disposing of several grades of goats at the same time, make sure different grades are marked with different identifiers such as different colours. It is also essential that the agent knows what all the identifiers mean.

### 8.7.8 Agreed load size

It is very important for farmers to deliver the number and type of goats that were agreed to be sold. Agents organize their purchases to match deliveries along the meat supply chain. Carcasses cannot be stored for extended lengths of time and will deteriorate. Delivering too many or too few goats or not delivering on time causes disruption to orderly marketing arrangements.

### 8.7.9 Timeliness

Goats should be ready for loading when the livestock transport arrives. Transport drivers do not appreciate long delays while they wait for farmers to move animals. Usually, transport drivers have complex timetables to meet, in both collecting and delivering animals to a range of locations. Be considerate of the driver and the next farmers by having all animals nearby at the agreed time.

### 8.7.10 Paper work

Livestock transporters do not want to wait while farmers search for or fill in any necessary forms. Prepare all forms the night before a consignment is to be loaded. Be organized and keep a supply of the correct forms. If no one will be present when the transport arrives, arrange for a safe and dry place for the paper work to await collection by the driver.

### 8.7.11 Timetable for selling

Table 8.12 gives the outline of routine activities to be completed before the dispatch of goats for slaughter. This table can be used as a checklist by ticking each activity when completed.

## 8.7 Conclusions

The marketing of goats that meet the specifications of the meat buyers is essential for goat meat farmers. Goat meat farmers are strongly advised to improve their skills in live-weight assessment and body-condition scoring. For meat production from goats:

- Live weight is the best single indicator of the carcass weight and boneless meat yield.
- Farmers aiming to market goats should weigh and inspect their goats regularly.

**Table 8.12.** A suggested list of routine activities to be completed before the sale of goats for meat. The table can be used as checklist by ticking each activity when completed.

| Time before sale | Activity | √ |
|---|---|---|
| 6 months | Research suitable markets | □ |
| | Contact agents to determine market specifications | □ |
| 5 months | Implement correct nutrition and husbandry practices | □ |
| | Record chemical usage | □ |
| 6 weeks | Ensure compliance with any withholding periods | □ |
| | Organize shearing or crutching if needed | □ |
| 4–5 weeks | Contact agent to reconfirm marketing arrangements | □ |
| | Inspect and evaluate (weigh and condition score) all potential goats for suitability for market | □ |
| | Adjust nutrition as needed | □ |
| 3 weeks | Shearing and crutching must be completed | □ |
| 2–7 days | Inspect and evaluate each animal (weight and condition score, health) for compliance with market specification. Reject animals not meeting specification | □ |
| | Move to paddocks near yards | □ |
| 0–1 day | Draft suitable animals into the agreed size sale lines | □ |
| | Identify different sale lots | □ |
| | Adult goats given 12 h fast | □ |
| 0–1 day | Fill in all required paper work and forms | □ |
| | If wet, put goats into clean undercover shedding | □ |
| Day of transport | Wean kids before transport arrives | □ |
| | Driver signs and takes copy of required forms | □ |
| After marketing | Make contact with agent or marketing group to obtain feedback on sale | □ |

- Farmers should use body-condition scoring to monitor nutritional management and commercial suitability of goats prior to slaughter.
- Farmers should ensure that they prepare their goats to meet market specifications and other marketing, transport and regulatory requirements.

## References


Amin, M.R., Husain, S.S. and Islam, A.B.M.M. (2000) Evaluation of Black Bengal goats and their cross with the Jamunapari breed for carcass characteristics. *Small Ruminant Research* 38, 211–215.

Anon. (2001) *Code of Accepted Farming Practice for the Welfare of Goats*. Department of Primary Industries, Melbourne, Australia.

Anon. (2002) *Code of Practice for Welfare of Farm Animals During Transportation*. Department of Primary Industries, Melbourne, Australia.

Anon. (2009) *Goat ESIs and WHPs 2009*. Meat and Livestock Australia Ltd, Sydney, Australia <http://www.mla.com.au/Meat-safety-and-traceability/On-farm-assurance/LPA/Requirements/>.

Aumont, G., Poisot, F., Saminadin, G., Borel, H. and Alexandre, G. (1994) Body condition score and adipose cell-size determination for in-vivo assessment of body-composition and postmortem predictors of carcass components of Creole goats. *Small Ruminant Research* 15, 77–85.

Aus-Meat (2001) *Goatmeat Language*. Aus-Meat Ltd, Woolloongabba, Queensland, Australia.

Colomer-Rocher, F., Morand-Fehr, P. and Kirton, A.H. (1987) Standard methods and procedures for goat carcass evaluation, jointing and tissue separation. *Livestock Production Science* 17, 149–159.

Eggen, N.R., Smith, G.C., Carpenter, Z.L., Berry, B.W. and Shelton, M. (1973) Composition of Angora goat carcasses. *Journal of Animal Science* 37, 259–260.

Ferguson, M.B. and McGregor, B.A. (2005) *Selecting High Performing Angoras*. RIRDC Research Report No. 05/188. Rural Industries Research and Development Corporation, Barton, ACT, Australia.

Gall, C. (1981) *Goat Production*. Academic Press, London, UK.

Gall, C.F. (1983) Carcass composition. In: *Proceedings of the Third International Conference on Goat Production and Disease*. Dairy Goat Journal Publishing Co., Scottsdale, Arizona, pp. 472–487.

Holst, P.J. (1999) Recording and on-farm evaluations and monitoring: breeding and selection. *Small Ruminant Research* 34, 197–202.

Holst, P.J. and Denny, G.D. (1980) The value of dentition for determining the age of goats. *International Goat and Sheep Research* 1, 41–47.

Honhold, N., Petit, H. and Halliwell, R.W. (1989) Condition scoring scheme for small East-African goats in Zimbabwe. *Tropical Animal Health and Production* 21, 121–127.

Hopkins, D.L., Stanley, D.F. and Ponnampalam, E.N. (2007) Relationship between real-time ultrasound and carcass measures and composition in heavy sheep. *Australian Journal of Experimental Agriculture* 47, 1304–1308.

Jefferies, B.C. (1961) Body condition scoring and its use in management. *Tasmanian Journal of Agriculture* 32, 19–21.

KidPlan (2008) Meat and Livestock Australia, Armidale, Australia <http://kidplan.mla.com.au/>.

Kwantes, L.J. (1994) Aging of Omani small ruminants by permanent incisor growth. *Tropical Animal Health and Production* 26, 210–212.

LambPlan (2008) Meat and Livestock Australia and Australian Wool Innovation, Armidale, Australia. <http://www.sheepgenetics.org.au/lambplan/>.

Matika, O., Sibanda, R. and Beffa, M.L. (1992) Eruption of permanent incisors in indigenous goats and sheep. In: *Proceedings of the First Biennial Conference of the African Small Ruminant Research Network*. ILRAD, Nairobi, Kenya, 10–14 December 1990, pp. 499–504 <http://www.fao.org/wairdocs/ilri/x5520b/x5520b1e.htm>.

McClymont, G.L. and Lambourne, J. (1958) Interactions between planes of nutrition during early and late pregnancy. *Proceedings of the Australian Society of Animal Production* 2, 135–140.

McGregor, B.A. (1982) Growth of organ and body components of grazing goats. *Proceedings of the Australian Society of Animal Production* 14, 487–490.

McGregor, B.A. (1983) Assessing the carcasses of goats using condition scores. *Mohair Australia* 13, 26–27.

McGregor, B.A. (1985) Growth, development and carcass composition of goats: a review. In: Copeland, J.W. (ed.) *Goat Production and Research in the Tropics*. Australian Centre for International Agricultural Research, Canberra, Australia, pp. 82–90.

McGregor, B.A. (1988) Effects of different nutritional regimens on the productivity of Australian cashmere goats and the partitioning of nutrients between cashmere and hair growth. *Australian Journal of Experimental Agriculture* 28, 459–467.

McGregor, B.A. (1990) Boneless meat yields and prediction equations from carcass parameters of Australian cashmere goats. *Small Ruminant Research* 3, 465–473.

McGregor, B.A. (1992) Body composition, body condition scores and carcass and organ components of grazing Angora goats. *Proceedings of the Australian Society of Animal Production* 19, 273–276.

McGregor, B.A. (1996) Carcass quality and commercial acceptance of Angora goat kids fed supplementary energy and slaughtered at 5 months of age. *Proceedings of the Australian Society of Animal Production* 21, 135–138.

McGregor, B.A. (2005) *Nutrition and Management of Goats in Drought*. RIRDC Research Report No. 05/188. Rural Industries Research and Development Corporation, Barton, Australian Capital Territory, Australia.

McGregor, B. (2007a) *Assessment Skills for Goat Meat Marketing*. Agriculture Note AG0997, Department of Primary Industries, Melbourne, Australia.

McGregor, B. (2007b) *Preparing Meat Goats for Sale*. Agriculture Note AG1006, Department of Primary Industries, Melbourne, Australia.

McGregor, B.A. (2010a) The influence of stocking rate and mixed grazing of Angora goats and Merino sheep on animal and pasture production in southern Australia. 2. Live weight, body condition, carcass yield and mortality. *Animal Production Science* 50, 149–157.

McGregor, B.A. (2010b) *Benchmarking Mohair Production in Australia*. Rural Industries Research and Development Corporation Research Report 09/171, Barton, Australian Capital Territory, Australia.

McGregor, B.A. and Butler, K.L. (2008) The effect of body condition score, live weight, stocking rate and grazing system on the mortality from hypothermia of Angora goats and the application of results in the assessment of welfare risks to individual and flocks of Angora goats. *Australian Veterinary Journal* 86, 12–17.

McGregor, B.A. and Butler, K.L. (2011) Determinants of permanent first incisor eruption in grazing Australian Angora goats. *Australian Veterinary Journal* (in press). DOI: 10.1111/j/1751-0813.2011.00842.

McGregor, B.A., Wolde-Michael, T. and Holmes, J.H.G. (1988) The influence of energy supplementation and zeranol implants on growth and carcass characteristics of Australian feral goat kids. *Proceedings of the Australian Society of Animal Production* 17, 234–237.

Mekasha, Y., Tegegne, A., Abera, A. and Rodriguez-Martinez, H. (2008) Body size and testicular traits of tropically-adapted bucks raised under extensive husbandry in Ethiopia. *Reproduction in Domestic Animals* 43, 196–206.

Mellado, M., Valdez, R., Lara, L.M. and Garcia, J.E. (2004) Risk factors involved in conception, abortion, and kidding rates of goats under extensive conditions. *Small Ruminant Research* 55, 191–198.

Mitchell, T.D. (1983) *Condition Scoring Goats.* Agfact A7.2.3. Agriculture New South Wales, Sydney, Australia.

Mohammed, I.D. and Amin, J.D. (1997) Estimating body weight from morphometric measurements of Sahel (Borno white) goats. *Small Ruminant Research* 24, 1–5.

Morand-Fehr, P. (2005) Recent developments in goat nutrition and application: a review. *Small Ruminant Research* 60, 25–43.

Morand-Fehr, P., Amaro, R.P., Rubino, R., Branca, A., Santucci, P.M. and Hadjipanayitou, M. (1992) Assessment of goat body condition and its use for feeding management. In: Lokeshwar, R.R. (ed.) *Proceedings of the 5th International Conference on Goats.* Indian Council of Agricultural Research Publishers, New Delhi, India, Vol. 2, pp. 212–223.

Nsoso, S.J., Aganga, A.A., Moganetsi, B.P. and Tshwenyane, S.O. (2003) Body weight, body condition score and heart girth in indigenous Tswana goats during the dry and wet seasons in southeast Botswana. *Livestock Research for Rural Development* 15(4) <http://lrrd.cipav.org.co/lrrd15/4/nsos154.htm>.

Owen, J.E., Norman, G.A., Fisher, I.L. and Frost, R.A. (1977) Studies on the meat production characteristics of Botswana goats and sheep. Part I: Sampling, methods and materials, and measurements on the live animals. *Meat Science* 1, 63–85.

Pomroy, W.E., Chalmers, K. and Charleston, W.A.G. (1987) The relationship of heart-girth to liveweight of female goats in New Zealand. *New Zealand Veterinary Journal* 35, 167–169.

Pugh, D.G. (2002) *Sheep and Goat Medicine*, 1st edn. Saunders, Philadelphia, PA.

Radostits, O.M., Gay, C.C., Hinchcliff, K.W. and Constable, P.D. (2007) *Veterinary Medicine*, 10th edn. Saunders, London, UK.

Slippers, S.C., Letty, B.A. and de Villiers, J.F. (2000) Prediction of the body weight of Nguni goats. *South African Journal of Animal Science* 30 (Suppl. 1), 127–128.

Smith, G.C., Berry, B.W. and Carpenter, Z.L. (1972) *Comparative Boning Yields for Goat Carcasses.* Texas Agricultural Experimental Station Report PR-3027. Texas Agricultural and Mechanical University, College Station, TX, USA.

Smith, M.C. and Sherman, D.M. (2009) *Goat Medicine*, 2nd edn. Wiley-Blackwell, Ames, IA, USA.

Stanford, K., McAllister, T.A., MacDougall, M. and Bailey, D.R.C. (1995) Use of ultrasound for the prediction of carcass characteristics in Alpine goats. *Small Ruminant Research* 15, 195–201.

Teixeira, A., Joy, M. and Delfa, R. (2008) In vivo estimation of goat carcass composition and body fat partition by real-time ultrasonography. *Journal of Animal Science* 86, 2369–2376.

TruckCare (2009) Australian Livestock Transporters Association, Canberra <http://www.alta.org.au/>.

Tulloh, N.M. (1963) Carcass composition of sheep, cattle and pigs as functions of body weight. In: Tribe, D.E. (ed.) *Symposium on Carcass Composition and Appraisal of Meat Animals.* CSIRO, Melbourne, Australia, Chapter 5:1.

Warmington, B.G. and Kirton, A.H. (1990) Genetic and non-genetic influences on growth and carcass traits of goats. *Small Ruminant Research* 3, 147–165.

Wilson, P.N. (1958) The effect of plane of nutrition on the growth and development of the East African dwarf goat. *Journal of Agricultural Science* 51, 4–21.

Wilson, R.T. and Durkin, J.W. (1984) Age at permanent incisor eruption in indigenous goats and sheep in semi-arid Africa. *Livestock Production Science* 11, 451–455.

Wood, J.D. and Fisher, A.V. (eds) (1990) *Reducing Fat in Meat Animals.* Elsevier Science Publishers, Barking, UK.

# 9 Tissue Distribution in the Goat Carcass

O. Mahgoub[1], I.T. Kadim[1] and E.C. Webb[2]
[1]*Department of Animal & Veterinary Sciences, College of Agricultural & Marine Sciences, Sultan Qaboos University, Sultanate of Oman;* [2]*Department of Animal and Wildlife Sciences, University of Pretoria, South Africa*

## 9.1 Abstract

This chapter discusses the distribution of the major tissues (muscle, bone and fat) in the goat carcass. Tissue distribution is extremely important in determining carcass quality, value and marketability. Carcass tissue distribution is influenced by stage of maturity, sex, breed and nutrition. It changes with age but is principally a function of body weight or, more precisely, empty body weight (EBW). Mature body weight is affected by breed or type, sex and nutrition, so it follows that these factors indirectly affect carcass tissue distribution. The most variable animal body tissue is fat. Bone and muscle proportions are less variable. Tissues are deposited at different rates in various sites of the body resulting in carcass cuts that contain different proportions of lean, bone and fat. Male goats have more lean, especially in the front quarter, whereas female carcasses contain more lean in the hind quarters. Goats are leaner than sheep with more fat distributed in the body cavity and less subcutaneous fat. About 50% of the carcass bone is found in the axial skeleton, with the remaining proportion being divided almost equally between the fore- and hindlimbs.

## 9.2 Introduction

The importance of goats as meat-producing animals is increasing worldwide. However, their meat-production characteristics are not well studied, unlike the other red meat-producing animals, such as beef cattle and sheep. Goats vary in size according to breed/types, ranging from small tropical breeds to large European dairy goats. The specialized meat-producing Boer goat is the largest, reaching up to 100 kg body weight (Warmington and Kirton, 1990). The goat is a much leaner animal than other meat-producing animals, especially sheep (Gaili *et al.*, 1972; Owen and Norman, 1977; Kirton, 1982; Gallo *et al.*, 1996; Sen *et al.*, 2004). The proportions and location of fat, muscle and bone in the carcasses of meat animals are important because they affect carcass quality. The effects of breed, body weight and sex on carcass composition and tissue distribution are important. For instance, smaller goats have a higher proportion of muscle but a lower proportion of bone in the carcass than larger goats (Mahgoub and Lu, 1998). Similar reports in sheep have indicated that the Dorset Horn, a smaller meat-type sheep, had higher proportions of muscle in the carcass than a larger sheep, the Merino (Butterfield, 1988).

Sex is an important factor in muscle and fat distribution in the carcass (Mahgoub *et al.*, 2004, 2005). This chapter aims to review information on carcass tissue distribution and the factors affecting it in goats, with emphasis on economic implications.

## 9.3 Carcass Tissue Distribution

The major tissues in the animal carcass are muscle, fat and bone. Muscle and to some extent fat are edible. Changes in the proportions of one tissue in the carcass influence those of other components. Carcass tissue distribution is affected by several factors including stage of maturity, nutrition, body size, breed and sex. The body composition of goats changes markedly during growth, with muscle and fat increasing and bone decreasing with progress of maturity. Allometric growth analysis has indicated that fat grows much faster than lean during postnatal life (Warmington and Kirton, 1990; Mahgoub *et al.*, 2005). This results in significant effects on meat production from the goat. Lean:bone, edible tissue:bone and fat:lean ratios increase as chronological age and body weight increase. For instance, Zimerman *et al.* (2008) found that the muscle:bone ratio increased from 2.13 to 2.65 and the muscle:total body fat ratio decreased from 6.20 to 3.84 in Argentinean Criollo goats between 3 and 5–7 months of age.

The proportions of tissues varied in the carcass as a result of several factors, especially carcass weight (Warmington and Kirton, 1990; Table 9.1). For instance, lean varied between 51.5 and 71.5%; fat ranged between 4.2 and 33.7% and bone ranged between 12.0 and 28.6% with increasing carcass weight. Body weight, the major determinant of carcass tissue distribution, is influenced by age, breed, sex and level of nutrition.

## 9.4 Goat Carcass Muscle

Muscle growth is isometric in relation to carcass size and therefore its proportion in the carcass remains unchanged, unlike the other components, fat and bone. Being the most edible tissue, muscle distribution in the carcass is important as it affects the value of commercial cuts. There is a lack of information on individual muscle distribution in goats. Some reports on the subject include those of Mahgoub (1997), Mahgoub and Lodge (1996) and Mahgoub *et al.* (2005) in Omani goats. Goat meat is usually sold in bulk in most parts of the world. However, for the modern supermarket marketing systems, carcass cutting needs to be addressed.

### 9.4.1 Muscle distribution in goat carcass

Several researchers have studied tissue distribution in goat carcasses using dissection of individual carcass cuts as an indicator of tissue distribution (Cameron *et al.*, 2001). Muscle content varied according to its location in the carcass. For instance, the highest muscle content (70%) was in the long leg, whereas the lowest (58.3%) was in the flank of chevon goats (Dhanda *et al.*, 2003). Studies on muscle distribution in goats also used muscle grouping, similar to that used for sheep by Butterfield (1988). A summary of the findings of Mahgoub *et al.* (2005) on muscle distribution in Omani Jebel Akhdar goats is given in Tables 9.2–9.4. Muscle group (MG) 1 (proximal hindlimb) comprised the highest proportions of the weight of one side of the carcass (25.7–28.7%). Together with MG3 (surrounding the spinal column) and MG5 (proximal forelimb), they comprised about 53–56% of the total muscle weight of the carcass side (Table 9.2). These muscle groups are known as expensive muscle groups (EMGs) because they represent the high-priced carcass cuts (Butterfield, 1988). Between 35 and 39% of the total side muscle weight was found in the forequarter of the carcass, which includes muscle groups of the proximal forelimb (MG5), muscles connecting the thorax to forelimb (MG7), muscles connecting the neck to forelimb (MG8) and intrinsic muscles of the neck and shoulder (MG9). Another major group of muscles is the

**Table 9.1.** Dissectible carcass tissues in goats of various sexes and breeds (from Warmington and Kirton, 1990).

| Breed | Sex[a] | Carcass weight (kg) | Lean (%) | Fat (%) | Bone (%) | Lean:bone ratio | Reference |
|---|---|---|---|---|---|---|---|
| West African Dwarf | W | 8 | 64.1 | 10.1 | 20.3 | 3.16 | Amegee (1996); Vidyadaran *et al.* (1984) |
| | F | 8 | 71.5 | 4.2 | 20.9 | 3.42 | |
| Chernequiera | M | 9 | 60.0 | 9.3 | 28.5 | 2.11 | Fonesca (1987) |
| Raiana Serpentima | M | 10 | 59.1 | 9.1 | 28.1 | 2.10 | Fonesca (1987) |
| Australian feral | M | 5 | 63.3 | 5.8 | 28.6 | 2.21 | Ash and Norton (1987) |
| | F | 5 | 64.0 | 10.0 | 22.7 | 2.82 | |
| | M | 11 | 64.1 | 13.2 | 19.3 | 3.32 | |
| | F | 11 | 59.5 | 22.6 | 15.8 | 3.70 | |
| Alpine | M | 8 | 67.3 | 5.1 | 24.4 | 2.76 | Fehr *et al.* (1976) |
| | | 10 | 67.8 | 6.3 | 22.8 | 2.97 | |
| | | 11 | 67.2 | 6.9 | 22.9 | 2.93 | |
| | | 13 | 68.6 | 7.1 | 21.9 | 3.13 | |
| Boer | M | 4 | 70.0 | 9.2 | 20.8 | 3.37 | Casey and Naude (cited in Casey, 1987) |
| | | 12 | 68.1 | 17.8 | 13.8 | 4.93 | |
| | | 17 | 64.5 | 21.8 | 12.6 | 5.12 | |
| | | 22 | 63.3 | 24.1 | 12.0 | 5.28 | |
| Saanen | M | 5 | 60.9 | 9.9 | 25.6 | 2.38 | F. Colomer-Rocher and A.H. Kirton (personal communication, 1988) |
| | | 20 | 60.1 | 14.0 | 21.5 | 2.80 | |
| | | 50 | 59.7 | 17.6 | 19.2 | 3.11 | |
| | F | 10 | 61.8 | 10.6 | 24.7 | 2.50 | |
| | | 20 | 55.1 | 22.0 | 17.4 | 3.17 | |
| | | 30 | 514 | 33.7 | 14.1 | 3.65 | |

[a]M, Male; F, female; W, wether.

**Table 9.2.** Least square means (± SEM) of weights of muscle groups (MGs) of the left side of the carcass (as % of total side muscle weight) in buck, wether and doe Omani Jebel Akhdar goats slaughtered at 11, 18 or 28 kg body weight (from Mahgoub *et al.*, 2005).

| | Buck | | | Wether | | | Doe | | | | Effect[a] | | |
|---|---|---|---|---|---|---|---|---|---|---|---|---|---|
| Muscle group | 11 kg | 18 kg | 28 kg | 11 kg | 18 kg | 28 kg | 11 kg | 18 kg | 28 kg | SEM | Sex | SLWT | Sex × SLWT |
| Proximal hindlimb (MG1) | 26.60 | 26.39 | 26.76 | 25.68 | 27.62 | 28.14 | 28.14 | 27.67 | 28.72 | 0.59 | * | | |
| Distal hindlimb (MG2) | 5.98 | 5.58 | 5.16 | 6.01 | 5.63 | 5.69 | 5.96 | 5.40 | 5.03 | 0.12 | * | *** | |
| Surrounding spinal column (MG3) | 14.69 | 15.30 | 14.39 | 13.64 | 14.30 | 14.68 | 14.49 | 14.47 | 15.48 | 0.32 | | | * |
| Abdominal wall (MG4) | 10.65 | 11.13 | 12.12 | 11.16 | 10.95 | 10.78 | 11.06 | 12.22 | 11.54 | 0.42 | | | |
| Proximal forelimb (MG5) | 13.51 | 12.76 | 12.28 | 13.84 | 13.09 | 12.36 | 13.17 | 12.14 | 11.84 | 0.25 | ** | *** | |
| Distal forelimb (MG6) | 4.18 | 3.50 | 3.26 | 3.96 | 3.82 | 3.45 | 3.66 | 3.44 | 3.08 | 0.09 | *** | *** | * |
| Connecting thorax to forelimb (MG7) | 8.16 | 8.80 | 9.09 | 8.39 | 8.70 | 9.14 | 7.94 | 8.85 | 9.07 | 0.23 | | *** | |
| Connecting neck to forelimb (MG8) | 4.58 | 4.82 | 5.23 | 4.83 | 4.51 | 4.96 | 4.44 | 4.45 | 4.77 | 0.17 | | * | |
| Intrinsic neck and thorax (MG9) | 10.70 | 11.13 | 11.26 | 12.17 | 11.05 | 10.37 | 10.18 | 10.55 | 9.59 | 0.34 | ** | | * |
| Expensive muscle group[b] | 54.80 | 54.45 | 53.43 | 53.16 | 55.02 | 55.26 | 55.70 | 54.38 | 56.04 | 0.75 | | | |
| Forequarter[c] | 36.46 | 36.97 | 37.23 | 38.68 | 36.86 | 36.30 | 35.24 | 35.44 | 34.73 | 0.55 | ** | | * |

SEM, Standard error of the mean.
[a]Effects of sex and slaughter weight (SLWT) are different: *, $P < 0.05$; **, $P < 0.01$; ***, $P < 0.001$.
[b]Expensive muscle group = MG1, MG3 and MG5.
[c]Forequarter = MG5, MG7, MG8 and MG9.

**Table 9.3.** Least square means (± SEM) of weights of some individual muscles of the left half of the carcass (as % of side total muscle) in Omani Jebel Akhdar bucks, wethers and does slaughtered at 11, 18 or 28 kg body weight (Mahgoub *et al.*, 2005).

| | Buck | | | Wether | | | Doe | | | | Effects[a] | | |
|---|---|---|---|---|---|---|---|---|---|---|---|---|---|
| Muscle group | 11 kg | 18 kg | 28 kg | 11 kg | 18 kg | 28 kg | 11 kg | 18 kg | 28 kg | SEM | Sex | SLWT | Sex × SLWT |
| Muscle group 1 | | | | | | | | | | | | | |
| M. biceps femoris | 4.58 | 7.76 | 5.15 | 4.18 | 4.96 | 5.37 | 4.70 | 5.19 | 5.55 | 0.18 | | *** | |
| M. gluteus medius | 2.49 | 2.55 | 2.69 | 2.29 | 2.57 | 2.89 | 2.52 | 2.80 | 3.09 | 0.12 | * | *** | |
| M. semitendinosus | 1.49 | 1.74 | 1.59 | 1.54 | 1.70 | 1.74 | 1.66 | 1.68 | 1.70 | 0.07 | | | |
| M. semimembranosus | 4.13 | 3.90 | 4.08 | 3.75 | 4.06 | 4.12 | 4.17 | 4.21 | 4.59 | 0.15 | * | | |
| Muscle group 2 | | | | | | | | | | | | | |
| M. gastrocnemius et. Soleus | 3.13 | 3.10 | 2.73 | 3.06 | 3.01 | 3.00 | 3.11 | 2.82 | 2.76 | 0.07 | | *** | |
| Mm. extensors | 1.13 | 1.09 | 1.07 | 1.17 | 1.23 | 1.16 | 1.18 | 1.09 | 1.02 | 0.07 | | | |
| Mm. flexors | 1.67 | 1.40 | 1.36 | 1.78 | 1.39 | 1.53 | 1.66 | 1.49 | 1.25 | 0.09 | | *** | |
| Muscle group 3 | | | | | | | | | | | | | |
| M. psoas major | 1.54 | 1.50 | 1.52 | 1.31 | 1.49 | 1.68 | 1.57 | 1.62 | 1.63 | 0.07 | | | |
| M. longissimus thoracis et. lumborum | 7.76 | 8.39 | 8.05 | 6.80 | 7.74 | 8.45 | 7.80 | 8.13 | 9.42 | 0.33 | * | *** | |
| M. spinalis et spinalis | 1.56 | 1.54 | 1.37 | 1.63 | 1.45 | 1.30 | 1.33 | 1.21 | 1.24 | 0.09 | ** | * | |
| Muscle group 4 | | | | | | | | | | | | | |
| M. obliquus externus abdominis | 1.85 | 2.19 | 2.42 | 2.21 | 1.99 | 2.02 | 1.95 | 2.22 | 2.37 | 0.11 | | * | * |
| M. rectus abdominis | 2.52 | 2.55 | 2.70 | 2.86 | 2.72 | 2.83 | 2.65 | 2.24 | 2.86 | 0.12 | * | | |
| Muscle group 5 | | | | | | | | | | | | | |
| M. infraspinatus | 2.33 | 2.41 | 2.24 | 2.51 | 2.48 | 2.29 | 2.33 | 2.04 | 2.05 | 0.08 | ** | * | |
| M. triceps brachii (caput longum) | 2.69 | 2.47 | 2.49 | 2.75 | 2.52 | 2.49 | 2.70 | 2.48 | 2.47 | 0.08 | | *** | |
| M. supraspinatus | 2.79 | 2.51 | 2.41 | 2.71 | 2.67 | 2.39 | 2.69 | 2.52 | 2.52 | 0.08 | | *** | |
| Muscle group 6 | | | | | | | | | | | | | |
| Mm. extensors | 1.06 | 0.93 | 0.93 | 1.15 | 1.07 | 0.93 | 1.04 | 1.08 | 0.83 | 0.04 | | *** | |
| Mm. flexors | 1.81 | 1.51 | 1.34 | 1.71 | 1.65 | 1.49 | 1.51 | 1.38 | 1.29 | 0.06 | *** | *** | |

*Continued*

**Table 9.3.** Continued.

| | | | | | | | | | | | | | |
|---|---|---|---|---|---|---|---|---|---|---|---|---|---|
| Muscle group 7 | | | | | | | | | | | | | |
| M. serratus ventralis thoracis | 2.84 | 3.05 | 3.05 | 3.12 | 2.93 | 3.03 | 2.95 | 3.06 | 3.04 | 0.07 | | | |
| M. pectoralis superficialis | 2.31 | 2.58 | 2.71 | 2.32 | 2.55 | 2.80 | 2.24 | 2.68 | 2.80 | 0.09 | | *** | |
| Muscle group 8 | | | | | | | | | | | | | |
| M. rhomboidious | 0.80 | 0.76 | 0.77 | 0.79 | 0.74 | 0.67 | 0.67 | 0.66 | 0.68 | 0.04 | * | | |
| M. serratus ventralis cervicis | 1.57 | 1.61 | 1.41 | 1.58 | 1.49 | 1.43 | 1.49 | 1.39 | 1.43 | 0.05 | | * | |
| M. brachiocephalicus | 1.07 | 1.25 | 1.59 | 1.30 | 1.09 | 1.57 | 1.13 | 1.21 | 1.42 | 0.10 | | *** | |
| Muscle group 9 | | | | | | | | | | | | | |
| M. intercostalis (externi et interni) | 3.12 | 2.94 | 3.21 | 3.37 | 3.32 | 3.12 | 2.99 | 3.32 | 3.02 | 0.16 | | | |
| M. splenius | 0.31 | 0.40 | 0.59 | 0.34 | 0.33 | 0.35 | 0.27 | 0.29 | 0.31 | 0.02 | *** | *** | *** |
| M. longissimus capitis et atlantis | 0.54 | 0.66 | 0.70 | 0.63 | 0.58 | 0.53 | 0.44 | 0.44 | 0.41 | 0.04 | *** | | * |
| M. complexus | 1.37 | 1.33 | 1.35 | 1.55 | 1.29 | 1.03 | 1.17 | 1.13 | 1.18 | 0.06 | ** | ** | *** |

SEM, Standard error of the mean.
[a]Effects of sex and slaughter weight (SLWT) are different: *, $P < 0.05$; **, $P < 0.01$; ***, $P < 0.001$.

**Table 9.4.** Growth coefficients (± SEM) of muscle groups relative to empty body weight in Omani Jebel Akhdar bucks, wethers and does slaughtered over a body weight range of 11–28 kg (from Mahgoub *et al.*, 2005).

| Muscle group | Pooled data | Buck | Wether | Doe |
|---|---|---|---|---|
| Proximal hindlimb (MG1) | 1.05a,b ± 0.02 | 1.01b ± 0.02 | 1.09a* ± 0.03 | 1.02b ± 0.02 |
| Distal hindlimb (MG2) | 0.89b** ± 0.02 | 0.86b*** ± 0.03 | 0.96a ± 0.03 | 0.84b*** ± 0.02 |
| Surrounding spinal column (MG3) | 1.06a*a ± 0.02 | 0.97b ± 0.02 | 1.09a* ± 0.03 | 1.08a** ± 0.02 |
| Abdominal wall (MG4) | 1.03a,b ± 0.03 | 1.11a* ± 0.04 | 0.95b ± 0.05 | 1.05a,b ± 0.05 |
| Proximal forelimb (MG5) | 0.90*** ± 0.01 | 0.92** ± 0.02 | 0.90** ± 0.02 | 0.89*** ± 0.03 |
| Distal forelimb (MG6) | 0.84a,b*** ± 0.02 | 0.78b*** ± 0.04 | 0.88a*** ± 0.02 | 0.85a,b*** ± 0.03 |
| Connecting thorax to forelimb (MG7) | 1.09* ± 0.02 | 1.11* ± 0.01 | 1.08 ± 0.03 | 1.13** ± 0.03 |
| Connecting neck to forelimb (MG8) | 1.05a,b ± 0.03 | 1.11a ± 0.05 | 1.02b ± 0.03 | 1.06a,b ± 0.05 |
| Intrinsic neck and thorax (MG9) | 0.92b,c* ± 0.03 | 1.03a ± 0.03 | 0.85c*** ± 0.03 | 0.94b ± 0.05 |
| Expensive muscle group[a] | 1.01a,b ± 0.01 | 0.98b ± 0.01 | 1.04a ± 0.02 | 1.00a,b ± 0.01 |
| Forequarter[b] | 0.97b ± 0.01 | 1.02a ± 0.02 | 0.93c* ± 0.02 | 0.98a,b ± 0.02 |

SEM, Standard error of the mean.
Coefficients on the same line denoted by the same or no letter (a, b, c) do not differ ($P > 0.05$).
Growth coefficient values that differ significantly from 1.0 are indicated: *, $P < 0.05$; **, $P < 0.01$; ***, $P < 0.001$.
[a]Expensive muscle group = MG1, MG3 and MG5.
[b]Forequarter = MG5, MG7, MG8 and MG9.

muscles surrounding the spinal column (MG3), which comprise about 14–16% of one side of the carcass and include the largest muscle of the body, the m. longissimus thoracis et lumborum. This muscle contributes the highest proportion of any individual muscle in the side (7–9%) followed by the m. biceps femoris (4–6%) and m. semimembranosus (4%), both in MG1. More than half (48/88) of the individual muscles dissected from the carcass side contribute <1.0% each of the muscles in one side of the carcass.

Warmington and Kirton (1990) stated that, although goats contain a higher proportion of total muscle, the distribution in high-priced muscle groups is less favourable than in sheep. However, the high proportion of high-priced cuts in the carcass (~53–56%) in the Jebel Akhdar goats indicates a good potential for meat production from goats compared with sheep. This was higher than the 51% for this group of muscles in sheep (Butterfield, 1988).

Comparison of Jebel Akhdar goats with Omani sheep (Mahgoub and Lodge, 1994) at 28 kg body weight indicates some differences in proportions of some muscle groups, as well as some individual muscles, between the two species. Proportions of MG3 were lower whereas those of MG4 and MG5 were higher in goats than in sheep. Likewise, the proportions of individual muscles within these groups followed this trend. Proportion of individual muscles of the neck was higher in Omani goats than sheep (m. splenius, m. longissimus capitis et atlantis). However, these differences were small in absolute terms. Therefore, differences in carcass conformation between sheep and goats are more likely to be caused by differences in levels of fatness, especially in subcutaneous fat, which is reported to be less well developed in goats than in sheep (Naudé and Hofmeyr, 1981).

Muscle distribution is affected by the differential rate of growth in individual muscles and muscle groups on the carcass (Table 9.4). Muscle groups situated in the hindquarters and those in distal limbs grew at a slower rate and decreased as a proportion of total muscle weight, indicating that they are early maturing

(Butterfield, 1988; Mahgoub and Lodge, 1994; Mahgoub *et al.*, 2005).

### Effect of body weight on muscle distribution

There are significant effects of slaughter weight on the proportions of muscle groups in the carcass. Muscles of MG2, MG5 and MG6 generally decreased with increasing slaughter weight, whereas those of MG7 and MG8 increased (Table 9.2). Individual muscles of these groups followed a similar trend of growth pattern to that of the muscle group itself (Table 9.3).

Work with Omani goats (Mahgoub and Lodge, 1996; Mahgoub, 1997; Mahgoub *et al.*, 2005) indicated that the degree of maturity influences muscle distribution in the carcass, resulting in differences in proportions of individual and groups of muscles. Similar results were reported in temperate (Butterfield, 1988) and tropical (Mahgoub and Lodge, 1994) sheep. Sheep studies (Hogg *et al.*, 1992) indicated that the magnitude of the differences in lean tissue distribution was small in absolute terms. When goats are slaughtered at an early age and lower body weight, the proportions of high-priced cuts are lower in the carcass. For instance, Santos *et al.* (2007) reported that high-priced cuts contributed ~44% in suckling goats slaughtered at 8–11 kg body weight. These figures are lower than the 53–56% reported for the same group of muscles in Omani goats (Table 9.2). Increasing the slaughter weight from 11 to 18 kg decreased the proportions of the muscles in the proximal and distal hindlimb, proximal forelimb, distal forelimb and the EMGs in the proximal hindlimb, around the spinal column and in the proximal forelimb (Mahgoub and Lu, 1998). However, increasing goat body weight increased the proportions of the muscles in the abdominal wall, connecting the thorax to the forelimb, in the intrinsic muscles of the neck and in the thorax and forequarter.

### Effect of breed type on muscle distribution

Several authors observed a difference in distribution of individual carcass tissue in various cuts of different breeds of goats, although in some cases differences were small. For instance, Cameron *et al.* (2001) reported that the percentage of lean was lowest in the shoulder of Boer × Angora goats, whereas that of bone was greatest in the leg of Spanish goats, while the rack contained the lowest fat. Tshabalala *et al.* (2003) reported that the improved Boer goat had higher lean in the neck, forelimb and ventral trunk and slightly higher amounts in the hind leg but lower lean in the dorsal trunk than unimproved indigenous goats. Mahgoub and Lu (1998) compared carcass tissue distribution in Omani goats of different sizes. The small-sized Dhofari goat had higher proportions of muscles in the proximal hindlimb, around the spinal column and in the abdominal wall but lower proportions in the proximal forelimb, distal forelimb, connecting the forelimb to thorax, in the intrinsic muscles of neck and thorax and in the total forequarter than the larger Batina goats. These findings indicate that the smaller Dhofari goat may be more suitable for meat production than larger breeds such as the Batina. The difference in muscle proportion between the Dhofari and Batina breeds was 3–4% at 18 kg body weight. This difference may be large enough to produce commercial implications in favour of the smaller breed, slaughtered between 11 and 18 kg body weight, especially if combined with the 2–3% lower proportion of carcass bone, the higher dressing percentage and the faster growth rate relative to body size (Mahgoub and Lu, 1998). The differences between goats of different body sizes in carcass muscle distribution are in agreement with findings in sheep. Butterfield (1988) reported differences between large Merino and small Dorset sheep breeds in carcass muscle distribution.

### Effect of sex on muscle distribution

Sex influences muscle distribution in goat carcasses. Males have a higher lean content in their carcasses, associated with more developed shoulder and neck, than does (Mahgoub and Lodge, 1996; Mahgoub, 1997; Mahgoub *et al.*, 2005). Males are reported to have heavier necks and forequarters whereas females have heavier hindquarters than males (Colomer-Rocher *et al.*, 1992; El Moula *et al.*, 1999). The male neck cut has more muscle than that of the female (Gallo *et al.*, 1996). This is similar to reports in sheep (Butterfield, 1988). An important economic implication is that male goats would have heavier meat cuts at the front of the carcass, whereas females have heavier cuts towards the rear of the carcass (Kirton, 1970). Castration, which had been practised in goats for a long time, may have a significant effect on muscle distribution. Hutchison (1964) found that castration resulted in higher proportions of loin and hindquarters in the carcass of crossbred Boer goats in Tanzania.

Expressed as a percentage of total side muscle weight, male goats, especially intact ones, had lower proportions of muscle in the proximal and distal hindlimbs (MG1 and MG2) but higher proportions of muscle in the groups of the proximal and distal forelimbs, intrinsic muscles of the neck and shoulder and forequarter than castrates and does (Table 9.3). Does also have higher proportions of muscles in the proximal hindlimb and EMGs. The effects of sex on individual muscles follows a similar trend as for the muscle groups (Table 9.4). Sex differences in the neck and shoulder regions are caused by more development of some intrinsic muscles in bucks than in does. These included m. rhomboidious, m. splenius, m. longissimus capitis et atlantis and m. complexus. These muscles were reported to be affected by male sex hormones (Butterfield, 1988). Jebel Akhdar bucks had better development of these muscles at 28 kg body weight, a weight that probably coincides with the onset of puberty in male kids, than at 11 kg body weight (Mahgoub *et al.*, 2005). Mahgoub and Lu (1998) also reported that male goats generally had lower proportions of muscle in the proximal hindlimb and abdominal wall but higher proportions in the proximal forelimb, distal forelimb, intrinsic muscles of the neck and thorax and the forequarter than females of small- and large-sized goat breeds.

Sex affects the rate of growth of individual muscles and muscle groups. For data pooled from animals of all sexes, growth rates of muscle groups relative to total side muscle weight were higher for MG1, MG3, MG4, MG7 and MG8 than those for MG2, MG5, MG6, MG9, EMGs and the forequarter (Table 9.4). When data were analysed separately, bucks generally had lower values for growth rates of muscle groups that are situated towards the hindquarters (MG1, MG3 and EMGs) but higher values for those towards the anterior of the body (MG8, MG9 and forequarter). These results indicated that differences in muscle distribution in goats due to sex are not very large and are unlikely to have a commercial impact on meat production from these goats if they were slaughtered over a low body weight range. This suggests that entire male goats, which have always been subject to prejudice because of the male goat odour, may be used for meat production equally with other sexes to utilize their higher potential for growth, especially if slaughtered at low weights before attaining sexual maturity. On the other hand, some reports have indicated that differences in the proportions of muscle groups due to the effects of sex or slaughter weight may reach up to 3% (Mahgoub *et al.*, 2005). The female goats had 1.5% higher proportions of EMGs but 2% lower proportions of forequarter in the carcass muscle than males. These differences are important, especially at higher carcass weights.

## 9.5 Goat Carcass Fat

### 9.5.1 Growth and partitioning of fat

It is well established that fat is the most variable tissue in the carcass. The proportions and locations of fat in the body are important in meat animals. Differences in the contents and properties of subcutaneous and intramuscular fat between and within breeds are important factors, resulting in differences in meat quality in goats (Tshabalala *et al.*, 2003). Fat is a late-growing body tissue and therefore proportions in the carcass greatly change with progress of growth. For instance, the proportion of fat in the West African dwarf goat increased from 3.5% at birth to 15.5% at 10 kg but decreased slightly in goats approaching maturity (Wilson, 1960).

Body fat depots were deposited at a higher rate in the carcass of the Omani Jebel Akhdar goats relative to the EBW (Mahgoub *et al.*, 2005). Consequently, their proportions in the body were higher at 28 kg than at 11 and 18 kg body weight. This increasing fat deposition rate with age is in line with findings in other Omani goats (Mahgoub and Lodge, 1996; Mahgoub, 1997) and sheep (Mahgoub and Lodge, 1994).

Fat is deposited at a different rate in various parts of the goat body. In general, carcass fat such as subcutaneous fat develops at a slower rate in goats compared with non-carcass fat. Carcasses of 30% fat may contain only 2–3.5 mm fat cover over the m. longissimus dorsi (Warmington and Kirton, 1990). The ascending order of fat deposition in goats was: subcutaneous, intermuscular, mesenteric, kidney knob and channel and omental fat (Teixeira *et al.*, 1995). Within the carcass, intermuscular fat was later developing than subcutaneous fat (Teixeira *et al.*, 1995). Therefore, proportions of intermuscular fat will increase with increasing body weight. The proportions of intermuscular fat were higher at 25 kg than at 6 kg slaughter weight (Marichal *et al.*, 2003). Colomer-Rocher *et al.* (1992), Teixeira *et al.* (1995), Mahgoub and Lu (1998) and Santos *et al.* (2007) reported that the intermuscular fat depot in goats was higher than the subcutaneous fat in goats of similar weight and the same sex. Sumarmono *et al.* (2001) reported that intermuscular fat was approximately fourfold higher than the level of subcutaneous fat. This is reflected in carcass distribution, as the loin region in feral goats, for instance, was devoid of subcutaneous fat (Kirton, 1970). Subcutaneous fat proportions were different in various carcass anatomical sites. Tshabalala *et al.* (2003) reported values of 0.88, 3.19, 4.39, 3.84 and 1.76% subcutaneous fat in the neck, forelimb, ventral trunk, dorsal trunk and hind leg of Boer goats, respectively.

The greatest part of the body fat in goats is deposited in the abdomen (40%), followed by subcutaneous fat (30%), intermuscular fat (23%) and mesenteric fat (6%) (Wilson, 1960). Internal fat such as omental and mesenteric fat develops faster in goats (McGregor, 1982; Thonney *et al.*, 1987). Kidney knob and channel and omental fat is deposited at a higher rate in relation to total carcass fat than subcutaneous and intermuscular fat (Teixeira *et al.*, 1995). Within the goat carcass, the sites in which subcutaneous fat is deposited last are the breast and chump, whereas intermuscular fat is deposited late in the breast and loin (Teixeira *et al.*, 1995).

Tropical goats appeared to have lower proportions of subcutaneous fat and subsequently higher proportions of intermuscular fat than those of sheep at the same body weight (Mahgoub and Lodge, 1994, 1996; Mahgoub, 1997; Mahgoub *et al.*, 2005). There are similar reports in temperate goats (Naudé and Hofmeyr, 1981). This low subcutaneous fat in goats indicates a negative effect on the storage properties of goat carcasses (Hogg *et al.*, 1992). Kirton (1970) reported a 5.3 and 6.1% cold storage loss in male and female feral goats, respectively, which was higher than the 4.5% for New Zealand lamb. He related this to the low fat cover and low fat carcass content.

In Omani Jebel Akhdar goats (Table 9.5), the weight of total body fat as a proportion of EBW in goats ranged between the lowest value of 7.6% in bucks at 11 kg body weight to the highest value of 23.5% in does at 28 kg body weight. Total non-carcass fat generally had lower proportions in the EBW and total body fat than total carcass fat. Among body fat depots, intermuscular fat contributed the highest proportions in EBW and TBF followed by subcutaneous, omental, kidney, mesenteric, scrotal/udder, pelvic and channel fats, respectively. For data pooled from all sexes, total non-carcass fat and total carcass fat were deposited at a rate higher than EBW and total body fat. Generally, proportions of carcass and non-carcass fats in EBW of all sexes increased with increasing body wieght (Table 9.5). For

**Table 9.5.** Least square means (± SEM) of percentages of carcass and non-carcass fat depots in the empty body weight and total body fat in Omani Jebel Akhdar buck, wether and doe goats slaughtered at 11, 28 and 28 kg BW (Mahgoub *et al.*, 2005).

| | Buck | | | Wether | | | Doe | | | | Effects[a] | | |
|---|---|---|---|---|---|---|---|---|---|---|---|---|---|
| Fat depot | 11 kg | 18 kg | 28 kg | 11 kg | 18 kg | 28 kg | 11 kg | 18 kg | 28 kg | SEM | Sex | SLWT | Sex × SLWT |
| In empty body weight (%) | | | | | | | | | | | | | |
| Omental | 0.77 | 1.48 | 2.68 | 1.13 | 2.15 | 2.39 | 1.57 | 2.46 | 4.47 | 0.26 | *** | *** | * |
| Mesenteric | 1.31 | 1.36 | 1.80 | 1.28 | 1.56 | 2.21 | 1.43 | 1.56 | 2.51 | 0.12 | ** | *** | |
| Scrotal/udder | 0.41 | 0.61 | 0.79 | 0.39 | 0.75 | 0.78 | 0.56 | 0.80 | 1.01 | 0.08 | * | ** | |
| Kidney | 0.83 | 1.00 | 1.50 | 0.99 | 1.29 | 1.56 | 1.13 | 1.36 | 2.83 | 0.20 | *** | *** | ** |
| Channel | 0.06 | 0.10 | 0.14 | 0.07 | 0.10 | 0.10 | 0.11 | 0.20 | 0.23 | 0.03 | ** | * | |
| Pelvic | 0.22 | 0.28 | 0.30 | 0.24 | 0.26 | 0.30 | 0.28 | 0.25 | 0.36 | 0.05 | | | |
| Total non-carcass fat | 3.61 | 4.82 | 7.20 | 4.10 | 6.11 | 7.32 | 5.82 | 6.63 | 11.41 | 0.56 | *** | *** | * |
| Subcutaneous | 1.32 | 2.49 | 3.22 | 1.81 | 3.32 | 4.29 | 1.48 | 3.79 | 5.53 | 0.41 | ** | *** | |
| Intermuscular | 2.71 | 3.54 | 5.21 | 4.14 | 4.70 | 5.53 | 3.07 | 4.61 | 6.57 | 0.34 | ** | *** | |
| Total carcass fat | 4.03 | 6.03 | 8.43 | 5.98 | 8.01 | 9.82 | 4.55 | 8.40 | 12.09 | 0.76 | | ** | |
| Total body fat | 7.64 | 10.85 | 15.63 | 10.62 | 14.12 | 17.27 | 11.11 | 15.03 | 23.50 | 1.18 | *** | *** | |
| In total body fat (%) | | | | | | | | | | | | | |
| Omental | 9.97 | 13.71 | 16.84 | 11.95 | 15.26 | 13.85 | 16.82 | 16.25 | 19.16 | 1.23 | *** | ** | |
| Mesenteric | 19.26 | 12.90 | 11.52 | 12.73 | 11.14 | 12.83 | 14.39 | 10.38 | 10.73 | 1.36 | | * | |
| Scrotal/udder | 5.26 | 5.62 | 5.13 | 4.63 | 5.33 | 4.57 | 5.05 | 5.30 | 4.29 | 0.35 | | * | |
| Kidney | 10.81 | 9.24 | 9.44 | 10.73 | 9.28 | 8.83 | 12.66 | 9.12 | 11.98 | 0.78 | * | ** | |
| Channel | 0.80 | 0.92 | 0.94 | 0.71 | 0.69 | 0.46 | 1.03 | 1.34 | 1.00 | 0.25 | | | |
| Pelvic | 3.10 | 2.58 | 1.91 | 2.69 | 1.84 | 1.74 | 2.61 | 1.38 | 1.54 | 0.25 | * | *** | |
| Total non-carcass fat | 48.94 | 45.00 | 45.79 | 43.44 | 43.55 | 42.27 | 52.27 | 44.07 | 48.68 | 1.85 | * | | |
| Subcutaneous | 15.44 | 22.42 | 20.68 | 15.56 | 23.09 | 25.15 | 17.00 | 24.57 | 23.36 | 1.80 | | *** | |
| Intermuscular | 35.62 | 32.60 | 33.53 | 41.00 | 33.36 | 32.58 | 30.43 | 31.06 | 27.95 | 2.02 | ** | * | |
| Total carcass fat | 51.05 | 55.02 | 54.21 | 56.56 | 56.45 | 57.73 | 47.43 | 55.93 | 51.32 | 1.85 | * | | |

SEM, Standard error of the mean.
[a]Effects of sex and slaughter weight (SLWT) are different: *, $P < 0.05$; **, $P < 0.01$; ***, $P < 0.001$.

individual fat depots, the highest rate of deposition was exhibited by omental, subcutaneous and kidney depots followed by intermuscular fat, and the lowest by pelvic and mesenteric fat (Table 9.6).

### 9.5.2 Effect of sex

Generally, female goats have higher fat content in their bodies than males and the proportions of fat increase as body weight increases (Mahgoub and Lu, 1998). These findings are consistent with findings in temperate goats (Morand-Fehr, 1981; Warmington and Kirton, 1990; Colomer-Rocher *et al.*, 1992), sheep (Butterfield, 1988) and cattle (Berg and Butterfield, 1976).

Castration, which is widely practised in goats to reduce the goat male odour, affects fat deposition and distribution in the carcass. Castrated male Boer goats had twice the amount of intermuscular and subcutaneous fat as entire male goats (Sumarmono *et al.*, 2001). Castrated males had higher total fat, carcass fat, internal fat and kidney fat and less lean than intact males (Ruvuna *et al.*, 1992).

Carcass and non-carcass fat in Jebel Akhdar, Batina and Dhofari Omani goats was deposited at a faster rate in does and to a lesser extent in castrates with increasing EBW than in the intact males (Mahgoub and Lodge, 1996; Mahgoub, 1997; Mahgoub *et al.*, 2005). Similar findings were reported in Omani sheep (Mahgoub and Lodge, 1994) and temperate sheep (Butterfield, 1988).

**Table 9.6.** Growth coefficients (± SEM) of fat depots relative to empty body weight in buck, wether and doe Omani Jebel Akhdar goats slaughtered over a body weight range of 11–28 kg (adapted from Mahgoub *et al.*, 2005).

| Fat depot | Pooled data | Buck | Wether | Doe |
|---|---|---|---|---|
| Relative to empty body weight | | | | |
| Omental | 2.28 ± 0.15 | 2.42 ± 0.23 | 2.07* ± 0.20 | 2.28 ± 0.19 |
| Mesenteric | 1.53* ± 0.08 | 1.34 ± 0.14 | 1.56** ± 0.09 | 1.68* ± 0.14 |
| Scrotal/udder | 1.95*** ± 0.14 | 1.81b ± 0.19 | 2.22a*** ± 0.29 | 1.61a,b** ± 0.19 |
| Kidney | 1.91* ± 0.14 | 1.73b ± 0.21 | 1.76***a ± 0.21 | 2.21a,b ± 0.23 |
| Channel | 1.80 ± 0.23 | 1.95 ± 0.47 | 1.39* ± 0.26 | 1.90 ± 0.34 |
| Pelvic | 1.36** ± 0.12 | 1.35b ± 0.17 | 1.31a*** ± 0.12 | 1.40a,b*** ± 0.34 |
| Total non-carcass fat | 1.84 ± 0.10 | 1.77b ± 0.15 | 1.79a*** ± 0.13 | 1.82a,b ± 0.13 |
| Subcutaneous | 2.46** ± 0.20 | 2.24 ± 0.32 | 2.29** ± 0.28 | 2.80** ± 0.37 |
| Intermuscular | 1.68* ± 0.10 | 1.78b ± 0.18 | 1.37a*** ± 0.08 | 1.87a*** ± 0.12 |
| Total carcass fat | 1.94** ± 0.12 | 1.91b ± 0.21 | 1.66a*** ± 0.14 | 2.16a,b*** ± 0.16 |
| Total body fat | 1.84*** ± 0.11 | 1.84b ± 0.17 | 1.66a*** ± 0.14 | 1.88a,b*** ± 0.10 |
| Relative to total body fat | | | | |
| Omental | 1.23*** ± 0.04 | 1.30b*** ± 0.05 | 1.16a ± 0.09 | 1.08a,b ± 0.06 |
| Mesenteric | 0.78*** ± 0.04 | 0.69 *** ± 0.07 | 0.92 ± 0.10 | 0.85 ± 0.06 |
| Scrotal/udder | 0.94 ± 0.04 | 0.98b ± 0.05 | 1.06a ± 0.08 | 0.86a,b ± 0.09 |
| Kidney | 0.99 ± 0.04 | 0.95 ± 0.06 | 0.90 ± 0.09 | 1.01 ± 0.08 |
| Channel | 1.01 ± 0.12 | 1.06b ± 0.22 | 0.74a ± 0.15 | 0.98a,b ± 0.22 |
| Pelvic | 0.69 *** ± 0.07 | 0.74*** ± 0.06 | 0.69** ± 0.08 | 0.77 ± 0.20 |
| Total non-carcass fat | 0.98 ± 0.02 | 0.95b ± 0.04 | 0.98a ± 0.04 | 0.97a,b ± 0.04 |
| Subcutaneous | 1.28** ± 0.05 | 1.26 * ± 0.09 | 1.45*** ± 0.09 | 1.17 ± 0.07 |
| Intermuscular | 0.90** ± 0.03 | 0.96b ± 0.04 | 0.82a* ± 0.04 | 0.93a,b ± 0.06 |
| Total carcass fat | 0.63*** ± 0.07 | 1.05 ± 0.03 | 1.02 ± 0.03 | 1.03 ± 0.04 |

SEM, Standard error of the mean.
Coefficients on the same line denoted by the same or no letter (a, b) do not differ significantly ($P > 0.05$).
Growth coefficient values that differ significantly from 1.0 are indicated: *, $P < 0.05$; **, $P < 0.01$; ***, $P < 0.001$.

This suggests that intact male goats should be more suited for meat production in spite of the male goat odour, although this is not well founded (Kirton, 1970; Gaili *et al.*, 1972).

Does have higher proportions of fat in their EBW than bucks and wethers, especially at higher body weight (Table 9.5). Mahgoub and Lu (1998) reported that females generally had higher proportions of carcass and non-carcass fats than males. These effects were significant for omental, mesenteric, scrotal or udder, total non carcass, total body, kidney, intermuscular and total carcass fat in Omani goats.

### 9.5.3 Effect of body size and breed on fat distribution

Body weight affects the proportions of fat in the goat body in absolute terms and relative to body weight. For instance, proportions of omental, total non-carcass, total body, kidney, intermuscular, total carcass, scrotal or udder and subcutaneous fat depots were higher at 18 kg than at 11 kg body weight in Omani goats (Mahgoub and Lu, 1998).

Goats vary in size to a large extent. There are significant differences in carcass and non-carcass fat distribution in goats of various sizes. For instance, as a proportion of EBW, small goats (Dhofari) had higher proportions of omental, kidney, total non-carcass, intermuscular and total body fat than large (Batina) goats (Mahgoub and Lu, 1998). Dhofari goats had higher total body fat and total non-carcass fat than Batina goats. The differences were more pronounced in females than in males and at higher than at lower body weights. This indicates an earlier maturity for the smaller goat breeds such as the Dhofari, which means that they enter the 'fattening stage' at lower weights than the larger goats. This phenomenon has economic implications for meat production. Early maturing breeds such as the Dhofari goats should be slaughtered at a lower body weight than larger breeds such as the Batina goats to avoid the need for trimming of excess carcass fat. The higher proportion of total body fat in Dhofari goats compared with Batina goats appears to be attributed more to the higher total and individual non-carcass fats (omental, kidney, etc.). This should add to the suitability of the small-sized Dhofari goat for meat production, as the non-carcass fat is readily separable at the time of slaughter.

The proportions of fat vary among breeds along the goat carcass. Tshabalala *et al.* (2003) reported that Boer goats had higher subcutaneous fat in the forelimb, ventral trunk and hind leg than indigenous goats.

## 9.6 Goat Carcass Bone

### 9.6.1 Bone growth and development

Bone is an important body and carcass component. During the animal's life, it gives stature and support for the animal, protects internal organs, provides movement in coordination with skeletal muscles and serves as a reservoir for minerals and trace elements. However, not being edible, bone is often overlooked, yet its proportion affects that of muscle and fat. Bone and muscle growth are strongly related (Mahgoub, 1988), and the muscle:bone ratio is regarded as an important attribute for carcass evaluation. Bone is an early-maturing carcass component so it grows at a slower rate during post-natal life, consequently decreasing with increasing body weight (Mahgoub, 1997; Marichal *et al.*, 2003; Liméa *et al.*, 2009).

Significant changes occur in the skeleton during pre-natal and post-natal life from birth to maturity. These changes are attributed to the differential rates of growth of different parts and tissue of the body (Hammond, 1932; Mahgoub, 1988). This has led to a general concept of skeletal growth and development in meat animals such as the axial craniocaudal gradient of increasing growth (Hammond, 1932).

### 9.6.2 Proportions of bone in goat carcass

The percentage of bone in the carcass decreases whereas that of fat increases with increasing body weight. Bone proportions in the goat carcass range between 12 and 28% depending on the factors influencing it such as breed and sex (Table 9.1). Bone contributed about 13% of EBW of Batina goats (Mahgoub and Lodge, 1996). Oman *et al.* (1999) reported a high bone proportion of 37%, whereas Sen *et al.* (2004) reported a value as low as 17.6%. There are even lower proportions of bone in the carcass reported for West African dwarf goats of 9.6% at a body weight of 20 kg (Attah *et al.*, 2006); however, this value appears to be doubtful as the total addition of carcass tissue did not add up to 100%.

Goats generally have higher levels of bone in the carcass than sheep (Gaili, 1976). These differences in conformation are attributed to the ability of goats to browse; hence, their necks and shoulders are more developed and adapted to browsing than those of sheep. Although Mahgoub and Lodge (1998) found no difference in the proportions of bone in the carcass, sheep had a higher proportion of axial skeleton but lower proportion of forelimb in carcass bone than goats.

The distribution of bone in the carcass is an important trait, especially when carcasses are sold in the form of wholesale or retail cuts. It is affected by breed and sex. Generally, the axial skeleton of the carcass side comprises ~50% of the total side bone weight of which 30% is in the vertebral column. The fore- and hindlimb constitute about 22% each of the total side bone (Mahgoub and Lodge, 1996; Mahgoub, 1997) (Table 9.7). The largest single bone was the femur, which contributed about 10% of the half carcass weight. However, goats appear to have higher proportions of bone in the limbs and lower proportions in the axial skeleton than sheep. At 28 kg body weight, bucks had 51, 23 and 22.3% side carcass in the axial skeleton, forelimb and hindlimbs, respectively, compared with 55, 21 and 24% in Omani rams (Mahgoub and Lodge, 1994) and 65.4, 17.4 and 17.2% in mature small-strain Merino rams (Butterfield, 1988).

Tshabalala *et al.* (2003) reported values of 19.23, 21.85, 14.1, 26.7 and 20.16% bone in the neck, forelimb, ventral trunk, dorsal trunk and hind leg of Boer goats, respectively.

#### *Effect of breed and size on bone distribution*

Breed effects are more manifest when comparing dairy- versus meat-goat breeds or improved versus non-improved breeds. For instance, goats with Boer blood have higher proportions of bone in their carcasses, and Boer goats had lower carcass bone proportions (20.6 versus 24.6%) than indigenous South African goat breeds (Webb *et al.*, 2005). Oman *et al.* (1999) found a significant difference in proportion of bone in the carcass between Boer × Spanish crosses and Spanish goats, with those raised on range having higher proportions of bone in the side than those raised in feedlots (36.9 and 36.5 versus 26.5 and 27.6%, respectively). Cameron *et al.* (2001) also found that carcass bone, fat and lean weight were significantly or numerically greater for Boer crossbreeds than for Spanish goats. The percentage of bone was higher for Boer × Spanish than for Spanish and Boer × Angora wethers, ranging between 26 and 29% of carcass weight. Pralomakran *et al.* (1995) found that Thai native goats had lower bone content in their carcasses than Anglo-Nubian cross male goats. Tshabalala *et al.* (2003) reported that Boer goats had lower bone content in the neck, forelimb, ventral trunk and hind leg but higher bone content in the dorsal trunk than indigenous goats.

Lower proportions of carcass bone are mainly due to higher proportions of carcass fat. In tropical, non-improved breeds, there was a wide variation in bone proportions in the carcass. Attah *et al.* (2006) reported a wide variation between West African dwarf goats and Red Sokoto goats in Nigeria. Red Sokoto goat carcasses contained higher weights and proportions of bone than the West African dwarf goats. However, within indigenous breeds, there were no effects

**Table 9.7.** Least square means (± SEM) of weights of some individual bones of the left half carcass (as % of total half carcass bone weight) in Omani Jebel Akhdar goats slaughtered at 11, 18 and 28 kg body weight (from Mahgoub *et al.*, 2005).

| | Buck | | | Wether | | | Doe | | | | Effects[a] | | |
|---|---|---|---|---|---|---|---|---|---|---|---|---|---|
| Bone or bone group | 11 kg | 18 kg | 28 kg | 11 kg | 18 kg | 28 kg | 11 kg | 18 kg | 28 kg | SEM | Sex | SLWT | Sex × SLWT |
| Cervical vertebrae | 11.10 | 11.76 | 11.81 | 13.23 | 11.80 | 12.95 | 10.10 | 10.68 | 10.08 | 0.70 | ** | | |
| Thoracic vertebrae | 12.20 | 11.09 | 10.27 | 11.36 | 11.55 | 10.10 | 11.37 | 10.78 | 9.60 | 0.56 | | ** | |
| Lumbar vertebrae | 8.04 | 8.83 | 7.47 | 8.00 | 7.8 | 7.58 | 9.82 | 9.01 | 8.80 | 0.46 | ** | | |
| Sacral vertebrae | 2.87 | 2.46 | 2.64 | 2.36 | 2.61 | 2.81 | 3.62 | 2.84 | 3.08 | 0.29 | * | | |
| Total vertebral column | 34.13 | 34.14 | 32.18 | 34.94 | 33.84 | 33.44 | 34.91 | 33.01 | 31.56 | 1.07 | | | |
| Ribs | 5.98 | 5.89 | 6.63 | 6.31 | 6.83 | 6.92 | 5.59 | 7.09 | 6.65 | 0.37 | | * | |
| 1st rib | 0.86 | 0.78 | 0.96 | 0.72 | 0.86 | 0.83 | 0.71 | 0.83 | 0.86 | 0.04 | | * | |
| 6th rib | 0.83 | 0.84 | 0.98 | 0.82 | 0.96 | 1.09 | 0.80 | 1.06 | 1.07 | 0.04 | * | *** | |
| 12th rib | 0.28 | 0.30 | 0.36 | 0.31 | 0.29 | 0.29 | 0.29 | 0.34 | 0.32 | 0.02 | | | |
| Total ribs | 7.95 | 7.80 | 8.93 | 8.16 | 8.93 | 9.13 | 7.39 | 9.32 | 8.91 | 0.40 | | ** | |
| Pelvis | 6.57 | 6.77 | 7.03 | 6.32 | 6.59 | 6.95 | 6.40 | 6.35 | 6.93 | 0.25 | | * | |
| Sternum | 2.30 | 2.97 | 2.81 | 2.84 | 2.54 | 2.94 | 2.40 | 2.39 | 3.22 | 0.33 | | | |
| Total axial skeleton | 51.05 | 51.68 | 50.95 | 52.26 | 51.90 | 52.46 | 51.09 | 51.07 | 50.61 | 1.00 | | | |
| Scapula | 4.74 | 5.51 | 5.26 | 4.47 | 4.66 | 4.86 | 4.68 | 5.03 | 5.36 | 0.22 | * | * | |
| Humerus | 8.88 | 9.00 | 9.32 | 8.96 | 8.98 | 8.7 | 9.08 | 9.11 | 9.15 | 0.22 | | | |
| Radio-ulna | 6.98 | 6.59 | 6.74 | 6.85 | 6.73 | 6.44 | 6.76 | 6.56 | 6.61 | 0.18 | | | |
| Carpus | 1.68 | 1.50 | 1.61 | 1.53 | 1.57 | 1.50 | 1.56 | 1.50 | 1.53 | 0.10 | | | |
| Total forelimb | 22.28 | 22.60 | 22.93 | 21.78 | 21.94 | 21.58 | 22.09 | 22.20 | 22.66 | 0.54 | | | |
| Femur | 10.42 | 10.21 | 10.36 | 10.39 | 10.41 | 10.28 | 10.61 | 10.51 | 10.07 | 0.26 | | | |
| Tibia | 8.34 | 8.16 | 8.29 | 8.03 | 8.49 | 8.46 | 8.53 | 8.47 | 8.48 | 0.21 | | | |
| Patella | 0.60 | 0.59 | 0.69 | 0.61 | 0.64 | 0.65 | 0.65 | 0.67 | 0.70 | 0.04 | | | |
| Tarsus | 3.46 | 3.21 | 3.00 | 3.28 | 3.13 | 2.91 | 3.37 | 3.31 | 3.59 | 0.25 | | | |
| Total hindlimb | 22.83 | 22.17 | 22.34 | 22.31 | 22.67 | 22.30 | 23.16 | 22.96 | 22.84 | 0.51 | | | |

SEM, Standard error of the mean.
[a]Effects of sex and slaughter weight (SLWT) that are significantly different are indicated: *, $P < 0.05$; **, $P < 0.01$; ***, $P < 0.001$.

(Santos *et al.*, 2007) on young goats (Serrana and Bravia and crosses).

Mahgoub and Lu (1998) found no major differences between small-sized (Dhofari) and large-sized goats (Batina) in individual bone distribution except for the forelimb where the Dhofari goat had a higher proportion of bone than the Batina goats. Butterfield (1988) stated that large sheep breeds do not need to have higher proportions of bone in the limbs to be able to carry the extra weight.

The proportions of total ribs, pelvis and scapula increased whereas that of the radio-ulna decreased between 11 and 18 kg body weight in Omani goats (Mahgoub and Lu, 1998). Smaller Dhofari goats had higher proportions of the forelimb and radio-ulna but lower proportions of the humerus and femur than large Batina goats.

### *Effect of sex on bone distribution*

Although bone proportions in the carcass are reported to be affected by body weight, breed and nutrition, reports on the effects of sex are scarse. Bone growth is affected by sex hormones (Mahgoub, 1988). Therefore, sexual dimorphism is evident in goats, with males being much larger and more muscular than females.

Castration affects bone growth and dimensions due to the lack of male sex hormones. Intact males have higher proportions of bone in the carcass compared with females. El Moula *et al.* (1999) reported that male Sudan Desert goats had slightly higher proportions of bone (25.3%) than females (23.2%), but the difference was not significant. There was also no effect of sex on the proportions of bone in Serrana or Bravia goats or their crosses with bone proportions being 20.7 and 21.2% for females and males, respectively (Santos *et al.*, 2007). However, some reports have indicated that bone proportions are significantly lower in females than males due to higher fat proportions (Peña *et al.*, 2007). Male goats had lower proportions of pelvis bone but higher proportions of the humerus than females (Mahgoub and Lu, 1998).

Most parts of the skeleton generally grow at a growth coefficients rate of <1.0, with males (intact and castrated) demonstrating higher growth coefficients than females (Table 9.8). The proportions of thoracic vertebrae decreased, whereas those of ribs, pelvis and scapula increased with increasing body weight.

### *Bone distributions in carcass cuts*

Bone proportions in carcass cuts vary greatly in published reports on goats. However, these reports should be evaluated carefully, as methods of carcass cutting vary to a great extent. Generally, bone contents in individual carcass cuts are similar to that of the whole carcass (20–30%). However, some carcass cuts have higher proportions of bone, especially the rack (25–40%). This is due to the high proportions of the thoracic vertebrae and ribs, with the m. longissimus dorsi being the major muscle. El Moula *et al.* (1999) reported the lowest proportion of bone in the loin (10.6%). Goats are more browsers than grazers, especially in the arid and semi-arid regions of the world, which requires an erect and extended neck posture with bipedal stance (Bhatta *et al.*, 2001). This may have contributed to differences between goats and other animal species such as sheep in bone proportions in this region.

There are some reports of sex effects on bone proportions in various cuts. El Moula *et al.* (1999) reported 10.6 versus 18.6% bone in the loin and 16.1 versus 19.8% in the breast cuts of male and female Sudan Desert goats, respectively. Generally, male goat carcasses had heavier bone in cuts such as single short forequarter, best end of neck and neck, while females had heavier bone in the leg and chump, loin and breast cuts (El Moula *et al.*, 1999). Males usually have better developed forequarters and neck due to the effects of male sex hormones (Colomer-Rocher *et al.*, 1992). Gallo *et al.* (1996) reported that males had higher proportions of bone in the shoulder (21.6 versus 19.8%) and thorax (18.6 versus 15.3%) but lower neck bone (22.2 versus 27.4%) than female Criollo Chilean goats.

**Table 9.8.** Growth coefficients (± SEM) of carcass and non-carcass fat depots relative to EBW in buck, wether and doe Omani Jebel Akhdar goats slaughtered over a body weight range of 11–28 kg (adapted from Mahgoub *et al.*, 2005).

| Skeletal part | Pooled data | Buck | Wether | Doe |
|---|---|---|---|---|
| Cervical vertebrae | 0.63 ± 0.11 | 0.67 ± 0.27 | 0.77* ± 0.14 | 0.59 ± 0.10 |
| Thoracic vertebrae | 0.53* ± 0.08 | 0.54 ± 0.14 | 0.66** ± 0.10 | 0.45* ± 0.13 |
| Lumbar vertebrae | 0.64*** ± 0.07 | 0.69b ± 0.19 | 0.77a*** ± 0.06 | 0.46a,b*** ± 0.11 |
| Sacral vertebrae | 0.65* ± 0.19 | 0.33b ± 0.79 | 1.05a*** ± 0.12 | 0.40a,b ± 0.14 |
| Total vertebral column | 0.60 ± 0.07 | 0.61 ± 0.18 | 0.76* ± 0.07 | 0.49 ± 0.07 |
| Ribs | 0.78** ± 0.08 | 0.78b ± 0.09 | 0.91a*** ± 0.06 | 0.72a,b*** ± 0.18 |
| 1st rib | 0.85 ± 0.07 | 0.72b ± 0.08 | 1.02a*** ± 0.07 | 0.80a,b ± 0.11 |
| 6th rib | 1.00** ± 0.05 | 0.89 ± 0.10 | 1.13*** ± 0.09 | 0.97** ± 0.05 |
| 12th rib | 0.82* ± 0.07 | 1.11b ± 0.19 | 0.72a*** ± 0.10 | 0.77a*** ± 0.08 |
| Total ribs | 0.82** ± 0.07 | 0.80b ± 0.07 | 0.94a*** ± 0.05 | 0.77a,b*** ± 0.14 |
| Pelvis | 0.80*** ± 0.07 | 0.80b ± 0.11 | 0.97a*** ± 0.07 | 0.69a,b*** ± 0.09 |
| Sternum | 0.24 ± 0.53 | 1.42 ± 1.17 | 0.69* ± 0.69 | 0.62 ± 0.94 |
| Total axial skeleton | 0.67* ± 0.07 | 0.69 ± 0.17 | 0.82** ± 0.05 | 0.56* ± 0.08 |
| Scapula | 0.83* ± 0.07 | 0.83b ± 0.25 | 0.90a*** ± 0.09 | 0.76a,b ± 0.03 |
| Humerus | 0.69 ± 0.04 | 0.77 ± 0.08 | 0.79* ± 0.05 | 0.59 ± 0.02 |
| Radio-ulna | 0.63** ± 0.28 | 0.61b ± 0.09 | 0.75a*** ± 0.05 | 0.57a,b*** ± 0.03 |
| Carpus | 0.65 ± 0.08 | 0.54b ± 0.14 | 0.81a*** ± 0.11 | 0.57a,b ± 0.11 |
| Total forelimb | 0.70** ± 0.05 | 0.72 ± 0.12 | 0.80*** ± 0.05 | 0.62** ± 0.02 |
| Femur | 0.72** ± 0.04 | 0.70b ± 0.07 | 0.89a*** ± 0.03 | 0.60a,b*** ± 0.04 |
| Tibia | 0.72*** ± 0.04 | 0.70b ± 0.07 | 0.89a*** ± 0.03 | 0.60a,b*** ± 0.04 |
| Patella | 0.80 ± 0.06 | 0.83 ± 0.09 | 0.88* ± 0.10 | 0.73 ± 0.11 |
| Tarsus | 0.63* ± 0.07 | 0.42 ± 0.22 | 0.70** ± 0.08 | 0.67* ± 0.08 |
| Total hindlimb | 0.70*** ± 0.04 | 0.67b ± 0.08 | 0.83a*** ± 0.04 | 0.61a,b*** ± 0.03 |

SEM, Standard error of the mean.
Coefficients on the same line denoted by the same or no letter (a, b) do not differ ($P > 0.05$).
Growth coefficient values that differ significantly from 1.0 are indicated: *, $P < 0.05$; **, $P < 0.01$; ***, $P < 0.001$.

## 9.7 Conclusions

The published literature indicates differences in carcass tissue distribution influenced by body weight, breed and sex. Fat is the most variable tissue in the carcass and increases at a rate higher than the carcass, whereas muscle grows at a similar rate and bone at a lower rate. Differences in bone and soft tissue distribution in the carcass affect carcass conformation, composition and the marketability of meat.

## References

Amegee, Y. (1996) Finishing performance and carcass quality of West African dwarf goat. *Revue d'élevage et de Médecine Vétérinaire des Pays Tropicaux* 39, 75–80.

Ash, A.J. and Norton, B.W. (1987) Studies with the Australian cashmere goat. II. Effects of dietary protein concentration and feeding level on body composition of male and female goats. *Australian Journal of Agricultural Research* 38, 971–982.

Attah, S., Omojola, A.B. and Adesehinwa, A.O.K. (2006) Yield and carcass composition of goats as affected by breed and slaughter weight. *World Applied Science Journal* 1, 8–11.

Berg, R. and Butterfield, R.M. (1976) *New Concepts in Cattle Growth.* University of Sydney Press, Sydney, Australia.

Bhatta, R., Sankhya, S.K., Shinde, A.K. and Verma, D.L. (2001) Seasonal changes in diet selectivity and grazing behavior of goats on semiarid rangeland. *Indian Journal of Animal Sciences* 71, 62–65.

Butterfield, R.M. (1988) *New Concepts of Sheep Growth.* Department of Veterinary Anatomy, University of Sydney, Sydney, Australia.

Cameron, M.R., Luo, J., Sahlu, T., Hart, S.P., Coleman, S.W. and Goetsch, A.L. (2001) Growth and slaughter traits of Boer Spanish, Boer Angora, and Spanish goats consuming a concentrate-base diet. *Journal of Animal Science* 79, 1423–1430.

Casey, N. (1987) Meat production and meat quality from Boer goats. In: *Proceedings of the IVth International Conference on Goats.* EMBRAPA and IGA, Brasilia, pp. 211–239.

Colomer-Rocher, F., Kirton, A.H., Mercer, G.J.K. and Duganzich, D.M. (1992) Carcass composition of New Zealand Saanen goats slaughtered at different weights. *Small Ruminant Research* 7, 161–173.

Dhanda, J.S., Taylor, D.G. and Murray, P.J. (2003) Part 2. Carcass composition and fatty acid profiles of adipose tissue of male goats: effects of genotype and liveweight at slaughter. *Small Ruminant Research* 50, 67–74.

El Moula, I.H.A., Babiker, S.A., Khidir, O.A. and Ibrahim, S.E. (1999) Meat production from female goat kids compared with males. *Journal of Agricultural Science* 133, 223–226.

Fehr, P.M., Sauvant, D., Delage, J., Dumont, B.L. and Roy, G. (1976) Effect of feeding methods and age at slaughter on growth performances and carcass characteristics of entire young male goats. *Livestock Production Science* 3, 183–194.

Fonesca, P.D. (1987) Quantitative and qualitative aspects of meat production by Raiana-Serpentima and Chernequiera goats. In: *Proceedings of the IVth International Conference on Goats.* EMBRAPA and IGA, Brasilia, p. 1465.

Gaili, E.S.E. (1976) A comparison of the development of body components in Sudan desert sheep and goats. *Tropical Animal Health and Production* 10, 103–108.

Gaili, E.S.E., Ghanem, Y.S. and Mukhtar, A.M.S. (1972) A comparative study of some carcass characteristics of Sudan desert sheep and goats. *Animal Production* 14, 351–357.

Gallo, C., Le Breton, Y., Wainnright, I. and Berkhoff, M. (1996) Body and carcass composition of male and female Criollo goats in the South Chile. *Small Ruminant Research* 23, 163–169.

Hammond, J. (1932) *Growth and Development of Mutton Qualities in the Sheep.* Oliver & Boyd, Edinburgh, UK.

Hogg, B.W., Mercer, G.J.K., Mortimer, B.J., Kirton, A.H. and Duganzich, D.M. (1992) Carcass and meat quality attributes of commercial goats in New Zealand. *Small Ruminant Research* 8, 243–256.

Hutchison, H.G. (1964) *4th Annual Report (1963).* Livestock Research Division, Research Division, Ministry of Agriculture, Tanganyika.

Kirton, A.H. (1970) Body and carcass composition and meat quality of the New Zealand feral goat (*Capra hircus*). *New Zealand Journal of Agricultural Research* 13, 167–181.

Kirton, A.H. (1982) Carcass and meat quality. In: Coop, I.E. (ed.) *Sheep and Goat Production.* Elsevier, Amsterdam, the Netherlands, pp, 259–272.

Liméa, L., Gobardham, J., Gravillon, G., Nepos, A. and Alexandre, G. (2009) Growth and carcass traits of Creole goats under different pre-weaning, fattening and slaughter conditions. *Tropical Animal Health and Production* 41, 61–70.

Mahgoub, O. (1988) Studies in normal and manipulated growth of sheep with special references to skeletal growth. PhD thesis, Lincoln College, University of Canterbury, New Zealand.

Mahgoub, O. (1997) Meat production from the Omani Dhofari goat. 2. Distribution of carcass tissue. *International Journal of Animal Science* 12, 31–38.

Mahgoub, O. and Lodge, G.A. (1994) Growth and development of Omani Local sheep. 2. Growth and distribution of the musculature and skeleton. *Animal Production* 58, 373–379.

Mahgoub, O. and Lodge, G.A. (1996) Growth and body composition in meat production of Omani Batina goats. *Small Ruminant Research* 19, 233–246.

Mahgoub, O. and Lodge, G.A. (1998) A comparative study on growth, body composition and carcass tissue distribution in Omani sheep and goats. *Journal of Agricultural Science* 131, 329–340.

Mahgoub, O. and Lu, C.D. (1998) Growth, body composition and carcass tissue distribution in goats of large and small sizes. *Small Ruminant Research* 27, 267–278.

Mahgoub, O., Kadim, I.T., Al-Saqri, N.M. and Al-Busaidi, R.M. (2004) Effects of body weight and sex on carcass tissue distribution in goats. *Meat Science* 67, 577–585.

Mahgoub, O., Kadim, I.T., Al-Saqri, N.M. and Al-Busaidi, R.M. (2005) Potential of the Omani Jebel Akhdar goat for meat production under feedlot. *Small Ruminant Research* 56, 223–230.

Marichal, A., Castro, N., Capote, J., Zamorano, M.J. and Argüello, A. (2003) Effects of live weight at slaughter (6, 10 and 25 kg) on kid carcass and meat quality. *Livestock Production Science* 83, 247–256.

McGregor, B.A. (1982) Growth of organs and body components of grazing goats. *Proceedings of the Australian Society of Animal Production* 14, 487–490.

Morand-Fehr, P. (1981) Growth. In: Gall, C. (ed.) *Goat Production*. Academic Press, London, UK, pp. 253–283.

Naudé, R.T. and Hofmeyr, H.S. (1981) Meat production. In: Gall, C. (ed.) *Goat Production*. Academic Press, London, pp. 285–307.

Oman, J.S., Waldron, D.F., Griffin, D.B. and Savell, J.W. (1999) Effect of breed-type and feeding regimen on goat carcass traits. *Journal of Animal Science* 77, 3215–3218.

Owen, J.E. and Norman, G.A. (1977) Studies on meat production characteristics of Botswana goats and sheep. II. General body composition, carcass measurements and joint composition. *Meat Science* 1, 283–306.

Peña, F., Perera, J., Garcia, A. and Acero, R. (2007) Effects of weight at slaughter and sex on the carcass characteristics of Florida sucking kids. *Meat Science* 75, 543–550.

Pralomakran, W., Saithanoo, S., Kochapakdee, S. and Norton, B.W. (1995) Effect of genotype and plane of nutrition on carcass characteristics of Thai native and Anglo-Nubian × Thai native male goats. *Small Ruminant Research* 16, 21–25.

Ruvuna, F., Taylor, J.F., Okeyo, M., Wanyoike, M. and Ahuya, C. (1992) Effects of breed and castration on slaughter weight and carcass composition of goats. *Small Ruminant Research* 7, 175–183.

Santos, V.A.C., Silva, A.O., Cardoso, J.V.F., Silvestre, A.J.D., Silva, S.R., Martins, C. and Azevedo, J.M.T. (2007) Genotype and sex effects on carcass and meat quality of suckling kids protected by the PGI 'Cabrito de Barroso'. *Meat Science* 75, 725–736.

Sen, A.R., Santra, A. and Karim, S.A. (2004) Carcass yield, composition and meat quality attributes of sheep and goats under semiarid conditions. *Meat Science* 66, 757–763.

Sumarmono, J., Pratiwi, N.M.W., Murray, P.M. and Taylor, D.G. (2001) Carcass composition of entire and castrated full blood improved Boer bucks. *Proceedings of the Nutrition Society of Australia* 25, S4.

Teixeira, A., Azevedo, J., Delfa, R., Morand-Fehr, P. and Costa, C. (1995) Growth and development of Serrana kids from Montesinho natural park (NE of Portugal). *Small Ruminant Research* 16, 263–269.

Thonney, M.L., Taylor St, C.S., Murray, J.I. and McCleland, T.H. (1987) Breed and sex differences in equally mature sheep and goats. 2. Body components at slaughter. *Animal Production* 45, 261–276.

Tshabalala, P.A., Strydom, P.E., Webb, E.C. and de Kock, H.L. (2003) Meat quality of designated South African indigenous goat and sheep breeds. *Meat Science* 65, 563–670.

Vidyadaran, M.K., Razak, K. and Ganesamurty, P. (1984) Carcass composition and muscle distribution of Kambing Katjang does. *Malaysia Applied Biology* 13, 45–52.

Warmington, B.G. and Kirton, A.H. (1990) Genetic and non-genetic influences on growth and carcass traits of goats. *Small Ruminant Research* 3, 147–165.

Webb, E.C., Casey, N.H. and Simela, L. (2005) Goat meat quality. *Small Ruminant Research* 60, 153–166.

Wilson, P.N. (1960) The effect of plane of nutrition on the growth and development of the East African dwarf goat. III. The effect of plane of nutrition and sex on the carcass composition of the kid at two stages of growth, 16 lb weight and 30 lb weight. *Journal of Agricultural Sciences* 54, 105–130.

Zimerman, M., Domingo, E. and Lanari, M.R. (2008) Carcass characteristics of Neuquén Criollo kids in Patagonia region, Argentina. *Meat Science* 79, 453–457.

# 10 Influences of Diets on Fatty Acid Composition of Edible Tissues of Meat Goat

J.H. Lee and G. Kannan
*Georgia Small Ruminant Research and Extension Center, Fort Valley State University, Georgia, USA*

## 10.1 Abstract

Goat meat (chevon) is attractive to health-conscious consumers because of its lower fat content compared with other traditional red meats. The fatty acid compositions of edible tissues of meat goats are influenced by intrinsic and extrinsic factors such as breed, gender, slaughter weight, age, and quality and quantity of feed. Dietary regime is the major factor that can alter the fatty acid composition of chevon. Goats deposit more internal fat and less intramuscular fat compared with other ruminants. As meat goats are fed with forages, additional protein and energy are required to maintain goat performances. Goats fed pastures have higher concentrations of linoleic (C18:2ω–6), linolenic (C18:3ω-3) and conjugated linoleic acids (C18:2; CLAs) than those fed concentrates. It is suggested that nutritional characteristics of chevon from a human health perspective can be improved through pasture feeding. However, this trend is not found in goats fed different pastures unless pastures have different fatty acid profiles. Because of the ruminal biohydrogenation, it is hard to increase the concentrations of polyunsaturated fatty acids (PUFAs) in edible tissues of goats supplemented within the same major fatty acid groups. Additional research is needed to investigate ways to further increase these nutritionally interesting PUFAs.

## 10.2 Introduction

Consumption of red meat has been criticized because of its high content of saturated fatty acids (SFAs), produced by hydrogenation of dietary unsaturated fatty acids (USFAs) in the rumen (Jenkins, 1994; Simopoulos, 1994; Vanerveen, 1996). Myristic (C14:0) and palmitic (C16:0) acids are hypercholesteraemic and associated with increased incidences of arteriosclerosis and coronary heart diseases (Noakes *et al.*, 1996). Because of the concerns of the medical community and health-conscious consumers, it is necessary to increase unsaturated fat concentration in meat from ruminants and to develop low-fat meat commodities, particularly in developed countries. Efforts to increase the concentration of unsaturated fat in ruminant meat and milk has had limited success as ruminal microorganisms biohydrogenate unsaturated fat during digestion (Gulati *et al.*, 1997; Scollan *et al.*, 2001). Because chevon has a lower fat content and higher USFA concentration compared with beef, lamb and pork, it is an excellent source for the production of low-fat meat products (James and Berry, 1997; McMillin and Brock, 2005). Data on the fatty acid composition of chevon lipids have been limited because of its relatively low commercial importance (Wood *et al.*, 2003; Dubeuf *et al.*, 2004).

Many factors influence red meat quality, and these can be classified into genetic or environmental (Priolo *et al.*, 2001). The fatty acid composition of muscle and adipose tissues can be influenced by breed of the animal, age, sex, quality and quantity of feed consumed, and body weight (Melton, 1990; Banskalieva *et al.*, 2000). Banskalieva *et al.* (2000) reviewed the available literature on fatty acid composition of muscle and fat depots of goats, mainly on genetic factors, including species, whereas the influence of diets on fatty acid profiles of chevon was not completed. In the present review, research on the effects of diets such as pasture and concentrate on fatty acid composition of chevon is emphasized.

## 10.3 Effect of Breed, Sex, Slaughter Weight and Age

Banskalieva *et al.* (2000) reported considerable differences among goat breeds in the contents of SFAs, monounsaturated fatty acids (MUFAs) and PUFAs, as well as individual fatty acids. However, the comparison was focused on dairy goats such as Alpine and Nubian goats in their review. Werdi-Pratiwi *et al.* (2006) conducted an experiment to determine the effects of breed, slaughter weight and castration on fatty acid profiles in longissimus thoracis muscles from Boer and Australian feral goats. These two meat-goat breeds were raised under identical feeding regimes. The authors observed that the fatty acid composition of the longissimus thoracis muscles was significantly affected ($P < 0.05$) by slaughter weight, but only certain fatty acids were affected ($P < 0.05$) by breed and castration. The proportions of fatty acids changed with slaughter weight, which is closely related to changes in age. Compared with the Australian feral goats, the fatty acid content of longissimus thoracis muscle from the Boer goats contained higher ($P < 0.05$) USFAs, MUFAs and PUFAs, which are considered to be desirable fatty acids. The effects of sex and body weight on fatty acid composition of chevon were investigated by other researchers (Johnson *et al.*, 1995; Mahgoub *et al.*, 2002). Mahgoub *et al.* (2002) fed Omani Jebel Akhdar bucks, wethers and doe kids with the same pelleted feed (16% crude protein) plus Rhode grass hay from weaning to slaughter and found that sex had significant effects on the fatty acid profiles of muscle tissues. Males had higher ($P < 0.05$) levels of pentadecanoic (C15:0), linoleic (C18:2ω-6) and linolenic (C18:3ω-3) acids but lower ($P < 0.05$) levels of C16:0, margaric (C17:0) and stearic (C18:0) acids than females. Johnson *et al.* (1995) also reported sex differences in fatty acid composition of chevon. Cooked leg slices from male carcasses had higher ($P < 0.05$) levels of palmitoleic (C16:1ω-7) acid and greater ($P < 0.05$) ratios of PUFAs:SFAs than wethers and female goats. The SFA (C14:0, C15:0 and C16:0) contents were higher ($P < 0.05$) in intact than in castrated kids (Banskalieva *et al.*, 2000).

Mahgoub *et al.* (2002) observed a significant effect of slaughter weight on fat depots of goats. The concentration of decanoic (C10:0), lauric (C12:0) and C14:0 acids of the kidney fat decreased ($P < 0.05$) with increasing slaughter weight. There was a trend of the C16 chain fatty acids decreasing and the C18 chain fatty acids increasing in kidney fat with body weight. The SFAs in kidney fat decreased ($P < 0.05$), whereas USFAs increased ($P < 0.05$) with increasing body weight. Similarly, Dhanda *et al.* (1999) also reported that the concentration of USFAs increased ($P < 0.05$) with increasing age in goats.

## 10.4 Effect of Diet

Compared with monogastric animals, the fatty acid composition of tissues from ruminants is generally less influenced by feeding sources because dietary lipids are hydrolysed to glycerol and fatty acid in the rumen (Jenkins, 1994). The dietary PUFAs are largely hydrogenated to SFAs and *trans*-fatty acids by the ruminal microbes. In ruminants, with a lipid intake containing 66% PUFAs from plant sources, only 4.4% PUFAs were found in duodenal lipid contents (Ward *et al.*, 1964). Furthermore, adding USFAs to lipid supplements for ruminants may cause digestive disturbances because of their antimicrobial effects and inhibition of ruminal fermentation (Jenkins, 1993). Consequently, hydrogenation of dietary fatty acid and the low fat content (2–6%) of ruminant diets are the reasons why the fats in ruminant tissues are highly saturated (Jenkins, 1994). Extensive studies have been conducted on the supplementation of the ruminant diet with a variety of fat sources, including saturated and unsaturated fats, oils and oilseeds, and lipids protected and unprotected from ruminal biohydrogenation (Grummer, 1991; Banskalieva *et al.*, 2000; Wood *et al.*, 2003). In addition, different dietary regimes can also modify the muscle lipid composition of ruminants (Banskalieva *et al.*, 2000; Wood *et al.*, 2003).

Feed cost makes up 50–70% of the total expenditure in livestock production (Wilkinson and Stark, 1987). Meat goats are fed forages to meet most of their nutrient requirements to increase economic returns to the producers; however, pasture-based production systems have a limitation because of the effect of seasonal variation in nutrient contents. This often means that pasture alone does not always provide adequate nutrition for fast-growing animals (Wilkinson and Stark, 1987). Because of this, additional protein and energy (lipid) are offered to maintain acceptable goat performance.

According to Atti *et al.* (2006), in arid and semi-arid regions, spineless cactus (*Opunita ficus indica f. intermis*) is readily available and considered as green forage. Atti *et al.* (2006) reported that intramuscular fat from meat goats on a cactus-supplemented diet contained more C18:2ω-6 (4.03 versus 2.34%) and CLAs (0.32 versus 0.19%) compared with goats on a control diet (barley plus soybean meal). Furthermore, cactus-fed goats showed a higher ($P < 0.05$) proportion of PUFAs (4.54 versus 2.73) and a higher PUFA:SFA ratio than goats in the control group. The results indicated that cactus, as green forage, produces high-quality goat meat in terms of nutritionally important fatty acids.

*Sericea lespedeza* (previously recognised as *Lespedeza cuneata*) has been recognized as a quality forage because of its high concentration of crude protein (Puchala *et al.*, 2005). However, the forage quality of *Sericea lespedeza* is generally considered to be limited by the relatively high concentration of condensed tannins. Lee *et al.* (2008a) fed intact male goats (Kiko × Spanish) with either 75% *Sericea lespedeza* or Bermuda grass (*Cynodon dactylon*) hay and 25% concentrate (Table 10.1). Twenty-two fatty acids were isolated and identified in total lipids of intramuscular fat of longissimus muscle from goats fed the experimental diets (Table 10.2), which consisted of 11 SFAs (C10:0; C12:0; C14:0; C15:0; C16:0, *iso*; C16:0; C17:0; C18:0; C20:0; C21:0 and C22:0), seven MUFAs (C14:1ω-5; C16:1 *trans*; C16:1ω-7; C17:1; C18:1 *trans*; C18:1ω-9 and C20:1ω-9) and four PUFAs (C18:2ω-6; C18:2 CLA; C18:3ω-3 and C20:4ω-6) fatty acids. Three major fatty acids, C16:0, C18:0 and oleic (C18:1ω-9) acids, made up 83.6% of the total lipids in the longissimus muscle of the goats (Table 10.2). No significant differences were found in the concentrations of total SFAs (46.9 or 48.7%), MUFAs (47.3 or 46.8%) and PUFAs (6.0 or 5.0%) in the longissimus muscle lipids from goats fed the *Sericea lespedeza* or Bermuda grass diet. However, the goats from the Bermuda grass group had higher ($P < 0.05$) levels of C17:0 and *trans*-7-hexadecenoic (C16:1 *trans*) acids in longissimus muscle compared with those from the *Sericea lespedeza* group. No significant differences were observed in any of the longissimus muscle PUFAs. Similar results were observed by Priolo *et al.* (2005), who reported that the longissimus muscle fatty acid profile of

**Table 10.1.** Chemical composition of *Sericea lespedeza* (SL), Bermuda grass (BG) hay and concentrate (CON)[a] (from Lee *et al.*, 2008a).

| | Ingredient | | |
|---|---|---|---|
| Item | SL | BG | CON |
| Chemical composition (%) | | | |
| Dry matter | 91.2 | 88.3 | 93.6 |
| Crude protein | 10.2 | 9.3 | 18.0 |
| Ether extract | 1.4 | 1.2 | 7.1 |
| Ash | 3.6 | 4.4 | 6.5 |
| Acid detergent fibre | 34.4 | 26.4 | 4.0 |
| Neutral detergent fibre | 45.1 | 63.3 | 20.0 |
| Fatty acid (%) | | | |
| C8:0 | 0.22 | – | – |
| C12:0 | 0.30 | 0.43 | 0.08 |
| C13:0 | 0.43 | – | – |
| C13:1ω-9 | 2.41 | 1.49 | 0.27 |
| C14:0 | 0.51 | 1.54 | 0.49 |
| C14:1ω-5 | 6.56 | 2.77 | 0.11 |
| C15:0 | – | 0.58 | 0.11 |
| C16:0 | 27.16 | 27.67 | 21.32 |
| C16:1ω-7 | 1.14 | 1.15 | 5.09 |
| C17:0 | 0.41 | 1.36 | 0.16 |
| C18:0 | 5.85 | 11.60 | 4.86 |
| C18:1ω-9 | 6.23 | 12.82 | 33.70 |
| C18:2ω-6 | 17.25 | 13.58 | 25.08 |
| C18:3ω-3 | 15.91 | 11.02 | 3.69 |
| C20:0 | 1.47 | 1.84 | 0.30 |
| C20:1ω-9 | 0.25 | 0.26 | 0.04 |
| C20:5ω-3 | 0.63 | 0.53 | 0.09 |
| C22:0 | 2.01 | 1.88 | 0.10 |
| C22:5ω-3 | 1.27 | 1.47 | 0.16 |

lambs fed sulla (condensed tannins) was not different from that of lambs fed the same diet, supplemented with polyethylene glycol. Lambs fed sulla only had lower ($P < 0.05$) concentrations of C16:1ω-7 but higher ($P < 0.05$) amounts of C18:3ω-3 in the longissimus muscle lipids compared with lambs fed polyethylene glycol supplement (Priolo *et al.*, 2005). However, compared with lambs fed carob pulp (45% as fed basis), lambs supplemented with polyethylene glycol had higher ($P < 0.05$) concentrations of *trans*-vaccenic (C18:1 *trans*-11) and isomer of *cis*-9 *trans*-11 of linoleic acid (CLA) in the longissimus muscle lipids (Vasta *et al.*, 2007). These results support the suggestion that tannins have a strong negative effect on the microorganisms that are responsible for ruminal biohydrogenation (Molan *et al.*, 2001). The CLA isomers and all the *trans*-vaccenic acids originated in the rumen during biohydrogenation, while CLA is also synthesized in tissues by the action of Δ-9 desaturase on *trans*-vaccenic acid (Bauman *et al.*, 2000). However, this trend of increasing *trans*-vaccenic acid and CLA isomers in the longissimus muscle from goats fed the lower amount of condensed tannins was not noticed by Lee *et al.* (2008a). A possible explanation for this is that the diet based on *Sericea lespedeza* (highly condensed tannins) could have negatively impacted the microorganisms responsible for ruminal biohydrogenation. It is also possible that a reduction in ruminal biohydrogenation

**Table 10.2.** Fatty acid composition (weight % of fatty acid methyl esters) of longissimus muscle (intramuscular fat) from goats fed either *Sericea lespedeza* (SL) or Bermuda grass (BG) hay supplemented with concentrate (adapted from Lee *et al.*, 2008a).

| | Diet | | |
|---|---|---|---|
| Fatty acid (%) | SL | BG | SEM |
| C10:0 | 0.10 | 0.10 | 0.027 |
| C12:0 | 0.10 | 0.18 | 0.026 |
| C14:0 | 1.68 | 1.80 | 0.143 |
| C14:1ω-5 | 0.13 | 0.20 | 0.024 |
| C15:0 | 0.43 | 0.50 | 0.047 |
| C16:0 *iso* | 0.94 | 0.82 | 0.123 |
| C16:0 | 22.27 | 22.19 | 0.628 |
| C16:1 *trans* | 0.72[b] | 0.85[a] | 0.023 |
| C16:1ω-7 | 1.76 | 1.68 | 0.142 |
| C17:0 | 1.44[b] | 1.64[a] | 0.042 |
| C17:1 | 1.28 | 1.31 | 0.136 |
| C18:0 | 19.22 | 20.69 | 0.945 |
| C18:1 *trans* | 1.27 | 1.33 | 0.117 |
| C18:1ω-9 | 41.64 | 40.76 | 0.787 |
| C18:2ω-6 | 4.38 | 3.40 | 0.509 |
| C18:2, CLA | 0.18 | 0.18 | 0.009 |
| C18:3ω-3 | 0.26 | 0.20 | 0.069 |
| C20:0 | 0.13 | 0.10 | 0.034 |
| C20:1ω-9 | 0.48 | 0.34 | 0.067 |
| C21:0 | 0.44 | 0.60 | 0.161 |
| C22:0 | 0.12 | 0.12 | 0.019 |
| C20:4ω-6 | 1.18 | 1.21 | 0.220 |

SEM, Standard error of the mean; CLA, conjugated linoleic acid.
Within a row, least squares means that do not have a common letter differ ($P < 0.05$).

induced by the *Sericea lespedeza* might not have been enough to increase the C18:3ω-3 in the longissimus muscle as in the other studies (Priolo *et al.*, 2005; Vasta *et al.*, 2007).

Cashew nut oil, which is rich in C18:1ω-9 (75% of total fatty acids), has been used in the animal diet to improve the performance of goats in Brazil. Santos-Filho *et al.* (2005) investigated the effects of cashew on the fatty acid composition of longissimus dorsi muscle from crossbred goats. No significant difference was found in the concentration of C18:1ω-9 in longissimus dorsi muscles from goats fed cashew nut (13%) plus maize (55.7%) or maize only (63.3%). One reason for this result was that the amount of C18:1 from the supplemented diet was not enough to bypass the biohydrogenation of rumen as reported by Andrews and Lewis (1970).

Lee *et al.* (2008b) investigated the effect of lucerne hay and concentrate on the fatty acid composition of edible tissues of meat goats. Crossbred (Boer × Spanish) intact male goats were fed lucerne (*Medicago sativa*) hay alone (H); an 18% crude protein concentrate diet (C), consisting predominantly of lucerne meal and yellow maize; or a combined diet, consisting of the hay diet for the first 45 days, followed by the concentrate diet (HC) (Table 10.3). The major fatty acids in the longissimus dorsi muscle lipids from H-, C- and HC-fed goats were C16:0, C18:0 and C18:1ω-9, which accounted for 73.1, 75.3 and 71.4% of total

**Table 10.3.** Ingredients and chemical composition of hay and concentrate diets (adapted from Lee *et al.*, 2008b).

| Item | Diet | |
|---|---|---|
| | Hay | Concentrate |
| Ingredient (%) | | |
| Lucerne hay | 100 | – |
| Lucerne meal | – | 50.2 |
| Yellow maize | – | 35.0 |
| Soybean meal (44%) | – | 8.8 |
| TM salt (red salt) | – | 0.50 |
| Vitamin premix | – | 0.50 |
| Poultry fat | – | 5.00 |
| Chemical composition (%) | | |
| Dry matter | 91.7 | 93.6 |
| Crude protein | 17.3 | 18.0 |
| Ether extract | 2.4 | 7.1 |
| Ash | 5.9 | 6.5 |
| Acid detergent fibre | 34.0 | 4.0 |
| Neutral detergent fibre | 45.0 | 29.0 |
| Fatty acid methyl ester (%) | | |
| C12:0 | 0.27 | 0.07 |
| C13:0 | 0.23 | – |
| C13:1ω-9 | 2.70 | 0.27 |
| C14:0 | 0.79 | 0.49 |
| C14:1ω-5 | 0.14 | 0.11 |
| C15:0 | 0.48 | 0.11 |
| C16:0 | 20.97 | 21.32 |
| C16:1ω-7 | 1.67 | 5.09 |
| C17:0 | 0.47 | 0.16 |
| C18:0 | 3.94 | 4.86 |
| C18:1ω-9 | 10.00 | 33.70 |
| C18:2ω-6 | 19.90 | 25.08 |
| C18:3ω-3 | 21.66 | 3.69 |
| C20:0 | 1.16 | 0.30 |
| C20:1ω-9 | 1.05 | 0.04 |
| C20:5ω-3 | 0.45 | 0.00 |
| C22:0 | 1.15 | 0.10 |
| C22:5ω-3 | 0.84 | 0.16 |

fatty acids, respectively (Table 10.4). In general, meat from pasture-fed animals contains a similar proportion of SFAs, a lower concentration of MUFAs and a higher percentage of PUFAs than that from concentrate-fed animals (Webb *et al.*, 2005; Marino *et al.*, 2006). In the study by Lee *et al.* (2008b), the longissimus dorsi muscle lipid from goats fed the C diet contained a lower ($P < 0.05$) concentration of saturated fat (36.4 versus 41.0%) and a higher ($P < 0.05$) concentration of monounsaturated fat (42.8 versus 48.0%) than those from H-fed goats. Rhee *et al.* (2000) also found that intramuscular fat from crossbred (Boer × Spanish) goats grazed on pasture (grasses, browses and forb) without any grain supplementation was more saturated than that from goats fed a grain diet (sorghum grain and cottonseed hulls). Furthermore, compared with goats fed the grain diet, goats grazed on pasture had increased ($P < 0.05$) proportions of C18:2ω-6, C18:3ω-3 and eicosatrienoic (C20:3ω-6) acids

**Table 10.4.** Fatty acid composition (%) of intramuscular fat of goats fed three different diets (adapted from Lee *et al.*, 2008b).

| Fatty acid | Diet[a] H | C | HC | SEM |
|---|---|---|---|---|
| C10:0 | 0.17 | 0.17 | 0.15 | 0.04 |
| C12:0 | 0.17 | 0.14 | 0.14 | 0.03 |
| C13:1ω-9 | 0.09 | 0.06 | 0.06 | 0.01 |
| C14:0 | 2.16 | 2.00 | 1.77 | 0.15 |
| C14:1ω-5 | 0.52 | 0.29 | 0.42 | 0.11 |
| C15:0 | 0.70a | 0.47b | 0.51b | 0.06 |
| C16:0 | 21.87 | 21.40 | 20.60 | 0.68 |
| C16:1ω-7 | 3.38 | 3.69 | 3.24 | 0.27 |
| C17:0 | 1.75a | 1.35b | 1.69a | 0.10 |
| C18:0 | 12.49 | 10.01 | 11.69 | 1.33 |
| C18:1ω-9 | 38.73b | 43.85a | 39.12b | 1.45 |
| C18:2ω-6 | 6.93 | 7.92 | 7.07 | 0.61 |
| C18:3ω-3 | 0.46a | 0.12b | 0.19b | 0.15 |
| C20:0 | 0.07 | 0.05 | 0.08 | 0.02 |
| C20:1ω-9 | 0.12 | 0.11 | 0.16 | 0.05 |
| C20:4ω-6 | 4.65 | 3.46 | 4.19 | 1.16 |
| C20:5ω-3 | 0.45 | 0.38 | 0.59 | 0.14 |
| C22:0 | 1.57 | 0.83 | 1.90 | 0.62 |
| C22:5ω-3 | 0.37 | 0.37 | 0.63 | 0.26 |

SEM, Standard error of the mean.
Within a row, least squares means that do not have a common letter (a, b) differ ($P < 0.05$).
[a]H, Hay for 45 days; C, concentrate for 45 days; HC, hay for 45 days plus concentrate for 45 days.

but decreased ($P < 0.05$) contents of C18:1ω-9 in intramuscular fat. Lee *et al.* (2008b) also reported that among the SFAs, goats fed the H diet had higher ($P < 0.05$) percentages of C15:0 and C17:0 in longissimus dorsi muscle lipids than those fed the C or HC diet (Table 10.4). However, no significant difference was found in C17:0 between goats fed the H and HC diets. SFAs such as C12:0, C14:0 and C16:0 raise the low-density lipoprotein (LDL)-cholesterol concentrations in blood, increasing the risk of cardiovascular diseases (Noakes *et al.*, 1996). These three LDL-cholesterol-increasing fatty acids made up 24.2, 23.5 and 22.8% of total fatty acids in the longissimus dorsi from goats fed the H, C and HC diets, respectively (Table 10.4). Of the MUFAs, the mean concentration of C18:1ω-9 of the longissimus dorsi muscle lipids of goats that consumed the C diet was higher ($P < 0.05$) than that of goats fed either the H or HC diet. No significant difference was detected in the C18:1ω-9 concentration of the longissimus dorsi muscle lipid from goats fed the H or HC diet. Goats fed the H diet had higher ($P < 0.05$) levels of C18:3ω-3 in longissimus dorsi muscle lipids than goats fed either the C or HC diet; however, there were no differences in the concentrations of C18:3ω-3 in the longissimus dorsi muscle lipids between the goats fed the C and HC diets. Current recommendations are that the PUFA:SFA ratio should be around 0.45 (Webb *et al.*, 2005). The PUFA:SFA ratios noticed by Lee *et al.* (2008b) shown in Table 10.4 were lower than the recommended ratio. Duckett *et al.* (1993) reported a higher PUFA:SFA ratio (0.26) for beef from grass-finished steers than for meat from concentrate-finished animals (0.07).

Pasture-fed goats had higher proportions of C18:2ω-6 and CLA (Rhee *et al.*, 2000; Atti *et al.*, 2006) or C18:3ω-3 (Rhee *et al.*, 2000; Lee *et al.*, 2008b) than concentrate-fed goats. These results confirm previous reports that finishing ruminants on pasture enhanced the PUFA profile of intramuscular fat, including CLA and ω-3 fatty acids. Results from these studies (Rhee *et al.*, 2000; Atti *et al.*, 2006; Lee *et al.*, 2008b) suggest that fatty acid composition of goat meat can be improved from a human health perspective through pasture feeding. However, the tendency of increasing USFAs and CLA isomers in chevon was not found in goats fed different grass hays (Lee *et al.*, 2008a). Furthermore, this trend also applied to goats fed different concentrate diets (Santos-Filho *et al.*, 2005). These differences are consequences of fatty acid composition of feed sources, C18:3ω-3 being the major fatty acid in plant materials, C18:2ω-6 the major fatty acid in grains and C18:1ω-9 the major fatty acid in nuts (Weiss, 1983). However, when these USFA sources are fed to ruminants, they undergo hydrogenation and degradation in the rumen (Jenkins, 1994). Consequently, a relatively low proportion of dietary USFAs is deposited in muscle tissues of ruminants. Because of the ruminal biohydrogenation, it is hard to increase the concentrations of PUFAs in edible tissue of goats supplemented with the same major fatty acid groups.

## 10.5 Conclusion

Many studies have been conducted to evaluate the effects of intrinsic factors (breed, gender, slaughter weight and age) on fatty acid composition of muscle and adipose tissues in goats. However, relatively few studies investigated the influence of dietary regime on fatty acid profiles of edible tissues of meat goats. Meat goats are fed most of their required nutrients from forages to increase the profit to producers; however, pasture-based production systems have a limitation because of the effect of seasonal variation on the nutrient contents. Therefore, additional protein and energy are offered to maintain the performance of goats at acceptable levels. In general, pasture-fed goats have higher proportions of linoleic acid, α-linolenic acid and CLAs than concentrate-fed goats. From a human dietary health viewpoint, it is suggested that the fatty acid composition of goat meat can be improved through pasture feeding. However, the tendency of increasing unsaturated fatty acid and CLA isomers in meat was not noticed in goats fed different pasture diets. Differences are the consequences of the fatty acid compositions of feed sources. In addition, a relatively low proportion of dietary unsaturated fatty acid is deposited in muscle tissues of meat goats because of ruminal biohydrogenation of dietary unsaturated fatty acids. However, feeding rumen-protective dietary polyunsaturated fat to ruminants could drastically increase these beneficial fatty acids in edible tissue. Several studies have been conducted to increase nutritionally important PUFAs in ruminant meats such as beef and lamb by feeding rumen-protective dietary supplements. However, these supplements have not yet been tested extensively in meat goats.

## References


Andrews, R.J. and Lewis, D. (1970) The utilization of dietary fats by ruminants. Part II. The effect of fatty acids chain length and unsaturation on digestibility. *Journal of Agricultural Science* 75, 55–60.

Atti, N., Mahouachi, M. and Rouissi, H. (2006) The effect of spineless cactus (*Opunita ficus-indica f. inermis*) supplementation on growth, carcass, meat quality and fatty acid composition of male goat kids. *Meat Science* 73, 229–235.

Banskalieva, V., Sahlu, K. and Goetsch, A.L. (2000) Fatty acid composition of goat muscles and fat depots: a review. *Small Ruminant Research* 37, 255–268.

Bauman, D.E., Baumgard, L.H., Corl, B.A. and Griinari, J.M. (2000) Biosynthesis of conjugated linoleic acid in ruminants. In: *Proceedings of the American Society of Animal Science 1999*, Indianapolis, USA.

Dhanda, J.S., Taylor, D.G., Murray, P.L. and McCosker, J.E. (1999) The influence of goat genotype on the production of Capretto and Chevon carcasses. 4. Chemical composition of muscle and fatty acid profiles of adipose tissue. *Meat Science* 52, 375–379.

Dubeuf, J.P., Morand-Fehr, P. and Rubino, R. (2004) Situation, changes and future of goat industry around the world. *Small Ruminant Research* 51, 165–173.

Duckett, S.K., Wagner, D.G., Yates, L.D., Dolezal, H.G. and May, S.G. (1993) Effect of time on feed on beef nutrition in beef nutrient composition. *Journal of Animal Science* 71, 2079–2088.

Grummer, R.R. (1991) Effect of feed on the composition of milk fat. *Journal of Dairy Science* 74, 3244–3257.

Gulati, S.K., Byers, E.B., Byers, Y.G., Ashes, J.R. and Scott, T.W. (1997) Effect of feeding different fat supplements on the fatty acid composition of goat milk. *Animal Feed Science and Technology* 66, 159–164.

James, N.A. and Berry, B.W. (1997) Use of chevon in the development of low-fat meat products. *Journal of Animal Science* 75, 571–577.

Jenkins, T.C. (1993) Lipid metabolism in the rumen. *Journal of Dairy Science* 76, 3851–3863.

Jenkins, T.C. (1994) Regulation of lipid metabolism in the rumen. *Journal of Nutrition* 124, S1372–S1376.

Johnson, D.D., Eastridge, J.S., Neubauer, D.R. and McGowan, C.H. (1995) Effect of sex class on nutrient content of meat from young goats. *Journal of Animal Science* 73, 296–301.

Lee, J.H., Kannan, G., Kouakou, B., Moore, D.A. and Terrill, T.H. (2008a) Influence of dietary condensed tannins in meat goats on fatty acid composition of carcasses. *Journal of Animal Science* 86 (Suppl. 2), 97.

Lee, J.H., Kouakou, B. and Kannan, G. (2008b) Chemical composition and quality characteristics of chevon from goats fed three different post-weaning diets. *Small Ruminant Research* 75, 177–184.

Mahgoub, O., Khan, A.J., Al-Maqbaly, R.S., Al-Sabahi, J.N., Annamalai, K. and Al-Sakry, N.M. (2002) Fatty acid composition of muscle and fat tissues of Omani Jebel Akhdar goats of different sexes and weights. *Meat Science* 61, 381–387.

Marino, R., Albenzio, M., Girolami, A., Muscio, A., Sevi, A. and Braghieri, A. (2006) Effect of forage to concentrate ratio on growth performance, and on carcass and meat quality of Podolian young bulls. *Meat Science* 72, 415–424.

McMillin, K.W. and Brock, A.P. (2005) Production practices and processing for value added goat meat. *Journal of Animal Science* 83, E57–E68.

Melton, S.L. (1990) Effects of feeds on flavor of red meat: a review. *Journal of Animal Science* 68, 4421–4435.

Molan, A.L., Attwood, G.T., Min, B.R. and McNabb, W.C. (2001) The effect of condensed tannins from *Lotus pedunculatus* and *Lotus corniculatus* on the growth of proteolytic rumen bacteria in vitro and their possible mode of action. *Canadian Journal of Microbiology* 47, 626–633.

Noakes, M.N., Nestle, P.J. and Clifton, T.M. (1996) Modifying the fatty acids profile of dairy products through feedlot technology lowers plasma cholesterol of humans consuming the products. *American Journal of Clinical Nutrition* 63, 42–46.

Priolo, A., Micol, D. and Agabriel, J. (2001) Effects of grass feeding systems on ruminant meat colour and flavor: a review. *Animal Research* 50, 185–200.

Priolo, A., Bella, M., Lanza, M., Galofaro, V., Biondi, L., Barbagallo, D., Ben Salem, H. and Pennisi, P. (2005) Carcass and meat quality of lambs fed fresh sulla (*Hedysarum coronarium* L.) with or without polyethylene glycol or concentrate. *Small Ruminant Research* 59, 281–288.

Puchala, R., Min, B.R., Goetsch, A.L. and Sahlu, T. (2005) The effect of a condensed tannin-containing forage on methane emission by goats. *Journal of Animal Science* 83, 182–186.

Rhee, K.S., Waldron, D.F., Ziprin, Y.A. and Rhee, K.C. (2000) Fatty acid composition of goat diets vs intramuscular fat. *Meat Science* 54, 313–318.

Santos-Filho, J.M., Morais, S.M., Rondina, D., Beserra, F.J., Neiva, J.N.M. and Magalhães, E.F. (2005) Effect of cashew nut supplemented diet, castration, and time of storage on fatty acid composition and cholesterol content of goat meat. *Small Ruminant Research* 57, 51–56.

Scollan, N.D., Dhanoa, M.S., Choi, N.J., Maeng, W.J., Enser, M. and Wood, J.D. (2001) Biohydrogenation and digestion of long chain fatty acids in steers fed on different source of lipids. *Journal of Agricultural Science* 136, 345–355.

Simopoulos, A.S. (1994) Fatty acids. In: Goldberg, I. (ed.) *Functional Foods*. Chapman & Hall, New York.

Vanerveen, J.E. (1996) Dietary recommendation for lipids and measures designed to facilitate implementation. In: McDonald, E.R. and Min, D.B. (ed.) *Food Lipids and Health.* Marcel Dekker, New York.

Vasta, V., Pennisi, P., Lanza, M., Barbagallo, D., Bella, M. and Priolo, A. (2007) Intramuscular fatty acid composition of lambs given a tanniniferous diet with or without polyethylene glycol supplementation. *Meat Science* 76, 739–745.

Ward, P.F.V., Scott, T.W. and Dawson, R.M.C. (1964) The hydrogenation of unsaturated fatty acids in the ovine digestive tract. *Biochemistry Journal* 92, 60–68.

Webb, E.C., Casey, N.H. and Simela, L. (2005) Goat meat quality. *Small Ruminant Research* 60, 153–166.

Weiss, T.J. (1983). *Food Oils and Their Uses*, 2nd edn. AVI Publishing, Westport, Connecticut.

Werdi-Pratiwi, N.W., Murray, P.J., Taylor, D.G. and Zhang, D. (2006) Comparison of breed, slaughter weight and castration on fatty acid profiles in longissimus thoracic muscle from male Boer and Australian feral goats. *Small Ruminant Research* 64, 94–100.

Wilkinson, J.M. and Stark, B.A. (1987) *Commercial Goat Production.* BSP Professional Books, London, UK.

Wood, J.D., Richardson, R.I., Nute, G.R., Fisher, A.V., Campo, M.M., Kasapidou, E., Sheard, P.R. and Enser, M. (2003) Effect of fatty acids on meat quality: a review. *Meat Science* 66, 21–32.

# 11 Mineral Composition of Goat Meat

N.H.I. Osman[1] and O. Mahgoub[2]
[1]Open University of Sudan, Obeid Khatim Street, Arkaweet, Khartoum, Sudan;
[2]Department of Animal & Veterinary Sciences, College of Agricultural & Marine Sciences, Sultan Qaboos University, Sultanate of Oman

## 11.1 Abstract

In this chapter, the literature on the mineral content of goat meat is reviewed. Goat meat is a rich source of various minerals. Chemical ash, which comprises ~3.5% of total body weight, represents the inorganic material and is composed mainly of minerals. The ash content of goat meat is influenced by several factors including breed, age, sex, litter size and cooking method. There are about 90 natural elements in the body, of which only about 40 are essential. Some of the elements, the trace elements, are present in the animal body in small proportions (≤50 mg/kg of dry matter). Others, which are found in larger quantities, are known as macroelements and comprise the major elements of ash. The most common macroelements in the body include calcium, phosphorus, potassium, sodium and magnesium. The major trace elements are iron, copper, zinc, selenium and manganese, plus several others (cobalt, cadmium, lead, nickel and vanadium). Factors that affect the mineral concentration in animal tissues include species, type of tissue, muscles, sex, age, breed, diet and cooking method. There are significant correlations between various minerals in goat meat.

## 11.2 Introduction

The animal body is made of organic and inorganic materials. Inorganic materials or minerals are required by all living organisms to ensure their normal functioning. The animal body is not equipped to synthesize minerals as it is with many other organic nutrients. Therefore, minerals need to be supplied in animal feeds. Chemical ash, which comprises ~3.5% of total body weight, represents the inorganic material and is composed mainly of minerals (Keeton and Eddy, 2004). There are about 90 natural elements in the body (Chesworth, 1992), of which only about 40 are essential (McDonald *et al.*, 2002). The term 'essential mineral elements' refers to minerals that have a metabolic role in mammalian tissue. Some of the elements, the trace elements or microminerals, are present in the animal body at low concentrations [≤50 mg/kg dry matter (DM)]. Others, which are found in larger quantities, are known as macrominerals or macroelements and comprise the major elements of ash. Research has shown that there is a highly complex interrelationship between the various mineral elements in the body. This affects the nutritional requirements of humans and animals.

 *Goat Meat Production and Quality* (eds O. Mahgoub, I.T. Kadim and E.C. Webb)

Minerals are an integral part of various organs and tissues of the animal body. These tissues contain a variety of elements such as calcium, magnesium, potassium, sodium, silicon, phosphorus, chromium, manganese, iron, nickel, copper, zinc, cobalt, cadmium, lead, vanadium and molybdenum. Macrominerals include calcium, magnesium, sodium and potassium as the principal cations and phosphorus, chlorine and sulfur as the principal anions. The microminerals include cobalt, copper, iodine, iron, manganese, molybdenum, selenium and zinc (Campbell *et al.*, 2003). Microminerals of significant nutritional importance in meat and meat products are iron, zinc and magnesium, and these are found in quantities of milligrams per 100 g DM, while selenium is found in microgram quantities per 100 g DM (McDonald *et al.*, 2002; Keeton and Eddy, 2004; Leth and Ertbjerg, 2004).

The mineral content in meat is measured using standard methods of analysis such as the atomic spectrometric techniques of atomic absorption spectrometry and inductively coupled plasma atomic emission or mass spectrometry. The first step in analysis is ashing in a furnace overnight at 450°C after drying in an oven, or digestion of samples by wet digestion, which is most commonly used. Microwave digestion, which is faster, is carried out using nitric acid and hydrogen peroxide as oxidants in sealed containers of an inert material (Leth and Ertbjerg, 2004).

Humans obtain their body mineral requirements from plant and animal sources. Meat, a major component of the human diet, is a rich source of both macro- and microelements that are essential for human health. Minerals are also found in various concentrations in other animal body organs including the liver, kidney, brain, heart and spleen (Wan Zahari and Abdul Wahid, 1985). Minerals of animal origin are more readily available for humans than those of non-animal origin (plants) (Mioč *et al.*, 2000). The mineral content in meat is affected by many factors such as animal species, breed, climate, mineral content of the diet and type of tissue. Within species and breed types, the concentration of minerals in muscles is affected by their position in the body and their function (Mioč *et al.*, 2000).

Mineral and trace element concentrations in animal muscles can affect meat quality. For instance, Satterlee *et al.* (1977) reported that the ratio of iron:zinc in beef was highly correlated with muscle tenderness. A low concentration of zinc in skeletal muscles prevents collagen cross-linking, which in turn influences muscle tenderness (Seideman *et al.*, 1984).

Goat meat, similar to other red meats, is a rich source of minerals. The muscles of the goat contain a variety of minerals and trace elements including calcium, magnesium, potassium, sodium, silicon, phosphorus, chromium, manganese, iron, nickel, copper, zinc, cobalt, cadmium, lead, vanadium and molybdenum, but the proportions of these elements vary greatly. The minerals that are most abundant in goat meat are potassium, phosphorus and magnesium (Sheridan *et al.*, 2003), which is similar to the situation in lamb meat (Hoffman *et al.*, 2003).

This chapter reviews the nutritive role and contents of the mineral component of goat meat and the factors affecting them.

## 11.3 Ash Content in Goat Meat

As meat is burned, most elements in the organic material (carbon, nitrogen, hydrogen and oxygen) will be lost in the form of carbon dioxide, water and dioxides of nitrogen (Keeton and Eddy, 2004). The remaining part comprises the minerals. It is usually analysed by burning in a muffle furnace at high temperatures and is expressed as percentage ash. Therefore, ash is an estimate of the total mineral content in the body that is in the form of oxides, sulfates, phosphates, nitrates, chlorides and other halides (Chesworth, 1992). The ash in lean meat is an estimate of total mineral content that makes up the cellular constituents (myoglobin, haemoglobin and enzymes) or together with bone (bone fragments, mechanically separated tissue and advanced meat recovery systems) in the whole minced carcass.

The ash content reported for fresh goat meat ranges between 0.8 and 1.7% (Table 11.1). The factors that cause the wide variation of ash content found in goat meat include breed (Beserra *et al.*, 2004), sex and castration (Madruga *et al.*, 1999), dietary supplementation (Solaiman *et al.*, 2006) and treatment of the meat, including the cooking method (Dzudie *et al.*, 2000). The ash content in cooked goat meat is at the higher end of the range of fresh meat values (Johnson *et al.*, 1995). Reports on ash content in goats using whole minced carcasses suggest a higher ash content (Hogg *et al.*, 1989). On a DM basis, goat carcasses contain a lower fat and ash content but higher protein content than sheep (Mahgoub and Lu, 2004).

### 11.3.1 Factors affecting ash contents in goat meat

*Breed*

There is a large body of literature on ash content in goat meat, but the values reported vary widely, indicating a clear effect of breed. Goats of the Moxotó Brazilian breed had a significantly higher ash content of 1.2 ± 0.02% (mean ± SEM) at 4–6 months and 2.7 ± 0.23% at 8–10 months of age than crosses of (Moxoto × Pardo Alpina) × Anglo-Nubian castrated male goats with 1.1 ± 0.01 and 2.3 ± 0.12%, respectively (Beserra *et al.*, 2004). Similar values of 1.05 ± 0.04 and 1.00 ± 0.07% were reported by Madruga *et al.* (1999) in castrated male 'Mestico' goats (breeds of 'Creole' × Anglo-Nubian, Saanen or British Alpine crossbred) at 175 and 220 days, respectively. Proportions of 1.06 and 1.09% in groups of young and old

**Table 11.1.** Ash content in the meat of goats of various breeds, sexes and ages reported by various authors.

| Ash content (%) | Breed | Comment | Reference |
|---|---|---|---|
| Uncooked meat | | | |
| 1.06–1.09 | Unspecified breed | | Turgut (1984) |
| 0.8 | Spanish goat | Chevon chops from 1-year-old | James and Berry (1997) |
| 1.00–1.05 | Mestiço | Wethers (175 and 220 days) | Madruga *et al.* (1999) |
| 1.01–1.06 | Saanen, Boer, Feral and Angora crosses | Capretto and chevon, males and wethers | Dhanda *et al.* (2003) |
| 0.9 | French goat | | Paleari *et al.* (2003) |
| 1.2–2.7 | Moxotó (Brazil) | Age 4–10 months | Beserra *et al.*, 2004 |
| 1.13 | White improved | | Niedziolka *et al.* (2005) |
| 1.05 | Lowland Polish | | Niedziolka *et al.* (2005) |
| 1.32–1.62 | Boer × Spanish | Intact males, age 4 months | Lee *et al.* (2008) |
| 1.2–1.63 | Pakistani mixed | Age <7 months to >11 months | Arain *et al.* (2010) |
| Cooked meat | | | |
| 1.4 | Florida native or F1 crosses of Florida natives with Nubian or Spanish goats | Intact male cooked meat | Johnson *et al.* (1995) |
| 1.3 | | Castrate | Johnson *et al.* (1995) |
| 1.2 | | Female | Johnson *et al.* (1995) |
| 0.97–1.07 | Angora | Longissimus muscle | Schonfeldt (1989) |
| 1.00–1.08 | Boer | Longissimus muscle | Schonfeldt (1989) |
| 5.3 | Feral yearling bucks | Whole minced half carcass | Hogg *et al.* (1989) |
| 5.2 | Saanen males | Whole minced half carcass | Hogg *et al.* (1989) |
| 5.4 | Saanen females | Whole minced half carcass | Hogg *et al.* (1989) |

goats, respectively, were reported by Turgut (1984) with an unspecified breed, which were comparable to those in other types of meat-producing animals in the same report, including cattle (1.05 and 1.06%), sheep (1.08 and 1.05%) and water buffalo (1.12 and 1.07%). Compared with lambs of a similar age (150 days), the ash content in meat from kids of White Improved goat breeds was 1.13%, and was 1.05% in meat from a lowland Polish breed (Niedziolka *et al.*, 2005). A higher percentage of 1.32–1.62% (± 0.15%) was reported in weaned, 4-month-old, intact Boer × Spanish male goats by Lee *et al.* (2008). A lower percentage of 0.9 ± 0.1%, however, was reported in French goat meat (Paleari *et al.*, 2003), which was also lower than other species in the same study, including deer (1.3 ± 0.04%), boar (1.1 ± 0.05%), horse (1.3 ± 0.09%) and cattle (1.3 ± 0.21%). A very low content of 0.8% in raw chevon chops from 1-year-old Spanish goats was reported by James and Berry (1997). The ash content in whole minced half carcasses of feral yearling bucks and male and female Saanen goats was 5.3, 5.2 and 5.4%, respectively (Hogg *et al.*, 1989). These values were higher than the range reported for goat meat because the minced whole half carcass includes bones. There were no breed effects on the ash contents of whole carcasses of Omani Batina and Dhofari goat breeds (Mahgoub *et al.*, 2005).

### Age

Reports on the effect of age on ash content are inconsistent. With increasing age and body weight, water, crude protein and ash concentrations were shown to decrease, whereas that of fat increased in carcasses of both Omani goats and sheep (Mahgoub and Lu, 2004). A value of 0.99% ash in the meat of goats at 175 days decreased to 0.9% at 220 days (7 months) and further to 0.88% at 265 days (8.5 months) but increased again to 0.97% at 310 days (10 months) (Madruga *et al.*, 1999). Differences between age groups per se, however, were not significant, although there was a significant age × sex interaction (Madruga *et al.*, 1999). Beserra *et al.* (2004), using different goat breed crosses, reported values of 1.2 ± 0.02 and 1.1 ± 0.01% at 4–6 months of age. The ash content was higher (2.7 ± 0.23 and 2.3 ± 0.12%) in animals of 8–10 months of age. Although the differences were not significant, there was a consistent trend similar to that of a previous report of Beserra *et al.* (2000). In contrast, Turgut (1984) reported values of 1.06 and 1.09% in two groups of goats of different ages. The ash content of Pakistani goat meat (<7 months to >11 months) increased linearly with age (Arain *et al.*, 2010).

### Sex

The effects of sex, i.e. male, female or castrated male, on ash content in meat have been contradictory. Meat from intact males at 175 and 220 days had a lower ash content (0.92 ± 0.10 and 0.81 ± 0.08%, respectively) than castrated male goats (1.05 ± 0.04 and 1.00 ± 0.07%, respectively) (Madruga *et al.*, 1999). At higher ages of 265 and 310 days, the ash content showed the opposite trend and was higher in castrates than in intact males (0.79 ± 0.22 and 0.92 ± 0.06% versus 0.96 ± 0.08 and 0.98 ± 0.08%, respectively) (Madruga *et al.*, 1999). The ash content of leg meat in male Saanen goats was 1.1 ± 0.1%, which was higher than in meat from females (0.9 ± 0.1%), although the difference was not significant (Hogg *et al.*, 1989). The ash percentage in cooked meat from male goats was also slightly higher but not significantly different between castrated and male and female goats (1.4, 1.3 and 1.2 g/100 g DM, respectively) (Johnson *et al.*, 1995). However, sex did not significantly affect the ash content of fat-free half carcasses in 4-month-old kids (4.4 and 4.6% in males and females, respectively) (Ash and Norton, 1987).

### Litter size

There are some reports indicating that litter size may also affect the ash content of goat meat. Meat from twin kids contained a significantly higher percentage of ash (5.29 ± 0.03%) than meat from a single-born kid (5.13 ± 0.03%) (Todaro *et al.*, 2006).

This may be attributed to the fact that single kids are usually fatter than twins.

*Cooking*

Cooking per se and the cooking method used significantly affect the ash content in goat meat. A significant effect of cooking was reported in loin sections of grassland African dwarf goats. Roast chops had the highest ash content of 6.76 ± 0.35%, followed by broiled chops at 5.18 ± 0.35%, while the lowest content was from water-bathed chops at 4.15 ± 0.31% (Dzudie *et al.*, 2000). Wet cooking such as boiling caused more water loss and consequently a higher mineral loss than dry cooking methods such as roasting. Comparing microwave cooking with broiling, James and Berry (1997) found no significant difference between cooking with a microwave at different powers and broiling. Microwaving at 100% power (1.1% ash) or 60% power (1.0% ash) produced a slightly higher mineral content than broiling (0.09% ash). However, all values were significantly higher than the ash content (0.8% ash) in raw meat. Differences in ash content from different methods of cooking are inversely proportional to the loss of moisture contents, except in water bathing, where minerals can be lost in the surrounding water.

*Type of diet*

Reports on effects of diet on goat meat mineral content are conflicting. There was no effect of energy level in the diet on mineral content in Boer goats (Sheridan *et al.*, 2003). However, Mahgoub and Lu (2004) reported a trend of increasing carcass fat content and decreasing protein and ash content with increasing levels of metabolizable energy in the diet causing increasing body weight in Omani goats and sheep.

The ash content in longissimus muscles of 4-month-old intact Boer × Spanish male goats was not affected by feeding kids either hay, concentrate, or hay plus concentrate for 45 days (Lee *et al.*, 2008). However, inclusion of certain minerals in the diet may influence mineral levels in the body. For instance, with an increase in the dietary copper supplementation to 100 and 200 mg/day, the ash content in the 9th to 11th ribs tended to increase linearly from 0.73 to 0.81 and 0.80%, respectively (Solaiman *et al.*, 2006). Dietary supplementation with 15 or 25% neem cake significantly increased the ash content from 1.10 ± 0.06 to 1.60 ± 0.11 and 1.40 ± 0.13%, respectively (Kesava *et al.*, 2003).

## 11.4 Mineral Content of the Goat Body

As with other red-meat animals, goat meat is a valuable source of minerals, especially in regions of the world where the goat is a major meat animal such as Asia and Africa. Goat meat contains different levels of various macro- and microelements with different concentrations in body organs such as liver, kidney, brain, muscle, spleen and heart.

McDonald *et al.* (2002) and Keeton and Eddy (2004) have reported the essential elements and their approximate concentrations in the whole body of farm animals and in muscles (Table 11.2). These included

**Table 11.2.** Nutritionally important minerals and their approximate concentration in the whole body and muscle of farm animals.

| Major minerals | Whole animal (g/kg DM)[a] | Muscle (mg/g DM)[b] |
|---|---|---|
| Calcium | 15 | 3–6 |
| Phosphorus | 10 | 167–216 |
| Potassium | 2 | 250–400 |
| Sodium | 1.6 | 55–94 |
| Chlorine | 1.1 | 0.65 |
| Sulfur | 1.6 | 2.5 |
| Magnesium | 0.4 | 22–29 |
| **Trace elements** | **Whole animal (mg/kg DM)[a]** | **Muscle (mg/g DM)[b]** |
| Iron | 20–80 | 1–3 |
| Zinc | 10–50 | 1–5 |
| Copper | 1–5 | 0.5–0.13 |
| Molybdenum | 1–4 | |
| Selenium | 1–2 | |
| Iodine | 0.3–0.6 | |
| Manganese | 0.2–0.5 | |
| Cobalt | 0.02–0.1 | |

[a]McDonald *et al.* (2002); [b]Keeton and Eddy (2004).

seven macroelements and eight microelements. The macromineral contents ranged from 0.4 g/kg DM for magnesium to 15 g/kg DM for calcium. The highest concentrations of macroelements in muscles were potassium (250–400 mg/g DM) and phosphorus (167–216 mg/g DM). The microelements with the highest concentrations in the whole animal body and muscle were iron and zinc, and the lowest was cobalt.

Table 11.3 gives the mean mineral concentrations in muscle and selected organs of goats in a study carried out by Wan Zahari and Abdul Wahid (1985). The mineral contents on a DM basis were highest in the liver, followed by muscle, brain, heart, spleen and kidney. Organs varied in their concentrations of individual minerals. For instance, the brain had a high concentration of many minerals including calcium, phosphorus, potassium, sodium and manganese. The liver contained high levels of phosphorus, magnesium, zinc and extremely high levels of copper. The spleen had the highest content of iron, which was many times higher than that of other organs. The kidneys had higher levels of most minerals, especially sodium. This renders the muscle and edible organs of the goat an excellent source of minerals for humans, especially in underprivileged parts of the world where most of the goat body is consumed. However, it should also be noted that extremely high concentrations of certain elements such as heavy metals (lead, mercury and cadmium) in meat are regarded as a potential human health hazard.

Macrominerals are more abundant than microminerals in goat meat. Sheridan *et al.* (2003) reported that the elements that contribute the largest proportions in goat meat are calcium, phosphorus, potassium and sodium (946, 653, 142 and 57 mg/100 g DM, respectively). Mahgoub (unpublished data) found that the most abundant minerals in goat meat are potassium, phosphorus, sodium, magnesium and calcium (757, 530, 108, 49 and 13 mg/100 g DM). It should be noted that reports from various studies vary greatly in values of minerals and trace elements. This may be attributed to a number of factors including sample collection, method of extraction and method of analysis, or to how results are expressed (e.g. wet weight or DM basis). For instance, muscle samples may include fat or connective tissue. There are several methods employed for extraction of minerals from meat samples using acids such as nitric acid. However, microwave extraction is the most effective. Atomic absorption is utilized more than inductive coupled plasma atomic emission spectrophotometry as it is more readily available in laboratories.

### 11.4.1 Macrominerals in goat meat

#### *Calcium*

Calcium contributes significantly to the hardness of bones and teeth. It is also important for bone development, neuromuscular activity, secretory functions, buffers, certain

**Table 11.3.** Mean mineral concentrations (mg/100 g dry matter) in muscle and selected organs of crossbred goats (from Wan Zahari and Abdul Wahid, 1985).

| Mineral | Muscle | Liver | Kidney | Heart | Spleen | Brain |
|---|---|---|---|---|---|---|
| Calcium | 11 | 10.06 | 13.58 | 7.7 | 11.47 | 46.99 |
| Phosphorus | 155.5 | 253.9 | 168.1 | 111.71 | 214.03 | 245.64 |
| Magnesium | 19.7 | 15.08 | 10.19 | 9.63 | 15.28 | 12.82 |
| Potassium | 350 | 188.55 | 122.26 | 100.15 | 194.9 | 277.68 |
| Sodium | 64.48 | 58.18 | 148.68 | 38.52 | 59.38 | 136.92 |
| Copper | 0.30 | 8.28 | 0.52 | 0.53 | 0.41 | 0.40 |
| Zinc | 3.51 | 2.99 | 2.61 | 1.41 | 2.19 | 1.40 |
| Iron | 4.37 | 7.82 | 9.78 | 4.40 | 34.79 | 3.07 |
| Manganese | 0.087 | 0.66 | 0.19 | 0.098 | 0.159 | 0.122 |
| Dry matter (%) | 21.90 | 25.14 | 16.98 | 19.26 | 19.11 | 21.36 |

coenzymes and nutrients for the nursed young (Casey, 1992), and has an essential role in blood clotting (Chesworth, 1992; McDonald *et al.*, 2002). It is directly involved in contraction of muscles and contributes to muscle fibre contraction post-mortem (Keeton and Eddy, 2004). Calcium deficiency in young growing animals causes rickets, symptoms of which include deformed bones, enlargement of the joints, lameness and stiffness. In adults, calcium deficiency produces osteomalacia, in which the calcium in the bone is withdrawn and not replaced, so that the bones become weak and easily broken (McDonald *et al*, 2002).

Calcium content varies in organs and tissues of goats, and was found to be highest in the brain (46.99 mg/100 g DM) followed by kidney (13.58 mg/100 g DM), spleen (11.47 mg/100 g DM) and liver (10.06 mg/100 g DM) and lowest in the heart (7.7 mg/100 g DM) in crossbred goats (Table 11.3). Calcium in the liver and kidney of Alpine and Saanen goats was 20.4 and 16.7 mg/100 g DM (Mioč *et al.*, 1998).

The calcium values reported for goat lean meat are quite variable. A value of 11 mg/100 g DM was found by Wan Zahari and Abdul Wahid (1985) (Table 11.3). Slightly higher values were found in Omani (12.72 mg/100 g DM) and Somali (14.83 mg/100 g DM) goat meat (Table 11.4). Oke *et al.* (2007) reported values of 0.98–23 g/100 g DM in goats ranging between 24 and 52 weeks of age. Much higher values (880–945 mg/100 g DM) were reported by Sheridan *et al.* (2003) in Boer goats of various levels of dietary energy. Madruga *et al.* (2006) reported levels of 134–213 mg/kg DM in goats raised under various management systems.

Several factors have been reported to affect calcium levels in goat meat including sex and age (Madruga *et al.*, 1999), cooking and processing (Johnson *et al.*, 1995), breed and system of management (Madruga *et al.*, 2006). Goat meat contains more calcium than chicken (Addrizzo, 2010) and sheep meat (Sheridan *et al.*, 2003).

### *Phosphorus*

Phosphorus is a major component in bone and is closely associated with calcium. It is essential for bone formation, enzymes and energy metabolism (Casey, 1992). It occurs

**Table 11.4.** Mineral and trace element content (mg/100 g dry matter) of Omani and Somali goats (O. Mahgoub, unpublished data).

| Element | Omani goat | Somali goat | PSEM | Significance |
|---|---|---|---|---|
| Calcium | 12.72 | 14.83 | 1.100 | NS |
| Magnesium | 49.19 | 59.02 | 1.763 | <0.0001 |
| Potassium | 756.61 | 941.07 | 25.76 | <0.0001 |
| Sodium | 108.26 | 126.88 | 6.393 | <0.0001 |
| Silicon | 0.21 | 0.03 | 0.020 | <0.0001 |
| Phosphorus | 529.96 | 650.36 | 18.18 | <0.0001 |
| Chromium | 0.01 | 0.11 | 0.007 | <0.0001 |
| Manganese | 0.02 | 0.13 | 0.009 | <0.0001 |
| Iron | 3.435 | 4.30 | 0.201 | 00.004 |
| Nickel | 0.016 | 0.08 | 0.019 | 00.014 |
| Copper | 0.151 | 0.349 | 0.020 | <0.0001 |
| Zinc | 6.461 | 6.518 | 0.321 | <0.0001 |
| Cobalt | 0.041 | 0.08 | 0.004 | <0.0001 |
| Cadmium | 0.004 | 0.001 | 0.001 | 0.049 |
| Lead | 0.014 | 0.06 | 0.010 | 0.001 |
| Vanadium | 0.003 | 0.02 | 0.002 | <0.0001 |
| Molybdenum | 0.130 | 0.16 | 0.007 | 0.003 |

PSEM, Pooled standard error of the mean; NS, Not significant.

in phosphoproteins, nucleic acids and phospholipids. It has a vital role in the formation of sugar phosphates and adenosine di- and triphosphates (McDonald *et al.*, 2002).

The content of phosphorus in farm animal bodies is about 10 g/kg DM and 167–216 mg/kg DM in muscle (Table 11.2). In goat body tissues, the highest level is found in the liver, followed by brain, spleen, kidney and muscle (Table 11.3). Muscle is rich in phosphorus. Wan Zahari and Abdul Wahid (1985) reported a value of 155.5 mg/100 g DM, with a value of 175.78 mg/100 g DM in intact males (Table 11.3). Oke *et al.* (2007) reported values of 52–197 mg/100 g DM in goats ranging between 24 and 52 weeks of age. Sheridan *et al.* (2003) reported higher levels of phosphorus (632–654 mg/100 g DM). Mahgoub (unpublished results) found similar levels in Omani and Somali goats (Table 11.4).

No significant effects of age or sex on phosphorus content were found by Madruga *et al.* (1999). Goat meat contains less phosphorus than beef (Johnson *et al.*, 1995). The system of rearing of goats affects the phosphorus contents of goat (Madruga *et al.*, 2006).

### Potassium

Potassium is directly involved in contraction of muscles, together with calcium and sodium (Keeton and Eddy, 2004). It is an important component in osmotic regulation of the body fluids and in the base balance in the animal (McDonald *et al.*, 2002). Farm animal bodies contain about 2 g/kg DM (Table 12.2). Keeton and Eddy (2004) reported potassium levels of 250–400 mg/g DM in meat (Table 11.2). Oke *et al.* (2007) reported values of 151–306 mg/100 g DM in goat bodies ranging between 24 and 52 weeks of age.

Goat body organs and tissues contain variable levels of potassium with the muscle having the highest levels followed by brain, spleen, liver, kidney and heart (Table 11.3). However, values in muscle vary. Wan Zahari and Abdul Wahid (1985) found that goat meat contained potassium levels of 350 mg/100 g DM (Table 11.3) and Mahgoub (unpublished results) reported a high value of 757–941 mg/100 g DM (Table 11.4), whereas Sheridan *et al.* (2003) reported a value of 130.9–141.6 mg/100 g DM.

### Sodium

Sodium is the major cation of blood plasma and other extracellular fluids of the body. It is essential for acid/base balance and osmotic regulation of body fluids (McDonald *et al.*, 2002). It is also involved in contraction of muscle (Keeton and Eddy, 2004).

The sodium level in farm animal bodies is ~1.6 g/kg DM (Table 11.2). Goat body organs and tissues contain various levels of sodium, with the kidney having the highest levels followed by brain, muscle, spleen, liver and heart (Table 11.3).

Keeton and Eddy (2004) reported a value of 55–94 mg/g DM in muscle (Table 11.2). Goat muscles were found to contain 64.48 mg/100 g DM (Table 11.3). Oke *et al.* (2007) reported values of 60–115 mg/100 g DM in goats of 24–52 weeks of age. Sheridan *et al.* (2003) reported a value of 49.8–56.7 mg/100 g DM. Mahgoub (Table 11.4) found much higher values for sodium in Omani and Somali goat meat (108 and 127 mg/100 g DM). Cooked goat meat has a much higher sodium content than beef (Johnson *et al.*, 1995; Addrizzo, 2010).

### Magnesium

Magnesium is directly involved in live muscle contraction and contributes to muscle fibre contraction post-mortem (Keeton and Eddy, 2004). While 70% of magnesium is found in hard tissue, the rest is found in soft tissues and body fluids (Chesworth, 1992; McDonald *et al.*, 2002). Magnesium is necessary for many essential biochemical reactions. It is the commonest enzyme activator and is important for activating transferases, decarboxylases and acyl transferases (McDonald *et al.*, 2002).

The bodies of farm animals contain magnesium at a level of ~0.4 g/kg DM (Table 11.2). Keeton and Eddy (2004) reported a magnesium value of 22–29 mg/g DM in muscle. Goat body organs and tissues contain various levels of magnesium, with the muscle having the highest levels followed by spleen, liver, brain, kidney and heart (Table 11.3).

Oke *et al.* (2007) reported values of 7–10 g/100 g DM in goats of 24–52 weeks of age. Sheridan *et al.* (2003) reported magnesium values of 32–35 mg/100 g DM. Mahgoub (unpublished results), who used microwave extraction, recorded higher magnesium values of 49–59 mg/100 g DM (Table 11.4). In other tissues, the highest levels of magnesium were in the spleen (15.28 mg/100 g DM) and liver (15.08 mg/100 g DM), followed by brain (12.82 mg/100 g DM), kidney (10.19 mg/100 g DM) and heart (9.63 mg/100) (Table 11.3).

Cooked goat meat contained magnesium at 29.2 mg/100 g DM (Johnson *et al.*, 1995), which was higher than beef at 22.4 mg/100 g DM. Cooked goat meat contains more magnesium than chicken meat (Addrizzo, 2010).

### 11.4.2 Microelements

#### *Iron*

Iron is directly involved in the use of oxygen in cells with its transport in blood (Chesworth, 1992). More than 90% of iron in the body is combined with proteins, the most important of which is haemoglobin, which contains 3.4 mg/kg DM (McDonald *et al.*, 2002). Meat is a very good source of iron. There are two types of iron in meat, haem and non-haem. About 50–60% of the iron in meat is haem. Haem iron in meat is in haemoglobin and myoglobin and has a high bioavailability (Mulvihill, 2004). Meat promotes the absorption of iron in itself and from other food consumed with meat at the same meal (Mulvihill, 2004).

The bodies of farm animals contain iron at about 20–80 g/kg DM (Table 11.2). Keeton and Eddy (2004) found that goat muscle contained 1–3 mg/kg DM (Table 11.2). Goat body organs and tissues contain various levels of iron, with spleen having the highest levels followed by kidney, liver, heart, muscle and brain (Table 11.3).

Reports on iron muscle content in goats vary greatly. Sheridan *et al.* (2003) reported values of 1.2–1.8 mg/100 g DM. Mahgoub (unpublished results) recorded values of 3.4 and 4.3 mg/100 g DM in Omani and Somali goats, respectively (Table 11.4). Park (1988) reported an iron range of 27–33.7 μg/g wet weight in the muscles of Alpine and Nubian goats. These differences are due to the effects of age, sex and breed. For instance, Madruga *et al.* (1999) reported that goat meat contains iron levels of 1.75–3.65 mg/100 g DM in goats of various sexes and ages. A value of 4.37 mg/100 g DM was reported by Wan Zahari and Abdul Wahid (1985) (Table 11.3).

Iron content was not affected by sex in the latter report, which was similar to the results reported by Johnson *et al.* (1995) in cooked meat. It was, however, affected by age × castration interaction in a study by Madruga *et al.* (1999). In the latter report, meat iron content was affected linearly by age (175–310 days).

Johnson *et al.* (1995) reported that iron content in goat meat was similar to that of cooked beef meat (2.6 mg/100 g DM). Cooked goat meat contains twice the level of iron of chicken meat (Addrizzo, 2010).

#### *Copper*

Copper is necessary for the uptake and transport of iron, the formation of collagen, body growth (Hart *et al.*, 1928; Harris, 1983), haemoglobin formation and the prevention of a wide range of clinical and pathological disorders in all types of farm animals (Underwood and Suttle, 2001).

The bodies of farm animals contain 1–5 g/kg DM (Table 11.2). Keeton and Eddy (2004) reported that meat contained 0.5–0.13 mg/g DM Cu (Table 11.2). Wan Zahari and Abdul Wahid (1985) reported that the highest level of copper in goat tissues was in liver (8.28 mg/100 g DM), followed by heart and kidney (0.53 and 0.52 mg/100 g DM, respectively) and speen and brain (0.41 and 0.40 mg/100 g DM, respectively), with muscles having the lowest values (0.30 mg/100 g DM) (Table 11.3). Values of 0.151 and 0.349 mg/100 g DM were reported for Omani and Somali goats, respectively (Table 11.4). Sheridan *et al.* (2003) reported a level of 0.14–0.20 mg/100 g DM in Boer goat meat. Fresh goat meat contained copper at 2.3–2.8 µg/g wet weight (Park, 1988).

High levels of copper in the animal body may be regarded as contamination of meat and meat products. For instance, Swaileh *et al.* (2009) reported a range of 1.03–217.9 µg/g DM in the liver, kidney and muscle of cattle, sheep, goat and poultry in the Israeli West Bank, with the highest concentrations in the kidney and liver and lowest concentration in the muscles.

Cooked goat meat contains higher levels of copper than beef with significant sex effects, with castrates having the highest values (Johnson *et al.*, 1995). The system of rearing of goats affects their copper contents. Madruga *et al.* (2006) reported a level of 0.68 mg/100 g DM in field-raised goats compared with 0.93 mg/100 g DM in goats raised under confinement. Cooked goat meat contains much more copper than chicken meat (Addrizzo, 2010).

### *Zinc*

Zinc is known to play a central role in the immune system, and zinc-deficient persons experience increased susceptibility to a variety of pathogens (Shankar and Prasad, 1998). Zinc is a constituent of many important enzymes and accumulates in bone (Chesworth, 1992). It also functions as an antioxidant and can stabilize membranes (Shankar and Prasad, 1998). Zinc plays an important role in effective utilization of vitamin A (Chesworth, 1992) and is essential for the functioning of a substantial number of enzyme systems (Miller, 1969).

The animal body contains zinc at a level of 10–50 g/kg DM (Table 11.2). Keeton and Eddy (2004) found that goat muscles contained zinc at 1–5 mg/g DM. The highest levels of zinc (3.51 mg/100 g DM) were found in the muscle, followed by liver (2.99 mg/100 g DM), kidney (2.61 mg/100 g DM) and spleen (2.19 mg/100 g DM), and the lowest were in the heart and brain (1.41 and 1.40 mg/100 g DM, respectively) (Table 11.3). Comparable values (Table 11.4) were reported for Omani and Somali goats (6.461 and 6.518 mg/100 g DM, respectively). Fresh goat meat contains zinc at an average of 40 µg/g wet weight (Park, 1988).

Levels of zinc in cooked goat meat (4.3 mg/100 g DM) were comparable to those in beef (5.0 mg/100 g DM) but there were no sex effects on levels of zinc in broiled goat meat (Table 11.5). Cooked goat meat contains more than twice the concentration of zinc than chicken meat

**Table 11.5.** Comparison of nutrient analysis of chicken, beef and sheep meat.

| Mineral | Goat (mg/100 g DM)[a] | Chicken (mg/100 g DM)[a] | Beef (ppm)[b] | Sheep (mg/100 g DM)[c] |
|---|---|---|---|---|
| Calcium | 25.3 | 12.8 | 168–229 | 4.6–6.5 |
| Iron | 2.2 | 1.1 | 127–150 | 1.4–1.9 |
| Sodium | 77.1 | 69.7 | 0.24–0.36 | 16.2–19.8 |
| Zinc | 4.3 | 1.7 | 127–323 | 2.8–3.6 |
| Magnesium | 23.7 | 20.0 | 703–906 | 18.0–22.0 |
| Potassium | 308.3 | 189.6 | 1.27–1.54 | 130.1–160.5 |
| Phosphorus | 57.8 | 154.7 | 0.68–0.78 | 125.7–139.9 |
| Copper | 1.7 | 0.06 | 5–22 | 0.09–0.14 |
| Manganese | | | 0.85–1.46 | |

DM, Dry matter.
[a]Nutrient profile information taken from USDA (1989) and Johnson *et al.* (1995) including castrates, intact males and females was dissected into separable components of bone and soft tissue.
[b]Ammerman *et al.* (1974); [c]Hoffman *et al.* (2003).

(Addrizzo, 2010). The muscles of cattle contained 126 ppm of zinc (Miller, 1969).

#### *Selenium*

Selenium is involved in the function of immune cells (McKenzie *et al.*, 1998). It is an essential constituent of a number of enzymes, some of which have antioxidant functions (Burk, 2002). Its role is closely related to its function as a cofactor for a number of biochemical systems that destroy peroxides, potentially toxic compounds (Chesworth, 1992). The concentration of selenium in the bodies of farm animals was estimated at 1–2 mg/kg DM (Table 11.2). There are not many available reports on selenium content in goat meat. However, lamb meat contained levels of 0.05–0.12 mg/100 g DM with significant sire and ewe breed effects (Hoffman *et al.*, 2003).

#### *Manganese*

The average farm animal body contains manganese at a concentration of 0.2–0.5 mg/kg DM (Table 11.2). Goat tissue was found to contain 0.66 mg/100 g DM in the liver, 0.19 mg/100 g DM in the kidney, 0.159 in the spleen, 0.122 mg/100 g DM in the brain, 0.098 mg/100 g DM in the heart and 0.087 mg/100 g DM in the muscles (Table 11.3). Values of 0.02 and 0.13 mg/100 g DM were reported for Omani and Somali goats (Table 11.4). Fresh goat meat contained manganese at 0.28–0.48 μg/g wet weight (Park, 1988) and Egyptian goat muscles contained 0.6 mg/kg DM wet weight (Abou-Arab, 2001).

#### *Other elements*

Farm animal bodies contain many other elements at various levels including molybdenum (1–4 g/kg DM), iodine (0.3–0.6 g/kg DM) and cobalt (0.02–0.1 mg/kg DM) (Table 11.2). Omani and Somali goat muscles contained chromium at 0.01 and 0.11 mg/100 g DM, molybdenum at 0.13 and 0.16 mg/100 g DM, cobalt at 0.041 and 0.08 mg/100 g DM, nickel at 0.016 and 0.08 mg/100 g DM, cadmium 0.004 and 0.001 mg/100 g DM, lead at 0.014 and 0.06 mg/100 g DM and vanadium at 0.003 and 0.02 mg/100 g DM, respectively (Table 11.4). Goat meat also contained iodine at 41–43 mg/100 g DM and lead at 0.013–0.016 mg/100 g DM (Sheridan *et al.*, 2003).

## 11.5 Factors Affecting Mineral Concentration in Goat Tissues

### 11.5.1 Species differences

The mineral content of goat meat is more comparable to other red meats than to white ones. Table 11.5 shows a comparison of meat from goat, chicken, beef and sheep. The red meats had higher levels of most minerals except phosphorus. Goat meat is a particularly good source of iron and calcium. Goat meat appeared to have higher levels of calcium compared with other species. Johnson *et al.* (1995) also reported higher calcium contents in goat meat (28.1 mg/g DM) than in beef (8.0 mg/g DM). Cooked goat meat was found to contain more calcium than chicken meat (Addrizzo, 2010).

Sheridan *et al.* (2003) carried out a study comparing the mineral content of goats and sheep raised on two levels of dietary energy (Table 11.6) and found that goat carcasses had higher calcium, potassium, magnesium, sodium, phosphorus and copper contents than sheep carcasses (Sheridan *et al.*, 2003). Johnson *et al.* (1995) reported that goat meat had a lower phosphorus content (103 mg/85 g cooked meat) than beef (173 mg/85 g cooked meat).

### 11.5.2 Tissue type

Liver usually has higher concentrations of minerals and trace elements. Park (1988) reported that liver contained 2.4, 11.0, 41.5 and 1.3 times higher levels of iron, manganese, copper and iron/zinc than muscles, respectively. This is because the liver stores most of these elements. Macro- and microminerals are also found in the brain, heart, kidney, spleen and muscle. Most of these tissues are consumed as well as the muscle

**Table 11.6** Mineral composition (mg/100 g DM) of the 8th to 10th rib cuts of Boer goat kids and Mutton Merino lambs (16 per treatment) receiving a low- or a high-energy feedlot diet (least squares mean ± SEM) (from Sheridan *et al.*, 2003).

| | Treatment | | | | | | | |
|---|---|---|---|---|---|---|---|---|
| | BGLE | | BGHE | | MMLE | | MMHE | |
| Mineral | Mean | SEM | Mean | SEM | Mean | SEM | Mean | SEM |
| Calcium | 880.84ab | 59.362 | 946.55a | 61.309 | 672.88c | 65.857 | 723.96bc | 59.362 |
| Iodine | 41.68ac | 2.048 | 43.38a | 2.115 | 36.56bc | 2.272 | 33.45b | 2.048 |
| Potassium | 141.57a | 6.476 | 130.88a | 6.689 | 95.46b | 7.185 | 86.16b | 6.476 |
| Magnesium | 32.51a | 1.594 | 35.36a | 1.646 | 24.90b | 1.594 | 24.32b | 1.768 |
| Sodium | 56.73a | 3.415 | 49.83ac | 3.527 | 42.39bc | 3.788 | 38.12b | 3.415 |
| Phosphorus | 631.97a | 37.685 | 653.69a | 38.920 | 485.39b | 41.807 | 510.38b | 37.685 |
| Copper | 0.20a | 0.027 | 0.14ab | 0.032 | 0.11b | 0.030 | 0.13ab | 0.027 |
| Iron | 1.19a | 0.142 | 1.78b | 0.147 | 1.19a | 0.158 | 1.01a | 0.142 |
| Lead | 0.013 | 0.0032 | 0.016 | 0.0029 | 0.014 | 0.0027 | 0.016 | 0.0039 |

SEM, Standard error of the mean; BGLE, low-energy diet given to Boer goats; BGHE, high-energy diet given to Boer goats; MMLE, low-energy diet given to Mutton Merino lambs; MMHE, high-energy diet given to Mutton Merino lambs. Means in the same row with different letters (a, b, c) differ significantly ($P < 0.05$).

in goat-meat-eating countries in Asia and Africa. Therefore, these organs form a useful source of minerals for the population.

### 11.5.3 Individual muscles

There are some variations in reports on mineral and trace element concentrations in goat tissues. The mineral concentration in muscles is related to physiological function, with red muscle tissue being richer in essential minerals than white muscles (Wagner *et al.*, 1976). Differences in concentrations of minerals in various carcass muscles were reported by Mioč *et al.* (2000). However, Park (1988) reported no differences in trace element concentrations in the biceps femoris and longissimus dorsi muscles in Alpine and Nubian goats. Reports from other species indicate differences in mineral content between carcass muscles. For instance, Marchello *et al.* (1985) reported a one- to twofold difference in mineral content in pork, with shoulder muscles containing higher levels of iron and zinc than leg or loin muscles. The copper, iron, zinc and magnesium contents in beef longissimus dorsi muscle were higher than those in the semimembranosus muscle (Doornenbal and Murray, 1981).

### 11.5.4 Sex

Park (1988) observed that differences in mineral content in goat tissues between sexes were less pronounced than between tissues, except for manganese, where females had higher manganese levels than males. Similar observations have been reported in other species. Doornenbal and Murray (1981) also observed that sex effects were less pronounced for mineral content in cattle.

The findings of Madruga *et al.* (1999) indicated an effect of age and sex on calcium levels in meat. Calcium concentration was highest in castrates at 175 days age (8.07 ± 1.24 mg/100 g wet weight) and lowest in intact males at 310 days of age (3.01 ± 0.71 mg/100 g wet weight). The effect of both age and castration was significant, with no effect of an age × sex interaction. In cooked meat, calcium values were 126.4, 136.2 and 128.3 mg/100 g DM in female, castrate and male broiled meat, respectively (Johnson *et al.*, 1995).

### 11.5.5 Age

Madruga *et al.* (1999) reported that calcium content decreased while iron content increased with age in goat meat of males. There were no differences in mineral content between Boer goats slaughtered at 28 or 56 days of age (Sheridan *et al.*, 2003). However, in other meat-producing species, age effects have been reported. For example, the age of cows affected the levels of iron, zinc, magnesium and sodium in muscles (Doornenbal and Murray, 1981).

### 11.5.6 Breed

Reports on the effects of breed on mineral content in goats vary. This is apparently because of differences in the analytical methods used and muscles studied. Mahgoub (unpublished data) found significant differences between native Oman and Somali goats in magnesium, potassium, sodium, silicon, phosphorus, chromium, manganese, iron, nickel, copper, zinc, cobalt, cadmium, lead, vanadium and molybdenum contents in the psoas major and minor muscles (Table 11.4). In contrast, Mioč *et al.* (2000) found no breed effect on macrominerals in goat meat (Table 11.7). However, they found significant breed effects on zinc, iron and copper in the back and shoulder muscles. Park (1988) observed no differences between breeds in iron, manganese and zinc content, regardless of sex, but there was an effect of breed on copper. Similar observations have been reported in other species. For example, Doornenbal and Murray (1981) also observed that breed effects were less pronounced for mineral content in cattle. However, Hoffman *et al.* (2003) reported an effect of breed on mineral composition in lambs. They found significant differences between six breed crosses in iron, potassium, magnesium and phosphorus content, with the dam effect being more significant than the sire effect.

### 11.5.7 Diet

There are few data on the effects of diet on macro- and micromineral content of goat tissues. However, results from other meat-producing animals may shed some light on this issue. Feeding high dietary iron increased iron concentrations in the liver, spleen, kidney and heart, whereas copper and zinc concentrations of the liver were decreased. Dietary iron treatments did not appear to affect the magnesium or manganese content of the liver, spleen, kidney, heart and muscle (Standish *et al.*, 1969).

**Table 11.7** Macrominerals and trace element content (mg/100 g) in Saanen and Alpine goats (Mioč *et al.*, 2000).

| | Saanen | | | | | | Alpine | | | | | |
|---|---|---|---|---|---|---|---|---|---|---|---|---|
| | Leg | | Back | | Shoulder | | Leg | | Back | | Shoulder | |
| Mineral | Mean | SEM | Mean | SEM | Mean | SEM | Mean | SEM | Mean | SEM | Mean | SEM |
| Calcium | 15.8 | 0.67 | 15.5 | 0.56 | 15.3 | 0.80 | 16.6 | 0.49 | 15.6 | 0.74 | 16.3 | 0.70 |
| Phosphorus | 68.6 | 1.32 | 65.8 | 0.97 | 64.4 | 3.87 | 62.2 | 1.79 | 31.9 | 1.68 | 60.6 | 1.79 |
| Potassium | 132.6 | 2.65 | 128.5 | 2.3 | 129.5 | 2.74 | 135.0 | 3.4 | 133.1 | 2.26 | 133.2 | 3.52 |
| Magnesium | 19.1 | 0.43 | 19.1 | 0.44 | 18.6 | 0.35 | 19.0 | 0.33 | 19.2 | 0.63 | 19.3 | 0.50 |
| Sodium | 78.9 | 2.73 | 77.9 | 2.64 | 78.3 | 3.12 | 75.0 | 1.89 | 72.2 | 1.40 | 72.8 | 2.67 |
| Zinc | 2.8 | 0.22 | 2.5 | 0.09 | 4.0 | 0.17 | 2.7 | 0.13 | 2.9 | 0.17 | 3.0 | 0.08 |
| Iron | 0.21 | 0.014 | 0.21 | 0.011 | 0.18 | 0.014 | 0.21 | 0.007 | 0.19 | 0.005 | 0.23 | 0.010 |
| Manganese | 0.06 | 0.002 | 0.06 | 0.004 | 0.06 | 0.002 | 0.06 | 0.004 | 0.06 | 0.003 | 0.07 | 0.005 |
| Copper | 0.37 | 0.068 | 0.31 | 0.046 | 0.31 | 0.033 | 0.17 | 0.001 | 0.23 | 0.019 | 0.15 | 0.017 |

Feeding 1600 ppm iron as either ferrous sulfate or ferric citrate resulted in increased iron in the kidney, liver and spleen. Levels of copper, zinc, magnesium and manganese in tissues were not influenced by the treatment. In general, increasing the dietary sulfate concentration resulted in tissue mineral concentrations similar to those for the basal diet, indicating that the effects of the dietary sulfate were due primarily to iron (Standish and Ammerman, 1971). Dietary iron influenced tissue mineral content in cattle (Standish *et al.*, 1969) and sheep (Standish and Ammerman, 1971).

### 11.5.8 Cooking

Cooking and processing affect the mineral content of goat meat, including that of calcium. Higher values of calcium were reported in broiled Florida native or F1 crosses of Florida native × Nubian or Spanish goat meat at 6–10 months by Johnson *et al.* (1995) (Table 11.8). Calcium values of broiler meat were much higher than values reported for uncooked meat (Madruga *et al.*, 1999), amounting to more than eightfold higher values in intact males and more than fivefold higher values than castrates of comparable ages. Cooked goat meat contained 28.1 mg calcium/g DM (Johnson *et al.*, 1995). The effect of cooking on calcium content is difficult to explain as the difference in cooking loss was similar (31.1 and 30.5%), and ash content in intact male meat was about 1.44 times than that reported by Madruga *et al.* (1999) (Table 11.8).

## 11.6 Correlation Between Minerals in Goat Meat

Significant correlations were found between minerals in goat meat and these correlations were more pronounced for macro- than for microminerals (Mioč *et al.*, 2000). These included significant positive correlations between the major macrominerals calcium, phosphorus, potassium and magnesium in the thigh muscle of Alpine goats (Mioč *et al.*, 2000). Park (1990) also reported positive correlations between phosphorus and potassium.

**Table 11.8.** Effect of sex on mineral analysis of 100 g composite of broiled and moisture content of cooked and uncooked goat meat and comparison of 85 g cooked goat and beef meat (adapted from Johnson *et al.*, 1995).

| | Broiled goat meat (100 g) | | | | Cooked meat (85 g) | |
|---|---|---|---|---|---|---|
| Component | Female | Castrate | Male | SEM | Goat | Beef |
| Mineral (mg) | | | | | | |
| Calcium | 30.6 | 32.7 | 37.9 | 6.22 | 28.1 (33.1) | 8.0 (9.4) |
| Magnesium | 29.6 | 28.8 | 29.8 | 1.5 | 24.8 (29.2) | 19.0 (22.4) |
| Potassium | 394.4 | 348.9 | 372.7 | 15.00 | 315.0 (370.6) | 266.0 (312.9) |
| Phosphorus | 126.4 | 136.2 | 128.3 | 7.87 | 103.0 (121.2) | 173.0 (204.5) |
| Sodium | 113.0 | 115.4 | 106.2 | 5.62 | 92.3 (108.6) | 52.0 (61.2) |
| Copper | 0.1 | 2.3 | 0.7 | 0.75 | 1.0 (1.2) | 0.1 (0.11) |
| Iron | 2.6 | 2.4 | 2.4 | 0.13 | 2.2 (2.6) | 2.2 (2.6) |
| Manganese | 0.04 | 0.04 | 0.00 | 0.00 | 0.04 (0.05) | 0.01 (0.01) |
| Zinc | 5.3 | 4.7 | 4.9 | 1.28 | 4.4 (5.2) | 5.0 (5.9) |
| Moisture (g) | | | | | | |
| Uncooked | 68.5 | 68.4 | 70.3 | 0.76 | | |
| Cooked | 57.4 | 59.5 | 59.3 | 0.61 | | |
| Cooked ash (g) | 1.2 | 1.3 | 1.4 | 0.08 | | |
| Cooked loss (%) | 33.5 | 30.5 | 31.1 | 0.80 | | |

SEM, Standard error of the mean.

## References


Abou-Arab, A.A. (2001) Heavy metal contents in Egyptian meat and the role of detergent washing on their levels. *Food Chemistry and Toxicology* 39, 593–599.

Addrizzo, J. (2010) Nutrient composition of goat meat. <http://onlinekambingan.webs.com/apps/forums/topics/show/4045401-nutrient-composition-of-goat-meat?page=last>.

Ammerman, C.B., Loaiza, J.M., Blue, W.G., Gamble, J.F. and Martin, F.G. (1974) Mineral composition of tissues from beef cattle under grazing conditions in Panama. *Journal of Animal Science* 38, 158–162.

Arain, M.A., Khaskheli, M., Rajput, I.R., Faraz, S., Rao, S., Umer, M. and Devrajani, K. (2010) Effect of slaughtering age on chemical composition of goat meat. *Pakistani Journal of Nutrition* 9, 404–408.

Ash, A.J. and Norton, B.W. (1987) Studies with the Australian cashmere goat. II Effects of dietary protein concentration and feeding level on body composition of male and female goats. *Australian Journal of Agricultural Research* 38, 971–982.

Beserra, F.J., Monte, A.L. de Sousa, Bezerra, L.C. Nogueira de Moraes and Nassu, R.T. (2000) Chemical characterization of kid meat from Moxotó goat and Pardo Alpina × Moxotó crossbreeds. *Pesquisa Agropecuária Brasileira* 35, 171–177.

Beserra, F.J., Madruga, M.S., Leite, A.M., da Silva, E.M.C. and Maiaa, E.L. (2004) Effect of age at slaughter on chemical composition of meat from Moxotó goats and their crosses. *Small Ruminant Research* 55, 177–181.

Burk, R.F. (2002) Selenium, an antioxidant nutrient. *Nutrition in Clinical Care* 5, 75–79.

Campbell, J.R., Kenealy, M.D. and Campbell, K.L. (2003) The nutritional contributions of minerals to humans and animals. In: *Animal Sciences: Biology, Care and Production of Domestic Animals.* McGraw Hill, Boston, USA, pp. 356–369.

Casey, N.H. (1992) Goat meat in human nutrition. In: *Proceedings of the Vth International Conference on Goats*, Vol. 2, Part II. Indian Council of Agricultural Research, New Delhi, India, pp. 582–596.

Chesworth, J.M. (1992) Minerals. In: *Ruminant Nutrition.* Macmillan Press, London, UK, pp. 25–40.

Dhanda, J.S., Taylor, D.G. and Murray, P.J. (2003) Part 2. Carcass composition and fatty acid profiles of adipose tissue of male goats: effect of genotype and liveweight at slaughter. *Small Ruminant Research* 50, 67–74.

Doornenbal, H. and Murray, A.C. (1981) Effects of age, breed, sex and muscle on certain mineral concentrations in cattle. *Journal of Food Science* 47, 55–58.

Dzudie, T., Ndjouenkeub, R. and Okubanjo, A. (2000) Effect of cooking methods and rigor state on the composition, tenderness and eating quality of cured goat loins. *Journal of Food Engineering* 44, 149–153.

Harris, E.D. (1983) Copper in human and animal health. In: Rose, J. (ed.) *Trace Elements in Health – A Review of Current Issues.* Butterworths, London, UK, pp. 44–73.

Hart, E.B, Steenbock, H., Waddell, J. and Elvehjem, C.A. (1928) Iron in nutrition, IV. Copper as a supplement to iron for hemoglobin building in the rat. *Journal of Biological Chemistry* 77, 797–812.

Hoffman, L.C., Muller, M., Cloete, S.W.P. and Schmidt, D. (2003) Comparison of six crossbred lamb types: sensory, physical and nutritional meat quality characteristics. *Meat Science* 65, 1265–1274.

Hogg, B.W., Catcheside, L.M., Mercer, G.J.K. and Duganzich, D.M. (1989) Meat yields and chemical composition of muscle in New Zealand goats. *Proceedings of the New Zealand Society of Animal Production* 49,153–156.

James, N.A. and Berry, B.W. (1997) Use of chevon in the development of low-fat meat products. *Journal of Animal Science* 75, 571–577.

Johnson, D.D., Eastrige, J.S., Neubauer, D.R. and McGowan, C.H. (1995) Effect of sex on nutrient content of meat from young goat. *Journal of Animal Science* 73, 296–301.

Keeton, J.T. and Eddy, S. (2004) Chemical and physical characteristics of meat. In: Jensen, W.K., Devine, C. and Dikeman, M. (eds) *Encyclopaedia of Meat Sciences*, Vol. 1. Elsevier Academic Press, Oxford, UK, pp. 210–218.

Kesava Rao, V., Kowale, B.N. and Verma, A.K. (2003) Effect of feeding water washed neem (*Azadirachta indica*) seed kernel cake on the quality, lipid profile and fatty acid composition of goat meat. *Small Ruminant Research* 47, 213–219.

Lee, J.H., Kouakou, B. and Kannan, G. (2008) Chemical composition and quality characteristics of chevon from goats fed three different post-weaning diets. *Small Ruminant Research* 75, 177–184.

Leth, T. and Ertbjerg, P. (2004) Micronutrients and other minor meat components. In: Jensen, W.K., Devine, C. and Dikeman, M. (eds) *Encyclopedia of Meat Sciences*, Vol. 1. Elsevier Academic Press, Oxford, UK, pp. 190–195.
Madruga, M.S., Arruda, S.G.B. and Nascimento, J.A. (1999) Castration and slaughter age effects on nutritive value of the 'Mestiço' goat meat. *Meat Science* 52, 119–125.
Madruga, M.S., Resosemito, F.S., Narain, N., Souza, W.H., Cunha, M.G.G and Ramos, J.L.F. (2006) Effect of raising conditions of goats on physic-chemical and chemical quality of its meat. *Ciencia y Tecnologia de los Alimentos* 5, 100–104.
Mahgoub, O. and Lu, C.D. (2004) Influence of various levels of metabolisable energy on chemical composition of whole carcass and non-carcass portion of goats and sheep. *South African Journal of Animal Science* 34 (Suppl. 1), 81–84.
Mahgoub, O., Lu, C.D., Hameed, M.S., Richie, A., Al-Halhali, A.S. and Annamalai K. (2005) Performance of Omani goats fed diets containing various metabolizable energy densities. *Small Ruminant Research* 58, 175–180
Marchello, M.H., Slanger, W.D. and Milne, D.G. (1985) Macro and micro minerals from selected muscles of pork. *Journal of Food Science* 50, 1375.
McDonald, P., Edwards, R.A., Greenhalgh, J.F.D. and Morgan, C.A. (2002) Minerals. In: *Animal Nutrition*, 6th edn. Pearson Education, Edinburgh, UK, pp. 108–145.
McKenzie, R.C., Rafferty, T.S. and Beckett, G.J. (1998) Selenium: an essential element for immune function. *Immunology Today* 19, 342–345.
Miller, W.J. (1969) Absorption, tissue distribution, endogenous excretion, and homeostatic control of zinc in ruminants. *American Journal of Clinical Nutrition* 22, 1323–1331.
Mioč, B., Pavič, V. and Kaps, M. (1998) Mineral composition of liver and kidneys in Alpine and Saanen kids. *Agricultural Conceptus Scientificus* 63, 61–66.
Mioč, B., Pavič, V., Ivanovič, A. and Havranek, D. (2000) Concentration of macro and microminerals in muscles of kids. *Czech Journal of Animal Science* 45, 533–538.
Mulvihill, B. (2004) Micronutrients in meat. In: Jensen, W.K., Devine, C. and Dikeman, M. (eds) *Encyclopedia of Meat Sciences*, Vol. 1. Elsevier Academic Press, Oxford, UK, pp. 612–623.
Niedziolka, R., Pieniak-Lendzion, K. and Horoszewicz, E. (2005) Comparison of the chemical composition and fatty acids of the intramuscular fat of goat kid and ram lambs meat. *Electronic Journal of Polish Agricultural Universities* 8, #11.
Oke, D.B., Oke, M.O. and Fasina, O.E. (2007) Mineral elements in the muscle groups of West African dwarf goats. *Research Journal of Animal Science* 1, 53–55.
Paleari, M.A., Moretti, V.M., Beretta, G. and Caprino, F. (2003) Chemical parameters, fatty acids and volatile compounds of salted and ripened goat thigh. *Small Ruminant Research* 74, 140–148.
Park, Y.W. (1988) Trace mineral contents and Fe/Zn ratio in goat meat. *Journal of Food Composition and Analysis* 1, 283–289.
Park, Y.W. (1990) Effect of breed, sex and tissues on concentrations of macrominerals in goat meat. *Journal of Food Science* 55, 308–311.
Satterlee, L.D., Arnold, R.G. and Anderson, P.C. (1977) Process for measuring tenderness of cooked meat. US patent no. 4.009.390.
Schonfeldt, H.C. (1989) A comparison of the quality characteristics of goat meat with that of sheep meat. MSc Dissertation, Department of Home Economics and Dietetics, Faculty of Science, University of Pretoria, South Africa.
Seideman, S.C., Cross, H.R. and Crouse, J.D. (1984) The effect of sex and age on the textural properties and mineral content of beef steaks. *Journal of Food Quality* 7, 91–96.
Shankar, A.H. and Prasad, A.S. (1998) Zinc and immune function: the biological basis of altered resistance to infection. *American Journal of Clinical Nutrition* 68, 447S–463S.
Sheridan, R., Hoffman, L.C. and Ferreira, A.V. (2003) Meat quality of Boer goat kids and Mutton Merino lambs. 1. Commercial yields and chemical composition. *Animal Science* 76, 63–71.
Solaiman, S.G., Shoemaker, C.E., Jones, W.R. and Kerth, C.R. (2006) The effects of high levels of supplemental copper on the serum lipid profile, carcass traits, and carcass composition of goat kids. *Journal of Animal Science* 84,171–177.
Standish, J.F. and Ammerman, C.B. (1971) Effect of excess dietary iron as ferrous sulfate and ferric citrate on tissue mineral composition of sheep. *Journal of Animal Science* 33, 481–484.

Standish J.F., Ammerman, C.B., Simpson, C.F., Neal, F.C. and Palmer, A.Z. (1969) Influence of graded levels of dietary iron, as ferrous sulfate, on performance and tissue mineral composition of steers. *Journal of Animal Science* 29, 496–503.

Swaileh, K.M., Abdulkhaliq, A., Hussein, R.M. and Matani, M. (2009) Distribution of toxic metals in organs of local cattle, sheep, goat and poultry from the West Bank, Palestinian Authority. *Bulletin of Environmental Contamination and Toxicology* 83, 265–268.

Todaro, M., Corrao, A., Barone, C.M.A., Alicata, M.L., Schinelli, R. and Giaccone, P. (2006) Use of weaning concentrate in the feeding of suckling kids: effects on meat quality. *Small Ruminant Research* 66, 44–50.

Turgut, H. (1984) Emulsifying capacity and stability effect of cooking methods and rigor state on the composition, tenderness and eating quality of cured goat loins, sheep and cattle muscle proteins. *Journal of Food Science* 49, 168–182.

Underwood, E.J. and Suttle, N.F. (2001) Copper. In: *The Mineral Nutrition of Livestock.* CABI Publishing, Wallingford, UK, pp. 238–342.

USDA (1989) *USDA Agriculture Handbook 8: Composition of Foods: Raw, Processed, Prepared.* US Department of Agriculture.

Wagner, K.H., Sarican, C., Ali, A. and Wanger-Hering, E. (1976) *Ein Beitrag zur verteilung von mineralstoffen und amonsauren in hellen und dunlem schweinfleisch (m. longissimus dorsi and diaphragm). Die Fleischwirtschaft* 11, 1651–1654.

Wan Zahari, M. and Abdul Wahid, S. (1985) Mineral concentration in blood plasma and in various soft tissues of local crossbred goats. *MARDI Research Bulletin* 13, 333–340.

# 12 Linear Body Measurements and Carcass Characteristics of Goats

I.T. Kadim and O. Mahgoub

*Department of Animal & Veterinary Sciences, College of Agricultural and Marine Sciences, Sultan Qaboos University, Sultanate of Oman*

## 12.1 Abstract

This chapter discusses the carcass characteristics of goats, with emphasis on live body weight and linear carcass measurements. Goats are a good potential source of meat as they yield good carcass weights under inexpensive management systems. The economic importance of goat carcasses in the world can be attributed to a demand for goat meat, which is believed to make up 80% of the total meat consumed in Asia and Africa. Age, breed, live weight and diet influence goat live body and carcass measurements. Linear measurements reflect the length of the body dimensions of the live animal to describe the changing body shape in order to predict both animal live weight and composition. Goat carcass characteristics can also be evaluated by assessing the conformation and distribution of muscle and fat in the carcass. Goat carcass weight is one of the most variable parameters, with the variation apparently due to condition, sex, breed and age at slaughter. Goat carcasses are thin and shallow but become thicker and more compact as carcass weight increases. Carcass weight is the best predictor of the meat content because of the lean nature of the carcass. The economic value of a goat carcass depends on its yield of reasonable carcass weight under inexpensive management systems. The dressing percentage of goats varies from 38.5 to 52.3% depending on sex, body condition and breed. The dressing percentage is lower for male than female goats, with this difference increasing with age. Goat carcasses are leaner than other meat animal carcasses because the fat tends to be concentrated around the viscera. The latter anatomical feature makes goats more adaptable to the environmental extremes of the tropics.

## 12.2 Introduction

Goats are one of the most widely domesticated animals, as evidenced by their wide distribution and utilization due to their high adaptability to a broad range of environments. They are also able to utilize marginal land to produce high-quality protein products. Goat meat is the primary source of protein in many parts of the world, especially Asia and Africa. The importance of goats is associated with the increasing number of goats globally (Morand-Fehr *et al.*, 2004), despite major changes in agriculture due to industrial merges, globalization and technological advances in developed countries (McMillin and Brock, 2005). There is also a worldwide tendency

 *Goat Meat Production and Quality* (eds O. Mahgoub, I.T. Kadim and E.C. Webb)

towards a rapid increase in the demand for goat meat (Gipson, 1998; Stankov *et al.*, 2002). Goat meat has an enormous market potential, as it could become an ideal choice for health-conscious consumers because it has a lower fat content than other types of red meat. This is an important factor in reducing the risk of cardiovascular diseases (van Niekerk and Casey, 1988; Park *et al.*, 1991; Colomer-Rocher *et al.*, 1992; Giese, 1992; Stankov *et al.*, 2002). Goat meat is also a good source of desirable fatty acids, as goats deposit relatively higher amounts of polyunsaturated fatty acids than other ruminants (Banskalieva *et al.*, 2000; Mahgoub *et al.*, 2002). Carcass quality, an important aspect in the marketing of meat, is not well defined. Elucidation of the available reports on goat carcass quality is complicated by the fact that there are so many goat breeds kept under such widely different conditions that comparisons of results are not always meaningful. Information on live body parameters and carcass characteristics from different breeds of goats can be evaluated properly. However, unfortunately much of the data cannot be used directly to evaluate carcass yields.

This chapter attempts to highlight the characteristics of goat carcasses with special emphasis on live and linear carcass measurements, dressing percentage and meat composition.

## 12.3 External Live Linear Body Measurements

Animal live linear body measurements are recognized as preferable objective measurements and have been used as a predictor of goat body composition (Kadim *et al.*, 2006). Live body dimensions can be measured with a caliper and flexible tapes. An accurate description of the body growth dimensions is important for a better understanding of the relationship between growth stage and body size (Lawrence and Fowler, 1997; Kamalzadeh *et al.*, 1998; Kadim *et al.*, 2006). Most of the linear measurements primarily reflect the lengths of the bones of the animal. Overall, when taken sequentially over a period of time, they indicate the way in which the animal body is changing shape, and have been used as predictors of both animal live weight and carcass composition. Examples of measurements that may be taken are given in Fig. 12.1 and Table 12.1. The five basic live body measurements in Table 12.1 are considered adequate to describe the goat's body composition, allowing comparisons to be made of the stage of growth reached by different parts of the body in the live animal. The purpose of taking these live body measurements is to provide a comparative

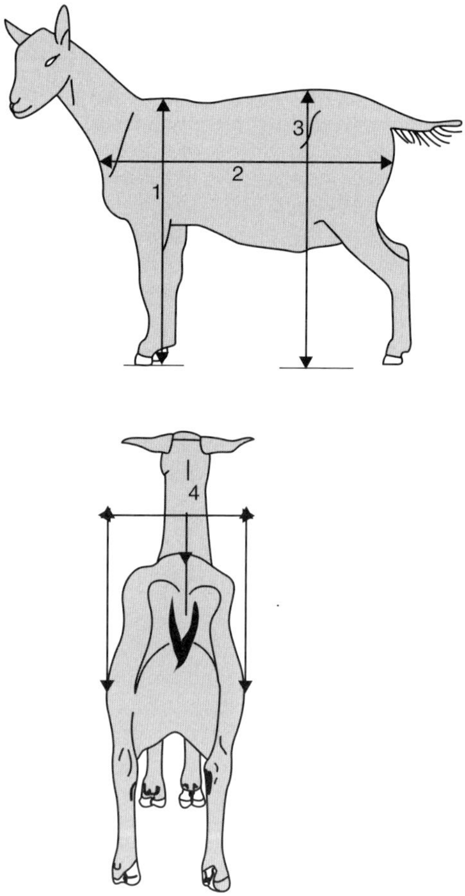


**Fig. 12.1.** Diagram indicating where measurements are taken on the live goat. 1, Wither height; 2, body length; 3, rump height; 4, rump width. See Table 12.1 for details.

**Table 12.1.** Definitions of live and carcass linear measurements shown in Fig. 12.1 (from Kadim, 1988).

| Measurement | Definition |
|---|---|
| Body length | Distance between the sciatic tuber (pinbone) and the most distal point of the major tuberosity of the humerus |
| Rump height | Highest point over the hip bone |
| Rump width | Widest horizontal width across the hips region |
| Wither height | Highest point over the scapulae vertically to the ground |
| Wither width | Widest horizontal width across the shoulder region |

description of the various groups of goats studied, rather than to provide a basis for predicting body weight or composition from such measurements, as many studies have aimed to do in the past (Kadim *et al*., 2006). Fisher (1975a) pointed out that there are three sources of error in taking body measurements: (i) correct identification; (ii) location of the end reference points in linear measurements; and (iii) anatomical distortion produced by the animal changing either position or posture. Errors involved in actually taking the measurement at any one position will be minimal for caliper measurements but greater for measurements using flexible tape over surfaces, particularly if they are concave. The accuracy of body measurements depends on the size of skeletal units and the development of both soft tissues and the skeleton or the development of soft tissues only (Lawrence and Fowler, 1997). The order of decreasing accuracy is from skeletal measurements to skeletal plus flesh measurements to soft tissue measurements. In goats, the prediction of soft tissue deposition and distribution within regions of the live body measurements is inadequate. Goat live body dimensions vary in both shape and size. Therefore, live body measurements cannot be used as accurate predictors of body weight. However, if the animal retained a constant shape and varied in size only, then, as long as the basic geometry of the goat was understood, dimensions or tissue volumes could be predicted from other dimensions or tissue volumes. Fisher (1975b) pointed out that, in addition to the range of variable sizes in animal bodies, there are also ranges in tissue shapes and proportions.

A series of live linear body measurements on three breeds of Omani goats (Jebel Akhdar, Batina and Dhofari) was carried out by Kadim *et al*. (2006). These measurements comprised body length, rump height, rump width, wither height and wither width (Fig. 12.2). The measurements were taken at weaning and then at 4-week intervals until 48 weeks of age. They found that the Jebel Akhdar breed had significantly greater linear body dimensions than the Batina and Dhofari breeds. This is in line with the breed effects on carcass linear measurements of the same three breeds of Omani goats reported by Kadim *et al.* (2004). The differences between the three breeds were more pronounced in length and height measurements than in bone width measurements, as bone length is more closely associated with early development than width.

The length of rump, loin and chine; chest and shoulder widths; chest depth; height at withers and at hip; barrel circumference; and heart girth were recorded on kids, yearlings and adults of Angora, Spanish, dairy and Boer × Spanish cross goats (Pinkerton and McMillin, 2000). Boer-cross kids had longer rump, loin and chine measurements and wider and deeper chest dimensions, as well as larger heart girth and barrel circumference measurements, than kid goats of the other breeds. Spanish yearling goats had slightly greater dimensions at each location, including height at withers and height at hip, than Boer-cross yearlings, which could indicate differences in maturation patterns between the two breeds. Angora goats had smaller dimensions at each age compared with the other breeds. Yearling goats of each breed generally had greater dimensions at each location than kid goats of the same breed, and

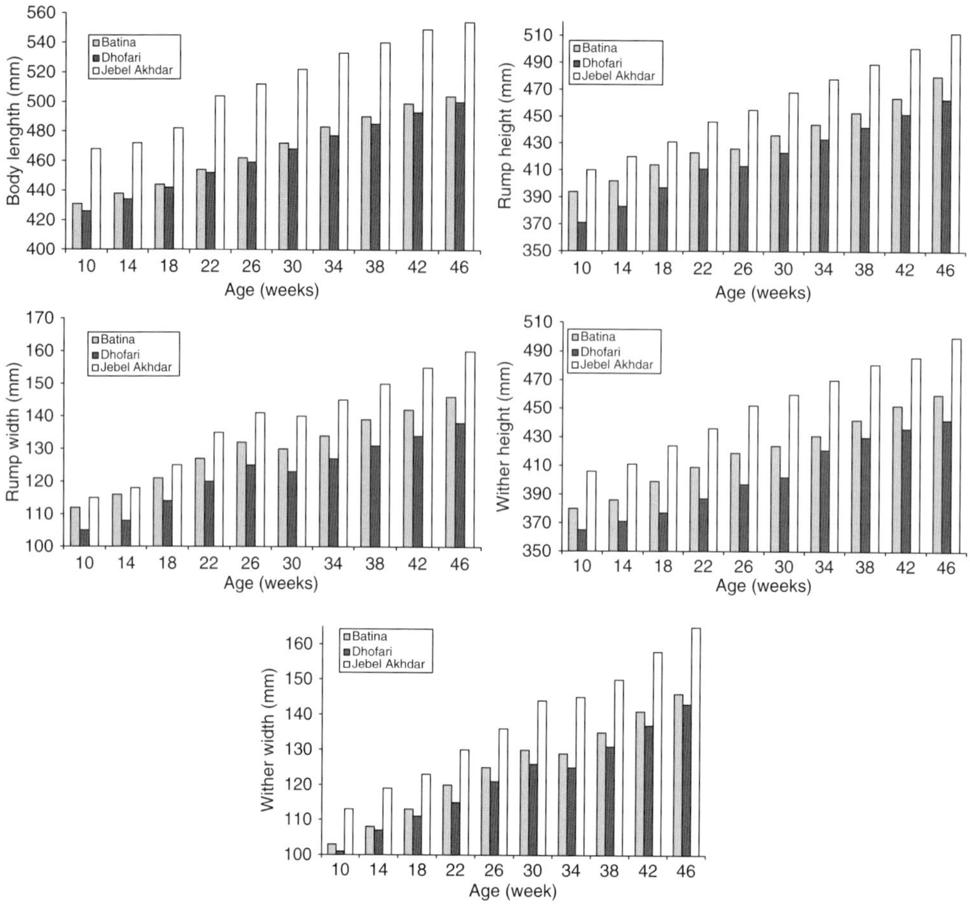


**Fig. 12.2.** Body length, rump height, rump width, wither height and wither width measurements of three local Omani goat breeds, showing increases with increasing age (Kadim *et al.*, 2006).

adult goats correspondingly had greater dimensions than the yearling goats. This indicates an obvious age effect. The live animal traits were generally highly correlated with one another, indicating that animals that are larger in one measurement are usually larger in others (Pinkerton and McMillin, 2000). Breed differences in body measurements of the Bornu White, Red Sokoto, Small East African, Sudanese Desert and West African Dwarf breeds of goat indigenous to the African continent have been reported (Quartermain, 1991). The width and shape of the Alpine and Rove breeds were significant due to heterotic effects (Mourad and Anous, 1998). The Anglo-Nubian breed was crossed with smaller indigenous goats to take advantage of body conformation in the exotic breed (Wilson *et al.*, 1980; Hussain *et al.* 1983).

## 12.4 Carcass Weight

In practice, goats are sold on a weight basis. Therefore, the relationship between the carcass weight and the live weight of the goat is very important. The economic value of goat carcass characteristics depends on its yield of meat, as well as the cutting and processing quality of the meat. Goats are a good potential source of meat as they yield reasonable carcass weights under inexpensive management systems. Goat carcass weight is one of the most variables parameters, with the variation apparently due to condition, sex, breed and age at slaughter. Carcass weight was found to be the best predictor of the meat content because of the lean component of carcasses (Hendrick, 1983; Simela *et al.*, 1999). This justifies the suggestion that carcass weight should be included in all goat carcass classification schemes (Prasad and Kirton, 1992). Generally, goat carcass weight ranges between 10 and 32 kg and increases with increasing body weight. The average carcass weight was 10.1 kg in Indian breeds, but much higher at 31.2 kg in South African goats (Ueckermann, 1969). The average Omani goat carcass weight was 11.7–12.1 kg (Kadim *et al.*, 2004). The main problem in recording carcass weight in goats is that of standardizing the time after slaughter at which the weight is recorded. Immediately after slaughter, the carcass is hot. The subsequent cooling and shrinking, which takes place for about 24 h in a chiller room, results in variable losses in carcass weight of up to 2%. Pinkerton and McMillin (2000) reported a highly significant simple correlation between hot and cold carcass weights in goats ($r$ = 0.96). Boer-cross kid goats raised for meat had heavier hot carcasses than dairy breeds of Angora or Spanish kid goats. Spanish × Boer cross yearling goats had heavier hot carcass weights than Angora and dairy yearling goats (Pinkerton and McMillin, 2000). The adult dairy goats had heavier hot carcasses than Angora and Spanish adult goats because their weight was much higher.

The effects of goat breed, sex and nutrition within breed on carcass weight and dressing percentage have been investigated (Kadim *et al.*, 2004; Mahgoub *et al.*, 2005; Safari *et al.*, 2009). The adult dairy goats had higher ($P < 0.05$) hot carcass weights than Angora and Spanish adult goats because the live weight was much heavier (Pinkerton and McMillin, 2000).

Carcasses of mature male goats were significantly heavier than female carcasses due to the heavier bone and forequarters of males (Simela *et al.*, 1999). The effects of concentrate supplementation on carcass quality of Small East African goats were assessed by Safari *et al.* (2009), who found that hot and cold carcass weights of goats that received 100 and 66% ad libitum concentrates were 3 kg heavier than those that did not receive concentrate.

## 12.5 Carcass Linear Measurements

Objective techniques are used to predict carcass quality according to the anticipated proportions of muscle, bone and fatty tissues. Most carcass linear techniques have aimed to describe carcass quality in the context of the measurable length of bone on the basis that this defines body conformation and may be an important predictor of carcass components. Some schemes have used back-fat thickness as a component for grading carcasses, while others have used a combination of subjective and objective techniques for carcass comparison predictions.

The weight and conformation of the goat carcass and its body composition are important to farmers and consumers. Carcass conformation is a product of the rates of deposition of the various carcass components. However, techniques used in research work vary widely. Therefore, uniform, fast, repeatable and precise techniques are needed for practical evaluation of carcasses in the field and to assess carcass quality. Carcass linear measurements can be used for carcass quality evaluation. A series of external linear measurements were suggested by Moxham and Brownlie (1976) including carcass length, leg length, gigot width, maximum shoulder width, depth from scapula to sternum and width behind the shoulders (Pálsson, 1939). The measurements that are used most commonly for sheep are summarized in Table 12.2.

Differences in carcass dimensions should be taken into consideration if this form of carcass grading is to be adopted for goats, as they affect carcass conformation. Carcass evaluation was carried out on 145 goats by Owen *et al.* (1977) to provide background data on various body and carcass characteristics. Although most of the linear measurements reflected the steady growth in goats with increasing age, certain parts of the goats did not show any increased growth in the older age groups (Owen *et al.*, 1977). The head and fore limbs are early maturing relative to the rest of the body, according to the theory of centripetal or heterogenic growth (Pálsson, 1955).

Carcass quality characteristics can be evaluated by assessing the conformation of the carcass and the amount and distribution of muscle and fat in a carcass. Goat carcasses are thin and shallow and not compact as in other meat-producing animals. However, goat carcasses become thicker and more compact as carcass weight increases. Safari *et al.* (2009) found that concentrate-supplemented Small East African goats displayed higher values for carcass length, chest depth and leg length than those of non-supplemented goats. The minimal difference in carcass fatness between concentrate-supplemented and non-supplemented groups (Safari *et al.*, 2009) could be attributed to the unique fattening pattern of goats, as they deposit most of their fat around viscera and less in the carcass (Babiker *et al.*, 1990; Webb *et al.*, 2005). However, increasing levels of carcass conformation with concentrate allowance suggest that goats respond to improved nutrition by accretion of more muscle protein (Sheridan *et al.*, 2003). Simela *et al.* (1999) reported that carcass length, chest depth, thigh circumference and eye muscle area significantly increased with age of male goats, while fat depth over the eye muscle did not vary with age.

There were no significant differences in carcass dimensions between different sexes of mature goats except for the eye muscle area, which was significantly greater in male than in female goats (Simela *et al.*, 1999). Omani Batina, Dofari and Jebel Akhdar goats were slaughtered at similar ages (Kadim *et al.*, 2006). Several carcass measurements were recorded and the results are presented in Table 12.3. The carcass and the leg lengths of the Jebel Akhdar goat were longer by 50 and 27 mm, respectively, than the Dhofari goats. The longer carcass of Jebel Akhdar goats compared with the other two breeds can be related to the larger size of the breed. The depth from scapula to sternum was significantly greater by 19 and

**Table 12.2.** Definitions of the carcass linear measurements shown in Fig. 12.3.

| Measurement | Definition |
|---|---|
| Carcass length | From the point where the gambrel is inserted through the Achilles tendon to a point just anterior to the point of the humerus (Moxham and Brownlie, 1976) |
| Leg length | From the distal end of the tarsals to the centre of the tuberosity of the tibia, which is visible on the ventral aspect of the hanging carcasses (Pálsson, 1939) |
| Gigot width (G) | Maximum width of the gigots, with the carcass suspended from a gambrel; the measurement is taken at right angles to the length of the carcass at a line level with the femoral trochanter (Pálsson, 1939) |
| Maximum shoulder width | Maximum width of the shoulder, measured at the level of the scapula from one lateral surface to the other (Pálsson, 1939) |
| Depth from scapula to sternum | Maximum depth of the chest taken behind the shoulders at a line cutting the posterior angles of the scapula and at right angles to the length of the carcass (Pálsson, 1939) |
| Width behind shoulders | Minimum width behind the scapulae (Pálsson, 1939) |

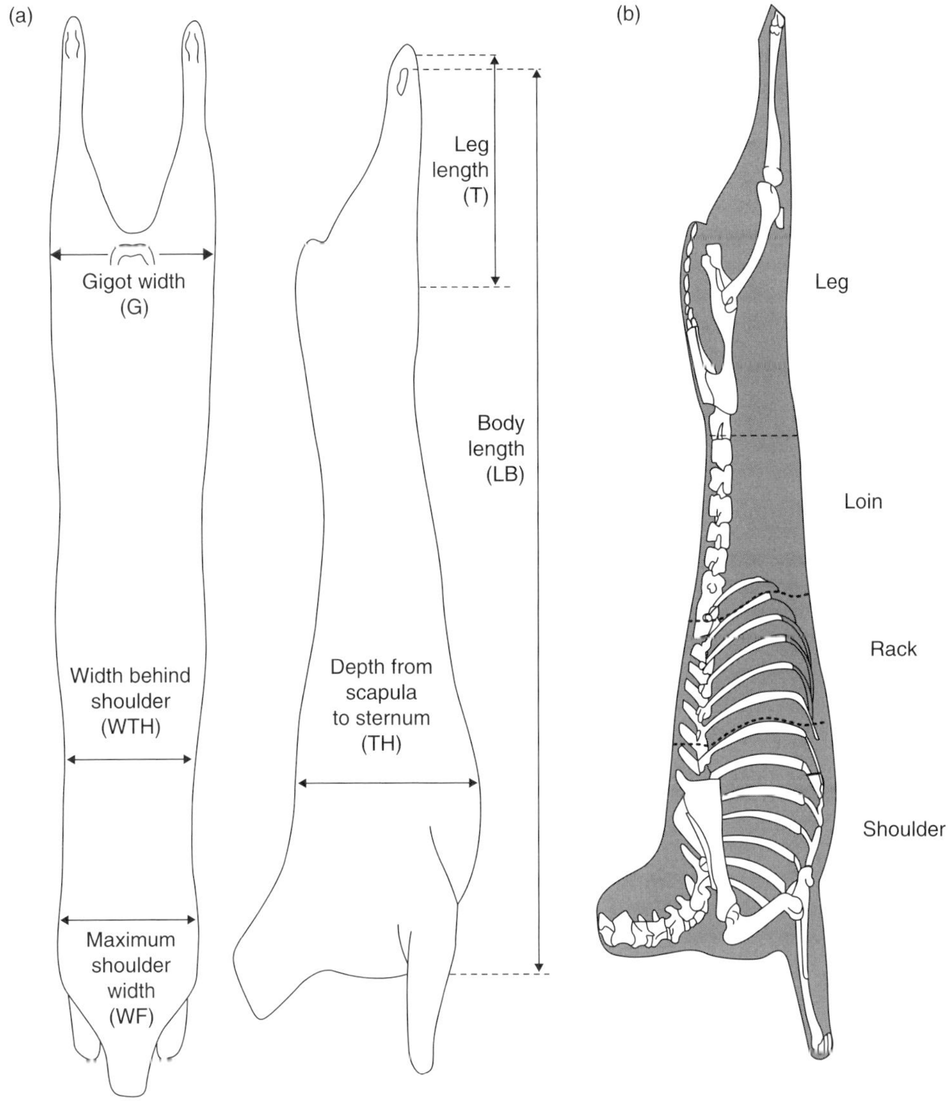


**Fig. 12.3.** (a) Diagram indicating where measurements are taken on a hanging carcass (see Table 12.2 for details). (b) Side view of a carcass showing the general position of wholesale cuts.

27 mm for the Jebel Akhdar carcass than the Batina and Dhofari carcasses, respectively. The shorter carcass of the Dhofari goat was accompanied by a significantly wider gigot (10 mm more than the Batina goat). Differences between breeds in carcass dimensions should be taken into consideration if carcass grading is adopted for Omani goats as well as other goats as they affect carcass conformation. The longissimus muscle area differed significantly between breeds, with the Jebel Akhdar goats having a significantly larger area by 24 mm$^2$ than the Batina breed (Table 12.3). The longissimus muscle area for the three Omani goats ranged from 106 to 130 mm$^2$. These values are comparable to the value of 112 mm$^2$ for Dhofari goats reported by El Hag and El Shargi (1996) but was much higher than the 68 and 72 mm$^2$ reported for other breeds by

**Table 12.3.** Slaughter weight, empty body weight, carcass weight, dressing percentage and carcass linear dimensions for three breeds of Omani goats (Kadim *et al.*, 2004).

| | Breed | | |
|---|---|---|---|
| Parameter | Batina | Dhofari | Jebel Akhdar |
| Body weight (kg) | 29.3 | 29.9 | 33.1 |
| Empty body weight (kg) | 21.9 | 22.1 | 24.6 |
| Hot carcass (kg) | 11.9 | 12.7 | 13.4 |
| Carcass (kg) | 11.7 | 12.5 | 13.1 |
| Dressing percentage[a] | 39.8 | 41.8 | 39.5 |
| Dressing percentage[b] | 53.4 | 56.6 | 53.3 |
| Carcass length (mm) | 1046 | 1025 | 1075 |
| Leg length (mm) | 269 | 253 | 280 |
| Th (mm)[c] | 267 | 259 | 286 |
| Gigot width (G) (mm) | 163 | 173 | 169 |
| WTH (mm)[d] | 198 | 217 | 201 |
| WF (mm)[e] | 132 | 137 | 139 |
| Longissimus depth (mm) | 56 | 57 | 60 |
| Longissimus width (mm) | 25 | 28 | 30 |
| Longissimus area (mm) | 106 | 119 | 130 |

[a]Based on full live body weight.
[b]Based on empty body weight.
[c]Th, Depth from scapula to sternum.
[d]WTH, Width behind shoulder.
[e]WF, Maximum shoulder width.

Potchoiba *et al.* (1990) and Dhanda *et al.* (1999a), respectively. These differences can be attributed to differences in the live weight of the goats used in the different studies. There were no significant breed differences for longissimus dorsi muscle depth and width. Boer × Spanish goats had carcasses with higher conformation scores and a larger leg circumference than carcasses from Spanish goats, but lean, bone and fat proportions were similar in the carcass and wholesale cuts within diet groups (Oman *et al.*, 1999).

Castrated indigenous male goats are heavier and larger in most linear measurements than goats in general and are generally larger in spite of a similar overall shape. This agrees with reports on sheep (Hammond, 1932, 1960). Hammond reported that Boer goats were heavier and larger in all respects than indigenous goats, except in the distal parts of the fore- and hindlegs and in shoulder height. Shorter fore cannons have been linked with earlier maturity and breed improvement in sheep (Pálsson, 1939, 1955). The bigger and blockier bodies of the Boer goats are also indicative of a relatively improved breed.

## 12.6 Dressing Percentage

The dressing percentage is a measure of carcass weight relative to live weight of the goat. It depends on the state of maturity, breed, sex and alimentary tract contents. The latter factor will vary depending on the period of fasting and the amount of feed consumed before slaughter. Dressing percentage, expressed in terms of empty body weight (EBW), usually rises as the animal increases in age and body weight by around 2–5% (Norman, 1991). Dividing the hot or cold carcass weight including kidney by live body weight will yield dressing percentage values in the 39–52% range in goats. The interactions among the individual factors affecting dressing percentage are great, making it difficult to accurately predict carcass yield or quality by visual examination of the live goat.

In the goat, the dressing percentage was found to vary from 38.5 to 52.3% (Owen, 1975; Kadim *et al.*, 2004; Safari *et al.*, 2009) depending on sex, body condition and breed. As with most livestock species, the age and sex of the goat influence carcass and dressing percentage. The dressing percentage is lower for male than female goats, with this difference increasing with age (Mahgoub *et al.*, 2005).

Average dressing percentages across age groups were 48–54% for Spanish × Boer cross goats and 43–49% for Angora and dairy goats in the study of Pinkerton and McMillin (2000). Dressing percentage (based on EBW) of goats ranged between 53 and 57%, with the Dhofari goats having a significantly higher value than the other two breeds (Kadim *et al.*, 2004). It is rather difficult to compare values of dressing percentages from different studies because different methods of slaughter are applied (e.g. kidney, kidney fat, head). Values varied between 44 and 55% (Gaili *et al.* 1972; Owen, 1975; Fehr *et al.*, 1976; Pinkerton and McMillin, 2000; Kadim *et al.*, 2004).

Sex class also influences carcass weight and dressing percentage. There are, however, some reports where the differences in dressing percentage of male and female goats were 53.5 and 55.3%, respectively, within differences being non-significant (Mahgoub *et al.*, 2005). Intact male goats and wethers had a lower dressing percentage than female goats, but the difference was not significant (Mahgoub *et al.*, 2005). Gut fill was greater in entire males than castrated males with little difference in dressing percentage (53.5 and 52.9%, respectively) when dressing percentage was calculated on an EBW basis (Mahgoub *et al.*, 2005). Mahgoub (1997) reported similar results and attributed the small difference in dressing percentage between males and females to the higher percentage of subcutaneous fat in the female carcasses. Furthermore, the heavier bone of the head and lower extremities of the male also has an adverse effect on the dressing percentage.

Differences among breeds in dressing percenatge have been widely investigated (Nagpal *et al.*, 1995; Dhanda *et al.*, 1999a,b; Kadim *et al.*, 2004; Oman *et al.*, 1999; Mushi, 2004; Mahgoub *et al.*, 2005). Dressing percentage based on full and empty live body weights were in the range of 39.5–41.8% and 53.3–56.6%, respectively, and dressing percentage significantly varied between three Omani breeds. Dhofari goat had a significantly higher dressing percentage. Jebel Akhdar goats had higher cold and hot carcass weights than Batina goats (Mahgoub *et al.*, 2005). This result is in agreement with Kadim *et al.* (2004), who showed significant differences among the same breeds of goat. The dressing percentage of Jebel Akhdar goats was higher than those reported for most tropical breeds (52 and 53%) such as Batina (on a fasted weight basis) at the same weight of 18 kg (Mahgoub and Lodge, 1996). Dhanda *et al.* (1999a) reported significant differences between various goat breeds for dressing percentage based on full body weight. They attributed these differences to variations in the weight of the digestive tract contents. A digestive tract content of 9.8–13.9% is comparable to 11–13% (of fasted slaughter weight) for Batina goats (Mahgoub and Lodge, 1996) but lower than the value of 17.8% found for some tropical breeds (Devendra and Burns, 1983). Variations for dressing percentage based on full body weight are generally less than for dressing percentage based on EBW. A dressing percentage (based on EBW) of goats ranging from 53.3 to 56.6% is in agreement with the reports of Potchoiba *et al.* (1990), El Hag and El Shargi (1996), Mahgoub and Lodge (1996) and Marinova *et al.* (2001). Breed differences in dressing percentage or lack of them are influenced by the degree of full gut at slaughter. The dressing percentage was higher in young intact males of Spanish breeds compared with Angora breeds (Nagpal *et al.*, 1995).

The dressing percentage of goats is influenced by nutritional status and diet, and was found to increase from 40.4–43.0% in unfattened goats to 54.1% in fattened goats (Gaili *et al.*, 1972). Animals kept on high nutritional plane diets produced higher dressing percentages (Gaili *et al.*, 1972). The energy density of diets fed to goats can influence carcass characteristics across various slaughter weights (Ueckermann, 1969). Dressing percentage rises with increasing slaughter weight and length of feeding period. The feed conversion ratio (kilograms of feed required for a kilogram of weight gain) decreases as heavier carcasses are produced. An all-roughage ration required more time on feed and required more kilograms of hay per kilogram of weight gain than rations containing concentrates. The responses to ratios of 60:40 and 40:60 roughage:concentrate were rather similar. The relative costs of hay and concentrates and the reduced time on feed at the higher concentrate level will determine the economic choice among roughage:concentrate ratios. Difference in feeding regimes between studies is a possible factor for the observed discrepancies. Shahjalal *et al.* (1992) reported increased carcass weight and dressing percentage of British Angora goats with increasing levels of high-energy concentrate diets. Safari *et al.* (2009) stated that dressing percentage increased with levels of concentrate supplementation in a curvilinear fashion. Yayneshet *et al.* (2008) studied the effect of different levels of *Acacia etbaica* and *Dichrostachys cinera* fruits on the dressing percentage of Abergelle goats. The data indicated that dressing percentage increased with an increased level of either type of fruit supplement, the highest (51.8%) being achieved at the 1.5% *D. cinera* level.

## 12.7 Carcass Components

Goat carcasses are considerably leaner than other livestock carcasses, chiefly because the fat tends to be concentrated around the viscera and is separated as offal at slaughter (Owen *et al.*, 1978). Goat carcasses usually have patch coverage of fat (less than 2–3 mm) and consequently they have a high lean content. Mahgoub *et al.* (2004) showed that, as a proportion of total fat, goats deposited relatively high levels of fat in the body cavity. Total non-carcass fat accounted for 42–52% of the total body fat in the goat (Mahgoub *et al.*, 2005) with more intermuscular fat (28–41%) than subcutaneous fat (15–25%). The apparent ability of goats to store fat internally around the viscera is a demonstration of their adaptability to the environmental extremes of the tropics, as indigenous African wildlife has similar characteristics.

As with most livestock species, carcass weight, breed, sex and nutritional systems of the goat influence carcass components. Goat carcass weight has a remarkable influence on carcass components, which is reflected in the goat meat markets. Increased carcass weight resulted in increased carcass fat percentage in both sexes (Mahgoub *et al.*, 2005; Table 12.4). Similar conclusions were reported by Dhanda *et al.* (2003a). Other studies reported that increased carcass weight increased the proportions of lean to fat and bone (Ruvuna *et al.*, 1992). In contrast, Simela *et al.* (1999) found that carcass lean and bone weights increased with age of male goats but fat content and proportions of the tissues did not vary with age. The lack of variation in subcutaneous and intermuscular fat measurements in goats with age is due to the inherent low priority for fat deposition in the carcass depot (Simela *et al.*, 1999). In addition, development of this issue is limited by the anti-lipogenic effect of androgens in males and the off-range diet, which does not enhance fat deposition (Casey and van Niekerk, 1985).

Sex also influenced goat carcass composition, with fat tissue being the most affected (Mahgoub *et al.*, 2004, 2005). Intact males at 28 kg live weight had a higher lean to fat to bone ratio (64:16:16) than females (62:21:13) (Mahgoub *et al.*, 2005). Similar conclusions were reported by Ruvuna *et al.* (1992). These data substantiated the findings of Ruvuna *et al.* (1992) that castrated males had lower levels of carcass fat than females. Carcasses from intact males had higher contents of muscle and lower contents of fat than carcasses from females (Colomer-Rocher *et al.*, 1992), whereas carcasses of castrated male kid goats had higher percentages of lean and lower amounts of

**Table 12.4.** Carcass weight and body components of Jebel Akhdar goats (from Mahgoub *et al.*, 2005).

| | Buck | | | Wether | | | Doe | | |
|---|---|---|---|---|---|---|---|---|---|
| Item | 11 kg | 18 kg | 28 kg | 11 kg | 18 kg | 28 kg | 11 kg | 18 kg | 28 kg |
| Cold carcass wt. (kg) | 4.7 | 7.9 | 13.6 | 4.2 | 7.9 | 13.2 | 5.1 | 4.8 | 13.9 |
| Carcass muscle (%) | 62.7 | 63.2 | 64.0 | 60.1 | 61.0 | 62.5 | 63.3 | 62.4 | 61.5 |
| Carcass bone (%) | 23.6 | 20.2 | 15.6 | 21.5 | 18.7 | 16.0 | 19.5 | 15.8 | 13.1 |
| Carcass fat (%) | 8.5 | 11.0 | 10.1 | 13.1 | 16.0 | 17.7 | 11.1 | 16.6 | 21.3 |
| Muscle:bone ratio | 2.7 | 3.2 | 4.1 | 2.6 | 3.3 | 3.9 | 3.3 | 4.0 | 4.7 |
| Muscle:fat ratio | 7.4 | 5.6 | 4.1 | 4.8 | 3.8 | 3.6 | 5.8 | 3.8 | 3.0 |

carcass fat than carcasses from female kid goats (Hogg *et al.*, 1992; Mahgoub *et al.*, 2005). Those findings were similar to earlier reports on goats. Wilson (1960) reported that female East African Dwarf kid goats had higher fat levels and less bone in the carcass than male goats, with differences increasing with age. Johnson *et al.* (1995) also reported that carcasses of female kid goats had less bone and more fat than those of males, which in turn had less bone and less fat than carcasses of castrated males. Mahgoub *et al.* (2004, 2005) reported that weight at slaughter influences composition, with wethers having more total carcass fat than intact males or females at an 11 kg slaughter weight, whereas females had more total body and carcass fat than wethers, which had more fat than intact males, at 18 and 28 kg slaughter weights. Goats at a lighter carcass weight of 16 kg were used in the study of Hogg *et al.* (1992), whereas a range of carcass weights from 2 to 52 kg of 37 male and female Saanen goats was reported in the study by Colomer-Rocher *et al.* (1992). This may explain some of the differences in results. Litter size did not affect body composition in goats (Todaro *et al.*, 2004).

Muscling was not found to be different in goats of five different genotypes (Boer × Angora, Boer × Saanen, feral × feral, Saanen × Angora and Saanen × feral) at the same live weight (Dhanda *et al.*, 1999a,b). In other studies, however, fat thickness at the 12th to 13th rib was different among male goats of these genotypes (Dhanda *et al.*, 2003a). Although there were no differences in percentages of the carcass cuts with genotype, percentages of muscle in the shoulder and leg were higher from goats with feral genotypes (Dhanda *et al.*, 2003b). Johnson (2000) found that 14–20 kg capretto carcasses from Boer × cashmere and cashmere male goats had more subcutaneous and intermuscular fat than from Boer × feral male kid goats. These results were generally reinforced by the findings of Hussain *et al.* (1983), where, in 11 genotypes, goats with some feral breeding had higher percentages of muscle and lower percentages of fat in carcasses than goats from established breed genotypes. Imported feral goat carcasses from Australia had a superior conformation with the same amount of external fat, higher percentages of total primal cuts and lower percentages of total boneless meat than did carcasses from goats raised on pasture (Nuti *et al.*, 2003).

Goats on a high plane of nutrition had heavier carcass weights with higher levels of fat than did goats on a lower plane of nutrition (Wilson, 1960; Haddad, 2005). Feedlot finishing of Boer × Spanish and Spanish goats with 80% concentrate diets ad libitum resulted in increased carcass fat thickness and increased fat percentage in primal cuts compared with goats raised on rangeland with no supplemental feeding (Table 12.5; Oman *et al.*, 1999).

Boer-cross kid goats received some grain and had slightly higher levels of external fat than the Boer-cross yearling goats that were raised only on pasture. It was anticipated that external fat would increase in goats of older ages as a normal

**Table 12.5.** Selected characteristics of carcasses from Boer × Spanish and Spanish goats from feedlot or range regimens (from Oman *et al.*, 1999).

| | Boer × Spanish | | Spanish | |
|---|---|---|---|---|
| Item | Feedlot | Range | Feedlot | Range |
| Hot carcass weight (kg) | 21.7 | 10.0 | 19.2 | 8.8 |
| Dressing percentage | 56.9 | 48.8 | 57.4 | 47.5 |
| Adjusted fat thickness (cm) | 0.16 | 0.04 | 0.11 | 0.04 |
| Leg circumference (cm) | 54.9 | 44.0 | 52.0 | 42.6 |
| Carcass tissue | | | | |
| Lean (%) | 57.8 | 55.9 | 57.6 | 55.3 |
| Bone (%) | 26.5 | 36.89 | 27.6 | 36.5 |
| Fat (%) | 15.7 | 7.3 | 13.4 | 8.2 |

indication of increased animal maturity. The external fat score was correlated with estimated kidney and pelvic fat (0.60) and actual kidney and pelvic fat (0.56). Readers should be aware that growing/ageing goats deposit fat first (and in the greatest quantity) in the kidney/pelvic region and secondly over the rib cage. Only rarely does a goat get enough extra feed to lay down fat along the top of its back, similar to marbling (the intramuscular deposition of fat). Other red-meat species are specifically fattened in feedlots to have excessive external fat that is deposited somewhat uniformly along the back and over the ribs before marbling is deposited inside the muscle.

## 12.8 Conclusions

Goats can produce good-quality meat from few resources, which may be important with the increased global demand for meat. Age, breed and nutrition can influence goat body weight and carcass quality characteristics. Goat carcass characteristics can be evaluated by assessing the carcass components, which can be used to predict meat content.

## References

Babiker, S.A., El Khider, I.A. and Shafie, S.A. (1990) Chemical composition and quality attributes of goat meat and lamb. *Meat Science* 28, 273–277.

Banskalieva, V., Sahlu, T. and Goetsch, A.L. (2000) Fatty acid composition of goat muscles and fat depots: a review. *Small Ruminant Research* 37, 255–268.

Casey, N.H. and van Niekerk, W.A. (1985) Fatty acid composition of subcutaneous and kidney fat depots of Boer goats and the response to varying levels of maize meal. *South Africa Journal of Animal Science* 15, 60–62.

Colomer-Rocher, F., Kirton, A.H., Mercer, G.J.K. and Duganzich, D.M. (1992) Carcass composition of New Zealand Saanen goats slaughtered at different weights. *Small Ruminant Research* 7, 161–173.

Devendra, C. and Burns, M. (1983) *Goat Production in the Tropics*. Commonwealth Agriculture Bureaux, Farnham Royal, UK.

Dhanda, J.S., Taylor, D.G., McCosker, J.E. and Murray, P.J. (1999a) The influence of goat genotype on the production of Capretto and Chevon carcasses. 1. Growth and carcass characteristics. *Meat Science* 52, 355–361.

Dhanda, J.S., Taylor, D.G., Murray, P.J. and McCosker, J.E. (1999b) The influence of goat genotype on the production of Capretto and Chevon carcasses. 2. Meat quality. *Meat Science* 52, 363–367.

Dhanda, J.S., Taylor, D.G. and Murray, P.J. (2003a) Growth, carcass and meat quality parameters of male goats: Part 1: Effects of genotype and liveweight at slaughter. *Small Ruminant Research* 50, 57–66.

Dhanda, J.S., Taylor, D.G. and Murray, P.J. (2003b) Carcass composition and fatty acid profiles of adipose tissue of male goats. Part 2: Effects of genotype and live weight at slaughter. *Small Ruminant Research* 50, 67–74.

El Hag, M.G. and El Shargi, K.M. (1996) Feedlot performance and carcass characteristics of local (Dhofari) and exotic (Cashmere) goats fed on a high-fiber by-products diet supplemented with fish sardine. *Asian-Australian Journal of Animal Science* 9, 389–396.

Fehr, P.M., Sauvant, D., Delage, L., Dumont, B.L. and Ray, G. (1976) Effect of feeding methods and age of slaughter on growth performances and carcass characteristics of entire young male goats. *Livestock Production Science* 3, 183–194.

Fisher, A.V. (1975a) The accuracy of some body measurements on live beef steers. *Livestock Production Science* 2, 357–366.

Fisher, A.V. (1975b) Live animal measurements as a means of evaluating animals in beef production experiments. In: *EEC Seminar on Criteria and Methods for Assessment of Carcass and Meat Characteristics in Beef Production Experiments.* Zeist, the Netherlands, pp. 43–55.

Gaili, E.S.E., Ghanem, Y.S. and Mukhtar, A. M.S. (1972) A comparative study of some carcass characteristics of Sudan desert sheep and goats. *Animal Production* 14, 351–357.

Giese, J. (1992) Developing low fat meat products. *Food Technology* 46, 100–105.

Gipson, T.A. (1998) Current market trends and potential for meat goat production. *Journal of Animal Science* 76 (Suppl. 1), 110.

Haddad, S.G. (2005) Effect of dietary forage:concentrate ratio on growth performance and carcass characteristics of growing Baladi kids. *Small Ruminant Research* 57, 43–49.

Hammond, J. (1932) In: Crewe, F.A.E. and Cutler, D.W. (eds) *Growth and Development of Mutton Qualities in the Sheep.* Oliver & Boyd, Edinburgh, pp. 13, 40, 72, 238, 254, 355.

Hammond, J. (1960) Growth in size and body proportions in farm animals. In: Zarrow, M.X. (ed.) *Growth in Living Systems.* Basic Books, New York, pp. 321–334.

Hendrick, H.B. (1983) Methods of estimating live animal and carcass composition. *Journal of Animal Science* 57, 1316–1327.

Hogg, B.W., Mercer, G.J.K., Mortimer, B.J., Kirton, A.H. and Duganzich, D.M. (1992) Carcass and meat quality attributes of commercial goats in New Zealand. *Small Ruminant Research* 8, 234–256.

Hussain, M.Z., Naidu, R., Tuvuki, I. and Singh, R. (1983) Goat production and development in Fiji. *World Animal Review* 48, 25–38.

Johnson, D.D., McGowan, C.H., Nurse, G. and Anous, M.R. (1995) Breed type and sex effects on carcass traits, composition and tenderness of young goats. *Small Ruminant Research* 17, 57–63.

Johnson, T.J. (2000) Evaluation of capretto carcasses from Boer cross and Cashmere goats in the Mediterranean climate of Western Australia. In: *Proceedings of the 7th International Conference on Goats,* Tours, France. International Goat Association, Little Rock, AR, USA, p. 219.

Kadim, I.T. (1988) A study of the carcass composition and meat quality of Southdown sheep selected for differences in backfat depth. PhD thesis, Massey University, New Zealand.

Kadim, I.T., Mahgoub, O., Al-Ajmi, D.S., Al-Maqbaly, R.S., Al-Saqri, N.M. and Ritchie, A. (2004) An evaluation of the growth, carcass and meat quality characteristics of Omani goat breeds. *Meat Science* 66, 203–210.

Kadim, I.T., Mahgoub, O., Al-Ajmi, D., Al-Habsi, K.R. and Johnson, E.H. (2006) Comparative effects of low levels of dietary cobalt and parenteral injections of vitamin $B_{12}$ on body dimensions in different breeds of Omani goats. *Small Ruminant Research* 66, 244–252.

Kamalzadeh, A., Koops, W.J., van Bruchem, J., Tamminga, S. and Zwart, D. (1998) Feed quality restriction and compensatory growth in growing sheep: development of body organs. *Small Ruminant Research* 29, 71–82.

Lawrence, T.J. and Fowler, V.R. (1997) Measuring growth. In: *Growth of Farm Animals.* CABI Publishing, Wallingford, UK, pp. 271–313.

Mahgoub, O. (1997) Meat production from the Omani Dhofari goat. 1. Live weight growth and body composition. *International Journal of Animal Science* 12, 25–30.

Mahgoub, O. and Lodge, G.A. (1996) Growth and body composition in meat production of Omani Batina goats. *Small Ruminant Research* 19, 233–246.

Mahgoub, O., Khan, A.J., Al-Maqbaly, R.S, Al-Sabahi, J.N., Annamalai, K. and Al-Sakry, N.M. (2002) Fatty acid composition of muscle and fat tissues of Omani Jebel Akhdar goats of different sexes and weights. *Meat Science* 61, 381–387.

Mahgoub, O., Kadim, I.T., Al-Saqry, N.M. and Al-Busaidi, R.M. (2004) Effects of body weight and sex on carcass tissue distribution in goats. *Meat Science* 67, 577–585.

Mahgoub, O., Kadim, I.T., Al-Saqry, N.M. and Al-Busaidi, R.M. (2005) Potential of Omani Jebel Akhdar goat for meat production under feedlot conditions. *Small Ruminant Research* 56, 223–230.

Marinova, P., Banskalieva, V., Alexandrov, S., Tzvetkova, V. and Stanchev, H. (2001) Carcass composition and meat quality of kids fed sunflower oil supplemented diet. *Small Ruminant Research* 42, 219–227.

McMillin, K.W. and Brock, A.P. (2005) Production practice and processing for value-added goat meat. *Journal of Animal Science* 83 (Suppl.), E57–E68.

Morand-Fehr, P., Boutonnet, J.P., Devendra, C., Dubeuf, J.P., Haenlein, J.F.W., Holst, P., Mowlem, L. and J. Capote (2004) Strategy for goat farming in the 21st century. *Small Ruminant Research* 51, 175–183.

Mourad, M. and Anous, M.R. (1998) Estimates of genetic and phenotypic parameters of some growth traits in common African and Alpine crossbred goats. *Small Ruminant Research* 27, 197–202.

Moxham, R.W. and Brownlie, L.E. (1976) Sheep carcass grading and classification in Australia. *Wool Technology and Sheep Breeding* 23, 17–25.

Mushi, D.E. (2004) Studies on marketing on meat goats and carcass composition – a case study of Gairo auction markets. MSc dissertation, Sokine University of Agriculture, Morogoro, Tanzania.

Nagpal, A.K., Singh, D., Prasad, V.S.S. and Jain, P.C. (1995) Effect of weaning age and feeding system on growth performance and carcass traits of male kids in three breeds in India. *Small Ruminant Research* 17, 45–50.

Norman, G.A. (1991) The potential of meat from the goat. In: Lawrie, R. (ed.) *Developments in Meat Science*, Vol. 5. Elsevier Science Publishers, London, pp. 57–98.

Nuti, L., Pinkerton, F. and McMillin, K. (2003) Goat production and marketing in Australia and New Zealand. *Goat Rancher* August 24–25, 27–28.

Oman, J.S., Waldron, D.F., Griffin, D.B. and Savell, J.W. (1999) Effect of breed-type and feeding regimen on goat carcass traits. *Journal of Animal Science* 77, 3215–3218.

Owen, G.E. (1975) The meat producing characteristics of the indigenous Malawi goat. *Tropical Science* 17, 123–138.

Owen, J.E., Norman, G.A., Fisher, I.L. and Frost, R.A. (1977) Studies on the meat production characteristics of Botswana goats and sheep. Part 1: Sampling, methods and materials, and measurements on the live animals. *Meat Science* 1, 63–85.

Owen, J.E, Norman, G.A., Philbrooks, C.A. and Jones, N.S. (1978) Studies on the meat production characteristics of Botswana goats and sheep. Part III: Carcass tissue composition and distribution. *Meat Science* 2, 59–74.

Pálsson, H. (1939) Meat qualities in the sheep with special reference to Scottish breeds and crosses. *Journal of Agricultural Science* 29, 544–626.

Pálsson, H. (1955) In: Hammond, J. (ed.) *Progress in the Physiology of Farm Animals*, Vol. 2. Butterworth, London, p. 340.

Park, Y.W., Kouassi, M.A. and Chin, K.B. (1991) Moisture, total fat and cholesterol in goat organs and muscle meat. *Journal of Food Science* 56, 1191–1193.

Pinkerton, F. and McMillin, K. (2000) Evaluation of slaughter goat selection and goat meat classification systems. *Goat Rancher* February, 7–10.

Potchoiba, M.J., Lu, C.D., Pinkerton, F. and Sahlu, T. (1990) Effects of all-milk diet on weight gain, organ development, carcass characteristics and tissue composition, including fatty acid and cholesterol contents of growing male goats. *Small Ruminant Research* 3, 583–592.

Prasad, V.S.S. and Kirton, A.H. (1992) Evaluation and classification of live goats, their carcasses and cuts. In: Lokeshwar, R.R. (ed.) *Proceedings of the Vth International Conference on Goats*, Vol. II, part II, New Delhi, India, pp. 440–450.

Quartermain, A.R. (1991) Evaluation and utilization of goat breeds. In: Maijala, K. (ed.) *World Animal Science. B8. Disciplinary Approach.* Elsevier, Amsterdam, the Netherlands, pp. 451–469.

Ruvuna, F., Taylor, J.F., Okeyo, M., Wanyoike, M. and Ahuya, C. (1992) Effects of breed and castration on slaughter weight and carcass composition of goats. *Small Ruminant Research* 7, 175–183.

Safari, J., Mushi, D.E., Mtenga, L.A., Kifaro, G.C. and Eik. L.O. (2009) Effects of concentrate supplementation on carcass and meat quality attributes of feedlot finished Small East African goats. *Livestock Science* 125, 266–274.

Shahjalal, M., Galbraith, H. and Topps, J.H. (1992) The effect of changes in dietary protein and energy on growth, body composition and mohair fiber characteristics of British Angora goats. *Animal Production* 54, 405–412.

Sheridan, R., Hoffman, L.C. and Ferreira, A.V. (2003) Meat quality of Boer goat kids and Mutton Merino lambs. 1. Commercial yields and chemical composition. *Animal Science* 76, 63–71.

Simela, L., Ndlovu, L.R. and Sibanda, L.M. (1999) Carcass characteristics of the marketed Matebele goat from south-western Zimbabwe. *Small Ruminant Research* 32, 173–179.

Stankov, I.K., Todorov, N.A., Mitev, J.E. and Miteva, T.M. (2002) Study on some qualitative features of meat from young goat of Bulgarian breeds and crossbreeds of goats slaughtered at various ages. *Asian-Australian Journal of Animal Science* 15, 283–289.

Todaro, M., Corrao, A., Alicata, M.L., Schinelli, R., Giaccone, P. and Priolo, A. (2004) Effects of litter size and sex on meat quality traits of kid meat. *Small Ruminant Research* 54, 191–196.

Ueckermann, L. (1969) MSc Agriculture thesis, University of Pretoria, South Africa.

van Niekerk, W.A. and Casey, N.H. (1988) The Boer goat. II. Growth, nutrient requirements, carcass and meat quality. *Small Ruminant Research* 1, 355–368.

Webb, E.C., Casey, N.H. and Simela, L. (2005) Goat meat quality. *Small Ruminant Research* 60, 153–166.

Wilson, I.L., Katsigianis, T.S., Dorsett, A.A., Cathoppulis, T.E., Greaves, A.G. and Baylor, J.E. (1980) Performance of native and Anglo-Nubian crosses and observation on improved pastures for goats in the Bahamas. *Tropical Agriculture Trinidad* 57, 183–190.

Wilson, P.N. (1960) The effect of plane of nutrition on the growth and development of the East African Dwarf goat. III. The effect of plane of nutrition and sex on the carcass composition of the kid at two stages of growth, 16 lb weight and 30 lb weight. *Journal of Agricultural Science* 54, 105–130.

Yayneshet, T., Eik, L.O. and Moe, S.R. (2008) Feeding *Acacia etbaica* and *Dichrostachys cinera* fruits to smallholder goats in northern Ethiopia improves their performance during the dry season. *Livestock Science* 119, 31–41.

# 13 Nutritive Value and Quality Characteristics of Goat Meat

I.T. Kadim and O. Mahgoub
*Department of Animal & Veterinary Sciences, College of Agricultural and Marine Science, Sultan Qaboos University Muscat, Sultanate of Oman*

## 13.1 Abstract

Goat is one of the most common domesticated animals, both in its distribution and its utilization, due to its adaptation to a wide range of environments. There has been a worldwide tendency towards a rapid increase in the demand for goat meat, indicating an enormous market potential. Carcass quality characteristics of goats have received some attention, but products from goats, an important aspect in the marketing of goat meat, are not well defined. Goat meat quality parameters are extremely important for consumers to select meat. They are also important to upgrade meat quality through processing procedures, thus extending the shelf-life and distribution of products more effectively. Goat carcasses are considerably leaner than other livestock carcasses because their fat tends to be concentrated around the viscera and is discarded as offal during post-slaughter processing. The economic value of a carcass and its meat-quality characteristics depend on meat yield, as well as carcass cutting and processing. Goat meat has been recommended as an ideal meat for health-conscious people due to its low fat content. Goat meat is also a good source of desirable fatty acids, as goats deposit relatively higher amounts of polyunsaturated fatty acids than other ruminants. Goat meat thus provides high-quality nutrients but there are important palatability parameters. Pre- and post-mortem factors should be considered carefully to improve goat meat quality. New technologies can be used to improve goat meat quality by electrical stimulation, ageing and chilling.

## 13.2 Introduction

Goat meat is one of the most widely consumed meats in the world (Stankov *et al.*, 2002). The increasing economic importance of goat production in the world can be attributed to the increase in demand for its meat. This is because goats are widely distributed and well adapted to various environmental conditions and scarce feed resources. They are also able to utilize marginal land to produce high-quality animal protein products. Goat meat has an immense market potential, as it could become an excellent choice for health-conscious consumers because of its low fat content compared with other red meat and chicken (Colomer-Rocher *et al.*, 1992; Pratiwi *et al.*,

2007). Generally, goat carcasses contain between 10 and 19% more lean and between 47 and 54% less fat than cattle and sheep. Besides being low in fat and cholesterol, goat meat is an excellent source of protein, B vitamins, iron and zinc. Goat meat is also a good source of desirable fatty acids (FAs), as goats deposit relatively higher amounts of polyunsaturated fatty acids (PUFAs) than other ruminants (Banskalieva *et al.*, 2000; Mahgoub *et al.*, 2002; Pratiwi *et al.*, 2007). This is an important factor in reducing the risk of cardiovascular diseases (van Niekerk and Casey, 1988; Park *et al.*, 1991; Giese, 1992; Stankov *et al.*, 2002).

The conventional approach for studying the quality characteristics of goat meat starts with an understanding of the structure and physiology of living muscle. Goat skeletal muscle is made up of thousands of cylindrical muscle fibres, often running all the way from origin to insertion. The fibres are bound together by connective tissue through which run blood vessels and nerves (Fig. 13.1). Goat muscle biology, like that of other muscles, is complex due to its multi-functions such as contraction, composition and protection (Hocquette *et al.*, 1998). Muscle metabolism plays a significant role in the transformation of muscles to meat (Cortright *et al.*, 1997; Geay *et al.*, 2001). As quality has been recognized as one of the most important economic challenges for meat producers around the world, goat muscle characteristics are important because of their relationship with palatability. Animal physiology also generally plays an important role in controlling the changes that occur in the post-mortem conversion of muscle to meat, thereby affecting meat supply for human consumption (Cortright *et al.*, 1997; Hwang *et al.*, 2004a). The rate and extent of goat muscle post-mortem metabolism are dependent on the availability of glycogen at slaughter (Janz *et al.*, 2001), the temperature of the carcass (Newbold, 1996) and other post-mortem factors. Initially, during rigor mortis, muscles become stiff and hard, but they gain some softness after hanging and conditioning (ageing).

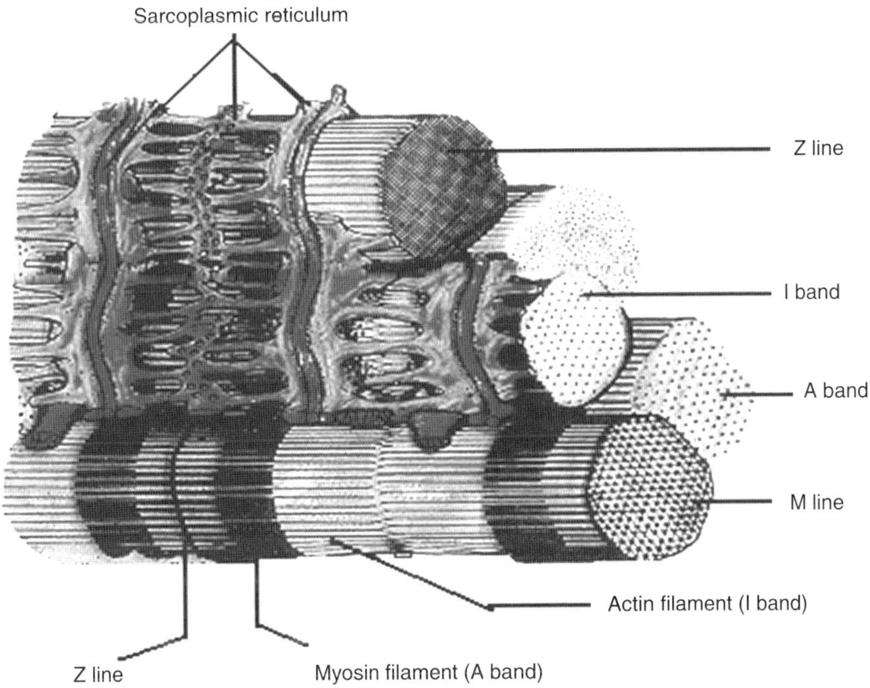


**Fig. 13.1.** Microstructure of animal muscle.

This chapter aims to summarize recent research to enable an understanding of the nutritive value and meat-quality characteristics of goat meat. The first part of the chapter covers the relationships between muscle ante-mortem changes and meat-quality traits. Next, the development of rigor mortis and the relationships between post-rigor changes and meat-quality traits are addressed. The third part describes the relationships between chilling, freezing and ageing processes and quality traits of goat meat. Lastly, the relationship between muscle fibre types and goat meat quality is discussed.

## 13.3 Muscle Structure, Physiology and Biochemistry

Each goat skeletal muscle cell is a filamentous, multinucleated structure composed of up to 1000 fibres known as myofibrils. A proper understanding of muscle structure architecture is essential for understanding the relationship of this contractile unit to muscle growth and development and ultimately meat-quality characteristics. The pivotal event in contraction of muscle is the precise assembly of the sarcomere, a highly ordered and complex array of numerous proteins. This section will discuss the major components in the muscle structure with an emphasis on those components expected to contribute significantly to meat quality.

Figure 13.1 shows a muscle fibre, the basic cellular unit of living muscle structure. The arrangement of myofibrils and myofilaments creates the meat texture (Swatland, 1984). The sarcosomes regulate calcium concentration to control the muscle contractions. Following the death of the animal, biochemical components necessary for anaerobic metabolism in muscles are still functional and, consequently, glycolysis proceeds until the glycogen is depleted and metabolism ceases. The adenosine triphosphate (ATP) necessary for muscle contraction and for the formation of actin–myosin crossbridges becomes completely depleted and rigor mortis becomes established. During the pre-rigor process, several changes occur in the concentrations of glycolytic substrates and products. The concentration of ATP does not decrease immediately post-mortem but instead remains at physiological levels for a short period before declining as a result of the regeneration of ATP from creatine phosphate during anaerobic glycolysis (Honikel *et al.*, 1983). When the ATP is exhausted, the thick and thin filaments remain locked to one another, causing the stiff nature of muscle in rigor. The characteristics of rigor mortis are elevated via loss of extensibility, muscle shortening (Honikel *et al.*, 1983), tension development (Nuss and Wolfe, 1981), resistance to strain (Lepetit *et al.*, 1998) and the combination of muscle tension and shortening (Olsson *et al.*, 1994).

Throughout the rigor mortis process, the production of $H^+$ leads to a more acidic environment leading to a decrease in meat pH. Fast-glycolysing muscles yield lower pH scores compared with slow-glycolysing muscle (Rosenvold and Andersen, 2003). The stress caused to goats before slaughter causes a depletion of muscle glycogen and therefore limits post-mortem glycolysis, resulting in high ultimate muscle pH (Warriss *et al.*, 1989). The rate of muscle biochemical reactions is influenced by meat temperature and ultimate pH. Muscle temperature and declining pH interactions during the onset of rigor influence meat quality via effects on protein denaturation and myofibrillar shrinkage (Rosenvold and Andersen, 2003). The calpain enzyme system may have an influence on myofibril-related palatability and proteolytic activity, and its function is affected as a result of the interactions between temperature and pH of muscle (Rees *et al.*, 2002; Hwang *et al.*, 2003). Muscle temperature at a pH above 6.0 has an important influence on meat quality, with an increase in 10°C resulting in a doubling of reaction rate. The temperature of the carcass at slaughter is 38–40°C. After processing, the carcass is usually placed in a chiller at 3–4°C. Carcass subcutaneous fat will act as an insulator and can significantly slow the rate of post-mortem temperature decline in the carcass. However, in goats, the subcutaneous fat is usually thin; therefore, the rate of post-mortem temperature decline is fast causing a slower decline in the rate of glycolysis. It is important to note that, within a given goat carcass, various muscles will display different cooling rates based on their location within the carcass.

## 13.4 Meat Composition

In many countries, fat is an unpopular constituent of meat for consumption, being considered unhealthy to the consumer. However, fat and FAs, whether in adipose tissue or muscle, contribute significantly to various aspects of meat quality and are essential for the nutritional value of meat. Goat meat varies in composition due to breed, age, sex, nutrition, conformation and site on the carcass. Water content differs only slightly among species, while differences in fat content are more marked (Mahgoub *et al.*, 2004). Goat meat contains 77% moisture (Dhanda *et al.*, 2003; Mahgoub *et al.*, 2004). This level is higher than that in the meat of other farm animals (Table 13.1). Goat meat is also a good source of protein, containing about 17.6–18.1% (Mahgoub *et al.*, 2004), comparable to the meat of other species (Table 13.1).

Fat serves as an energy store, providing a survival buffer against periodic food scarcity such as during periods of drought (Negussie *et al.*, 2003). Research findings show that the onset of fattening and the distribution of fat in domestic animals are affected by breed and plane of nutrition (Warren *et al.*, 2008; Mushi *et al.*, 2009b). An increase in non-carcass fat with increasing levels of concentrate supplementation is chiefly due to the increase in energy intake (Mushi *et al.*, 2009b). Animals bred for milk production deposit more fat internally around the viscera, while those bred for meat production deposit more in the carcass fat depots (Negussie *et al.*, 2003). Animals fattened on pasture generally have less body fat than those fed concentrates (Mushi *et al.*, 2009b). Fat in the carcass has beneficial roles with respect to reducing dehydration and cold shortening during the cooling process (Louvandini *et al.*, 2006). Goat meat has a fat content 50–65% lower than beef but a similar protein content. It has between 42 and 59% less fat than lamb, and is about the same to 25% lower than veal (James *et al.*, 1990). Goat intramuscular fat ranges between 1.19 and 2.34% (Mahgoub *et al.*, 2004; Banon *et al.*, 2006). A greater intermuscular fat content occurs in heavier carcasses and in older animals, decreasing the relative content of other nutrients. In Boer goat castrates, subcutaneous fat and intermuscular fat increased over a 10.4 kg growth range at exponential rates of 1.68 and 1.64 against empty body mass, respectively (Casey and Naude, 1984). Goats preferentially deposit fat internally as omental fat (Mushi *et al.* 2009b). Genotype has been recognized to have an effect on the chemical composition of goat meat (Dhanda *et al.*, 2003), and dif-

**Table 13.1.** Comparison of the composition of goat meat with meat from other species.

| Species | Moisture (%) | Protein (%) | Fat (%) | Ash (%) | Muscle | Reference |
|---|---|---|---|---|---|---|
| Camel | 71.0 | 21.4 | 4.4 | 1.1 | Longissimus | Kadim *et al.* (2006) |
| Beef | 71.5 | 21.5 | 5.5 | 0.9 | Longissimus | Mills *et al.* (1992) |
| Sheep | 68.9 | 21.0 | 8.5 | 1.2 | Longissimus | Sen *et al.* (2004) |
| Goat | 76.5 | 20.8 | 1.6 | 0.87 | Longissimus | Marinova *et al.* (2001) |
| Pig | – | 19.7 | 4.8 | 5.7 | Longissimus | Rosenvold *et al.* (2001) |
| Broiler | 75.5 | 22.4 | 1.5 | 0.6 | Pectoralis major | Castellini *et al.* (2002) |
| Duck | 76.8 | 21.0 | 1.68 | 1.0 | Pectoralis major | Baeza *et al.* (2002) |

**Table 13.2.** Comparison of cooked goat meat with other meats.

| 85 g Cooked (roasted) | Calories (g) | Fat (g) | Saturated fat (mg) | Protein (g) | Iron (g) |
|---|---|---|---|---|---|
| Goat[a] | 122 | 2.58 | 0.79 | 23 | 3.2 |
| Beef[b] | 245 | 16.00 | 6.80 | 23 | 2.9 |
| Pork[b] | 310 | 24.00 | 8.70 | 21 | 2.7 |
| Lamb[b] | 235 | 16.00 | 7.30 | 22 | 1.4 |
| Chicken[b] | 120 | 3.50 | 1.10 | 21 | 1.5 |

[a]USDA (1986); [b]Gebhardt and Thomas (2002).

**Table 13.3.** Mean composition of cooked muscle of lamb and Angora and Boer goats (Schonfeldt *et al.*, 1993).

| | Longissimus muscle | | | Semimembranosus muscle | | |
|---|---|---|---|---|---|---|
| Component (%) | Lamb | Angora | Boer | Lamb | Angora | Boer |
| Moisture | 64.6 | 64.7 | 65.4 | 63.9 | 64.2 | 64.4 |
| Protein | 26.6 | 26.8 | 27.2 | 29.4 | 29.2 | 29.1 |
| Fat | 7.1 | 7.0 | 6.2 | 4.7 | 4.7 | 4.4 |
| Ash | 1.06 | 1.07 | 1.08 | 0.99 | 0.97 | 1.00 |
| Dry matter | 35.3 | 35.3 | 34.4 | 36.0 | 35.8 | 35.8 |

ferent feeding regimes also contribute. Meat from Omani goats contained significantly more chemical fat and slightly less protein than Somali goats (Mahgoub *et al.*, 2004). Although meat with less fat might be preferred by consumers for health reasons, less carcass fat might also affect meat-keeping and quality attributes.

Cooked goat meat is also a good source of high-quality protein (Table 13.2), as well as being a rich source of iron. Its low-calorie content is comparable to that of chicken, mainly attributed to its low fat content (Table 13.2).

Decreasing moisture and fat content of meat during cooking results in an increase in the protein fraction, which also increases the amino acid content (Webb, 1991; Schonfeldt *et al.*, 1993). Both subcutaneous and intermuscular fat will affect this increase in the protein fraction. Different muscles on the carcass differ in chemical compositions, especially in terms of fat, as has been shown for cooked Angora and Boer goat meat and lamb (Table 13.3)

## 13.5 Nutritive Value of Goat Meat

The nutritive value of meat is becoming increasingly important for human health. It is not enough to determine the nutrient value of meat through chemical composition and bioavailability studies. Subsequently, it is not possible to apply such information as a norm for human nutrition without considering the role of meat as a supplementary nutrient source in most human diets. Human eating preferences, which are affected by palatability, flavour, taste and texture, tend to dictate dietary composition. In view of the lack of a complete dietary analysis, goat meat will be discussed from the basis of chemical composition. Generally, meat protein has a digestibility coefficient of 0.97, giving ingested meat a heat combustion of 17.87 kJ (Gopalan *et al.*, 1971). The average biological value (a measure of the proportion of absorbed protein that becomes incorporated into the proteins of the body) of goat meat reported by Mitra and Mitra (1945) was 60.4% compared with 68.6% for beef.

The nutritive value of meat lies in the extent to which the protein – and specifically the indispensable amino acid – requirements of humans are satisfied. The general composition of adult mammalian muscle is 75% water, 19% protein, 2.5% lipid, 1.2% carbohydrate, 0.65% minerals and <0.1% vitamins (Lawrie, 1985). The composition of muscle varies among muscles as a result of their greater or lesser amounts of connective tissue and intramuscular fat. As proteins accumulate and muscle hypertrophy occurs, the water:protein ratio changes. For instance, castrated Boer goat kids at 9.1% total body fat had a water:protein ratio in the buttock of 4.28 (Casey, 1982).

In general, the wealth of documentable evidence indicates that goat meat, regardless of age, breed or region, is a high-quality protein source along with a healthy fat level (high unsaturated fat:saturated fat ratio) with a minimal cholesterol intake risk. In addition, chevon contains comparatively higher values of iron, potassium and thiamine associated with a low sodium level (Eastridge and Johnson, 1990). All essential amino acids are present and a low caloric per serving value is available. Consequently, goat meat should be designated as a naturally occurring healthy meat.

### 13.5.1 Amino acid composition

The amino acid profile is very important in terms of meat composition, especially that of the indispensable amino acids. The amino acid profile of goat muscle shows a close resemblance to that of beef, pork and lamb (Table 13.4). Goat meat is a rich source of amino acids. Srinivasan and Moorjani (1974) reported that goat meat contains higher levels of arginine, leucine and isoleucine than mutton. Boer goat meat had significantly higher concentrations of 11 of the 18 measured amino acids than mutton (Sheridan *et al.*, 2003; Table 13.4). The usual limiting amino acids in different diets in various areas of the world are lysine, total sulfur amino acids, threonine and tryptophan. Pellett and Young (1990) noted that, by expressing amino acids in mg of meat, the supply of amino acids is determined largely by the amount of protein in a particular cut of meat. Meat is an important source of lysine, as 100 g of lean meat would provide 30–50% of the total protein needs of an adult and 60–100% of the estimated lysine needs.

On a lean-meat basis (muscle tissue), the amino acid composition variation between species is small. However, on a whole-meat basis (bone, fat and connective tissue), the amount of amino acids can be considerably different. Cuts of meat within species differ in composition according to the degree of fatness, age and sex.

### 13.5.2 Fat and fatty acids composition

The nutritive value of goat meat is becoming increasingly important for human health, particularly because of its leanness. The total lipid content of muscle (intramuscular fat) influences the tenderness and juiciness of cooked meat, although the strength of the correlation varies considerably between studies. Some studies have shown an important role for intramuscular fat, while others showed only a weak relationship with sensory traits. Goat meat obtained with green forages has less fat than that produced with concentrates and conserved forages (Atti *et al.*, 2006). There has been more emphasis on muscle composition because of its greater significance as a food and an increasing aversion to visible fat at retail. Muscle also contains higher concentrations of the long-chain ω-6 FAs, the importance of which in human nutrition has been recognized recently. Separation and identification procedures for low levels of unsaturated fatty acids (USFAs) in muscle have also greatly improved in recent years.

Saturated fatty acids (SFAs) are regarded as harmful to human health in contrast to PUFAs, which play a favourable role in the prevention of some human artery diseases (Mercier *et al.*, 2004). Although meat from monogastric animals contains

**Table 13.4.** Amino acid composition of muscle proteins of goat, beef, pork and lamb.

| Amino acid | Goat[a] | Mutton[a] | Goat[b] | Beef[b] | Pork[b] | Lamb[b] |
|---|---|---|---|---|---|---|
| Aspartic acid | 2.03 | 1.89 | | 88 | 89 | 85 |
| Threonine | 0.91 | 0.83 | 48 | 40 | 51 | 49 |
| Serine | 0.58 | 0.51 | | 38 | 40 | 39 |
| Glutamic acid | 3.16 | 2.91 | | 144 | 145 | 144 |
| Proline | 0.74 | 0.54 | | 54 | 46 | 48 |
| Glycine | 1.68 | 1.67 | | 71 | 61 | 67 |
| Alamine | 1.28 | 1.14 | | 64 | 63 | 63 |
| Valine | 1.19 | 1.04 | 54 | 57 | 50 | 52 |
| Methionine | 0.49 | 0.44 | 27 | 23 | 25 | 23 |
| Cystine | 0.30 | 0.30 | | 14 | 13 | 13 |
| Isoleucine | 1.03 | 0.92 | 51 | 51 | 49 | 48 |
| Leucine | 1.75 | 1.59 | 84 | 84 | 75 | 74 |
| Tyrosine | 0.63 | 0.56 | | 32 | 30 | 32 |
| Phenylalanine | 0.91 | 0.83 | 35 | 40 | 41 | 39 |
| Histidine | 0.63 | 0.55 | 21 | 29 | 32 | 27 |
| Lysine | 1.76 | 1.61 | 74 | 84 | 78 | 76 |
| Arginine | 1.44 | 1.38 | 75 | 66 | 64 | 69 |
| Tryptophan | 0.22 | 0.31 | 15 | 11 | 13 | 13 |

[a]Sheridan *et al.* (2003) (g/100 g); [b]Srinivasan and Moorjani (1974) (mg/g).

high levels of USFAs relative to ruminants, its meat is susceptible to oxidation (Leskanich *et al.*, 1997). Microorganisms in the rumen hydrogenate fat and increase the degree of saturated fat (Wood *et al.*, 1999). Meat FA composition is influenced by muscle type and its oxidation (Wood and Enser, 1997; Geay *et al.*, 2001). PUFAs are susceptible to rancidity because they contain double bonds. Meat with high concentrations of PUFAs can develop a rancid flavour faster than meat with fewer PUFAs. The interaction of oxygen with PUFAs is a non-enzymatic process. Vacuum packaging of meat products therefore provides a longer shelf life by excluding oxygen from the packaging. The FA composition of muscle affects its oxidative stability during processing and retail display, the PUFAs in phospholipids being liable to oxidative breakdown at this stage. A standard test for lipid oxidative stability in foods is the thiobarbituric acid reactive substances (TBARS) test of Tarladgis *et al.* (1960), which measures the oxidation product malondialdehyde. Values above 5 are considered critical, as they indicate a level of lipid oxidation products that produces a rancid odour and taste, which can easily be detected by consumers.

Goat meat is not only lower in total fat and cholesterol but is also lower in SFAs compared with other meats. FAs vary in length according to the number of carbon atoms that comprise their backbone, and may be saturated or unsaturated. SFAs are more solid at room temperature and contain no double bonds between carbon atoms. USFAs may contain one (monosaturated) or several (polyunsaturated) double bonds between the carbon atoms and are generally liquid at room temperature. It should be noted that the proportions of SFAs, monounsaturated FAs (MUFAs) and PUFAs in animal tissues vary depending on the species (Table 13.5). In monogastric species such as pigs and chickens, they may be influenced by diet.

The overall fat content of the animal and its muscle has an important impact on the overall FA composition because of the different FA compositions of neutral lipids and phospholipids. Phospholipids are essential components of cell membranes, and the amount of phospholipids remains fairly constant or increases slightly as the animal increases in fatness. In goats, the lower 18:1*cis*-9 and higher 18:2ω-6 content of phospholipids has a major influence on total muscle FA composition.

**Table 13.5.** Fatty acid content of muscle foods (from Hultin,1985).

| Species | Saturated (%) | Monounsaturated (%) | Polyunsaturated (%) |
|---|---|---|---|
| Beef | 55.5 | 52.0 | 3.0 |
| Pork | 44.0 | 56.5 | 10.5 |
| Mutton | 55.0 | 41.5 | 4.0 |
| Poultry | 30.5 | 45 | 18.5 |
| Fish | 30.0 | 33.0 | 37.0 |
| Goats | 51.3 | 43.5 | 5.09 |

Several factors can influence the FA composition of meat. Nutrition can affect the FA composition of muscle by improving the nutritional balance in ruminants to increase the level of PUFAs (Wood and Enser, 1997). It has been reported by Banskalieva *et al.* (2000), Mahgoub *et al.* (2002) and Pratiwi *et al.* (2007) that the FA composition of goat fat depots tends to change with age. Pratiwi *et al.* (2007) found more SFAs in the longissimus dorsi muscles taken from younger goats compared with those from older goats. This may be explained by the fact that young goats are suckling milk from their mother and the composition of FAs in their muscles is dependent on the composition of FAs in the consumed milk fat. Milk is rich in SFAs, which make up to 66% of total FAs (Devendra, 1980; Zygoyiannis *et al.*, 1992). As the rumen of young goats is not well developed, their diet would be influenced by the composition of FAs that exists in the milk consumed (Chilliard, 1993). In ruminants, the FAs in concentrate feedstuffs such as grains and oilseeds is degraded into MUFAs and SFAs in the rumen by microbial biohydrogenation, and only 10% will be incorporated into tissue lipids (Wood *et al.*, 2008). Muscle contains a significant proportion of long-chain (C20–C22) PUFAs, which are formed from 18:2ω-6 and 18:3ω-3 by the action of Δ5 and Δ6 desaturase and elongase enzymes. Important products are arachidonic acid (20:4ω-6) and eicosapentaenoic acid (20:5ω-3), which have various metabolic end products including eicosanoic acid.

The percentage of saturated fat in goat meat is 40% less than chicken (without skin), being far below beef, pork and lamb by 850, 1100 and 900%, respectively (USDA, 1989). According to Devendra (1988), the FA and protein values are usually constant with an intramuscular fat level of 0.94–1.4% in the Indian subcontinent breeds compared with Alpine, Toggenburg and Nubian Saanen goats (2.01%). Devendra (1988) also noted that USFAs predominate in goat meat, up to 70%, similar to the values in the USDA Handbook (1986) with a value of 69%, which is higher than the 50% reported by Eastridge and Johnson (1990). Lauric (2%), myristic (2.6%) and palmitic (27.6%) acids are SFAs of the hypercholesterolaemic group found in goat meat that elevate plasma cholesterol levels. The non-hypercholesterolaemic group of FAs consists of one SFA, stearic acid (C18:0, 14–16.6%) and the USFAs oleic (C18:1, 30.1–37%), linoleic (C18:2, 13.4%) and linolenic (C18:3 0.4%) acids.

The profile of the long-chain FAs of goat meat show oleic acid (C18:1) to be the most abundant, with levels of palmitic (C16:0) and stearic acid (C18:0) being relatively high (Casey and van Niekerk, 1985; Kuhne *et al.*, 1986; Casey *et al*, 1988). Although nutritional influences on the FA profile of ruminants are less than with monogastric animals, they can cause subtle changes in goats (Casey and van Niekerk, 1985). The high variance of each FA in goat kids can be ascribed to the monogastric characteristic of suckling animals, which makes them sensitive to nutritional influences. In adult Boer goat castrates, stearic acid and oleic acid in subcutaneous and kidney fat responded to five different energy levels (7.5, 8.4, 9.3, 10.3 and 11.2 MJ metabolizable energy/kg dry matter) fed for 90 days: stearic acid decreased by 41% and oleic acid increased by 21% (Casey and van Niekerk, 1985). Similarly, feeding eight types of pasture for 84 days influenced the levels of myristic, heptadecanoic, linoleic and stearic acids of subcutaneous fat of mutton (Casey *et al.*, 1988). In goats, oleic acid made up the greatest proportion (43%) of the subcutaneous fat, followed by palmitic acid (24%) and stearic acid (15%). Clearly, a range occurs in the FA profile of goat meat; for example, levels of stearic acid range from 12 to 26% and oleic acid from 21 to 46%.

Goat meat has a low ω-6 and ω-9 polyunsaturated fat content (Gimenenz *et al.*, 1985), which may have health-related nutritional implications for humans, particularly for the immune system (Wan *et al.*, 1989). The levels of C18:2ω-6, C20:1ω-9 and C20:4ω-6 in goat subcutaneous fat in goats fed four different diets were 3.12, 0.89 and 1.18%, respectively. Visceral fats are more saturated than subcutaneous fats, as is illustrated in the differences between FAs of the subcutaneous and kidney depots. Fat from the triceps brachii, biceps femoris and obliquus internus abdominis muscles of goats contained 57% C18:1 and 25% C16:0 (Ha *et al.*, 1986).

The FA profile of Jebel Akhdar Omani goats determined by Mahgoub *et al.* (2002) is presented in Table 13.6. The muscle tissue of the Jebel Akhdar goats contained an average of 51.3 and 48.7% of SFAs and USFAs, respectively (Table 13.6). These figures are in line with those reported elsewhere for goat meat. Potchoiba *et al.* (1990) reported a value of 50.6 and 49.4%, respectively, with Alpine kids. Johnson *et al.* (1995) reported a slightly higher ratio of USFAs to SFAs, but these results were from broiled rather than

**Table 13.6.** Mean (± SEM) fatty acid composition of muscle, subcutaneous and kidney fat tissue of Omani goats of pooled sexes and weights (from Mahgoub *et al.*, 2002).

| | Muscle | | Subcutaneous fat | | Kidney fat | |
|---|---|---|---|---|---|---|
| Group | Mean | SEM | Mean | SEM | Mean | SEM |
| C10 | 0.19b | 0.02 | 0.41a | 0.04 | 0.53a | 0.04 |
| C12 | 0.36b | 0.05 | 0.84a | 0.08 | 0.77a | 0.10 |
| C14 | 4.31c | 0.36 | 9.22a | 0.58 | 7.66b | 0.48 |
| C15 | 1.04b | 0.19 | 2.01a | 0.12 | 1.72a | 0.12 |
| C15:1 | 0.87a | 0.03 | 0.33b | 0.07 | 0.15c | 0.04 |
| C16 | 24.74a | 0.69 | 20.69b | 0.80 | 26.16a | 0.51 |
| C16:1ω7 | 6.88b | 0.73 | 9.65a | 0.50 | 3.67c | 0.31 |
| C17 | 1.91c | 0.13 | 5.28a | 0.51 | 3.59b | 0.32 |
| C18 | 18.72b | 0.67 | 16.06c | 0.80 | 28.82a | 1.09 |
| C18:1ω9 | 36.51a | 1.21 | 27.40b | 1.15 | 23.06 | 1.10 |
| C18:2ω6 | 4.08b | 0.66 | 6.70a | 0.47 | 3.05b | 0.22 |
| C:18:3ω3 | 0.17c | 0.03 | 0.35a | 0.06 | 0.30a | 0.17 |
| C19 | 0 | 0 | 0.73a | 0.06 | 0.27b | 0.04 |
| C20 | 0 | 0 | 0.40a | 0.02 | 0.20b | 0.04 |
| C20:2ω6 | 0.03 | 0.02 | 0 | 0 | 0.02 | 0.01 |
| C20:3ω3 | 0.16 | 0.07 | 0 | 0 | 0 | 0 |
| C20:4ω6 | 0.82a | 0.14 | 0.19b | 0.02 | 0.04c | 0.01 |
| SFAs | 51.27c | 1.22 | 55.37b | 0.82 | 69.72a | 0.93 |
| MUFAs | 43.47a | 1.08 | 37.38b | 1.02 | 26.88c | 1.13 |
| PUFAs | 5.09a | 0.67 | 6.90a | 0.48 | 3.10b | 0.22 |
| UFAs | 48.73a | 1.22 | 44.63b | 0.82 | 30.28c | 0.93 |
| UFA:SFA ratio | 1.00a | 0.06 | 0.82b | 0.03 | 0.44c | 0.02 |

SFAs, Saturated fatty acids; MUFAs, monounsaturated fatty acids; PUFAs, polyunsaturated fatty acids; UFAs, unsaturated fatty acids.
Means on the same row within sex or body weight without or with the same letter (a, b, c) do not significantly differ ($P > 0.05$).

raw meat. Generally, ruminants have a higher saturated:unsaturated ratio compared with monogastric animals (Wood, 1984). This is because USFAs in the diet are hydrogenated in the rumen to saturated fats to be absorbed in a more saturated form, in contrast to monogastric animals in which USFAs are absorbed directly from the intestine into the bloodstream. MUFAs in Jebel Akhdar goats were present at an average of 43.5%, which is very close to that reported for other goat breeds (45% by Johnson *et al.*, 1995) and for lamb (46.2%) and beef (47.1%) (Seman and McKenzie-Parnell, 1989).

### 13.5.3 Cholesterol

The cholesterol content of goat meat is associated with its fat content, which means that fattier meat contains more cholesterol than leaner meat. Pratiwi *et al.* (2007) reported that total cholesterol concentration in goat muscle decreased as carcass weight increased. Brown *et al.* (1990) and Huskey *et al.* (1993) reported that the amount of intramuscular fat is not always related to the cholesterol content of the muscles. Some variation in cholesterol content between different goat muscles was also reported. The differences may be due to variation in anatomical sites and physiological functions and the development of muscle types, which may reflect the rate of cholesterol synthesis among different muscles (Lawrie, 1985; Wheeler *et al.*, 1987; Chizzolini *et al.*, 1999; Pratiwi *et al.*, 2007).

The cholesterol content of chevon is controversially similar to that of beef, lamb, pork and chicken and much lower than some dairy and poultry products and some seafoods. Other studies have indicated goat meat cholesterol levels of 76 mg/100 g compared with 70 mg/100 g for beef, fish and lamb and 60 mg/100 g for pork and chicken (Pond and Maner, 1984; Potchoiba *et al.*, 1990). Cholesterol levels of uncooked beef meat range from 36–46 to 78.2 mg/100 g (Terrell *et al.*, 1969) compared with 58–70 mg/100 g for chevon (Park *et al.*, 1991). Sheridan *et al.* (2003) reported that meat from Boer goats on a high-energy diet contained lower levels of cholesterol than that from mutton on a high-energy diet (66.8 versus 99.3 mg/100 g).

Santos-Filho *et al.* (2005) reported a value of 54.4 mg/100 g for cholesterol in goat meat, with castration of male goats increasing the cholesterol levels. Similar effects of castration in goats were reported by Madruga *et al.* (2001). Cholesterol levels in goat meat were significantly affected by breed and age, with levels increasing dramatically with slaughter age (Madruga *et al.*, 1999; Beserra *et al.*, 2004).

### 13.5.4 Macro- and microelements

Goat meat is an excellent source of minerals required for normal growth and good health. The macro- and micromineral contents of goat meat are reviewed in detail in Chapter 11 of this book. Calcium is an important element in the body required for bone development, neuromuscular activity, secretory functions, buffers, certain coenzymes and nutrients for the nursed young. Although an essential dietary component, lean meat has a low calcium content that is insufficient to provide the recommended daily allowance (RDA). Lean goat meat was found to contain 11–12 mg calcium/100 g (McCance and Widdowson, 1960). Mechanically deboned meat tends to have a higher calcium content, in the range of 0.05–0.75% (Kolbye and Nelson, 1977). Phosphorus is essential for bone formation, enzymes and energy metabolism. The RDA for phosphorus, based on the need to provide a dietary calcium:phosphorus ratio of 1, is 800 mg. Protein foods are good sources of phosphorus, with goat muscle containing 157 mg/100 g (Wan Zahari and Wahid, 1985). Goat liver (259 mg/100 g) and brain (246 mg/100 g) had about 100 mg more phosphorus than muscle. Fresh goat meat is also a good source of potassium (359 mg/100 g), contributing to total potassium intake (Wan Zahari and Wahid, 1985). Goat muscle contains 55–77 mg of sodium per 100 mg (Wan Zahari and Wahid, 1985). The magnesium level in goat muscle is 19.7 mg/100 g. Of the organs, the spleen had the highest content (15.28 mg/100 g) followed by the liver (15.08 mg/100 g) and brain (12.82 mg/100 g). Trace minerals such as copper, manganese and zinc in goat meat have a high bioavailability as meat does not contain inhibitors. A 100 g portion of goat muscle would provide 0.28–0.35 mg copper, 0.059–0.145 mg manganese and 2.79–4.21 mg zinc (Wan Zahari and Wahid, 1985). Minerals in meat are not affected by normal cooking procedures, but salting and curing can increase sodium levels dramatically. Red meat muscle has a high myoglobin content and provides a high level of bioavailable iron (Worthington-Roberts and Monsen, 1990), the haem iron being 5–10% more available than non-haem iron and this appears to enhance the absorption of non-haem iron from other foods. A value of 2.1 mg iron/100 g was reported by Abdon *et al.* (1980) for lean goat meat.

### 13.5.5 Vitamins

Goat meat is a good source of several vitamins such as thiamine (0.11 mg/100 g), riboflavin (0.49 mg/100 g), niacin (3.75 mg/100 g), folate (5 μg/100 g) and vitamin $B_{12}$ (1.13 μg/100 g) (Collins, 2008). Each 170 g of goat meat contains 61%, which is 33% of the daily recommended intake of riboflavin and vitamin $B_{12}$, respectively. According to Abdon *et al.* (1980), thiamine, riboflavin and niacin levels in lean goat meat are comparable with those in lean beef, lamb and veal (Table 13.7). Gopalan *et al.* (1971) reported 4.5 mg total folic acid and 2.8 mg vitamin $B_{12}$ per 100 g of goat meat.

Goat meat contains several vitamins (Table 13.8). However, as for other meats, it is devoid of ascorbic acid but is rich in thiamine, riboflavin and folate (Johnson *et al.*, 1995).

**Table 13.7.** Thiamine, riboflavin and niacin content of goat, lean beef, veal and lamb.

| Species | Thiamine | Riboflavin | Niacin |
|---|---|---|---|
| Goat[a] | 0.1 | 0.56 | 3.6 |
| Beef[b] | 0.082 | 0.218 | 3.6 |
| Lamb[c] | 0.088 | 0.234 | 5.33 |
| Veal[d] | 0.06 | 0.3 | 7.6 |

[a]Abdon *et al.* (1980); [b]USDA, 1986; [c]Ono *et al.* (1984); [d]Ono *et al.* (1986)

**Table 13.8.** Comparison of nutrient analysis of an 85 g cooked portion of carcass composite meat from goat, beef and chicken (from Johnson *et al.*, 1995).

| Nutrient | Goat | Beef | Chicken |
|---|---|---|---|
| Ascorbic acid (mg) | 0 | 0 | 0 |
| Thiamine (mg) | 0.20 | 0.07 | 0.05 |
| Riboflavin (mg) | 0.23 | 0.18 | 0.14 |
| Niacin (mg) | 2.96 | 3.10 | 7.2 |
| Pantothenic acid (mg) | 0.32 | 0.30 | 0.88 |
| Vitamin $B_6$ (mg) | 0.18 | 0.28 | 0.34 |
| Folate (μg) | 3.78 | 6.0 | 4.3 |
| Vitamin $B_{12}$ (μg) | 0.90 | 2.1 | 0.26 |
| Vitamin A (IU) | 22.0 | 0 | 137 |

## 13.6 Meat-quality Characteristics

Meat-quality characteristics have an influence on meat consumption where the mere availability of meat is not a factor. Meat-quality parameters become more important when meat is selected and the customer has a free choice between meat types. Quality parameters are also important for processing procedures to upgrade poor-quality meat, extending shelf life or stabilization, and allowing the product to be distributed and merchandised more effectively. Meat quality is a combination of the attributes of flavour, juiciness, texture, tenderness and appearance that contribute to the edibility or desirability of the products. The consumer relates to quality in terms of the tenderness, juiciness and flavour of the cooked products. Tenderness, muscle pH, water-holding capacity and the MFI will be considered in this section. Skeletal muscle of goat meat not only provides nutrients but also contains the most important palatability parameters.

### 13.6.1 Muscle pH

Glycogen contained within goat muscle at slaughter is metabolized through an anaerobic process leading to the formation of lactic acid and consequently to a lowering of the pH. Meat from goat will usually achieve ultimate pH values of around 5.5–5.6 but high ultimate pH values (6.0 or above) may be found if muscle glycogen stores have been depleted prior to slaughter.

The ultimate pH of muscle is regarded as one of the important parameters affecting meat-quality characteristics. Muscle from goat has s high myoglobin content, contains both type I and type II muscle fibre types and undergoes the same post-mortem biochemical changes as other meat animals (Heffron and Dreyer, 1975; Lawrie, 1985; Kadim *et al.*, 2010). The decline in muscle pH follows a pattern typical of red meat carcasses, stabilizing at around pH 5.5 (Owen *et al.*, 1978; Breukink and Casey, 1989). Variations in the ultimate pH of goat muscles may be due to individual muscles, sex or pre-mortem stress. The ultimate goat muscle pH was 5.69 for longissimus dorsi, 5.93 for biceps femoris, 5.96 for semitendinosus and 5.73 for semimembranosus muscles (Kadim *et al.*, 2004). Such differences might be explained by the fact that individual muscles differ in their proportions of red and white muscle fibre types and therefore also differ in patterns of energy metabolism, both ante- and post-mortem (Swatland, 1982). Variations among goat muscles in ultimate pH values were also reported by Gonzalez *et al.* (1983) and Marinova *et al.* (2001). Goat muscle pH values were relatively higher compared with other meat animals (Atti *et al.*, 2006; Pratiwi *et al.*, 2007; Mushi *et al.*, 2009a), which generally implies a lower meat quality (Lawrie, 1985). A high ultimate pH generally reflects depletion of muscle glycogen due to stress or other factors (Kadim *et al.*, 2006, 2010). Exhausting pre-mortem stress yields dark, firm and dry meat with a high ultimate pH (pH >6.0). Post-mortem biochemical changes are associated with the loss of water-holding capacity as the pH reaches the isoelectric point of the muscle proteins, the onset of rigor mortis, and the release and activation of proteolytic enzymes, notably cathepsins, responsible for the ripening of goat meat. Muscles from Omani goats had significantly ($P < 0.01$) higher ultimate pH value than those of Somali goats (Mahgoub *et al.*, 2004). This may be attributed to the stressful conditions that deplete muscle glycogen reserves before slaughter, resulting in low residual levels of glucose and consequently increasing the ultimate pH of meat. In Oman, the majority of local animals are delivered daily to the central slaughterhouse from within a radius of a few hundred kilometres in open trucks. This exposes them to a variety of physical and psychological stimuli, many of which are novel and some of which are aversive. The pH values from heavier goats were usually higher than those from lighter goats (Pratiwi *et al.*, 2007).

### 13.6.2 Water-holding capacity

Water retention of meat is caused primarily by immobilization of tissue water within the myofibrillar system. Applying pressure can cause a shift of water from the intercellular to the extracellular space and then on to the meat surface as a result of structural alterations at the level of the sarcomeres or of the myofilaments structure. Water-holding capacity values vary according to goat breed and site of muscle on the carcass. Kadim *et al.* (2004) found that expressed juice from three Omani goat breeds ranged from 33.3 to 41.0%, with some differences between the three breeds and across four muscles (longissimus dorsi, semimembranosus, semitendinosus and biceps femoris). A similar range of expressed juice for other breeds of goat was reported by Gonzalez *et al.* (1983). Sen *et al.* (2004) reported higher values of 57% for water-holding capacity in Indian goats. Lower values of 24.3% at 90 days of age and 24.0% at 180 days of age were reported in Polish goats (Pieniak-Lendzion *et al.*, 2008).

### 13.6.3 Shear force values

Scientists and consumers both recognize that tenderness is one of the most important sensory attributes of all types of meat. The meat industry in many industrial countries has responded to consumer desire for better tenderness using a variety of techniques including ageing, electrical stimulation, and mechanical, chemical and enzymatic tenderization of muscle. Physiologists, muscle-cell biologists, biochemists and proteomists continue to investigate the muscle cell to learn more about tenderness indices.

The measurement of meat tenderness is extremely difficult because meat is not a simple one-component system. It is the result of two structural components, muscle fibres and connective tissue, and is further complicated by the presence of fat interspersed within these structural elements. A variety of instruments have been developed to measure tenderness of meat, usually based on a shearing, penetrating, biting, stretching, breaking or compressing action. The Warner–Bratzelr shear device has been widely used for the evaluation of meat tenderness.

Consumers believe that tenderness is a driver of eating satisfaction; however, several studies have indicated that goat meat is inherently less tender than that of sheep (Schonfeldt *et al.*, 1993; Roeder *et al.*, 1999). The difference has been attributed to the muscles of goat having a higher collagen content with lower collagen solubility and less intramuscular fat than sheep (Heinze *et al.*, 1986). Schonfeldt *et al.* (1993) and Sen *et al.* (2004) found lamb and mutton to be tenderer, with less fibrous tissue residue than Angora and Boer goat meat. Roeder *et al.* (1999) showed that meat from Boer-cross goats is less tender than meat from other breeds. The attributed toughness of goat meat has been ascribed to slaughtering of mature animals, in which the collagen in the connective tissue has a decreased ability to gelatinize under the influence of heat and moisture. However, goat muscle fibres are thicker and the fibre bundles larger than those of sheep, giving goat meat a characteristic coarser grain (Gaili and Ali, 1985). Such changes could also be due to species difference in carcass fat. Intramuscular fat content may also contribute to the observed differences in tenderness.

The meat of kids, or capretto, is a tender delicacy. However, Smith *et al.* (1978) reported that goats of 3–4 months of age produced meat that was less tender than those of 6 months. They attributed these differences to the rapid chilling of small, lean carcasses, which are potentially susceptible to reduced proteolysis and cold-induced sarcomere shortening. The shear force of goat meat tends to be greater for heavier than for lighter carcasses (Pratiwi *et al.*, 2007). Webb *et al.* (2005) suggested that goat meat does not readily attain a highly acceptable degree of tenderness. In contrast, Dhanda *et al.* (1999) found no difference in the sensory panel tenderness scores between meat from very young and older goats. Locker and Hagyard (1963) demonstrated a relationship between chilling temperature and sarcomere shortening, and reported that muscle shortening was minimal when the muscle entered rigor at approximately 15°C, and, as muscle temperature at the onset of rigor decreased, greater sarcomere shortening occurred. Goat breeds may also differ in meat quality. Meat from Angora goats is tenderer than meat of Boer goats. This could be due to lower collagen content and better collagen solubility. Evaluation of collagen alone, however, would be insufficient for conclusions on differences in tenderness. Other factors may be involved, especially muscle fibre size, the type of matrix formed by collagen and the state of muscle contraction.

### 13.6.4 Myofibrillar fragmentation index

The structural changes occurring in meat after slaughter are generally caused by interactions of myofibrillar proteins in the muscle (Nagaraj *et al.*, 2006). Degradation of the structural protein and integrity alterations of meat can be evaluated using myofibrillar fragmentation index (MFI) determination. Viewing myofibrillar length changes under a microscope or determining turbidity changes can be used as an indicator of post-mortem proteolysis in various meat species (Lametsch *et al.*, 2007), which accounts for differences in the rate of post-mortem tenderization of meat (King *et al.*, 2004; Nagaraj *et al.*, 2005, Kadim *et al.*, 2010). The MFI is also used as an indicator of the weakening of key cytoskeletal proteins, particularly titin and nebulin (Taylor *et al.*, 1995). There is a proportionate increase in MFI as the post-mortem ageing of meat advances (King *et al.*, 2004). The magnitude of MFI values varies among goat breeds. Kadim *et al.* (2010) found an average value of 77 for the MFI in Omani goats, which is much higher than the value of 30 observed by King *et al.* (2004) in capretto meat.

The metabolism of goat meat can be hastened by increasing the MFI through ageing or stimulation processes (Kadim *et al.*, 2010). The effect of low-voltage electrical stimulation on the MFI of goat meat was studied by Kadim *et al.* (2010). They found that stimulation improved tenderness through alteration of the MFI of longissimus dorsi myofibrillar protein in goat. The high MFI in stimulated Omani goat carcasses appeared to be caused by readily breaking the myofibrils into shorter segments in comparison with the non-stimulated group. However, King *et al.* (2004) found that electrical stimulation did not affect the MFI at 1, 3 or 14 days of ageing of goat meat. According to Kadim *et al.* (2010), the MFI can contribute up to 50% of the variation in goat meat tenderness. A similar conclusion was reported by Olson and Parrish (1977) for beef meat. Transportation of goat at a high ambient temperature (42°C) significantly decreases the MFI by 5.5%.

### 13.6.5 Meat colour

Meat colour measurements involve two basic methods: visual appraisal and instrumental analysis. Both methods inherently involve an assessment of the concentration and chemical form of myoglobin, the morphology of the muscle structure and the ability of the muscle to absorb or scatter incident light.

Meat colour is one of the most important sensory characteristics according to which consumers make judgements about meat quality. It is influenced by the pigment content, the chemical form of the pigment, meat pH and the meat structure (Lindahl *et al.*, 2001). Some residual blood may also be present in meat, but it is generally minimal and is of little practical important concern in considerations of meat colour. The degree of meat pigmentation is directly related to the chemical structure of myoglobin. In general, myoglobin concentration within a given muscle will differ according to the species or age and is dependent on muscle fibre type proportion (Lawrie, 1985). Muscle comprised predominantly of red fibre type contains more myoglobin than muscles with a high white fibre type content. Dark goat meat may be due to metmyoglobin, the oxidized form of myoglobin, and will be viewed as old and undesirable for consumption (Pratiwi *et al.*, 2007).

The haem group contains a centrally located iron atom that has six coordination sites available for chemical bonds. Four of these sites bond the iron atom within the haem structure, while the fifth bond links the iron atom to the amino acid chain. The sixth site bonds the iron atom to the haem chemical group that determines the meat colour. Proportions of deoxymyoglobin, oxymyoglobin and metmyoglobin in the meat depend on oxygen availability and determine the colour of fresh meat (Leeward, 1992; Lindahl *et al.*, 2001). According to Leeward (1992), oxygen availability depends on the oxygen partial pressure, penetration and consumption rate of the muscle (Leeward, 1992). The penetration depth of light decreases as an effect of light scattering due to an increased amount of myofibrillar water and meat pH (Feldhusen, 1994). The meat colour depends on the rate of pH decline, the ultimate pH and the extent of protein denaturation (Bendall and Swatland, 1988; Feldhusen, 1994). During post-mortem glycolysis, the muscle proteins become denatured, resulting in increased light scattering and less light penetration (Bendall and Swatland, 1988; Feldhusen, 1994), increased lightness of the meat (Joo *et al.*, 1999), decreased penetration depth of light and changes in the selective light absorption through chromophores such as myoglobin and haemoglobin (Feldhusen, 1994).

Meat colour is affected by several factors such as breed, type of muscle, meat pH and ageing. Breed differences in the lightness ($L^*$), redness ($a^*$) and yellowness ($b^*$) colour values (Hunter scale) of the biceps femoris, semitendinosus and longissimus dorsi muscles were reported (Kadim *et al.*, 2004). The longissimus dorsi muscle from Jebel Akhdar goats had significantly higher $L^*$ ($P < 0.05$), while Batina goats had significantly lower $L^*$ ($P < 0.05$) in the semimembranosus muscle than other breeds. Ageing of muscles for 6 days produced a significant effect on meat colour compared with meat aged for 1 day for the longissimus dorsi, semimembranosus, semitendinosus and biceps femoris muscles in goats (Kadim *et al.*, 2004). Kannan *et al.* (2001) reported that the average value of $a^*$ of chevon cuts was high at day 0 and low at day 8. The length of post-mortem glycolysis can modify the perceived colour independently of meat metmyoglobin formation (Leeward, 1985), which explained the differences between the values at days 1 and 6 of ageing.

## 13.7 Post-mortem Treatment and Meat Quality

Although the exact point of conversion of muscle to meat is not easy to determine, the metabolic activity of the skeletal muscle will not stop when the functional role of muscle is lost and rigor has been established. Many biochemical reactions have significant implications for the quality characteristics of goat meat. This section will discuss refrigeration, freezing, ageing, cold shortening and electrical stimulation regarding the quality characteristics of goat meat.

### 13.7.1 Refrigeration and freezing

Refrigeration is lowering the temperature of the carcass to slow down the rate of glycolysis of muscle. Controlling airflow inside chillers is important for efficiency and homogeneity of carcass chilling (Mirade and Picgirard, 2001). Blast chilling of goat meat, which has a thin layer of subcutaneous fat, may induce toughening and therefore compromise meat quality. According to Sheridan *et al.* (1998), at 35°C, almost 80% of the enzyme activity was lost during rigor development, while about 20% of the activity was lost when the meat was exposed to 15°C. Rapid chilling of carcasses (–20°C) was reported to produce meat as tender as those chilled at 4°C and reduced evaporative weight losses by 0.5–1% (McGeehin *et al.*, 2002). Therefore, blast chilling may be used in conjunction with electrical stimulation to accelerate the onset of rigor mortis in order to avoid the development of cold shortening. In contrast, freezing is a common practice in preserving meat quality for an extended time and offers several advantages. These indicate insignificant alterations in product dimensions and minimum deterioration in meat colour, flavour and texture. The disadvantages of frozen storage include freezer burn, dehydration, rancidity, drip loss and product bleaching. Many meat products go directly from the freezer to cooking. As the cooked appearance of frozen cuts does not differ from fresh cuts (Obuz and Dikeman, 2003), the consumer is not able to differentiate between the two. The shelf-life extension and the purchasing and inventory flexibility offered by frozen meat items are valuable assets in the food service industry.

Freezing of meat has been widely researched to enable lowering of the amount of drip loss on thawing. Drip loss is one of the main problems in frozen meat (Kropf and Bowers, 1992). Drip loss from thawing meat includes proteins, vitamins and other nutrients, in addition to moisture, and results in decreased cooked yields and juiciness. The loss of fluid generally reduces the eating quality, binding ability and weight of the meat. The volume of drip produced on thawing has been related to the rate of freezing, size and location of ice crystals in frozen meat. The loss of moisture in frozen–thawed meat may reach up to 85% of the water in muscle tissue. This water is located intracellularly in the myofibrils, whereas the remaining 15% is located in the extracellular space (Hamm, 1975). The main body of water is held by capillary action (Offer and Trinick, 1983) and a small amount (4–5%) is restricted in motion because of the proximity of protein molecules (Hamm, 1975). When the muscle is frozen, water associated with protein is replaced with protein (Fennema, 1982), which leads to a decrease in water-holding capacity after thawing (Wagner and Anon, 1985). The most critical temperature in thawing meat is between –10 and –2°C; therefore, meat must pass through this range rapidly (Calvelo, 1981). In rapid chilling of meat, calcium is released into the sarcosome, which causes remarkable muscle contraction in the presence of ATP. The temperature difference between the inside and outside of chilled muscles determines the cooling rate of the meat. This value decreases towards the centre of the muscle. This is important in a practical sense in carcasses from animals with less fat coverage, such as goats.

The rate of freezing and subsequent thaw drip loss may reduce the nutrient content of goat meat. Drip losses of cuts of Angora and Boer goat longissimus muscles frozen at –20°C and thawed at 10°C for 24 h were 3.68 and 3.19%, respectively. Drip losses of the semimembranosus muscles of Angora and Boer goats were about 3.5 times greater at 14.41 and 15.51%, respectively (Schonfeldt *et al.*, 1993).

Freezing causes physical and chemical changes in meat that lead to deterioration in the quality of the meat through ice crystal formation (Honikel *et al.*, 1986). The functionality of meat is adversely affected by long-term frozen storage, and most of the vitamins are lost in the dripping water (Miller *et al.*, 1980). Protein denaturation at low temperatures leads to loss of water. Slow freezing causes the water to separate from the tissue into pools that form large crystals, which may result in greater structural damage associated with larger intercellular ice crystals (Farouk *et al.*, 2004). Rapid freezing is more suitable for goat carcasses due to the small carcass size, small muscle thickness and low level of subcutaneous fat. Rapid freezing results in very little water separation, resulting in the formation of small crystals. The drip loss in fast-frozen meat is less than from slow-frozen meat. Fluctuations in temperature that occur during storage cause recrystallization phenomena that may explain the deterioration in meat quality during frozen storage (Bevilacqua and Zaritzky, 1982). The solubility of myofibrillar proteins is lower in slowly frozen meat compared with fast frozen meat (Farouk *et al.*, 2004).

### 13.7.2 Cold shortening

Cold shortening is a phenomenon that occurs in pre-rigor muscle and results in tough meat. 'Shortening' refers to the short sarcomere length of highly contracted muscle with protein denaturation and water loss (Devine *et al.*, 1999). Rapid chilling may have a detrimental effect through cold shortening, which may result in a drastic decrease in tenderness (Marsh *et al.*, 1974). The degree of overlap between myosin and actin filaments contributes primarily to meat toughening (Tornberg, 1996). Changes in the angles of criss-cross connective tissue lattice and crimp length are responsible in part for the relationship between sarcomere length and meat tenderness (Renerre *et al.*, 1999). However, the toughness of cold-shortened muscle is largely affected by an endogenous enzymatic tenderization mechanism rather than shortened sarcomere length (Hwang *et al.*, 2004b). There may be a possibility of a more direct cold-shortening/toughening relationship in lean goat carcasses exposed to rapid chilling early post-mortem. This would be more likely for lean carcasses with localized subcutaneous fat deposition (Koch *et al.*, 1995). The effect of shortening sarcomeres on shear force is significantly detrimental when proteolysis is relatively slow.

If the meat is frozen prior to rigor onset and subsequently thawed, it will dramatically shorten and become extremely tough. This phenomenon is referred to as 'thaw shortening'. The process of pre-rigor freezing can damage the sarcoplasmic reticulum and destroy its ability to regulate calcium concentrations within the myofibre. Both calpains and myosin ATPase require free $Ca^{2+}$ ions in the cytoplasm for their activities (Celio *et al.*, 1996). It has been shown that calcium-reserving organelles lose their function at abnormal cellular temperature (Cornforth *et al.*, 1980). During thawing, all components necessary for muscle contraction are still present but control of the reactions is lost. As a result, anaerobic metabolism proceeds at a very rapid rate and is concomitant with severe contraction.

### 13.7.3 Electrical stimulation

$Ca^{2+}$ ions and ATP content are the major factors that govern the degree of muscle contraction. The post-mortem release of $Ca^{2+}$ ions from the sarcoplasmic reticulum at high ATP levels in muscle results in a significant level of shortening. However, if $Ca^{2+}$ ions are released after some depletion of ATP from muscle has taken place, only a minor amount of shortening will occur. This suggests that depletion of ATP to minimum levels by increasing the rate of post-mortem glycolysis to exhaust the glycogen while the carcass temperature is still high will minimize cold shortening. Electrical stimulation increases the rate of post-mortem glycolysis, reducing the time for the onset of rigor mortis.

Electrical stimulation involves passing an electric current through the freshly slaughtered goat carcass in a series of short impulses, each causing the muscles to contract violently; however, in between impulses, the muscles return to their normal relaxed state. Goat carcasses have a poor insulating subcutaneous fat cover. This makes them susceptible to muscle toughening through the effects of cold shortening during the chilling process. This can be countered by application of a low-voltage electric current to the lean carcasses immediately after slaughter to accelerate post-mortem glycolysis (Kadim *et al.*, 2010). Electrical stimulation increases the rate of post-mortem glycolysis, depleting the ATP energy source for muscle contraction due to an anaerobic state (Hwang *et al.*, 2003). The variability in overall ATP post-mortem is primarily responsible for the variability in post-mortem pH fall in muscle. The time that it takes for muscle pH values to decline to 6.0 is a reflection of the onset of earlier rigor. With the acceleration of post-mortem glycolysis, a rapid build-up of lactic acid occurs and, in some cases, the pH of electrically stimulated goat muscle can reach a pH of 6.0 in 2–3 h instead of the 10–14 h that may be required for non-stimulated muscles (Gadiyaram *et al.*, 2008; Kadim *et al.*, 2010). A reduction in the time required for muscles to reach a pH of 6.0 is of practical importance as it determines the period of delay necessary before the muscle temperature can be dropped below 10°C if cold shortening is to be avoided (Chrystall *et al.*, 1984). The residual contractile properties of muscle are reduced, rigor mortis is advanced and the enzymes associated with the conditioning of meat are accelerated.

The proposed mechanism of electrical stimulation is the prevention of cold shortening by acceleration of rigor mortis onset while internal muscle temperature remains outside the cold-shortening risk zone (Swatland, 1981). Additionally, electrical stimulation causes muscular contraction sufficient to cause physical disruption of tissue (Ho *et al.*, 1996; Kadim *et al.*, 2009, 2010) Acceleration of proteolysis could be classified as a secondary effect mediated through time/temperature/pH interaction, affecting enzyme stability and activity (Hwang *et al.*, 2003). According to Breukink and Casey (1989) and Kadim *et al.* (2010), low-voltage electrical stimulation is most effective when cold shortening is an actual risk due to the low chilling temperatures applied in the early post-mortem period and/or when lean goat carcasses result in rapid heat dissipation. In this respect, it has been demonstrated that electrical stimulation can improve goat meat quality characteristics (Kadim *et al.*, 2010). However, improvements in meat quality would not result from electrical stimulation unless it markedly accelerated post-mortem glycolysis. Changes in glycolysis resulting from electrical stimulation should be followed by measuring the rate of pH decline, after stimulation and post-mortem of muscles, to a pH of 6.0. Figure 13.2 shows that the rate of pH fall in goat carcasses was significantly affected by electrical stimulation (Kadim *et al.*, 2010). At 40 min post-mortem, the average pH decline values in stimulated goats were 0.20–0.28 units below the non-stimulated group (King *et al.*, 2004; Gadiyaram *et al.*, 2008; Kadim *et al.*, 2010). After a relatively fast fall within the first 4 h, the mean pH values of goat muscles underwent a slow decline until the ultimate pH at 24 h post-mortem (Gadiyaram *et al.*, 2008; Kadim *et al.*, 2010).

Goat meat retailers benefit from the use of electrically stimulated carcasses because of the improved appearance of retail cuts. Meat from electrically stimulated goat carcasses had a brighter colour, less surface discoloration and a more desirable overall appearance than non-stimulated carcasses (Kadim *et al.*, 2010). The increased muscle brightness of the stimulated carcasses suggested that the early post-mortem conditions of these muscles favoured protein denaturation (Warriss and Brown, 1987). High muscle temperatures combined with low muscle pH values in early post-mortem are associated with increased protein denaturation. This was supported by significant differences in pH drop.

Muscles from stimulated goat carcasses have a significantly lower shear force value compared with non-stimulated carcasses (Savell *et al.*, 1977; King *et al.* 2004; Gadiyaram *et al.*, 2008; Kadim *et al.*, 2010;). The main mechanism through which stimulation improves tenderness is its effects on physical alteration and/or acceleration of energy turnover during and after the stimulation (Luo *et al.*, 2008). Physical disruption lowers the resistance to mechanical shearing force and therefore increases tenderness (Hopkins and Thompson, 2002). According to Kadim *et al.* (2010) and Savell *et al.* (1977), stimulated goat carcasses had significantly longer sarcomere lengths than non-stimulated muscles.

Electrical stimulation negatively affected the myofibrillar expressed juice of the goat muscle (Kadim *et al.*, 2010), being higher for stimulated than for non-stimulated muscle samples. Similarly, den Hertog-Meischke *et al.* (1997) found that filter-expressed juice was significantly higher for stimulated bovine muscles than for non-stimulated ones. In contrast, Whiting *et al.* (1981) found that electrical stimulation had no effect on the water-holding capacity of lamb longissimus dorsi muscle. The difference between the stimulated and non-stimulated muscle samples may be the result of shrinkage of the myofibrils due to the pH fall post-mortem, attachment of cross-bridges between thick and thin filaments at the onset of rigor, and denaturation of myosin (Offer and Knight, 1988). Denaturation of myosin takes place when a carcass experiences a low pH at high temperature. Moreover, expressed juice is affected by the integrity of the muscle cell

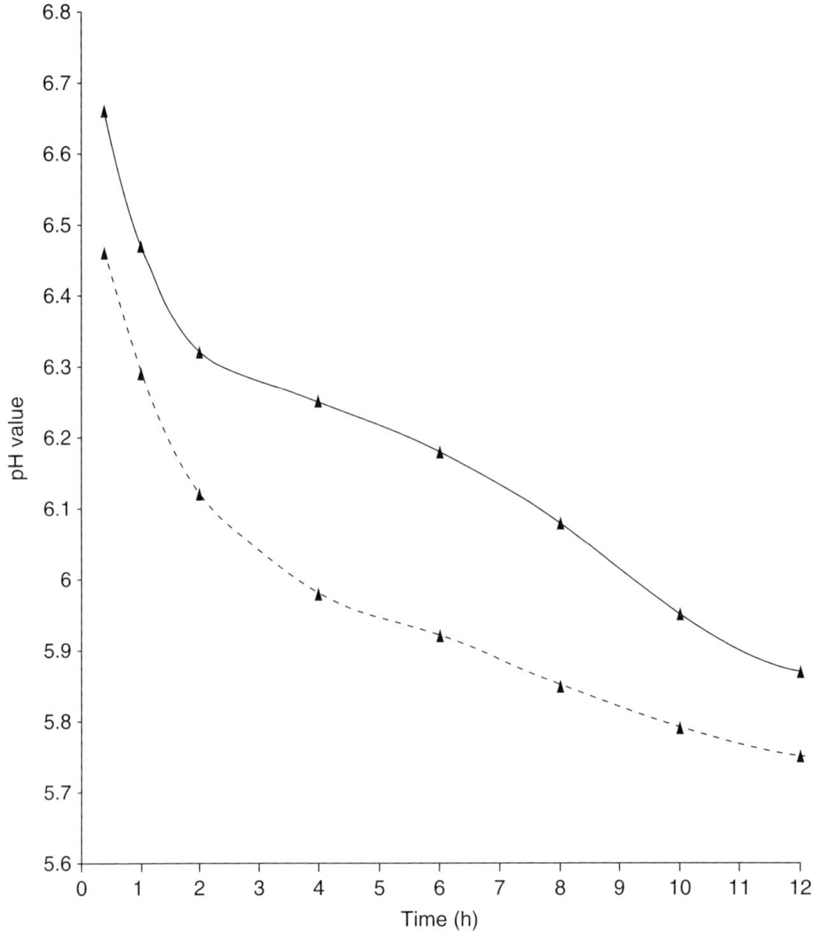


**Fig. 13.2.** Mean changes in pH within the longissimus dorsi muscle of two breeds of goats electrically stimulated (---) or non-stimulated (—) (Kadim *et al.*, 2010).

membranes and the rate of fluid migration within the muscle (den Hertog-Meischke *et al.*, 1997).

### 13.7.4 Ageing

Tenderness is the predominant quality determinant and probably the most important organoleptic characteristic of meat. Historically, meat has been aged to improve its quality characteristics by storage at a certain temperature for a period of time. Ageing is necessary as meat is usually unacceptably tough immediately following rigor onset. The time required for the ageing process varies depending on the type of meat. Although high-temperature conditioning of meat promotes bacterial growth, the ageing process may be accelerated by keeping carcasses above 15°C (Pearson and Dutson, 1985). This type of conditioning can be applied in the pre- or post-rigor state and is effective for improving meat quality. It has been suggested that, during the ageing process, tenderization occurs as a result of protein degradation. The ageing processes originate within the myofibre and are responsible for degradation of cellular constituents.

Analysis of muscle proteins along with meat-quality traits during ageing is crucial in understanding the biological basis of changes in meat quality. The proteolytic enzymes in meat that have been studied the most are the cathepsins and calpains.

Ageing was shown to have significant effects on Hunter $L^*$ (lightness) values, shear force values and expressed juice of goat meat (Kadim *et al.*, 2004). The most relevant consequence of ageing is an improvement in meat tenderness (Janz *et al.*, 2001; Ruiz de Huidobro *et al.*, 2003; Kadim *et al.*, 2004). Table 13.9 shows that ageing of goat muscles for 1 day resulted in significantly higher shear force values than in those aged for 6 days (Kadim *et al.*, 2004). The shear force values for the three Omani goat breeds Batina, Dhofari and Jebel Akhdar were reduced by 15–31, 17–28 and 17–29%, respectively, between 1 and 6 days of ageing for four muscles. The longissimus dorsi muscle had the highest reduction (31%) while the semitendinosus muscle had the lowest reduction (16%), and the shear force values of the other two were reduced by 27% for the biceps femoris and 25% for the semimembranosus muscles. Generally, longissimus dorsi muscles from goats had the lowest shear force values, while semimembranosus had the highest (Kadim *et al.*, 2004). These variations might be due to differences in connective tissue content or sarcomere length.

As ageing time increases, tenderness will improve as a result of complex changes in muscle metabolism, which are dependent on animal breed, metabolic status, rearing system and prior slaughter stress. During ageing, the structure of the myofibrillar and other associated proteins undergoes some modifications, and collagen is weakened, although to a lesser extent (Christensen *et al.*, 2004). The proteolytic enzymes in meat play a significant role in improving meat quality by the degradation of actin and/or actin-relevant peptides during ageing (Hwang *et al.*, 2004b). Enzymes require specific conditions such as temperature and pH for optimal activity, and, if these can be determined and maximized in meat, improvements in meat tenderness can be achieved. Cathepsins within lysosomes operate at a pH of <5.2 to produce better meat quality by degrading myofibrillar proteins. Calpains are proteases that require $Ca^{2+}$ ions and a pH of <5.6 for activity. The amount of calcium available in normal muscle cells is a major contributor to meat tenderization. The pH value required for optimal enzyme functioning is substantially higher than the pH 5.6 of normal meat; therefore, maximal activity of calpains is

**Table 13.9.** Means (± SD) for a range of quality characteristics for the longissimus dorsi muscle of three Omani goats breeds at 1 or 6 days ageing (from Kadim *et al.*, 2004).

| | Breed | | | | | |
|---|---|---|---|---|---|---|
| | Batina | | Dhofari | | Jebel Akhdar | |
| Parameters | 1 day | 6 days | 1 day | 6 days | 1 day | 6 days |
| Ultimate pH | 5.75 ± 0.1 | 5.78 ± 0.1 | 5.56 ± 0.0 | 5.60 ± 0.0 | 5.64 ± 0.1 | 5.67 ± 0.07 |
| Expressed juice[a] | 36.8 ± 3.7 | 35.6 ± 3.4 | 36.3 ± 1.8 | 35.0 ± 1.8 | 36.5 ± 4.2 | 35.5 ± 3.7 |
| Cooking loss (%) | 21.9 ± 3.7 | 21.3 ± 3.6 | 25.3 ± 2.2 | 23.8 ± 2.13 | 24.8 ± 1.7 | 23.4 ± 1.8 |
| WB values (kg)[b] | 7.2 ± 1.8 | 4.5 ± 2.2 | 7.4 ± 2.4 | 5.3 ± 1.4 | 7.7 ± 1.4 | 5.4 ± 1.2 |
| $L^*$ (lightness) | 40.7 ± 1.1 | 38.9 ± 1.5 | 40.1 ± 1.3 | 39.9 ± 1.7 | 42.1 ± 2.8 | 40.3 ± 3.5 |
| $a^*$ (redness) | 23.2 ± 0.7 | 23.4 ± 1.5 | 23.6 ± 1.2 | 23.1 ± 1.4 | 23.5 ± 1.8 | 23.4 ± 1.5 |
| $b^*$ (yellowness) | 4.7 ± 0.7 | 4.3 ± 0.8 | 4.8b ± 0.7 | 4.4 ± 1.0 | 5.7 ± 1.5 | 4.4 ± 0.8 |

SD, Standard deviation.
[a]Expressed juice = water area ($cm^2$)/sample weight (g).
[b]Warner–Bratzler shear force value.

most likely to occur during the early post-mortem conditions.

The influence of post-mortem ageing on meat water-holding capacity is of practical interest. Degradation of skeletal muscle proteins such as desmin, vinculin, titin and nebulin was considered to be responsible for changes in water-holding capacity during ageing (Kristensen and Purslow, 2001; Baron *et al.*, 2004; Lametsch *et al.*, 2004). Formation of drip is generally considered a result of denaturation of contractile proteins and shrinkage of myofibrils during rigor development (Bertram *et al.*, 2004; Hwang *et al.*, 2004b). However, reduced drip loss has also been related to the 'leak-out' effect, and ageing itself did not improve the water-holding capacity. A higher rigor temperature accelerated drip loss during vacuum-packed storage and drip loss increased at a high pH of 6.2 as ageing time lengthened (Hwang *et al.*, 2004b). Ageing of goat muscles had significantly lower water-holding capacity than non-aged samples (Moller *et al.*, 1983; Kim *et al.*, 1993; Joo *et al.*, 1995; Kadim *et al.*, 2004). This difference might be explained by the 'leaking-out' hypothesis, which states that water is lost by evaporation or dripping during the ageing period. Slight differences in expressed juice between different muscles were reported by Gonzalez *et al.* (1983), Marinova *et al.* (2001) and Kadim *et al.* (2004). Ageing of the biceps femoris, semitendinosus and semimembranosus muscles for 6 days had significantly lower percentage cooking losses than muscles aged for 1 day (Kadim *et al.*, 2004). Similarly, Kannan *et al.* (2001) found that the percentage cooking loss was higher at day 0 than at days 4, 8 or 12 of display for goat steaks. According to Trout (1988), cooking loss is more dependent on ultimate pH, sarcomere length and cooking conditions. Bouton *et al.* (1972) suggested that the myofibrillar protein changes structurally with ageing, resulting in significantly reducing cooking loss for aged rather than non-aged muscles. Goat semimembranosus muscle had the highest percentage cooking loss, while the longissimus dorsi had the lowest and the biceps femoris and semitendinosus muscles were intermediate (Kannan *et al.*, 2001; Kadim *et al.*, 2004). These authors reported that the average cooking loss was high in leg cuts and low in loin cuts.

Ageing of goat muscles for 6 days produced significantly lower $L^*$ values than those aged for 1 day (Kadim *et al.*, 2004). Kannan *et al.* (2001) reported that the average value of $a^*$ (redness) for chevon cuts was high at day 0 and low at day 8. The colour of the meat surface depends not only on the quantity of myoglobin but also on the relative proportions of the three main states of myoglobin on the surface. Ultimate pH can influence colour independently of meat myoglobin content (Leeward, 1985), which explains the differences between aged and non-aged samples.

### 13.7.5 Muscle fibre types and meat quality

Skeletal muscle features are complex due to their role in movement, deposition of protein, protection and transformation of muscles to meat (Hocquette *et al.*, 1998; Geay *et al.* 2001). Skeletal muscle is composed of a large number of different types of muscle fibre that contribute to a variety of functional capabilities and metabolic enzymes (Schiaffino and Reggiani, 1996). These fibre types differ according to their molecular, metabolic, structural and contractile properties (Pette and Staron, 1990). Fibre-type diversity is usually defined by the isoform of the myosin heavy chain present (Pette and Staron, 1990). Myosin heavy-chain composition and skeletal muscle fibre types are two of the most important determinants in meat quality and meat products, as quality is currently an important social and economic challenge for meat producers and retailers (Xiong, 1994). The molecular diversity of skeletal muscle fibre types is species specific (Pette and Staron, 1990). Physiological differences between species related to body size have been reported by Rome *et al.* (1990). The most useful schemes to describe skeletal fibre types are based on specific myosin profiles to provide greater insights into the molecular and functional diversity, versatility and adaptability of muscle fibres. Phenotypic profiles of skeletal muscle fibre types are affected by innervation/neuromuscular activity, exercise training, loading/unloading, hormones and ageing (Pette and Staron, 1990). According to the above information, the quality of meat is therefore determined by the muscle architecture, attachment of fibres to connective tissue and post-mortem changes in these structures. The quality characteristics of goat meat such as flavour, colour, juiciness and tenderness are influenced by many factors, among which fibre type is important. Muscle structures, the conversion of muscle to meat and the phenomenon of rigor mortis have been discussed above.

Goat skeletal muscle cells are classified into several specialized classes, termed fibre types, which show variations in contractile and metabolic properties. The isoforms of the myosin heavy-chain molecule represent the best markers of muscle fibre diversity (Pette and Staron, 1990). It is thought that both myosin heavy-chain composition and skeletal muscle fibre types are two of the most important determinants in meat quality and meat products (Xiong, 1994). Different combinations of myosin heavy-chain isoforms may occur within the same fibre, but the predominant isoform is the main determinant of the fibre's functional properties such as speed of contraction and fatigue resistance (Schiaffino and Reggiani, 1996). The molecular diversity of

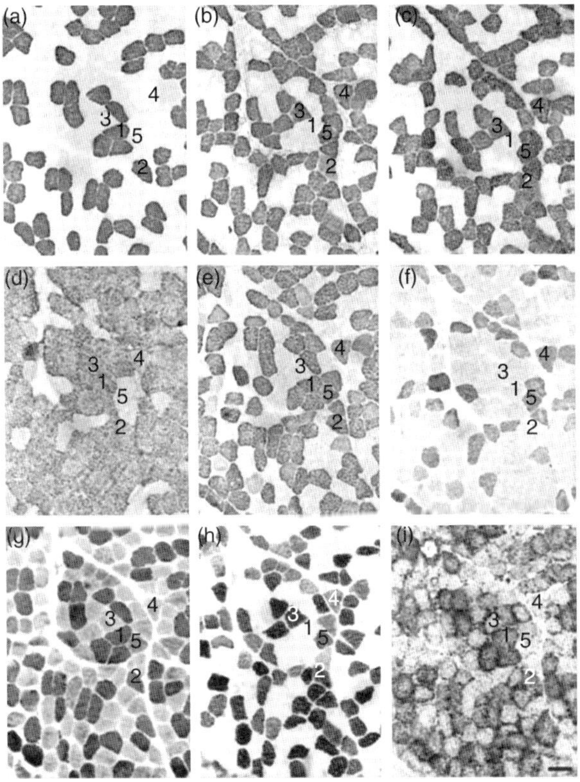

**Fig. 13.3.** Serial frozen sections of adult goat semitendinosus muscle stained for immunohistochemistry with monoclonal antibodies raised against specific myosin heavy-chain isoforms (a–f) and by enzyme histochemistry for myofibrillar ATPase and quantitative succinic dehydrogenase (g–i). The sections in a–f were stained with monoclonal antibodies specific for isoforms I, IIA, IIB and IIX. The fibres labelled 1, 3 and 5 are pure fibres containing isoforms I, IIA and IIX, respectively; fibres 2 and 4 are hybrid fibres containing isoforms I plus IIA, and IIA plus IIX, respectively. (g, h) Myofibrillar ATPase activity after pre-incubation at pH 4.5 (g) and pH 10.5 (h). (i) Succinate dehydrogenase activity. Bar, 50 μm (from Argüello *et al.*, 2001).

adult skeletal muscle fibres is species specific (Pette and Staron, 1990), and important physiological differences between species related to body size have already been reported (Rome *et al.*, 1990). Figure 13.3 demonstrates the presence of three different muscle fibre types in goat semitendinosus muscle containing a unique myosin heavy-chain isoform: one slow-twitch fibre type (type I) and two fast-twitch fibre types (types IIA and IIB) (Argüello *et al.*, 2001).

The quality of goat meat is the main factor of importance in the provision of meat, and is affected by muscle structure and muscle fibre types. Enhancing meat safety involves the application of measures to delay or prevent microbiological, chemical and/or physical changes that make meat unhealthy for human consumption. Recent advances have helped our understanding of meat structure and its effect on the meat-quality characteristics of goats.

Three myosin heavy-chain isoform proteins have been found in goat longissimus dorsi muscles: slow-twitch oxidative (type I), fast-twitch oxidative (type IIA) and fast-twitch (type IIB) (Kadim *et al.*, 2010). The constant staining pattern of muscle fibres for myosin ATPase is consistent with the goat muscle study by Argüello *et al.* (2001) (Fig. 13.3). It has been reported that the fast-twitch fibre type occurs at a significantly higher frequency than the slow-twitch oxidative or fast-twitch oxidative fibre types, while the slow-twitch oxidative fibre types occur at a significantly higher frequency than fast-twitch oxidative fibre types (Kadim *et al.*, 2010). The latter authors compared the diameters of slow-twitch oxidative, fast-twitch oxidative and fast-twitch muscle fibres and found that fast-twitch fibres were larger than fast-twitch oxidative fibres, and fast-twitch oxidative fibres were larger than slow-twitch oxidative fibres (Table 13.10). The effect of muscle fibre type on meat quality of goat meat may be due to the muscle fibre size – the larger the size, the tougher the meat. Fast-contracting fibres with a glycolytic metabolism (fast-twitch fibres) are larger than slow and oxidative red fibres. The red fibres are rich in lipid and red in colour, therefore contributing to taste and colour quality, and they are also related to metabolic differences.

## 13.8 Processing of Meat

Goat meat processing is important for human nutrition and refers to applying technology to improve or maintain quality and add value, and to preserve the meat and produce suitable high-quality meat products to be used at different times and places for consumption. Meat processing is an extensive subject and only some aspects of goat meat will be highlighted. Goat meat is preserved by drying, curing with salts or smoking, or is manufactured into reconstituted products. Goat meat is processed not only as a means of preserving but also to

**Table 13.10.** Muscle fibre type parameters (means ± SD) in longissimus dorsi muscles from two breeds of goats (from Kadim *et al.*, 2010).

| | Breed | | |
|---|---|---|---|
| Parameter | Dhofari | Batina | SD |
| Proportion | | | |
| Type I | 19.2 | 18.9 | 0.66 |
| Type IIA | 9.9 | 9.6 | 0.42 |
| Type IIB | 71.0 | 71.4 | 0.74 |
| Diameter (µm) | | | |
| Type I | 56.3 | 52.9 | 1.25 |
| Type IIA | 61.1 | 56.7 | 1.36 |
| Type IIB | 58.0 | 56.6 | 1.04 |

SD, Standard deviation.

produce consumer-acceptable products, compatible with marketing and lifestyles and related to human health. Therefore, the processing has to reflect the image and saleability of the envisaged products. The decision-making process of the consumer needs to be defined in terms of real and perceived values, the convenience the product offers and its palatability.

Although goat meat has a less desirable flavour and aroma than the meat of other species, objective evaluation of meat quality found that substitution of up to 40% of beef by goat meat was acceptable in frankfurters (Table 13.11) (Marshall *et al.*, 1977). The goat frankfurters had good physical attributes, being firm, resilient and springy under forefinger pressure and a firm bite, a desirable textural attribute in quality emulsified sausages. The goat frankfurters maintained their form and shape during peeling, indexing and packaging operations. Sausages made from mature does' meat had different quality parameters with higher shear force values than sausages made from beef. Differences in physical, chemical, structural and quality attributes between goat and beef sausages might be related to the characteristics of the raw meat. Cured and smoked leg muscles of low-voltage electrically stimulated carcasses of goats compared with the counterpart of beef silverside received higher ratings ($P < 0.01$) in terms of aroma, tenderness, juiciness and tastiness (Table 13.12) (Breukink and Casey, 1989). The authors concluded that smoked and cured goat leg meat has the potential to be a delicacy and could compete comfortably with other products such as smoked beef.

Pre-rigor goat muscle has a higher water-holding capacity, better fat-emulsifying properties and produces sausages with less moisture loss and rendering out when cooked. According to Padda *et al.* (1988), patties manufactured from hot goat carcasses (around 3–4 h post-mortem) had significantly lower cooking yields than patties prepared from chilled meat (24 h post-mortem). However, reducing the post-mortem time interval to processing to 1–2 h improved the yield. Reheating pre-cooked, frozen patties significantly reduced sensory scores. The fat content of patties has a significant influence on cooking loss, flavour, texture and overall acceptability (Padda *et al.*, 1985). According to sensory scores, a 20% fat content would be the optimal. The quality of warm minced goat meat (3 h post-mortem) could be improved

**Table 13.11.** Sensory panel rating for frankfurters in which beef was substituted by 40% goat, mutton and pork (from Marshall *et al.*, 1977).

| Formulation | Flavour | Juiciness | Texture | Overall |
|---|---|---|---|---|
| Control | 4.4 | 5.0 | 4.6 | 4.5 |
| Old age goat | 5.3 | 5.8 | 5.3 | 5.4 |
| Young age goat | 5.1 | 5.3 | 5.1 | 5.2 |
| Mutton | 5.1 | 5.5 | 4.8 | 5.1 |
| Pork | 3.1 | 3.2 | 1.8 | 2.4 |

**Table 13.12.** Sensory parameters (mean ± SD) of cured and smoked meat from goat and beef (from Breukink and Casey, 1989).

| Parameter | Goat | Beef |
|---|---|---|
| Aroma | 4.05 ± 0.27 | 3.89 ± 0.29 |
| Tenderness | 2.95 ± 0.50 | 2.37 ± 0.49 |
| Juiciness | 3.16 ± 0.32 | 1.88 ± 0.64 |
| Overall | 3.54 ± 0.60 | 2.96 ± 0.97 |

SD, Standard deviation.

by the addition of 2.5% sodium chloride and 1% tetrasodium pyrophosphate (Kondaiah *et al.*, 1985). These salts resulted in significantly increased pH, water-holding capacity and level of water-soluble proteins, decreased cooking loss and improved redness and overall appearance. The effects on the emulsifying capacity and salt-soluble protein concentration were also significant.

## 13.9 Conclusions

Goat meat is an important nutrient source to a large proportion of the world population. Goats are well adapted to a variety of environments and few feed resources as they are able to utilize marginal land to produce high-level protein products. There is a worldwide tendency towards a rapid increase in the demand for goat meat due to health reasons because of its lower fat content. This is an important factor in reducing the risk of cardiovascular diseases. Goat meat is also a good source of desirable fatty acids, as goats deposit relatively higher amounts of PUFAs than other ruminants. Goat meat is a nutrient-dense food, but the complementary role of goat meat in local diets, taking lifestyles and customs into consideration, should be quantified. The quality of goat meat can be improved through appropriate technology: processing can extend the range of products, improve the shelf life and give added value to products.

Pre- and post-mortem factors should be considered carefully to improve meat-quality characteristics. Pre-slaughter transportation may cause significant responses in goat meat quality and transportation under high ambient temperatures should be avoided or goats should be allowed to rest prior to slaughter in order to reduce economic losses. To counter post-slaughter conditions, technology has been used to improve goat meat quality through electrical stimulation, ageing and chilling.

## References


Abdon, I., Del Rosario, I.F. and Olga, L.G. (1980) *Food Composition Tables Recommended for Use in the Philippines, Handbook 1*, 5th revised edn. Food and Nutrition Research Institute, Manila, The Philippines.

Argüello, A., Lopez-Fernandez, J.L. and Rivero, J.L.L. (2001) Limb myosin heavy chain isoproteins and muscle fiber types in the adult goat (*Capra hircus*). *Anatomical Record* 264, 284–293.

Atti, N., Mahouachi, M. and Rouissi, H. (2006) The effect of spineless cactus (*Opuntia ficus-indica f. inermis*) supplementation on growth, carcass, meat quality and fatty acid composition of male goat kids. *Meat Science* 73, 229–235.

Babiker, S.A., El Khider, I.A. and Shafie, S.A. (1990) Chemical composition and quality attributes of goat meat and lamb. *Meat Science* 28, 273–277.

Baeza, E., Dessay, C., Wacrenier, N., Marche, G. and Iistra, A. (2002) Effect of selection for improved body weight and composition on muscle and meat characteristics in Muscovy duck. *British Poultry Science* 43, 560–568.

Banon, S., Vila, R., Price, A., Ferrandini, E. and Garrido, M.D. (2006) Effects of goat milk or milk replacer diet on meat quality and fat composition of suckling goat kids. *Meat Science* 72, 216–221.

Banskalieva, V., Sahlu, T. and Goetsch, A.L. (2000) Fatty acid composition of goat muscles and fat depots: a review. *Small Ruminant Research* 37, 255–268.

Baron, C.P., Jacobsen, S. and Purslow, P.P. (2004) Cleavage of desmin by cysteine proteases: calpains and cathepsin B. *Meat Science* 68, 447–456.

Bendall, J.R. and Swatland, H.J. (1988) A review of the relationships of pH with physical aspects of pork quality. *Meat Science* 24, 85–126.

Bertram, H.C., Chafer, A., Rosenvold, K. and Andersen, H.J. (2004) Physical changes of significance of early post mortem water distribution in porcine *M. longissimus*. *Meat Science* 66, 915–924.

Beserra, F.J., Madruga, M.S., Leite, A.M., da Silva, E.M.C. and Maiaa, E.L. (2004) Effect of age at slaughter on chemical composition of meat from Moxotó goats and their crosses. *Small Ruminant Research* 55, 177–181.

Bevilacqua, A.E. and Zaritzky, N.E. (1982) Ice recrystallization in frozen beef. *Journal of Food Science* 47, 1410–1414.

Bouton, P.E., Harris, P.V. and Shorthose, W.R. (1972) The effects of ultimate pH on ovine muscle: water-holding capacity. *Journal of Food Science* 37, 351–355.

Breukink, H.R. and Casey, N.H. (1989) Assessing the acceptability of processed goat meat. *South African Journal of Animal Science* 19, 76–80.

Brown, M.A., Huffman, D.L., Egbert, W.R. and Jungst, S.B. (1990) Physical and compositional characteristics of beef carcasses selected for leanness. *Journal of Food Science* 55, 9–14.

Calvelo, A. (1981) Recent studies on meat freezing. In: Lawrie, R.A. (ed.) *Developments in Meat Science.* Applied Science Publishers, Englewood, New Jersey, pp. 125–159.

Casey, N.H. (1982) Carcass and growth characteristics of four South African sheep breeds and the Boer goat. DSc thesis, Department of Livestock Science, Faculty of Agriculture, University of Pretoria, Republic of South Africa.

Casey, N.H. and Naude, R.T. (1984) Differential growth profiles of muscle and fat depots. In: *Proceedings of the Second International Conference on Cattle and Sheep Breeding*, Pretoria, Republic of South Africa, Vol. II, p. 12.

Casey, N.H. and van Niekerk, W.A. (1985) Fatty acid composition of subcutaneous and kidney fat depots of Boer goats and the response to varying levels of maize meal. *South African Journal of Animal Science* 15, 60–62.

Casey, N.H., van Niekerk, W.A. and Spreeth. E.B. (1988) Fatty acid composition of subcutaneous fat of sheep grazed on eight different pastures. *Meat Science* 23, 55–63.

Castellini, C., Mugnai, C. and Dal Bosco, A. (2002) Effect of organic production system on broiler carcass and meat quality. *Meat Science* 60, 219–225.

Celio, M.R., Pauls, T.L. and Schwaller, B. (eds) (1996) *Guidebook to the Calcium-binding Proteins.* Oxford University Press, New York.

Chilliard, Y. (1993) Dietary fat and adipose tissue metabolism in ruminants, pigs and rodents: a review. *Journal of Dairy Science* 76, 3897–3931.

Chizzolini, R., Zanardi, E., Dorigoni, V. and Ghidini, S. (1999) Calorific value and cholesterol content of normal and low fat meat and meat products. *Review in Trends in Food Science and Technology* 10, 119–128.

Christensen, M., Larsen, L.M., Ertbjerg, P. and Purslow, P.P. (2004) Effect of proteolytic enzyme activity and heating on the mechanical properties of bovine single muscle fibres. *Meat Science* 66, 361–369.

Chrystall, B.B., Devine, C.E., Ellery, S. and Wade, L. (1984) Low voltage electrical stimulation of lamb: its effect on muscle pH and tenderness. *New Zealand of Agricultural Research* 27, 513–523.

Collins, A. (2008) Nutrition and calories in goat's meat <http://www.calorie-counter.net/meat-calories/goats-meat.htm>.

Colomer-Rocher, F., Kirton, A.H., Mercer, G.J.K. and Duganzich, D.M. (1992) Carcass composition of New Zealand Saanen goats slaughtered at different weights. *Small Ruminant Research* 7, 161–173.

Cornforth, M.R., Pearson, A.M. and Merkel, R.A. (1980) Relationship of mitochondria and sarcoplasmic reticulum to cold shortening. *Meat Science* 4, 103–121.

Cortright, R.N., Muoio, D.M. and Dohm, G.L. (1997) Skeletal muscle lipid metabolism: a frontier for new insights into fuel homeostasis. *Journal of Nutrition Biochemistry* 8, 228–245.

den Hertog-Meischke, M.J.A., Smulders, F.J.M., van Logtestijn, J.G. and van Knapen, F. (1997) The effect of electrical stimulation on the water-holding capacity and protein denaturation of two bovine muscles. *Journal of Animal Science* 75, 118–124.

Devendra, C. (1980) Milk production in goats compared to buffaloes and cattle in the humid tropics. *Journal of Dairy Science* 63, 1755–1767.

Devendra, C. (1988) The nutritional value of goat. In: *Goat Meat Production in Asia.* Proceedings of a workshop held in Tando Jam, Pakistan, 13–18 March, pp. 76–86.

Devine, C.E., Wahlgren, N.M. and Tornberg, E. (1999) Effect of rigor temperature on muscle shortening and tenderization of restrained and unrestrained beef *m. longissimus thoracis et lumborum. Meat Science* 51, 61–72.

Dhanda, J.S., Taylor, D.G., Murray, P.J. and McCosker, J.E. (1999) The influence of goat genotype on the production of Cappretto and Chevon carcasses. 2. Meat quality. *Meat Science* 52, 363–367.

Dhanda, J.S., Taylor, D.G. and Murray, P.J. (2003) Part 1. Growth, carcass and meat quality parameters of male goats: effects of genotype and live weight at slaughter. *Small Ruminant Research* 50, 56–66.

Eastridge, J.S. and Johnson, D.D. (1990) The effect of sex class on nutrient composition of goat meat. In: *Proceedings of the International Goat Production Symposium*, Tallahassee, Florida, 22–26 October, pp. 143–146.
Farouk, M.M., Wielicko, K.J. and Merts, I. (2004) Ultra-fast freezing and low storage temperatures are not necessary to maintain the functional properties of manufacturing beef. *Meat Science* 66, 171–179.
Feldhusen, F. (1994) Einflüsse auf die postmortale Farbveränderung der Oberfläche von Schweinemuskulatur. *Fleischwirtschaft* 74, 989–991.
Fennema, O. (1982) Behaviour of proteins at low temperatures. In: Cherry, J.P. (ed.) *Food Protein Deterioration Mechanisms and Functionality.* ACS Symposium Series 206. American Chemical Society Washington, DC, USA, pp. 109–133.
Gadiyaram, K.M., Kannan, G., Pringle, T.D., Kouakou, B., McMillin, K.W. and Park, T.W. (2008) Effects of postmortem carcass electrical stimulation on goat meat-quality characteristics. *Small Ruminant Research* 78, 106–114.
Gaili, E.S. and Ali, A.E. (1985) Meat from Sudan desert sheep and goats. Part 2. Composition of the muscular and fatty tissues. *Meat Science* 13, 229–236.
Geay, Y., Bauchart, D., Hocquette, J.F. and Culioli, J. (2001) Effect of nutritional factors on biochemical, structural and metabolic characteristics of muscles in ruminants, consequences on dietetic value and sensorial qualities of meat. *Reproduction Nutrition Development* 41, 1–26.
Gebhardt, S.F. and Thomas, R.G. (2002) *Nutritive Value of Foods.* Home and Garden Bulletin No. 72, USDA, Washington, DC, USA.
Giese, J. (1992). Developing low fat meat products. *Food Technology* 46, 100–105.
Gimenenz, M.S., Baudini, O.M., Ojeda, M.S., Molins de Pedernera, M. and Gimenez, L.A. (1985) Some physicochemical properties and composition of adipose tissue of goats fed with different diets. *Nutrition Reports International* 32, 389–397.
Gonzalez, F.A.N., Owen, J.E. and Cereceres, M.T.A. (1983) Studies on the Criollo goat of northern Mexico: Part 2. Physical and chemical characteristics of the musculature. *Meat Science* 9, 305–314.
Gopalan, C., Rama Sastri, B.V. and Balasubramaniam, S.C. (1971) *Nutritive Value of Indian Foods.* National Institute of Nutrition, Hyderabad, India.
Ha, J.K., Ahn, B.H., Lee, Y.Y., Kang, D.H. and Kim, J.K. (1986) Study on the lipids and fatty acid composition of Korean native goat meat. *Korean Journal of Animal Science* 23, 666–672.
Hamm, R. (1975) Water-holding capacity of meat. In: Cole, D.A.J. and Lawrie, R.A. (eds) *Meat.* Butterworths, London, UK, pp. 321–328.
Heffron, J.J.A. and Dreyer, J.H. (1975) Postmortem glycolytic metabolism in the skeletal muscles of anaesthetised and stunned Boer goats. *South African Journal of Animal Science* 5, 61–65.
Heinze, P.J., Smith, M.C., Naude, R.T. and Boccard, R.L (1986) Influence of breed and age on collagen content and solubility of some ovine and goat muscles. Paper presented at the *32nd Meeting of European Research Works Workers*, 24–29 August 1986, Ghent, Belgium.
Ho, C.Y., Stromer, M.H. and Robson, R.M. (1996) Effect of electrical stimulation on postmortem titin, nebulin, desmin and troponin T degradation ultrastructural changes in bovine *longissimus* muscle. *Journal of Animal Science* 74, 1563–1575.
Hocquette, J.F., Ortigues-Marty, I., Pethick, D.W., Herpin, P. and Fernandez, X. (1998) Nutritional and hormonal regulation of energy metabolism in skeletal muscles of meat-producing animals. *Livestock Production Science* 56, 115–143.
Honikel, K.O., Roncales, P. and Hamm, R. (1983) The influence of temperature on shortening and rigor onset in beef muscle. *Meat Science* 8, 221–241.
Honikel, K.O., Kim, C.J., Hamm, R. and Roncales, P. (1986) Sarcomere shortening of pre-rigor muscles and its influence on drip loss. *Meat Science* 16, 267–282.
Hopkins, D.L. and Thompson, J.M. (2002) The relationship between post-mortem calcium concentration or pH and indicators of proteolysis in ovine muscle. *Meat Science* 61, 411–414.
Hultin, H.O. (1985) Characteristics of muscle tissue. In: Fennema, O.R. (ed.) *Food Chemistry.* Marcel Dekker, New York, p. 245.
Huskey, L.L., Brown, H.G., Lewis, P.K.J., Brown, A.H.J., Johnson, Z.B. and Perkins, J.L. (1993) Effects of end-point temperature and fat level on the residual cholesterol in ground beef patties. *Journal of Food Quality* 16, 187–196.
Hwang, I.H., Devine, C.E. and Hopkins, D.L. (2003) The biochemical and physical effects of electrical stimulation on beef and sheep meat tenderness. *Meat Science* 65, 677–691.

Hwang, I.H., Park, B.Y., Kim, J.H., Cho, S.H. and Lee, J.M. (2004a) Assessment of postmortem proteolysis by gel-based proteome analysis and its relationship to meat quality traits in pig *longissimus*. *Meat Science* 69, 79–91.

Hwang, I.H., Park, B.Y., Kim, J.H., Cho, S.H. and Lee, J.M. (2004b) Effects of muscle shortening and proteolysis on Warner–Bratzler shear force in beef *longissimus* and *semitendinosus*. *Meat Science* 68, 497–505.

James, N.A., Berry, B.W., Kotula, A.W., Lamikanra, V.T. and Ono, K. (1990) Physical separation and proximate analysis of raw and cooked cuts of chevon. In: *Proceedings of the International Goat Production Symposium*, Tallahassee, Florida, 22–26 October, p. 22.

Janz, J.A.M., Aalhus, J.L. and Price, M.A. (2001) Blast chilling and low voltage electrical stimulation influences on bison (*Bison bison bison*) meat quality. *Meat Science* 57, 403–411.

Johnson, D.D., Eastrige, J.S., Neubauer, D.R. and McGowan, C.H. (1995) Effect of sex on nutrient content of meat from young goat. *Journal of Animal Science* 73, 296–301.

Joo, S.T., Kaufman, R.G., Lee, S., Kim, B.C., Kim, C.J. and Greaser, M.L. (1995) Variation in water loss of PSE pork musculature over time. In: *Proceedings of the 41st International Congress of Meat Science and Technology*, San Antonio, Texas, 20–25 August, pp. 658–659.

Joo, S.T., Kauffman, R.G., Kim, B.C. and Park, G.B. (1999) The relationship of sarcoplasmic and myofibrillar protein solubility to color and water-holding capacity in porcine *longissimus* muscle. *Meat Science* 52, 291–297.

Kadim, I.T., Mahgoub, O., Al-Ajmi, D.S., Al-Maqbaly, R.S., Al-Saqri, N.M. and Ritchie, A. (2004) An evaluation of the growth, carcass and meat quality characteristics of Omani goat breeds. *Meat Science* 66, 203–210.

Kadim, I.T., Mahgoub, O., Al-Kindi, A.Y.A., Al-Marzooqi, W. and Al-Saqri, N.M. (2006) Effects of transportation at high ambient temperatures on physiological responses, carcass and meat quality characteristics of three breeds of Omani goats. *Meat Science* 73, 626–634.

Kadim, I.T., Mahgoub, O., Al-Marzooqi, O.W. and Khalaf, S.K. (2009) Effect of low voltage electrical stimulation and splitting the carcass on histochemical and meat quality characteristics of *Longissimus thoracis* muscle from the one-humped camel (*Camelus dromedarius*). *Journal of Camelid Science* 2, 30–40.

Kadim, I.T., Mahgoub, O., Al-Marzooqi, W., Khalaf, S.K., Al-Sinawi, S.S.H. and Al-Amri, I.S. (2010) Effects of transportation during the hot season, breed and electrical stimulation on histochemical and meat quality characteristics of goat longissimus muscle. *Animal Science Journal* 81, 352–361

Kannan, G., Kouakou, B. and Gelaye, S. (2001) Colour changes reflecting myoglobin and lipid oxidation in chevon cuts during refrigerated display. *Small Ruminant Research* 42, 67–75.

Kim, B.C., Warner, R.D. and Kaufman, R.G. (1993) Changes in expressible fluid losses of porcine musculature at different times post-rigor. In: *Proceedings of the 39th International Congress of Meat Science and Technology*, Calgary, Alberta, Canada, S3 P12.

King, D.A., Voges, K.L., Hale, D.S., Waldron, D.F., Taylor, C.A. and Savell, J.W. (2004) High voltage electrical stimulation enhances muscle tenderness, increases aging response and improves muscle color from capretto carcasses. *Meat Science* 68, 529–535.

Koch, R.M., Jung, H.G., Crouse, J.D., Varel, V.H. and Cundiff, L.V. (1995) Digestive capability, carcass and meat characteristics of *Bison bison*, *Bos taurus* and *Bos* × *Bison*. *Journal of Animal Science* 73, 1271–1281.

Kolbye, A. and Nelson, M.A. (1977) Health and safety aspects of the use of mechanical deboned meat. In: *Meat Poultry Inspection Program*, Vol. 2. Food Safety and Quality Service, USDA, Washington, DC, USA.

Kondaiah, N., Anjaneyulu, A.S.R., Sharma, N. and Josh, H.B. (1985) Effects of salt and phosphate on quality of buffalo and goat meats. *Meat Science* 15, 183–192.

Kristensen, L. and Purslow, P.P. (2001) The effect of ageing on the water-holding capacity of pork: role of cytoskeletal proteins. *Meat Science* 58, 17–23.

Kropf, D.H. and Bowers, J.A. (1992) Meat and meat products. In: Bowers, J. (ed.) *Food Theory and Applications*. Macmillan, NY, USA, pp. 22–29.

Kuhne, D., Freudnreich, P. and Ristic, M. (1986) Fettsauremuster verschiedener Tierarten. II. Mitteilung: Fette von Wiederkauern, Kaninschen und Hahncben. *Fleischwirtschaft* 66, 403–406.

Lametsch, R., Roepstorff, P., Møller, H.S. and Bendixen, E. (2004) Identification of myofibrillar substrates for μ-calpain. *Meat Science* 68, 515–521.

Lametsch, R., Knudsen, J.C., Ertbjerg, P., Oksbjerg, N. and Therkildsen, M. (2007) Novel method for determination of myofibril fragmentation post-mortem. *Meat Science* 75, 719–724.

Lawrie, R.A. (1985) *Meat Science*, 4th edn. Pergamon Press, Oxford, UK, pp. 173–175.
Leeward, D.A. (1985) Post-slaughter influences on the formation of metmyoglobin in beef muscles. *Meat Science* 15, 149–171.
Leeward, D.A. (1992) Colour of raw and cooked meat. In: Johnston, D.E., Knight, M.K. and Leeward, D.A. (eds) *The Chemistry of Muscle-based Foods.* The Royal Society of Chemistry, Cambridge, UK, pp. 128–144.
Lepetit, J., Canistro, J. and Favier, R. (1998) Rigor strength and temperature in rabbit muscle. In: *Very Fast Chilling in Beef*, Vol. 2. *Muscle to Meat.* Bristol University Press, Bristol, UK, pp. 95–103.
Leskanich, C.O., Mathews, K.R., Warkup, C.C., Noble, R.C. and Hazzledine, M. (1997) The effects of dietary oil containing (n-3) fatty acids on the fatty acid, physicochemical and organoleptic characteristics of pig meat and fat. *Journal of Animal Science* 75, 673–683.
Lindahl, G., Lundström, K. and Tornberg, E. (2001) Contribution of pigment content, myoglobin forms and internal reflectance to the colour of pork loin and ham from pure breed pigs. *Meat Science* 59, 141–151.
Locker, R.H. and Hagyard, C.J. (1963) A cold shortening effect in beef muscles. *Journal of the Science of Food and Agriculture* 14, 787–793.
Louvandini, H., McManus, C., Dallago, B.S., Machado, B.D. and Antunes, D.A. (2006) Evaluation of carcass traits, non-carcass components and 12th rib analysis of hair sheep supplemented with phosphorus. *Rev. Bras. Zootec. Brazilian Journal of Animal Science* 35, 550–554.
Luo, X., Zhu, Y. and Zhou, G. (2008) Electron microscopy of contractile bands in low voltage electrical stimulation beef. *Meat Science* 80, 948–952.
Madruga, M.S., Arruda, S.G.B. and Nascimento, J.A. (1999) Castration and slaughter age effects on nutritive value of the 'Mestiço' goat meat. *Meat Science* 52, 119–125.
Madruga, M.S., Souza, J.G. and Narain, J.G. (2001) Castration and slaughter age effects on fat components of the 'Mestiço' goat meat. *Small Ruminant Research* 42, 77–82.
Mahgoub, O., Khan, A.J., Al-Maqbaly, R.S., Al-Sabahi, J.N., Annamalai, K. and Al-Sakry, N.M (2002) Fatty acid composition of muscle and fat tissue of Omani Jebel Akhdar goats of different sexes and weights. *Meat Science* 61, 381–387.
Mahgoub, O., Kadim, I.T., Mothershaw, A., Al Zadjali, S.A., Annamalai, K. and Al-Mabsli, F.A. (2004) Chemical composition and quality attributes of meat from native and imported goats slaughtered at the central slaughter house in the Sultanate of Oman. In: *Proceedings of the 8th International Conference on Goats*, Pretoria, Republic of South Africa, 4–9 July 2004.
Marinova, P., Banskalieva, V., Alexandrov, S., Tzvetkova, V. and Stanchev, H. (2001) Carcass composition and meat quality of kids fed sunflower oil supplemented diet. *Small Ruminant Research* 42, 219–227.
Marsh, B.B., Leet, N.G. and Dickson, M.R. (1974) The ultrastructure and tenderness of highly cold-shortened muscle. *Journal of Food Technology* 9, 141–147.
Marshall, W.H., Smith, G.C., Dutson, J.R. and Carpenter, Z.L. (1977) Mechanically deboned goat, mutton and pork in frankfurters. *Journal of Food Science* 42, 193–196.
McCance, R.A. and Widdowson, E.M. (1960) *The Composition of Foods.* MRC Special Report. No. 297, HMSO, London, UK.
McGeehin, B., Sheridan, J.J. and Butler, F. (2002) Optimizing a rapid chilling system for lamb carcasses. *Journal of Food Engineering* 52, 75–81.
Mercier, Y., Garellier, P. and Renerre, M. (2004) Lipid and protein oxidation in vitro and antioxidant potential in meat from Charolaise cows finished on pasture or mixed diet. *Meat Science* 66, 467–473.
Miller, A.J., Ackerman, S.A. and Palumbo, S.A. (1980) Effects of frozen storage on functionality of meat for further processing. *Journal of Food Science* 45, 1466–1468.
Mills, E.W., Comerford, J.W., Hollender, R., Harpster, H.W., House, B. and Henning, W.R. (1992) Meat composition and palatability of Holstein and beef steers as influenced by forage type and protein source. *Journal of Animal Science* 70, 2446–2451.
Mirade, P.S. and Picgirard, L. (2001) Assessment of airflow patterns inside six industrial beef carcass chillers. *International Journal of Food Science and Technology* 35, 463–475.
Mitra, K. and Mitra, H.C. (1945) Biological value of proteins from muscle meat of cow, buffalo and goat. *Indian Journal of Medical Research* 33, 87–91.
Moller, A.J., Bouton, P.E., Harris, P.V. and Jones, P.N. (1983) Effect of electrical stimulation on the tenderization of mutton by ageing. *Journal of Food Science* 48, 874–877.
Mushi, D.E., Safari, J., Mtenga, L.A., Kifaro, G.C. and Eik, L.O. (2009a) Effects of concentrate levels of fattening performance, carcass and meat quality of small east African × Norwegian crossbred goats fed low quality grass hay. *Livestock Science* 124, 148–155.

Mushi, D.E., Safari, J., Mtenga, L.A., Kifaro, G.C. and Eik, L.O. (2009b) Growth and distribution of non-carcass components of Small East African and F1 Norwegian crossbred goats under concentrate diets. *Livestock Science* 126, 80–86.

Nagaraj, N.S., Anilakumar, K.R. and Santhanam, K. (2005) Post-mortem changes in myofibrillar proteins of goat skeletal muscles. *Journal of Food Biochemistry* 29, 152–170.

Nagaraj, N.S., Anilakumar, K.R. and Santhanam, K. (2006) Biochemical and physicochemical changes in goat meat during postmortem aging. *Journal of Muscle Food* 17, 198–213.

Negussie, E., Rottmann, O.J., Pirchner, F. and Rege, J.E.O. (2003) Patterns of growth and partitioning of fat depots in tropical fat-tailed Menz and Horro sheep breeds. *Meat Science* 54, 491–498.

Newbold, R.P. (1996) Changes associated with rigor mortis. In: Briskey, E.J., Cassens, R.G. and Trautmann, J.C. (eds) *The Physiology and Biochemistry of Muscle as Food.* University of Wisconsin Press, Madison, WI, USA.

Nuss, J.I. and Wolfe, H. (1981) Effect of post-mortem storage temperatures on isometric tension, pH, ATP, glycogen and glucose-6-phosphate for selected bovine muscles. *Meat Science* 5, 201–213.

Obuz, E. and Dikeman, M.E. (2003) Effects of cooking beef muscles from frozen or thawed states on cooking traits and palatability. *Meat Science* 65, 993–997.

Offer, G. and Knight, P. (1988) The structural basis of water-holding in meat. In: Lawrie, R.A. (ed.) *Developments in Meat Science*, Vol. 4. Elsevier Applied Science, London, UK, p. 63.

Offer, G. and Trinick, J. (1983) On the mechanism of water holding in meat: the swelling and shrinking of myofibrilis. *Meat Science* 8, 245–281.

Olson, D.G. and Parrish, F.C. Jr (1977) Relationship of myofibril fragmentation index to measures of beefsteak tenderness. *Journal of Food Science* 42, 506–509.

Olsson, U., Hertzman, C. and Tornberg, E. (1994) The influence of low temperature, type of muscle and electrical stimulation on the course of rigor mortis, ageing and tenderness of beef muscle. *Meat Science* 37, 115–131.

Ono, K., Berry, B., Johnson, H.K., Russek, E., Parker, C.F., Cahill, V. and Althouse, P. (1984) Nutrient composition of lamb of two age groups. *Journal of Food Science* 49, 1233–1239.

Ono, K., Berry, B. and Douglas, L. (1986) Nutrient composition of some fresh and retail cuts of veal. *Journal of Food Science* 51, 1352–1357.

Owen, I.E., Norman, G.A., Philbrooks, CA. and Jones, N.S.D. (1978) Studies on the meat production characteristics of Botswana goats and sheep. 3. Carcass tissue composition and distribution. *Meat Science* 2, 59–74.

Padda, G.S., Keshri, R.C., Sharma, B.D. and Sharma, N. (1985) Effect of different fat levels on the organoleptic acceptability of chevon (goat meat) patties. *Cheiron* 14, 183–187.

Padda, G.S., Keshri, R.C., Sharma, N., Sharma, B.D. and Murthy, T.R.K. (1988) Physico-chemical and organoleptic properties of patties from hot, chilled and frozen goat meat. *Meat Science* 22, 245–253.

Park, Y.W., Kouassi, M.A. and Chin, K.B. (1991) Moisture, total fat and cholesterol in goat organs and muscle meat. *Journal of Food Science* 56, 1191–1193.

Pearson, A.M. and Dutson, T.R. (1985) Scientific basis for electrical stimulation. In: Pearson, D.H. and Dutson, T.R. (eds) *Advances in Meat Research – Electrical Stimulation*, Vol. 1. AVI Publishing Company, Westport, Connecticut, pp. 185–218.

Pellett, P.L. and Young, V.R. (1990) Role of meat as a source of protein and essential amino acids in human nutrition. In: Pearson, A.M. and Dutson, T.R. (eds) *Advances in Meat Research – Meat and Health*, Vol. 6. Elsevier Applied Science, London, UK.

Pette, C. and Staron, S. (1990) Cellular and molecular diversities of mammalian skeletal muscle fibers. *Reviews of Physiology, Biochemistry and Pharmacology* 116, 1–76.

Pieniak-Lendzion, K., Niedziolka, R., Horoszewicz, E. and Borkowska, T. (2008) Evaluation of slaughter value and physicochemical attributes of goat meat. *Electronic Journal of Polish Agricultural Universities* 11, # 1.

Pond, W.G. and Maner, J.H. (1984) *Swine Production and Nutrition.* AVI Publishing Company, Westport, Connecticut.

Potchoiba, M.J., Lu, C.D., Pinkerton, F. and Sahlu, T. (1990) Effects of all-milk diet on weight gain, organ development, carcass characteristics and tissue composition, including fatty acid and cholesterol contents of growing male goats. *Small Ruminant Research* 3, 583–592.

Pratiwi, N.M.W., Murray, P.J. and Taylor, D.G. (2007) Feral goats in Australia: a study on the quality and nutritive value of their meat. *Meat Science* 75, 168–177.

Rees, M.P., Trout, G.R. and Warner, R.D. (2002) Tenderness, ageing rate and meat quality of pork *M. longissimus thoracis lumborum* after accelerated boning. *Meat Science* 60, 113–124.

Renerre, M., Poncet, K., Mercier, Y., Gatellier, P. and Metro, B. (1999) Influence of dietary fat and vitamin E on antioxidant status of muscles of turkey. *Journal of Agriculture and Food Chemistry* 47, 237–244.

Roeder, B., Ramsey, W.S., Hafley, B.S., Miller, R.K., Griffin, D.B., Davis, E.E. and Branson, R. (1999) *Consumer Acceptance and Quality Profile of Goat Meat.* Final Report to the Texas Department of Agriculture, Austin, Texas.

Rome, L.C., Sosnicke, A.A. and Goble, D.O. (1990). Maximum velocity of shortening of three fiber types from horse soleus muscle: implications for scaling with body size. *Journal of Physiology* 431, 173–185.

Rosenvold, K. and Andersen, H.J. (2003) Factors of significance for pork quality – a review. *Meat Science* 64, 219–237.

Rosenvold, K., Petersen, J.S., Laerke, H.N., Jensen, S.K., Therkildsen, M. and Karlsson, A.H. (2001) Muscle glycogen stores and meat quality as affected by strategic finishing feeding of slaughter pigs. *Journal of Animal Science* 79, 382–391.

Ruiz de Huidobro, F., Miguel, E., Onega, E. and Blazquez, B. (2003) Changes in meat quality characteristics of bovine meat during the first 6 days post mortem. *Meat Science* 65, 1439–1466.

Santos-Filho, J.M., Morais, S.M., Rondina, D., Beserra, F.J., Neiva, J.N.M. and Magalhães, E.F. (2005) Effect of cashew nut supplemented diet, castration, and time of storage of fatty acid composition and cholesterol content of goat meat. *Small Ruminant Research* 57, 51–56.

Savell, J.W., Smith, G.C., Dutson, T.R., Carpenter, Z.L. and Suter, D.A. (1977) Effect of electrical stimulation on palatability of beef, lamb and goat meat. *Journal of Food Science* 42, 702–706.

Schiaffino, S. and Reggiani, C. (1996) Molecular diversity of myofibrillar proteins: gene regulation and functional significance. *Physiological Review* 76, 371–423.

Schonfeldt, H.C., Naude, R.T., Bok, W., van Heerden. S.M., Sowden, L. and Boshoff, E. (1993) Cooking and juiciness related quality characteristics of goat and sheep meat. *Meat Science* 34, 381–394.

Seman, D.L and McKenzie-Parnell, J.M. (1989) The nutritive value of meat as food. In: *Meat Production and Processing*, Occasional Publication No. 11. New Zealand Society of Animal Production, Hamilton, New Zealand, pp. 13–28.

Sen, A.R., Santra, A. and Karim, A.A. (2004) Carcass yield, composition and meat quality attributes of sheep and goat under semiarid conditions. *Meat Science* 66, 757–763.

Sheridan, J.J., McGeehin, B. and Butler, F. (1998) Effects of ultra-rapid chilling and electrical stimulation on the tenderness of lamb carcass muscle. *Journal of Muscle Foods* 9, 403–417.

Sheridan, R., Hoffman, L.C. and Ferreira, A.V. (2003) Meat quality of Boer goat kids and mutton Merino lambs. 1. Commercial yields and chemical composition. *Animal Science* 76, 63–71.

Smith, G.C., Carpenter, Z.L. and Shelton, M. (1978) Effect of age and quality level on the palatability of goat meat. *Journal of Animal Science* 46, 1229–1235.

Srinivasan, K.S. and Moorjani, M.N. (1974) Essential amino acid content of goat meat in comparison with other meats. *Journal of Food Science and Technology* 11, 123–124.

Stankov, I.K., Todorov, N.A., Mitev, J.E. and Miteva, T.M. (2002) Study on some qualitative features of meat from young goat of Bulgarian breeds and crossbreeds of goats slaughtered at various ages. *Asian-Australian Journal of Animal Science* 15, 283–289.

Swatland, H.J. (1981) Cellular heterogeneity in the response of beef to electrical stimulation. *Meat Science* 5, 451–455.

Swatland, H.J. (1982) The challenges of improving meat quality. *Canadian Journal of Animal Science* 62, 15–24.

Swatland, H.J. (1984) The structure and properties of meat. In: *Structure and Development of Meat Animals.* Prentice-Hall, Englewood Cliffs, New Jersey, pp. 153–200.

Tarladgis, B.G., Watts, B.M., Younathan, N.T. and Dugan, L. (1960) A distillation method for the quantitative determination of malonaldehyde in rancid foods. *Journal of the American Oil Chemists Society* 37, 44–48.

Taylor, R.G., Geesink, G.H., Thompson, V.F., Koohmaraie, M. and Goll, D.E. (1995) Is Z-disk degradation responsible for postmortem tenderisation? *Journal of Animal Science* 73, 1351–1367.

Terrell, R.N., Suess, G.G. and Bray, R.W. (1969) Influence of sex, live-weight and anatomical location on bovine lipids. 2. Lipid components and subjective scores of six muscles. *Journal of Animal Science* 28, 454.

Tornberg, E. (1996) Biophysical aspects of meat tenderness. *Meat Science* 43, 175–191.

Trout, G.R. (1988) Techniques for measuring water-binding capacity in muscle foods: a review of methodology. *Meat Science* 23, 235–242.

USDA (1986) *Composition of Foods. Beef Products Fresh, Processed, Prepared.* USDA, Human Nutrition Information Service Agricultural Handbook no. 8–13, US Government Printing Office, Washington, DC, USA.

USDA (1989) *Official United States Standards for Grades of Carcass Beef.* Agricultural Marketing Service, USDA, Washington, DC, USA.

van Niekerk, W.A. and Casey, N.H. (1988) The Boer goat. II. Growth, nutrient requirements, carcass and meat quality. *Small Ruminant Research* 1, 355–368.

Wagner, J.R. and Ason, M.C. (1985) Effect of freezing rate on the denaturation of myofibrillar proteins. *Journal of Food Technology* 20, 735–744.

Wan, J.M.F., Haw, M.P. and Blackburn, G.L. (1989) Nutrition, immune function, and inflammation: an overview. *Proceedings of the Nutrition Society* 43, 315–335.

Wan Zahari, W.M. and Wahid, S.A. (1985) Mineral concentrations in the blood plasma and various tissues of local crossbred goats. *MARDI Research Bulletin* 13, 333–340

Warren, H.E., Scollan, N.D., Enser, M., Huges, S.I., Richardson, R.I. and Wood, J.D. (2008) Effects of breed and a concentrate or grass silage diet on beef quality in cattle of 3 ages. I: animal performance, carcass quality and muscle fatty acid composition. *Meat Science* 78, 256–269.

Warriss, P.D. and Brown, S.N. (1987) The relationship between initial pH, reflectance and exudation in pig muscle. *Meat Science* 20, 65–74.

Warriss, P.D., Bevis, E.A. and Ekins, P.J. (1989) The relationships between glycogen stores and muscle ultimate pH in commercially slaughtered pigs. *British Veterinary Journal* 145, 378–383.

Webb, E.C. (1991) The study and quantification of the influence of high energy nutrition on meat quality through the effects of carcass fat depots and fatty acids. MSc(Agric.) dissertation, University of Pretoria, Republic of South Africa.

Webb, E.C., Casey, N.H. and Simela, L. (2005) Goat meat quality. *Small Ruminant Research* 60, 153–166.

Wheeler, T.L, Davis, G.W., Stoeker, B.J. and Harmon, C.J. (1987) Cholesterol concentration of longissimus muscle, subcutaneous fat and serum of two beef cattle breed types. *Journal of Animal Science* 65, 1531–1537.

Whiting, R.C., Strange, E.D., Miller, A.J., Benedict, R.C., Mozersky, S.M and Swift, C.E. (1981) Effects of electrical stimulation on the functional properties of lamb muscle. *Journal of Food Science* 46, 434–487.

Wood, J.D. (1984) Fat deposition and the quality of fat tissue in meat animals. In: Wiseman, J. (ed.) *Fats in Animal Nutrition.* Butterworths, London, UK, pp. 407–435.

Wood, J.D. and Enser, M. (1997) Factors influencing fatty acids in meat and the role of antioxidation in improving meat quality. *British Journal of Nutrition* 78, S49–S60.

Wood, J.D., Enser, M., Fisher, A.V., Nute, G.R., Richardson, R.I. and Sheard, P.R. (1999) Manipulating meat quality and composition. *Proceedings of the Nutritional Society* 58, 363–370.

Wood, J.D., Enser, M., Fisher, M., Nute, G.R., Sheard, P.R., Richardson, R.I., Hughes, S.I. and Whittington, F.M. (2008) Fat deposition, fatty acid composition and meat quality: a review. *Meat Science* 78, 343–358.

Worthington-Roberts, B. and Monsen, E.R. (1990) Iron. In: Pearson, A.M. and Dutson, T.R. (eds) *Advances in Meat Research – Meat and Health*, Vol. 6. Elsevier Applied Science, London.

Xiong, Y.L. (1994) Myofibrillar protein from different muscle fiber types: implications of biochemical and functional properties in meat processing. *Critical Review Food Science and Nutrition* 34, 293–320.

Zygoyiannis, D., Kufidis, D., Katsaounis, N. and Philips, P. (1992) Fatty acid composition of carcass fat of indigenous (*Capra prisca*) suckled Greek kids and milk of their does. *Small Ruminant Research* 8, 83–95.

# 14 Effect of Early Nutrition on Carcass and Meat Quality of Young Goats Under Milk Production Systems

A. Argüello[1], N. Castro[1], D. Sánchez-Macías[1] and J. Capote[2]

[1]*Animal Science Department, Veterinary Faculty, Universidad de Las Palmas de Gran Canaria, Las Palmas, Spain;* [2]*Instituto Canario de Investigaciones Agrarias, La Laguna, Tenerife, Spain*

## 14.1 Abstract

Goat kids in intensive dairy goat farms are usually reared on milk replacers, due to farmers' desire to use all the milk for commercial purposes or cheese-making. In this chapter, growth, carcass and meat quality characteristics of goat kids fed milk replacers are studied in depth in comparison with kids reared naturally on goat milk.

## 14.2 Introduction

Different management methods have been proposed for goat kid rearing on dairy goat farms. These include natural suckling (NS), restricted natural suckling (RNS), ad libitum artificial rearing (ALAR) on milk replacers and restricted artificial rearing (RAR). These are summarized as follows:

- The NS system implies that kids remain with their dams from birth until the end of the suckling period with free access to goat milk 24 h a day.
- RNS is a system in which kids have access to their dams for a limited period of time per day.
- ALAR implies that kids are hand-fed colostrum during the first 2 days after birth. After the colostrum feeding period, the kids usually are accommodated in artificial rearing pens with at least 0.3 $m^2$ floor space per kid. Centrally heated pens have a temperature of around 20ºC. The animals are trained to suckle from an artificial teat fitted to a unit for feeding liquid diets. The milk replacer is continuously mixed and offered ad libitum on a 24 h basis.
- Under the RAR system, kids are reared as for the ALAR group except that milk replacers are made available twice a day for a limited period of time.

Goat milk and milk replacers differ in composition. The protein source of milk replacers is often milk protein concentrates and whey proteins (Beserra *et al.*, 2003). The casein content of milk replacers is lower than in goat milk, which may produce some problems with curd formation in the kid abomasum (Sanz Sampelayo *et al.*, 1990). Hashimoto *et al.* (2007) included soy as a protein source in milk replacers, but the effects on growth and meat quality are yet to be clearly defined. Carbohydrates are higher in milk replacers due to the overall lactose content. The main raw components in milk replacers are cow skimmed milk and cheese whey, and both are very rich in lactose. A high lactose content in milk replacers has been related to osmotic diarrhoea in kids (Argüello *et al.*, 1999). Some experiments have been carried out using starch as a raw material in milk replacers (Nitsan *et al.*, 1990), but diarrhoea was a major problem. A major difference in chemical composition between goat milk and milk replacers is also the fat source. Milk replacers are based on cow skimmed milk and cheese whey, and both ingredients have a low fat content. Therefore, vegetable fats are added to milk replacers as the main fat source. These include mainly palm or coconut oils. Bañón *et al.* (2006) recently reported substantial differences in fatty acid composition between goat milk and milk replacers. The main fatty acids in goat milk are C16:0 (30%), C18:1 (22%) and C18:0 (14%), whereas in milk replacers they are C12:0 (29%), C16:0 (23%) and C18:1 (16%). New advances in milk replacer formulations are currently under way. Tacchini *et al.* (2006) reduced the use of cow milk to 15%, and our group is introducing seaweed into formulations with encouraging preliminary results.

## 14.3 Goat Kid Growth Under Goat Milk or Milk Replacer Diets

The goat is an important source of meat in Africa, Asia and the Far East. It is now emerging as an alternative and attractive source of meat in other parts of the world. With milk goat breeds, kids are usually reared on milk replacers, so that all milk produced can be sold as fresh milk or processed into dairy products as cheese or yogurt. Young kids have to be routinely fed milk replacers.

### 14.3.1 Growth curves

Growth in farm animals in general, and goats in particular, is usually represented by an exponential curve. However, observations during the first month of life have shown a better statistical fit to a linear regression (Argüello *et al.*, 2004).

Growth curves of kids reared during the first month of life under NS, ALAR and RAR systems are shown in Fig. 14.1 (Argüello *et al.*, 2004). NS kids have a significantly higher average daily gain (ADG) than ALAR and RAR kids (Pérez *et al.*, 2001; Argüello *et al.*, 2004), and ALAR kids have a significantly higher ADG than RAR kids (Argüello *et al.* 2004). The higher ADG in NS kids is caused by the higher digestibility of components in goat milk than in milk replacers because the goat milk curd stays longer in the abomasum than the milk replacer curd (Sanz Sampelayo *et al.*, 1990;

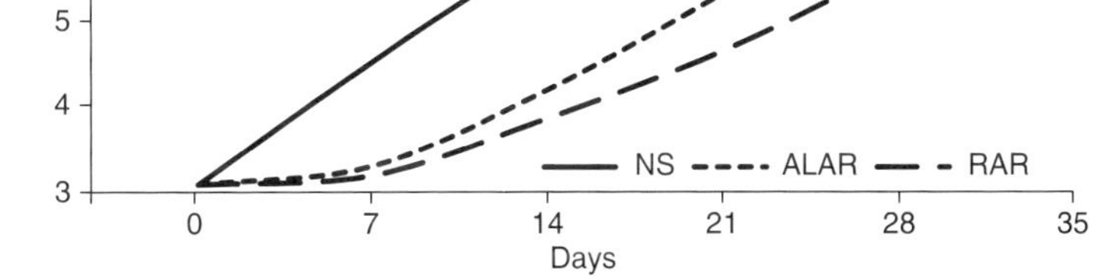


**Fig. 14.1.** Growth curves of goat kids raised under different feeding management systems (Argüello *et al.*, 2004). NS, natural suckling; ALAR, ad libitum artificial rearing; RAR, restricted artificial rearing.

Baumrucker and Blum, 1993). Baumrucker and Blum (1993) found that dams' milk has a growth promoter that is not present in milk replacers, which could explain the higher ADG in NS kids. NS kids' growth rates ranged from 140 to 200 g/day in different breeds such as Majorera (Argüello *et al.*, 2004), Verata (Fariña *et al.*, 1989) and Damascus (Louca *et al.*, 1977). Some authors have attributed the higher growth rate to a higher feed intake capacity in kids raised ad libitum (Sanz Sampelayo *et al.*, 1987; Yan *et al.*, 1993).

There are many factors that affect growth curves of farm animals, including birth weight, feed conversion efficiency and RNS, each of which is discussed in more detail.

### *Birth weight*

Birth weight has an important effect on the first month's growth pattern for goat kids (Argüello *et al.*, 2004). Table 14.1 shows the results for Majorera goat kids where the Pearson correlation coefficients between birth weight and weight at *n* days were statistically significant ($P < 0.01$) throughout the first 28 days of life for all groups. While the significance of the correlation coefficients observed with the ALAR and RAR methods lasted throughout the whole experiment, in the NS method the correlation lasted only until day 28. This behaviour is the opposite of what we expected, as the constant availability of food should have minimized the effect of birth weight.

### *Feed conversion efficiency*

From a financial point of view, it is also necessary to evaluate the feed conversion efficiency (FCE), which is a measure of an animal's efficiency in converting feed mass into increased body mass. Argüello (2000) reported similar FCEs for both male and female animals, with higher FCE values observed at the beginning of the experiment compared with the final period (Fig. 14.2). Tejón *et al.* (1995) reported FCE levels of 1060.19 and 1115.11 g/kg for males and females, respectively, in Guadarrama goat kids reared by artificial rearing for 0–21 days. The differences displayed between initial and final FCE levels are probably due to the lesser development of the digestive tract in kids at an early age. This becomes more evident in terms of milk assimilation and transformation as the animal grows older. The results obtained for FCE in males and females indicate that females are more efficient in terms of milk replacer transformation. Nevertheless, such differences are slight and are probably due to the greater voracity of males, which makes them tend to consume greater amounts of feed regardless of age.

### *Restricted natural suckling*

In special circumstances, kids are reared under RNS management. When the milk is highly valuable, farmers try to minimize the amount of milk used for feeding kids and allow the kids to access their mothers for

**Table 14.1.** Correlation matrix for birth and live weights of kids at different ages, as influenced by rearing methods (from Argüello *et al.*, 2004).

| | Birth weight | | | |
|---|---|---|---|---|
| Weight at: | NS | ALAR | RAR | NS+ALAR+RAR |
| 7 days | 0.92* | 0.79* | 0.90* | 0.78* |
| 14 days | 0.73* | 0.73* | 0.76* | 0.64* |
| 21 days | 0.83* | 0.68* | 0.65* | 0.60* |
| 28 days | 0.73* | 0.60* | 0.62* | 0.52* |
| 35 days | NS | 0.54* | 0.55* | 0.39* |

NS, Natural suckling; ALAR, ad libitum artificial rearing; RAR, restricted artificial rearing.
*, $P < 0.01$; NS, no significant differences.

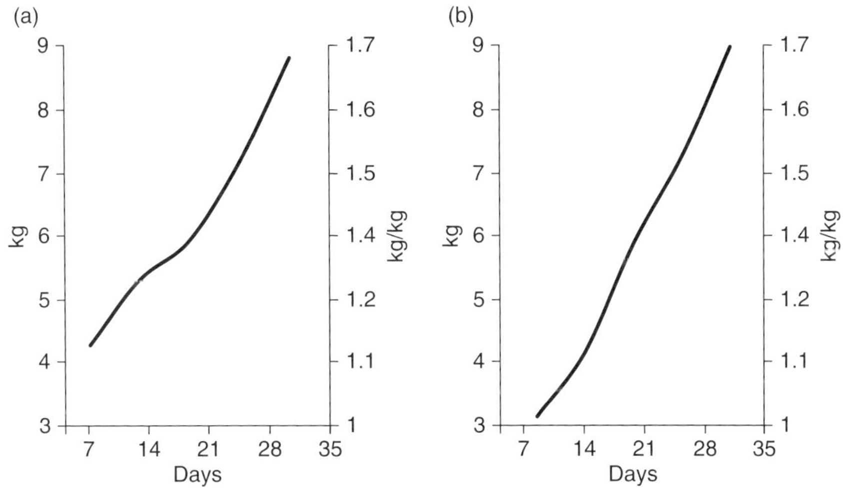


**Fig. 14.2.** Male (a) and female (b) goat kid feed conversion efficiency (efficiency of conversion of feed mass (kg) into increased body mass (kg)) under the ad libitum artificial rearing method (Argüello, 2000).

only a few hours a day. This system of management reduces the ADG in kids (Genandoy *et al.*, 2002) compared with NS management due to the lower milk intake. Therefore, RNS is not an adequate management system for meat production and must be recommended only when milk has a high price.

## 14.4 Effect of the Diet (Goat Milk or Milk Replacers) on Kid Carcass Quality

Goats are ruminant animals that produce useful products such as fibre, meat, milk and leather. In some regions of the world such as southern Europe, goats have been selected primarily for milk production (Harvey and Rigg, 1964). In such systems, few goats are raised for meat production as the major selection criterion in regions where milk production is the primary focus. Kid carcasses from dairy goats have little fat (Kirton, 1988). Traditionally, in non-specialized dairy goat herds, goat kids are reared with their dams, which results in reduced milk yield and thus less milk for cheese manufacturing. Therefore, goat keepers remove the kids from their dams very early postnatally (15 days of age; 5–6 kg live weight). These kids are then harvested for meat. Unfortunately, carcasses from these kids are very light in weight (~3 kg) and therefore have little saleable meat. Furthermore, consumers in Mediterranean countries and other regions (e.g. Canary Islands) prefer meat from kids that have only been fed milk. All these factors contribute to the production of a very light live weight of kids at slaughter and carcasses with low meat quality and yield, resulting in goat keepers not earning significant financial profits.

### 14.4.1 Body weight, carcass yield and offal

Losses in goat kid carcasses due to chilling are not affected by diet (goat milk versus milk replacers) according to Argüello *et al.* (2007). In reference to carcass yield, some authors (Argüello *et al.*, 2007) did not observe differences between animals fed goat milk or milk replacers due to lack of differences in nutritive characteristics between diets. Table 14.2 shows weights and carcass yield in goat kids fed goat milk or milk replacers (Argüello *et al.*, 2007).

Argüello *et al.* (2007) studied the influence of diet on offal components in kids (Table 14.3). There was no significant effect of diet on the percentage of blood, skin, feet, gastrointestinal tract full and empty, gastrointestinal content, liver, urinary bladder, testicle plus penis, spleen, head, lungs plus trachea, heart and thymus of the live weight at slaughter. However, the weight of the right kidney was significantly higher in kids fed with milk replacers.

### 14.4.2 Carcass conformation

Argüello *et al.* (2007) studied carcass conformation measurements and indices in Majorera goat kids fed goat milk or milk replacers. These included: width between hips (G), depth at 6th rib (Th), carcass length (L), leg length (F), chest width (Wr), hips perimeter (B), long leg compact indices

**Table 14.2.** Carcass yield parameters from kids fed milk or milk replacers (from Argüello *et al.*, 2007).

| | Goat milk | Milk replacers | SEM |
|---|---|---|---|
| Number of animals | 20 | 20 | |
| Live weight slaughter (kg) | 8.31 | 8.10 | 0.33 |
| Empty body weight (kg) | 5.89 | 5.91 | 0.31 |
| Hot carcass weight (kg) | 2.97 | 3.01 | 0.18 |
| Cold carcass weight (kg) | 2.87 | 2.75 | 0.18 |
| Chilling losses (%) | 3.45 | 3.29 | 0.16 |
| Net carcass yield (%)[a] | 50.43 | 50.23 | 0.39 |

SEM, Standard error of the mean.
[a]Net carcass yield = (carcass weight/empty body weight) × 100.

**Table 14.3.** Offal (% live weight at slaughter) in kids fed milk or milk replacers (from Argüello *et al.*, 2007).

| | Goat milk | Milk replacers | SEM |
|---|---|---|---|
| Blood | 3.89 | 3.27 | 0.14 |
| Skin | 10.18 | 9.93 | 0.01 |
| Feet | 3.85 | 4.09 | 0.10 |
| GI tract (full) | 14.57 | 15.33 | 0.49 |
| GI tract (empty) | 8.95 | 8.99 | 0.15 |
| GI content | 5.62 | 6.34 | 0.41 |
| Liver | 2.95 | 2.73 | 0.01 |
| Urinary bladder | 0.48 | 0.24 | 0.01 |
| Testicle and penis | 0.24 | 0.26 | 0.01 |
| Spleen | 0.22 | 0.21 | 0.01 |
| Right kidney | 0.33a | 0.44b | 0.01 |
| Head | 8.09 | 8.87 | 0.16 |
| Lungs + trachea | 1.72 | 1.69 | 0.01 |
| Heart | 0.75 | 0.68 | 0.01 |
| Thymus | 0.54 | 0.57 | 0.01 |

SEM, Standard error of the mean; GI, gastrointestinal.
Values with different letters (a, b) on the same row are statistically different ($P < 0.05$).

(G/F and B/F), cold carcass weight (CCW) and carcass compactness index (CCW/L) (Table 14.4 and Fig. 14.3). These authors found differences ($P < 0.01$) between diets for L and F measurements. Rearing in small pens (milk replacer diet) could be the reason for these little differences found between kids on different diets. There were

**Table 14.4.** Carcass conformation and indices from kids fed with different diets (from Argüello *et al.*, 2007).

| | Goat milk | Milk replacers | SEM |
|---|---|---|---|
| F (cm) | 24.42a | 23.13b | 0.22 |
| L (cm) | 42.71a | 40.48b | 0.04 |
| G (cm) | 9.88 | 10.20 | 0.13 |
| Wr (cm) | 11.77 | 10.61 | 0.09 |
| B (cm) | 33.08 | 31.94 | 0.15 |
| Th (cm) | 16.79 | 16.84 | 0.14 |
| CCW/L | 93.55 | 93.16 | 0.21 |
| G/F | 0.41 | 0.44 | 0.01 |
| B/F | 1.36 | 1.38 | 0.01 |

SEM, Standard error of the mean; F, Leg length; L, carcass length; G, width between hips; Wr, chest width; B, hips perimeter; Th, depth at 6th rib; CCW/L, carcass compactness index (cold carcass weight/carcass length), G/F and B/F, long leg compact indices. Values with different letters (a, b) on the same row are statistically different ($P < 0.01$).

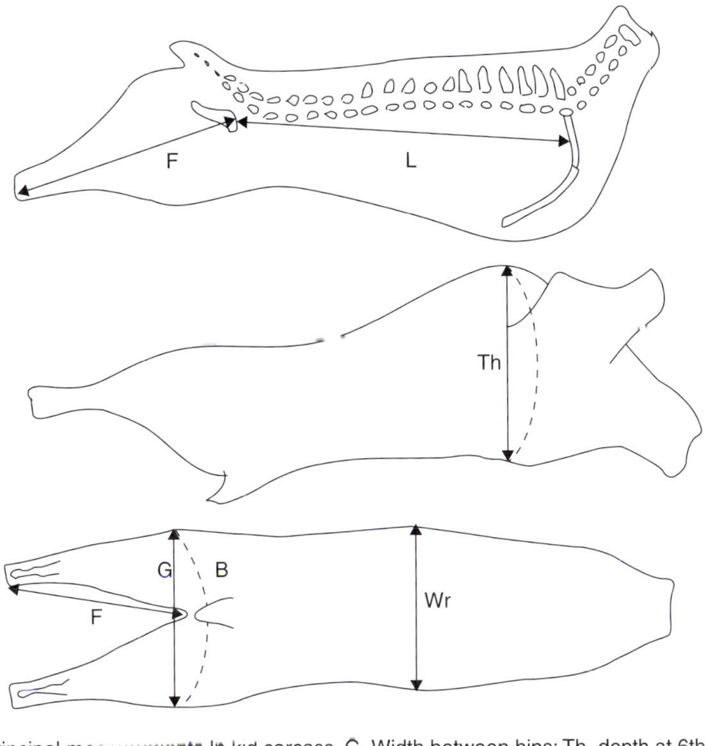


**Fig. 14.3.** Principal measurements in kid carcass. G, Width between hips; Th, depth at 6th rib; L, carcass length; F, leg length; Wr, chest width, B, hips perimeter.

no significant differences between diets in G, WR, B and TH measures and CCW/L, G/F and B/F indices. There were significant interactions between diet and live weight at slaughter for L and F measures and CCW/L, G/F and B/F indices. When kids' live weight at slaughter was 6 kg, higher CCW/L values in kids on the milk replacer were found, due to these animals being older.

### 14.4.3 Primal cut distribution

Some authors (Argüello *et al.*, 2007) found no differences in primal cut distribution in kids fed goat milk or milk replacers (Table 14.5). However, Sanz Sampelayo *et al.* (1987) found differences in lumbar rib percentages (~2%) between kids fed goat milk and kids fed milk replacers. These differences were probably a result of using different carcass jointing procedures.

### 14.4.4 Tissue composition

Argüello *et al.* (2007) studied the effect of diet (goat milk versus milk replacers) on tissue composition of kid carcasses (Table 14.6). There were significant effects of diet on subcutaneous, intermuscular and total fat in carcass and ribs, with the differences being higher in animals fed goat milk. Shoulder, long leg and flanks had lower percentages of intermuscular and total fat in kids fed milk replacers. There were no differences in fat content in neck. Morand-Fehr *et al.* (1986) previously reported similar results. They attributed this to a higher amount of fat fed in the NS system than in kids fed with milk replacers. The carcass total fat contents were lower than those reported by Gutiérrez *et al.* (1995) but were closer to those obtained by Colomer-Rocher *et al.* (1992). In the same breed, Argüello *et al.* (1997a,b,c) found that, while the amount of milk replacers increased, the total carcass fat percentage also increased. The bone and muscle tissue percentages did not differ as a result of diet.

## 14.5 Effect of Diet (Goat Milk or Milk Replacers) on Kid Meat Quality

The goat population in the world comprises four major types of goat: fibre goats (e.g. angora, cashmere), dairy goats (e.g. Saanen, Toggenburg and Nubian), meat goats (e.g. Boer) and feral goats (Naudé and Hofmeyr, 1981). The world's goat population was around 870 million in 2009,

**Table 14.5.** Contribution of organs and primal cuts to the carcass from kids fed dam milk or milk replacers (from Argüello *et al.*, 2007).

| | Goat milk | Milk replacers | SEM |
|---|---|---|---|
| Left kidney | 1.32 | 1.44 | 0.01 |
| Kidney and pelvic fat | 2.95 | 2.62 | 0.14 |
| Tail | 0.50 | 0.39 | 0.01 |
| Shoulder | 20.85 | 20.76 | 0.24 |
| Neck | 10.09 | 10.80 | 0.26 |
| Long leg | 32.93 | 33.86 | 0.39 |
| Flank | 9.64 | 9.54 | 0.20 |
| Ribs | 21.64 | 21.33 | 0.36 |
| By categories | | | |
| Extra | 54.58 | 55.20 | 0.45 |
| First | 20.85 | 20.76 | 0.24 |
| Second | 19.73 | 20.34 | 0.30 |

SEM, Standard error of the mean.

**Table 14.6.** Proportions of fat, bone, muscle and primal cuts of kids fed goat milk or milk replacers (from Argüello *et al.*, 2007).

| | Goat milk | Milk replacers | SEM |
|---|---|---|---|
| Carcass[a] | | | |
| Subcutaneous fat | 4.69a | 3.79b | 0.20 |
| Intermuscular fat | 3.71a | 2.56b | 0.22 |
| Total fat | 11.35a | 8.97b | 0.46 |
| Bone | 29.43 | 30.32 | 0.45 |
| Muscle | 55.03 | 55.70 | 0.46 |
| Losses | 1.08 | 2.31 | 0.26 |
| Shoulder[b] | | | |
| Subcutaneous fat | 3.20 | 2.52 | 0.20 |
| Intermuscular fat | 2.67a | 1.69b | 0.24 |
| Total fat | 5.87a | 4.21b | 0.32 |
| Bone | 30.87 | 31.69 | 0.46 |
| Muscle | 62.09 | 61.58 | 0.37 |
| Losses | 0.38 | 1.42 | 0.23 |
| Neck[b] | | | |
| Subcutaneous fat | 6.89 | 6.43 | 0.51 |
| Intermuscular fat | 4.22 | 3.15 | 0.44 |
| Total fat | 11.11 | 9.58 | 0.55 |
| Bone | 28.53 | 29.30 | 0.66 |
| Muscle | 55.18 | 50.67 | 1.08 |
| Losses | 4.09 | 8.57 | 1.12 |
| Long leg[b] | | | |
| Subcutaneous fat | 4.55 | 3.59 | 0.31 |
| Intermuscular fat | 3.64a | 2.46b | 0.23 |
| Total fat | 8.19a | 6.05b | 0.41 |
| Bone | 30.16 | 29.94 | 0.46 |
| Muscle | 60.26 | 61.93 | 0.40 |
| Losses | 0.49 | 1.03 | 0.13 |
| Flanks[b] | | | |
| Subcutaneous fat | 6.49 | 5.55 | 0.43 |
| Intermuscular fat | 6.95a | 4.70b | 0.57 |
| Total fat | 13.44a | 10.25b | 0.71 |
| Bone | 29.68 | 31.85 | 0.99 |
| Muscle | 54.19 | 54.34 | 1.01 |
| Losses | 0.47 | 0.43 | 0.12 |
| Ribs[b] | | | |
| Subcutaneous fat | 5.44a | 3.88b | 0.31 |
| Intermuscular fat | 3.96a | 2.66b | 0.32 |
| Total fat | 9.40a | 6.54b | 0.49 |
| Bone | 33.80 | 34.93 | 0.75 |
| Muscle | 53.18 | 52.97 | 0.63 |
| Losses | 1.61 | 3.06 | 0.30 |

SEM, Standard error of the mean.
Values with different letters (a, b) on the same row are statistically different.
[a]Proportions in carcass weight.
[b]Proportions in joint weight.

with annual meat production of around 4.9 million t (FAOSTAT, 2010). Consumers' preference for goat meat varies around the world. For instance, in India, the local community specifically seeks meat from mature goats, whereas, in France and Latin America, meat from young milk-fed kids is considered a traditional delicacy. The acceptability of meat is greatly influenced by local custom and preference, so it is not possible to apply a universal standard for the quality of goat meat (Naudé and Hofmeyr, 1981).

### 14.5.1 Physical attributes

Physical attributes of meat quality include: pH, colour, tenderness and water-holding capacity. Some authors have studied the effect of diet (goat milk or milk replacers) on the meat quality of young goats (Argüello *et al.*, 2005; Bañón *et al*, 2006). They did not find significant effects of diet on pH value or lightness, except for some slight differences in chroma values (a less intense red colour) (Tables 14.7, 14.8 and 14.9). Meat tenderness is considered one of the most important attributes in terms of consumer satisfaction. Diet significantly affected shear force values in the semimembranosus and triceps brachii muscles (Tables 14.8 and 14.9) with animals fed milk replacers having greater shear force values than animals fed goat milk. A similar trend was also observed for the longissimus dorsi muscle. These differences could be attributed to the fact that the animals were older and had consumed greater amounts of starter feed. This is in agreement with reports by Pisula *et al.* (1994), who found statistical differences between

**Table 14.7.** Effects of diet on longissimus dorsi muscle attributes (means ± SD) in kids fed milk or milk replacers (from Argüello *et al.*, 2005).

| | Goat milk | Milk replacers |
|---|---|---|
| pH[a] | 6.08 ± 0.24 | 6.30 ± 0.31 |
| pH[b] | 5.59 ± 0.18 | 5.73 ± 0.01 |
| Lightness[a] | 50.07 ± 3.92 | 49.53 ± 3.00 |
| Lightness[b] | 56.57 ± 4.82 | 56.93 ± 3.96 |
| Chroma[a] | 9.08 ± 1.72 | 10.45 ± 2.43 |
| Chroma[b] | 13.76 ± 3.99a | 16.11 ± 5.69b |
| Hue[a] | 26.79 ± 12.25 | 29.75 ± 8.94 |
| Hue[b] | 43.99 ± 7.67 | 42.08 ± 6.09 |
| Shear force (N) | 50.07 ± 14.93 | 55.71 ± 13.41 |
| Water-holding capacity (g) | 0.66 ± 0.11a | 0.46 ± 0.10b |
| Moisture (%) | 78.21 ± 0.38 | 78.40 ± 1.20 |
| Protein (%) | 18.67 ± 0.72 | 19.05 ± 1.74 |
| Fat (%) | 1.26 ± 0.41 | 0.96 ± 0.44 |
| Ash (%) | 1.15 ± 0.09 | 1.12 ± 0.05 |
| Collagen (%) | 0.60 ± 0.13 | 0.46 ± 0.16 |
| Collagen solubility (%) | 70.49 ± 8.47 | 85.62 ± 15.84 |
| Type I (%) | 24.00 ± 11.43 | 32.91 ± 22.67 |
| Type IIA (%) | 46.00 ± 10.70 | 35.50 ± 15.68 |
| Type IIB (%) | 30.00 ± 4.00 | 31.85 ± 19.30 |
| Type I (μm$^2$) | 484.27 ± 151.88 | 389.10 ± 123.79 |
| Type IIA (μm$^2$) | 541.23 ± 224.02 | 354.16 ± 164.43 |
| Type IIB (μm$^2$) | 472.49 ± 166.04 | 367.02 ± 86.79 |

SD, Standard deviation.
Values with different letters on the same row (a, b) are statistically different.
[a]At slaughter; [b]after chilling.

**Table 14.8.** Effects of diet on triceps brachii muscle attributes (means ± SD) in kids fed milk or milk replacers (from Argüello *et al.*, 2005).

| | Goat milk | Milk replacers |
|---|---|---|
| pH[a] | 6.34 ± 0.21 | 6.53 ± 0.27 |
| pH[b] | 5.82 ± 0.10 | 5.80 ± 0.14 |
| Lightness[a] | 53.08 ± 4.61 | 53.60 ± 3.95 |
| Lightness[b] | 56.33 ± 3.08 | 55.47 ± 4.98 |
| Chroma[a] | 12.50 ± 3.36 | 11.03 ± 1.41 |
| Chroma[b] | 13.99 ± 2.46a | 15.26 ± 1.76b |
| Hue[a] | 31.34 ± 8.93 | 31.87 ± 6.74 |
| Hue[b] | 39.82 ± 7.83 | 38.29 ± 10.36 |
| Shear force (N) | 83.18 ± 8.64a | 88.40 ± 6.85b |
| Water-holding capacity (g) | 0.41 ± 0.08a | 0.33 ± 0.07b |
| Moisture (%) | 78.38 ± 1.41 | 78.55 ± 0.40 |
| Protein (%) | 17.54 ± 2.07 | 18.53 ± 0.69 |
| Fat (%) | 0.84 ± 0.22 | 1.08 ± 0.51 |
| Ash (%) | 1.08 ± 0.07 | 1.16 ± 0.07 |
| Collagen (%) | 0.49 ± 0.06 | 0.40 ± 0.06 |
| Collagen solubility (%) | 83.04 ± 3.04 | 83.09 ± 12.14 |
| Type I (%) | 29.49 ± 8.63 | 17.53 ± 12.72 |
| Type IIA (%) | 40.13 ± 9.90 | 36.18 ± 19.26 |
| Type IIB (%) | 30.37 ± 6.51 | 46.28 ± 11.55 |
| Type I ($\mu m^2$) | 596.22 ± 126.13 | 570.73 ± 134.55 |
| Type IIA ($\mu m^2$) | 707.24 ± 238.84 | 636.02 ± 123.01 |
| Type IIB ($\mu m^2$) | 678.77 ± 254.48 | 640.60 ± 142.63 |

SD, Standard deviation.
Values with different letters on the same row (a, b) are statistically different.
[a]At slaughter; [b]after chilling.

kids slaughtered at 16 kg live weight and exclusively fed milk replacers and those that had consumed starter feed (35.7 versus 42.6 Newtons, respectively). The values obtained for water-holding capacity ranged between 0.31 g (6.2%) and 0.72 g (14.4%). The pH value and protein content play a fundamental role in the greater levels of expelled juice in animals fed goat milk (Argüello *et al.*, 2005). The average pH value for animals fed goat milk replacers after chilling was 5.65, while kids receiving milk replacers had an average pH value of 5.70.

### 14.5.2 Chemical composition and muscle characteristics

Argüello *et al.* (2005) reported protein values of ~17–20% in goat kids, which resembles that of very young animals, as well as low muscle fat (0.84–1.26%). Fat is a late-growing body tissue, and in goats it is deposited in the viscera more than in other animals (Chilliard *et al.*, 1981). The type of diet did not have a significant effect on the chemical composition of kid carcasses (Argüello *et al.*, 2005). This is in accordance with the observations of Mueller *et al.* (1985) using kids of similar weights and feed types. In contrast, Bañón *et al.* (2006) reported higher moisture in kids fed goat milk or milk replacers (77 and 76%, respectively) and less protein in kids fed milk replacers. Collagen percentages and solubility were not affected by diet (Argüello *et al.*, 2005; Bañón *et al.*, 2006). Diet did not affect muscle fibre areas (Tables 14.7, 14.8 and 14.9), following the muscle fibre classification of Argüello *et al.* (2001).

**Table 14.9.** Effects of diet on semimembranosus muscle attributes (means ± SD) in kids fed milk or milk replacers (from Argüello *et al.*, 2005).

| | Goat milk | Milk replacers |
|---|---|---|
| pH[a] | 6.09 ± 0.27 | 6.39 ± 0.22 |
| pH[b] | 5.58 ± 0.04 | 5.64 ± 0.07 |
| Lightness[a] | 47.13 ± 17.32 | 54.43 ± 3.11 |
| Lightness[b] | 53.61 ± 5.47 | 54.49 ± 2.11 |
| Chroma[a] | 9.73 ± 2.34 | 11.87 ± 2.61 |
| Chroma[b] | 12.43 ± 2.24a | 14.46 ± 3.64b |
| Hue[a] | 34.57 ± 13.71 | 32.28 ± 7.01 |
| Hue[b] | 44.93 ± 12.01 | 41.28 ± 6.62 |
| Shear force (N) | 32.64 ± 11.87a | 43.67 ± 6.24b |
| Water-holding capacity (g) | 0.72 ± 0.16a | 0.60 ± 0.15b |
| Moisture (%) | 78.46 ± 0.50 | 78.51 ± 0.88 |
| Protein (%) | 18.20 ± 0.99 | 18.10 ± 1.65 |
| Fat (%) | 0.91 ± 0.34 | 1.10 ± 0.51 |
| Ash (%) | 1.18 ± 0.06 | 1.18 ± 0.09 |
| Collagen (%) | 0.46 ± 0.05 | 0.42 ± 0.09 |
| Collagen solubility (%) | 81.52 ± 10.48 | 74.69 ± 8.77 |
| Type I (%) | 9.75 ± 5.68 | 31.40 ± 25.65 |
| Type IIA (%) | 84.25 ± 4.27 | 13.83 ± 9.24 |
| Type IIB (%) | 6.00 ± 1.82 | 54.77 ± 34.61 |
| Type I (μm²) | 508.45 ± 93.47 | 528.04 ± 106.56 |
| Type IIA (μm²) | 564.45 ± 198.00 | 602.65 ± 138.86 |
| Type IIB (μm²) | 565.55 ± 146.38 | 586.34 ± 129.95 |

SD, Standard deviation.
Values with different letters on the same row(a, b) are statistically different.
[a]At slaughter; [b]after chilling.

### 14.5.3 Fatty acid percentages

Bañón *et al.* (2006) investigated diet effects on fatty acid composition in perirenal fat. The major fatty acids were C18:1, C16:0 and C18:0. These authors reported higher percentages in goat milk-fed kids than in artificially reared kids for the fatty acids C10:0, C14:0, C15:0, C16:0 and C18:0. In contrast, the goat milk-fed kids had lower values of C12:0, C16:1, C17:0, C18:1, C18:2, C18:3, C20:0 and C:20:4. The main fatty acids in kids receiving goat milk were C16:0, C18:1 and C14:0, and the ratio of saturated:unsaturated fatty acids was 2.27. In animals receiving milk replacers, the main fatty acids were C18:1, C16:0 and C18:0, and the saturated: unsaturated ratio was 0.94. Together with previous results, Bañón *et al.* (2006) thus demonstrated a strong effect of diet on fatty acid content.

### 14.5.4 Sensorial quality

Bañón *et al.* (2006) observed that the goat milk or milk replacer diet had pronounced effects on the sensory quality of cooked meat. The milk replacer diet gave cooked meat a more intense characteristic odour and flavour, more tenderness and increased juiciness.

## 14.6 Conclusions

Rearing goat kids with milk replacers has significant repercussions on their growth, carcass and meat quality. Goat kids fed milk replacers grow at a slower rate and have leaner carcasses than kids fed goat milk. The meat from kids fed milk replacers is characterized by a more intense colour. Substantial differences have also been reported in fatty acid profile. However, although the kids produced using the two diets differ, rearing goat kids on intensive dairy goat farms on milk replacers is more profitable if there is a large enough price margin between goat milk and milk replacers.

## References


Argüello, A. (2000) Lactancia artificial de cabritos, encalostrado, crecimiento, calidad de la canal y de la carne. PhD thesis, Las Palmas de Gran Canaria University, Spain.

Argüello, A., Arjona, J., Piñán, J., Ginés, R., Capote, J. and López, J.L. (1997a) Características cárnicas de cabritos de la Agrupación Caprina Canaria (ACC: variedad tinerfeña) criados con lactancia artificial. In: *XXII Jornadas Científicas de la Sociedad Española de Ovinotecnia y Caprinotecnia*, Tenerife, Spain, pp. 407–415.

Argüello, A., Ginés, R., Afonso, J.M. and López, J.L. (1997b) Utilización de yogur en la lactancia artificial de cabritos de la Agrupación Caprina Canaria (ACC). In: *XXII Jornadas Científicas de la Sociedad Española de Ovinotecnia y Caprinotecnia*, Tenerife, Spain, pp. 423–430.

Argüello, A., Rodríguez, M., Darmanin, N., Afonso, J.M., Capote, J. and López, J.L. (1997c) Características cárnicas de cabritos de la Agrupación Caprina Canaria (ACC) criados con lactancia tradicional (media leche). In: *XXII Jornadas Científicas de la Sociedad Española de Ovinotecnia y Caprinotecnia*, Tenerife, Spain, p. 417.

Argüello, A., Ginés, R. and López, J.L. (1999) A note on yogurt utilisation in artificial rearing of kids. *Journal of Applied Animal Research* 15, 165–168.

Argüello, A., López-Fernández, J.L. and López-Rivero, J.L. (2001) Limb myosin heavy chain isoprotein and muscle fiber types in the adult goat (*Capra hircus*). *Anatomical Record* 264, 284–293.

Argüello, A., Castro, N. and Capote, J. (2004) Growth of milk replacer kids fed under three different managements. *Journal of Applied Animal Research* 25, 37–40.

Argüello, A., Castro, N., Capote, J. and Solomon, M. (2005) Effects of diet and live weight at slaughter in kid meat quality. *Meat Science* 70, 173–179.

Argüello, A., Castro, N., Capote, J. and Solomon, M.B. (2007) The influence of artificial rearing and live weight at slaughter on kid carcass characteristics. *Journal of Animal and Veterinary Advances* 6, 20–25.

Bañón, S., Vila, R., Price, A., Ferrandini, E. and Garrido, M.D. (2006) Effects of goat milk or milk replacer diet on meat quality and fat composition of suckling goat kids. *Meat Science* 72, 216–221.

Baumrucker, C.R. and Blum, J.R. (1993) Secretion of insulin-like growth factors in milk and their effect on the neonate. *Livestock Production Science* 35, 49–74.

Bezerra, F.J., Bezerra, L.C.N.M., Silva, E.M.C. and Silva, C.E.M. (2003) Influence of the replacement of cow milk by goat milk cheese whey on meat composition carcass characteristics of three cross suckling kids. *Ciencia Rural* 33, 929–935.

Chilliard, Y., Sauvant, D., Bas, P., Pascal, G. and Morand-Fehr, P. (1981) Importance relative at activités métaboliques des différent tissus adipeux de la chèvre laitière. In: *Nutrition and Systems of Goat Feeding*. ITOVIC-INRA, Paris, France, pp. 90–100.

Colomer-Rocher, F., Kirton, A.H., Mercer, G.J.K. and Duganzich, D.M. (1992) Carcass composition of New Zealand Saanen goats slaughtered at different weights. *Small Ruminant Research* 7, 161–193.

FAOSTAT (2010) Caprine heads and meat production. Live animals and livestock primary. US Food and Agriculture Organization <http://faostat.fao.org/site/339/default.aspx>.

Fariña, J., Martín, L., Rodríguez, P., Rojas, A., Rota, A. and Tovar, J. (1989) Estudio de los chivos veratos. Periodo de amamantamiento. *Archivos de Zootecnia* 38, 127–139.

Genandoy, H., Sahlu, T., Davis, J., Wang, R.J., Hart, S.P., Puchala, R. and Goetsch, A.L. (2002) Effects of different feeding methods on growth and harvest traits of young Alpine kids. *Small Ruminant Research* 44, 81–87.

Gutiérrez, M.J., Peña, F., Rodero, E. and Herrera, M. (1995) Producción ovina y caprina. In: *XX Jornadas Científicas de la Sociedad Española de Ovinotecnia y Caprinotecnia*, Madrid, Spain, pp. 421–425.

Harvey, D. and Rigg, J.C. (1964) Some aspects of goats as livestock. *Nutrition Abstracts and Reviews* 34, 641–645.

Hashimoto, J.H., Alcalde, C.R., Silva, K.T., Macedo, F.A.F., Mexia, A.A., Santello, G.A., Martin, E.N. and Matsushita, M. (2007) Caracteristicas de carcaça de caprinos Boer x Saanen confinados recebendo rações com casca do grão de soja em substituição ao milho. *Revista Brasileira de Zootecnia* 36, 165–173.

Kirton, A.H. (1988) Characteristics of goat meat including carcass quality and methods of slaughter. In: *Goat Meat Production in Asia. Proceedings of an International Workshop in Tando Jam, Pakistan.* International Development Research Centre, Ottawa, Canada, pp. 87–99.

Louca, A., Economides, S. and Hancock, J. (1977) Effect of castration on growth rate feed conversion efficiency and carcass quality in Damascus goats. *Animal Production* 24, 387–391.

Morand-Fehr, P., Bas, P., Schmidely, P. and Hervieu, J. (1986) Qualité des produits chez les ovins et les caprins. In: *10 Journées de la Recherche Ovine et Caprine.* INRA, pp. 236–247.

Mueller, R., Steinhart, H. and Scheper, J. (1985) Carcass composition and meat quality of kids. Influence of feeding. *Fleischwirtschaft* 65, 194–196.

Naudé, R.T. and Hofmeyr, H.S. (1981) Meat production. In: Gall, O. (ed.) *Goat Production.* Academic Press, London, UK, pp. 285–307.

Nitsan, Z., Golan, M. and Nir, I. (1990) Utilization of raw or heat-treated starch fed in liquid diet to preruminants 1. Kids. *Small Ruminant Research* 3, 325–339.

Pérez, P., Maino, M., Morales, M.S. and Soto, A. (2001) Effect of goat milk and milk substitutes and sex on productive parameters and carcass composition in Creole kids. *Small Ruminant Research* 42, 87–93.

Pisula, A., Slowinski, M., Pawlowski, P., Bidwel-Porebska, K. and Piotrowski, J. (1994) Chemical composition, physic-chemical properties and organoleptic quality of 'milk' kid meat reared to 16 kg of body weight. *Gospodarka-Miesna* 46, 15–17.

Sanz Sampelayo, M.R., Muñoz, F.J., Lara, L., Gil, F. and Boza, J. (1987) Effectos del nivel det alimentación, clase de leche y edad en el desarrollo de cabritos de raza Granadina. *Investigación Agraria, Producción y Sanidad Animal* 2, 93–103.

Sanz Sampelayo, M.R., Hernández-Clua, O.D., Naranjo, J.A., Gil, F. and Boza, J. (1990) Utilization of goat milk vs. milk replacer for Granadina goat kids. *Small Ruminant Research* 3, 37–46.

Tacchini, F., Rebora, C., van den Bosch, S., Gascón, A. and Pedrani, M. (2006) Formulation and testing of a whey-based kid goat's milk replacer. *Small Ruminant Research* 63, 274–281.

Tejón, D., López, C., Piñán, J., de la Fuente, J., Sanz, M., Fernández, A. and Rey, A. (1995) Contribución al estudio de las razas autóctonas de la C.A.M. Evaluación del crecimiento en relación con el sistema de cría de la raza Caprina del Guadarrama. In: *XX Jornadas Científicas de la Sociedad Española de Ovinotecnia y Caprinotecnia*, Madrid, Spain, pp. 389–400.

Yan, T., Cook, J.E., Gibb, M.J., Ivings, W.E. and Treacher, T.T. (1993) The effects of quantity and duration of milk feeding on the intake of concentrates and growth of castrated male Saanen kids to slaughter. *Animal Production* 56, 327–332.

# 15 Effects of Feeding System and Diet on Body Lipid Composition of Young Goats

P. Morand-Fehr[1], A. Araba[2], P. Bas[1] and A. El Aich[2]
[1]UMR 791 INRA/AgroParisTech, Paris, France;
[2]Institut Agronomique et Vétérinaire Hassan II, Rabat Instituts, Morocco

## 15.1 Abstract

The objective of this chapter is to present findings from our experimental work, as well as from other research institutes, related to the composition and quality of goat meat, particularly the lipid profile, with emphasis on dietary factors and feeding programmes. This topic is of current concern as the fatty acid (FA) composition of ruminant meats and fats contains essential FAs for health, conjugated linoleic acids (CLAs), which reduce the risks of cardiovascular diseases and cancers. It also plays a role in the organoleptic, nutritional and dietetic quality of meat. The FA compositions of kid and lamb meats are not very different. However, kid meat was observed to be a little richer in polyunsaturated fatty acids (PUFAs), ω-3 fatty acids and CLAs, although it was not clear whether this was due to a direct effect of species. In kids, as in other ruminants, muscle lipids are more unsaturated, richer in PUFAs and lower in saturated fatty acids (SFAs) than subcutaneous adipose tissues and especially internal fats. However, the subcutaneous fats are more sensitive to factors modifying their FA composition. Experimental information on the effect of factors modifying the lipid composition of meats and fats in kids is less available than in lambs and steers. The dietary factors and type of feeding system are the main factors, but it is difficult to evaluate their direct effects because of the influence of other factors such as genotype, sex and stage of growth. During the milk-feeding period, feeding systems based on dam grazing or supplementations with linseeds, fish oil or PUFA mixtures can increase the proportions of desirable FAs such as ω-3, PUFAs and CLAs, and decrease C18:1 and C16:0 fatty acids and SFAs in dams' milk and, to a lesser extent, in meat lipids and fats of suckling kids. During post-weaning periods, similar feeding systems with kids can have similar effects on meat and fats, although the results are limited and heterogeneous. Future progress in the commercialization of kid meat may occur but only if its quality, particularly dietetic aspects, is markedly good. Therefore, the dietetic quality of lipids present in meat or in consumed fats must be improved. Kid rearing systems such as under Argan tree forests in Morocco have been shown to reduce the proportion of fats in meat, increase the proportion of desirable FAs such as CLAs and the cholesterol content in meat, and decrease the ω-6:ω-3 ratio to about 5. Such findings allow optimism for the future in the production and consumption of goat meat.

## 15.2 Introduction

Dietary fat is one of the three major energy resource-providing macronutrient groups (mainly fatty acids, FAs). Humans require polyunsaturated fatty acids (PUFAs), which are essential for life. A component of these PUFAs, the conjugated linoleic acids (CLAs), may reduce the risk of various diseases, while other FAs, such as saturated FAs (SFAs), can increase the occurrence of cardiovascular diseases. Moreover, lipids enable the absorption of vitamins such as A, D, E and K. Ruminant meat flavour and consumer preferences are also influenced by FA composition (Melton, 1990).

The composition of meat lipids has been the subject of numerous investigations because of the effect of dietary lipids on human blood lipid composition, energy balance and the health consequences of cardiovascular and other metabolic diseases (Madsen *et al.*, 1992; Nürnberg *et al.*, 1998). The role of meat in human nutrition, particularly for goat meat, depends on the standards of living, cultural patrimony and socio-economic environment of consumers (Casey *et al.*, 2003). However, it is necessary to identify the characteristics of meat lipids and adipose tissues, as cardiovascular diseases are increasing in many countries such as in the Mediterranean Basin where much lamb meat is consumed. Therefore, it is essential to determine whether goat meat presents the same risk for human consumers and whether it is possible to modify goat-rearing techniques to improve the dietetic characteristics of goat meat lipids and consequently limit health risks.

The objective of this chapter is to present findings of our own experimental work and other studies on the composition of goat meat and adipose tissue lipids and the effects of diet and feeding programmes on their quality and composition, as feeding is the most practical method that can be applied by goat farmers.

## 15.3 Nutritional Role of Meat Lipids and Fats

The major role of meat lipids and fats is to meet human needs and consumers' satisfaction. Maintaining a good level of health depends on the frequency of meals including meat, on cultural and culinary traditions, and on the social position of consumers in society and their level of education.

### 15.3.1 Dietetic role of meat lipids and fat

In most industrialized as well as other countries, the lipid composition of ruminant meat is believed to contain high proportions of saturated FAs. This argument is used to explain the increase in cardiovascular diseases and other health problems such as colon cancer (Bauchart *et al.*, 2008). However, the lipid content and composition in ruminant meat and carcasses are quite variable in different cuts or muscles and depend on very numerous factors.

Recently, the positive and negative roles of FAs in atherogenesis and the risk of cardiovascular diseases are becoming better known, although there is still a lack of full knowledge in this area. Most FAs involved are present in goat meat and carcass fats. Recent information of the effects of FAs was reported by Ledoux (2006), Legrand (2008) and Lecerf (2008). Currently, the overconsumption of SFAs is considered to increase low-density lipoprotein (LDL)-cholesterol and results in other adverse effects. This effect is much more evident with palmitic acid (C16:0) than with stearic acid (C18:0), which is frequently considered not to have any negative effect. Information on the monounsaturated FAs (MUFAs), mainly represented by oleic acid (C18:1 *cis*), is currently insufficient to determine their role. Among PUFAs, the FAs of the ω-3 family, particularly linolenic acid (C18:3ω-3), eicosapentaenoic acid (C20:5ω-3; EPA), docosapentaenoic acid (C22:5ω-3; DPA) and docosahexaenoic acid (C22:6w-3; DHA), are the most efficient for decreasing cardiovascular risk. The effect of linoleic acid (C18:2ω-6) and other FAs of the ω-6 family may have a more moderate effect. Indeed, cardiovascular risks decrease with C18:2ω-6 excess or deficit, and an optimal level of 4% of total energy supply and a ratio of ω-6:ω-3 of around 5 or less are advised (Raes *et al.*, 2004). Moreover, the CLA and principally the main one, rumenic acid (C18:2 *cis*-9, *trans*-11), are said to be responsible for several health-promoting effects on consumers such as anti-carcinogenesis (particularly for breast cancer), anti-atherosclerosis, immunomodulation and shifting of the partitioning of energy towards protein instead of fat deposition (Enser, 2000; Webb *et al.*, 2005; Ledoux, 2006). Therefore, the lipids of red meat should have a high content of CLAs to improve its dietetic quality.

Some research workers group all the PUFAs (ω-3 and ω-6) and C18:0 FAs and refer to them as desirable FAs (DFAs). Some workers such as Banskalieva *et al.* (2000) and Webb *et al.* (2005) reported that the DFA content in goat meat ranged between 60 and 80%, with values tending to be higher than in other ruminants. The findings of our work are more moderate but in many experiments the effects of the studied factors on C18:0, ω-3 and ω-6 FAs and CLAs are different in lambs and kids.

These data on the link between FA composition and human diet and health emphasize the importance of FA content, particularly in small ruminant meat, which is one of the most saturated meats.

### 15.3.2 Effect of the lipid composition of goat meat and carcass fats on the organoleptic quality of meat

The physical properties and chemical composition of fat influence the conservation capacities and organoleptic qualities of meat.

Adipose deposits tend to appear later in kid carcasses than in lambs (Tshabalala *et al.*, 2003), as confirmed by allometric coefficients (Morand-Fehr, 1981). Consequently, when kids are slaughtered at an early age, their carcasses have little protection from subcutaneous adipose tissues. Therefore, conserving kid carcasses by refrigeration may lead to losses and dryness of external unprotected meat.

It was observed that the external carcass fat of lambs fed diets very rich in cereals and poor in fibre did not harden sufficiently during carcass cooling after slaughtering. The external fats remained soft and oily due to a modified fat composition rich in water, odd- and branched-chain FAs and unsaturated FAs (Molénat and Thériez, 1973; Bas *et al.*, 1980). The commercial value of these carcasses falls drastically. As kid carcasses are generally low in external fats, this aspect of carcass presentation is scarce in young kids, but was observed in 5–6-month-old kids fed a diet composed exclusively of concentrate feeds (Bas *et al.*, 1981).

Consumers' acceptability of red meats such as that of kids is determined by flavour (Melton,1990), which principally originates from lipids (Moody, 1983). Meat lipids act as a solvent for the volatile components that accumulate during the cooking of meat. The quantity and composition of fats in meats, particularly the marbled fat, play an essential role in meat succulence and flavour. The lamb flavour components were identified by Suzuki and Bailey (1985). The 'goaty' flavour is due to similar components in milk and meat. They are molecules with a chain of eight or nine carbons and with acid or aldehyde functions resulting from unsaturated FA (USFA) oxidation and with branched chains such as 4-ethyl-octanoic acid, 4-methyl-nanoic acid or 4-methyl-octanoic acid (Ha and Lindsay, 1990, 1991; Madruga *et al.*, 2000b; Martin *et al.*, 2005). In lambs, several FAs, particularly linoleic acid (C18:2), were also involved in the increase in lamb flavour (Ralph, 1989). When the flavour becomes stronger, the meat acceptability may increase or decrease according to consumers' taste. However, the meat flavour of small ruminants is a complicated issue due to various responses of consumers and the numerous factors that positively or negatively influence it, particularly conditions of cooking and dietary components added (Melton, 1990).

Similarly, intra- and intermuscular fats markedly influence the juiciness, succulence and tenderness of red meats. In particular, a little intramuscular fat significantly improves the succulence and tenderness. Indeed, small amounts of intermuscular fat are necessary to lubricate the muscle fibres and to increase the juiciness and flavour of cooked lamb meat (Beriain *et al.*, 2000).

Thus, lipids and fats play an essential role in the dietetic and organoleptic value of small ruminant meats; however, this topic has not yet been studied intensively in goat meat.

## 15.4 Characteristics of Goat Meat and Fat Lipids Composition

In Chapter 10, Lee and Kannan reviewed the nutritive value of FAs and fats in goat meat. However, the characteristics of the lipid composition of goat meat need to be highlighted to enable an understanding of the effects of feeding programmes on the dietetic and nutritive value of kid meat fats.

In comparison with steers and lambs (Morand-Fehr *et al.*, 1991), the FA composition of kid meat has not been investigated in much depth (Banskalieva *et al.*, 2000). However, since 2000, the amount of research on this topic has been expanding markedly.

In 2000, Banskalieva and colleagues published an extensive and informative review on the FA composition of goat muscle and fat deposits (Banskalieva *et al.*, 2000). They observed that experimental procedures and design, technical and statistical methodologies, sampling methods, breed and rearing methods, among other factors, varied among different researchers.

As reported for other ruminant species, the three major FAs in the muscle lipids of goats are oleic (C18:1 *cis*), palmitic (C16:0) and stearic (C18:0) acids, which represent 60–80% of total FAs, followed by linoleic acid (C18:2 *cis*). The other FAs are C14:0 in SFAs, C16:1 in MUFAs and C18:3ω-3,6,9 *cis*, C20:4ω-6,9,12,15 *cis* in PUFAs, and other FAs in lower concentrations such as C10:0, C12:0, C15:0, C15:1, C17:0, C17:1, C20:1, C20:3, C22:0, C24:0, C22:4, C22:5 and C22:6. The percentages of all these acids are quite variable from one study to another, for example: between 28 and 50% for C18:1, 15 and 31% for C16:0, 6 and 17% for C18:0, and 4 and 15% for C18:2. The average percentages of C16:0, C18:0, C18:1 and C18:2 in goat muscles are similar to those of other ruminant species, particularly sheep, but the concentration of C16:1 in goat muscles is frequently higher than in lambs, and goat muscle lipids can be a little richer in PUFAs (C18:2, C18:3 and C20:4) than lambs (Banskalieva *et al.*, 2000) with a higher ω6:ω-3 ratio (Sheridan *et al.*, 2003).

As with other ruminants, in goats, the anatomical location of adipose tissues is the main factor affecting the FA composition of adipose deposits. External fatty tissues such as subcutaneous tissues contain fewer SFAs and more USFAs, as well as minor FAs characterized by an odd number of carbons or a branched chain.

Regardless of the location of the adipose tissues in the body, the main FAs of goat fat deposits are C18:1, C18:0 and C16:0, followed by C14:0, C16:1, C17:0 and C18:2 (Banskalieva *et al.*, 2000). Other FAs such as C10:0, C12:0, C14:1, C15:0, C17:1 and C18:3 occur at the lowest levels, often making them unquantifiable. Goats deposit lipids that consist mainly of SFAs (30–71%) and MUFAs (20–57%) with PUFAs contributing <6% in some fat depots, although in most studies not all PUFAs were analysed.

Despite the high variation related to the anatomical location, it appears that, as in beef or lamb, internal deposits such as intermuscular or perirenal adipose tissues in young goats have more SFAs, particularly C18:0, fewer MUFAs, particularly 16:1, and much fewer PUFAs, especially C18:2, C18:3 and C20:4, than external deposits such as in subcutaneous tissues.

Table 15.1 shows data on the FA composition of muscle, lipids and carcass fats in goats and sheep from bibliographic references published after the review of Banskalieva *et al.* (2000), between 2000 and 2008.

**Table 15.1.** Fatty acid composition of meat and fat lipids in young sheep and goats.

| Animal | Breed | Nature of lipids | Age at slaughtering (weeks) | Fatty acid (molar %): 16:0 | 18:0 | 18:1 *cis*-9 | SFAs | USFAs | MUFAs | PUFAs | Reference |
|---|---|---|---|---|---|---|---|---|---|---|---|
| Sheep | Dorper | M+FLD | Unknown | 24.3 | 14.4 | 37.6 | 52.8 | 47.2 | 43.9 | 3.3 | Tshabalala *et al.* (2000) |
| Sheep | Damara | M+FLD | Unknown | 22.5 | 16.4 | 38.9 | 51.8 | 48.2 | 44.3 | 3.9 | Tshabalala *et al.* (2000) |
| Goat | Boer | M+FLD | Unknown | 21.0 | 20.4 | 36.7 | 54.7 | 45.3 | 41.9 | 3.4 | Tshabalala *et al.* (2000) |
| Goat | Indigenous | M+FLD | Unknown | 19.5 | 20.0 | 37.7 | 53.6 | 46.4 | 42.5 | 3.9 | Tshabalala *et al.* (2000) |
| Kid | Indigenous | IMLD | 54 | 22.2 | 15.1 | 44.5 | 39.0 | 56.0 | 50.3 | 5.7 | Bas *et al.* (2005) |
| Kid | Indigenous | PRAT | 54 | 30.5 | 32.6 | 19.0 | 67.3 | 26.3 | 24.9 | 1.4 | Bas *et al.* (2005) |
| Lamb | Timahdite | IMLD | 41 | 23.1 | 12.1 | 34.1 | 39.0 | 50.2 | 38.7 | 11.5 | Araba *et al.* (2008) |
| Lamb | Timahdite | PRAT | 41 | 18.8 | 30.2 | 28.6 | 53.5 | 39.2 | 35.5 | 3.9 | Araba *et al.* (2008) |
| Lamb | V | IM | V | 22.5 | 15.6 | 40.4 | 43.0 | 51.0 | 43.6 | 7.4 | Bas and Morand-Fehr (2000) |
| Lamb | V | PRAT | V | 20.6 | 26.1 | 36.7 | 52.6 | 45.8 | 39.4 | 6.4 | Bas and Morand-Fehr (2000) |
| Kid | Malaguena | SC leg | 7–8 | 25.5 | 11.5 | 43.4 | 46.6 | 52.7 | 47.0 | 5.7 | Fernandez-Navarro *et al.* (2005) |
| Kid | Malaguena | IM leg | 7–8 | 21.6 | 12.8 | 38.6 | 39.9 | 59.2 | 41.3 | 17.9 | Fernandez-Navarro *et al.* (2005) |
| Kid | Saanen | IM 3m | 4 | 22.6 | 12.0 | 36.3 | 40.8 | 59.0 | 42.4 | 16.2 | Mele *et al.* (2007) |
| Kid | Dutch White | PRAT | 6 | 24.4 | 17.5 | 35.7 | 51.2 | 47.3 | 40.7 | 8.9 | Yeom *et al.* (2002) |
| Sheep | Barbarine | SC leg | 36–40 | 22.2 | 10.6 | 45.1 | 41.8 | 55.1 | 51.9 | 3.2 | Atti *et al.* (2007) |
| Lamb | Meriniez-zata | IMLD | 14 | 17.8 | 10.6 | 21.1 | 30.6 | 69.5 | 23.8 | 45.8 | Camparra *et al.* (2007) |

SFAs, Saturated fatty acids; MUFAs, monounsaturated fatty acids; PUFAs, polyunsaturated fatty acids; USFAs, unsaturated fatty acids; M+FLD, lipids of meat (muscle and fat) in Longissimus dorsi; IMLD, intramuscular lipids in Longissimus dorsi; PRAT, lipids of perirenal adipose tissue; IM, intramuscular lipids; SC leg, lipids of leg subcutaneous fat; IM leg, leg intramuscular lipids; IM 3m, intramuscular lipids from three muscle mixtures (Longissimus dorsi, Triceps brachii and semimembranosus; V, very various.

There was an adequate level of methodological quality to allow comparisons of the results in this field during recent years.

When sheep and goats were reared under very close conditions, the compositions of meat lipids in these two species were similar (Tshabalala *et al.*, 2003), with goat meat lipids being a little higher in C18:0 and lower in C16:0 and MUFAs, with not much effect of breed. Bas *et al.* (2005) and Araba *et al.* (2008) carried out two experiments on lambs and goats with the same protocol under very similar experimental conditions. They collected samples from the same tissues and in the same location. Consequently, their results can be compared. In the two species, the perirenal adipose tissue (internal tissue) contained more SFAs, particularly C18:0, and fewer MUFAs and PUFAs than intramuscular lipids sampled on the longissimus dorsi muscle. However, lamb lipids tended to be higher in PUFAs than those of kids.

A meta-analysis of 979 observations from 108 references (Bas and Morand-Fehr, 2000) on lambs again showed that internal fats are higher in SFAs and C18:0, and lower in MUFAs and PUFAs. The results from other references (Yeom *et al.*, 2002; Fernandez-Navarro *et al.*, 2005; Atti *et al.*, 2007; Camparra *et al.*, 2007; Mele *et al.*, 2007) reflect the very high variability of FA composition in kid or lamb muscles and fats, probably due to various factors that can modify this composition.

There is a lack of information regarding the content of FAs likely to have a positive role on human health, particularly CLAs such as rumenic acid and ω-3 FAs such as EPA, DPA and DHA, as observed by Webb *et al.* (2005). It is frequently stated – although with no reliable evidence – that goat meat contains more CLAs or ω-3 FAs than sheep meat. However, in this regard, one should be aware of the large variability of findings (Webb *et al.*, 2005) due to differences in location of fat, breed, diet and more importantly the chromatographic method of assessment of these FAs. Similar variability has been reported in findings on beef (Bauchart *et al.*, 2008).

Table 15.2 presents recent results on the contents of ω-3 FAs and CLAs in intramuscular lipids of lamb and kid meat. As samples were collected from intramuscular lipids of the longissimus dorsi muscle, the results are likely to be comparable. In these studies, the authors generally attempted to improve the contents of CLAs or ω-3 FAs by modifying the diet or using different breeds. The percentage of rumenic acid (C18:2 *cis*-9, *trans*-11) is between 0.4 and 0.9% in kids. This confirms the values reported by Hausman and Snell (2000) of 0.4–0.8% for all fat locations in kids. Other values (Sikora and Borys, 2006; Mele *et al.*, 2007) are similar to those reported for lambs (Camparra *et al.*, 2007; Delmotte *et al.*, 2007; Vasta *et al.*, 2007; Araba *et al.*, 2008) but tend to be higher than in beef (Bauchart *et al.*, 2008).

The total content of ω-3 FAs varies between 1 and 2% in intramuscular lipids in kids, with values tending to be around 2% when the diet contains feeds supplying ω-3 FAs. The content of ω-3 FAs in intramuscular lipids in lambs is quite variable, with values similar to those of goats, although some values are much higher (Camparra *et al.*, 2007; Delmotte *et al.*, 2007). Similar to lambs, among the ω-3 FAs, the content of C22:5 (DPA) is always higher than the C20:5 (EPA) and C22:6 (DHA) content in kids.

It had been hypothesized that most of the CLAs in goat adipose tissues, as in the other ruminants, are synthesized by biohydrogenation and isomerization (*cis*/*trans* and non-conjugated/conjugated) of the PUFAs in feedstuffs in the rumen into C18:1 *trans*-11 (*trans* vaccenic acid). The C18:1 *trans*-11 is absorbed and accumulates in the adipose tissues, and is then converted into CLAs by desaturation, which explains the correlation ratio close to +1 between the contents of the *trans* vaccenic and rumenic acids in kid fats (Tsuneishi *et al.*, 2001).

To conclude, the FA composition of muscle lipids and adipose tissues, particularly concerning the FAs with positive or negative dietetic interest, is highly variable as a result of the interaction of numerous factors that may modify it. Generally, the lipid composition of goat fats is typical of

**Table 15.2.** Percentage of main ω-3 fatty acids and rumenic acid (a conjugated linoleic acid) and ω-6:ω-3 ratio in intramuscular lipids in lambs and kids.

| Animal | Fatty acid | | | | | | Reference |
|---|---|---|---|---|---|---|---|
| | 20:5ω-3 (EPA) | 22:5ω-3 (DPA) | 22:6ω-3 (DHA) | Total ω-3 | 18:2 *cis*-9, *trans*-11 (CLA) | ω-6:ω-3 ratio | |
| Lamb | 4.1–5.3 | 6.4–7.2 | 2.8–2.9 | 19.4–25.7 | 0.8 | 1.1–1.3 | Camparra *et al.* (2007) |
| Lamb | 1.0–1.4 | | 0.6–0.8 | 5.8 –6.7 | 0.7–1.3 | 2.8–6.6 | Delmotte *et al.* (2007) |
| Lamb | | | 2.1–3.2 | | 0.5–0.6 | | Vasta *et al.* (2007) |
| Lamb | | | | 1.3–2.8 | 0.6–0.7 | 2.8–7.6 | Arana *et al.* (2008) |
| Kid | 0.2 | 0.4–0.6 | 0.2–0.3 | 1.0–1.3 | 0.4–0.9 | 5.3–6.2 | Mele *et al.* (2007) |
| Kid | 0.2–0.8 | | 0.1–0.4 | 1.0–2.9 | | 6.1–9.4 | Sikora and Borys (2006) |
| Kid | 1.1 | | 1.1 | 3.3 | 0.8 | | Nudda *et al.* (2005) |
| Kid | 0.1–1.1 | 0.3–1.7 | 0.1–0.4 | 0.7–4.3 | | 2.9–8.2 | Bas *et al.* (2005) |

EPA, eicosapentaenoic acid; DPA, docosapentaenoic acid; DHA, docosahexaenoic acid; CLA, conjugated linoleic acid.

that of ruminants consuming a diet rich in fibre with intense rumen fermentation, with high contents of odd- and branched-chain FAs, SFAs and particularly stearic acid (C18:0) (Martin *et al.*, 2005).

## 15.5 Effect of Animal Factors and Diet on Composition of Muscle Lipids and Adipose Tissues

In addition to methodological reasons, the large variability in FA composition of goat muscle lipids and fats is a result of three factors:

1. Location of sampling, which takes into consideration whether the location is near blood vessels or not, and the metabolic specificity of the adipose tissue according to location and blood supply.
2. Animal factors, such as age, live weight, genotype and sex.
3. Exogenous factors, such as climatic and dietary factors. As subcutaneous fats must stay fluid in all climates, the subcutaneous tissues are higher in SFAs and lower in USFAs when the ambient temperature is high (Marchello *et al.*, 1967).

Dietary factors have the most significant effect on the composition of fat lipids. However, it is difficult to analyse these effects without evoking other factors because of the complexity of interactions. For example, a dietary factor can have a strong effect on one genotype and a more limited effect on another, as shown in beef between breeds with early and late growth (Bauchart, 2006). In lambs, Bas and Morand-Fehr (2000) also observed the risk of wrongly analysing results in a nutritional experiment if age, genotype, stage of growth and type of adipose tissue are not taken into account.

### 15.5.1 Location of sampling and type of adipose tissue

Bas *et al.* (1992) indicated the importance of sampling in goat adipose tissue at the same anatomically well-defined site because of the heterogeneity of the FA composition within the same adipose tissue. However, this factor is rarely taken into account in studies on goat fat composition, which may partially explain the variability in results.

The fat of the internal, omental, mesenteric, perirenal and pelvic depots is more saturated than the subcutaneous and intramuscular depots. This is attributed to the gradient in body temperature, which results in saturated fats with a higher melting point being deposited where the temperature is the highest (Marchello *et al.*, 1967; Beriain *et al.*, 2000). The PUFA content is highest in intramuscular lipids and lowest in subcutaneous adipose tissues, which also have the lowest C18:0 content and the highest odd- and branched-chain FA contents. In goats, several authors observed wide differences between FA composition in internal and external adipose tissues (Sauvant *et al.*, 1979; Bas *et al.*, 1996).

### 15.5.2 Effect of age or live weight at slaughter

It is not feasible to assess the effect of stage of growth (age or live weight at slaughtering) on the FA composition of goat fats because the FA composition depends on the composition of dietary lipids in diet feeds during suckling and post-weaning periods. During and immediately after weaning, the rate of rumen fermentation also increases progressively, and PUFAs are hydrogenated and transformed into conjugated and *trans* isomers (Bas *et al.*, 1991). Minor FAs such as branched- or odd-chain FAs appear. Consequently, the FA profile of tissues that reflects the milk-feeding period gradually disappears after weaning. During the milk-feeding period, Mandrefini *et al.* (1988), using a diet based on milk and concentrates, observed a decrease in SFAs and an increase in USFAs as live weight increased. With diets based on milk only, Bas *et al.* (1987) and Zygoyiannis *et al.* (1992) reported that the proportion of C18:0 FAs in fat depots decreased and the percentage of other FAs increased with increasing age of unweaned kids. Other authors studied the evolution of the FA composition of fats during kid weaning (Sauvant *et al.*, 1979; Sikora and Borys, 2006). Generally, SFAs and C18:0 FA concentrations in all fat depots tended to increase and the percentage of PUFAs decreased, while the MUFA content either decreased or increased. Similar changes were observed in lambs (Bas and Morand-Fehr, 2000; Beriain *et al.*, 2000). This is probably a result of the low proportions of short- and medium-chain FAs and high concentrations of C18 FAs in post-weaning diets in comparison with goat milk, and from hydrogenation of PUFAs in the rumen. The change in MUFAs is variable because it depends on the intensity of *trans*-isomerization in the rumen. Indeed, *trans* MUFAs are hydrogenated very slowly in the rumen. In addition, the cholesterol content in fats tends to increase after weaning.

During the growth period (1–2 months after weaning), cholesterol content is reported to increase continuously (Madruga *et al.*, 2000a; Beserra *et al.*, 2004) although this finding was not confirmed by other workers (Sikora and Borys, 2006). Several months after weaning, there was still an increase in the C18:0 and C18:1 FAs in kid muscle lipids and adipose tissues (Madruga *et al.*, 2000a; Dhanda *et al.*, 2003; Beserra *et al.*, 2004), but C14:0 and C16:0 proportions decreased. Nevertheless, Werdi Pratiwi *et al.* (2004) reported contradictory findings. The PUFA concentration is variable in kid tissues, probably because it depends on the intensity of rumen hydrogenation. Similar results were obtained in lambs (Webb and Casey, 1995; Nürnberg *et al.*, 1998; Bas and Morand-Fehr, 2000).

### 15.5.3 Effect of sex

The effect of sex on the FA composition of kid meat lipids and fats is limited. Johnson *et al.* (1995), Matsuoka *et al.* (1997) and Mahgoub *et al.* (2002) observed a trend of higher PUFA (C18:2 and C18:3) content in muscle lipids and a lower SFA (C14:0 and C8:0) content in male animals compared with females. Similar findings had been reported in lambs (Beriain *et al.*, 2000). However, other research workers (Rojas *et al.*, 1994 on perirenal fat; Todaro *et al.*, 2004 on pelvic fat) found no significant differences in the FA composition of fat between male and female kids. Similarly, Werdi Pratiwi *et al.* (2004) observed no difference in the composition of meat lipids and only minor changes in adipose tissues between entire and castrated Boer young bucks. In lambs, the concentration of PUFAs and C18:2 *cis* FAs in back fat decreases in the following order: male castrates > females > males (lowest in fat deposits), while SFAs increase (Nürnberg *et al.*, 1996).

It is also interesting to note an interaction between sex and diet affecting FA composition. Indeed, when kids are fed diets rich in grains and low in fibre, the subcutaneous fats are soft and oily, as has been mentioned above. In this case, these fats are softer in males than in females (Bas *et al.*, 1980).

### 15.5.4 Effect of breed genotypes

Banskalieva *et al.* (2000) reviewed a large body of experimental findings emphasizing the effect of breed on FA composition of kid muscles and fat depot lipids. This review suggested that other factors, particularly dietary factors, interact to produce this effect on FA composition, as indicated by Beriain *et al.* (2000) in sheep. Differences in fat composition due to breed should also be estimated at the same stage of maturity and fatness of young ruminants (Webb and Casey, 1995). Generally, different genotypes can have different growth potentials and levels of intake, feeding behaviour and feed preferences. For instance, Zygoyiannis *et al.* (1985) attributed their findings on the effects of various sheep genotypes to differences in milk intake in lambs as a result of differences in genotype. To determine whether findings are due to the effect of breed only, it is necessary to compare animals at the same level and composition of feed intake. Consequently, it is obvious that findings in kids varied because of differences in growth rates and the composition of feed intake. This is clearly observed when comparing improved breeds such as Saanen versus local breeds. Sikora and Borys (2006) reported that intramuscular lipids in Saanen kids have more ω-3 and C18:2 FAs and fewer MUFAs and C18:1 FAs than crossbreeds, while the contents of SFA, and C14:0, C16:0 and C18:0 FAs were similar in both genotypes. Similarly, differences appeared between crossbred Saanen or Boer and Feral or Angora goats (Dhanda *et al.*, 2003). In particular, adipose tissues in Saanen × Boer goats had a significantly higher USFA (mainly C18:1) content than the other crossbreeds. However, Beserra *et al.* (2004) observed that the variations in FA profiles in kid meat were not influenced by genetic groups such as Moxoto × Alpine or Anglo-Nubian crossbreeds. None the less, the specific genetic effects appeared to be of limited influence in sheep (Nürnberg *et al.*, 1998). The genetic effect is more apparent in subcutaneous adipose tissues than in internal tissues. Sauvant *et al.* (1979) emphasized that, in Alpine breeds, the effect of buck (male ascendance) is significant on the content of SFAs and minor FAs in various adipose tissues.

## 15.6 Effect of Feeding Programme and Type of Diet

Dietary factors have a very significant effect on the FA composition of muscle and fat lipids in ruminants, including goats. Their effects are of major interest, as goat farmers can easily modify the diets of kids and rapidly appreciate the effects of diet changes.

Young goat growth is composed of two physiologically distinguished periods: the milk-feeding period, when kids are pre-ruminants with their digestion similar to monogastric animals (i.e. without chemical modification of dietary lipids in the rumen), and the post-weaning period, when kids are true ruminants, with dietary lipids that are modified in the rumen.

### 15.6.1 Milk-feeding period

As in other ruminants, particularly lambs (Bas and Morand-Fehr, 2000), the FA composition of adipose tissues reflects the FA composition of milk ingested by kids (Banskalieva *et al.*, 2000). The period in which kids suckle their dams varies according to the production system. This period generally lasts for a short time in systems of goat-milk production but lasts longer in systems of goat meat production.

Table 15.3 clearly indicates that the FA composition of kids' muscle closely reflects the composition of dam milk and can be modified by the diet of dams, particularly by dietary fat changes (Nudda *et al.*, 2005). The dams received diets composed of hay and a concentrate with cottonseeds rich in C18:2ω-6 and C16:0 or linseeds rich in C18:3ω-3. With the supply of linseeds in the diet, the milk became richer in CLAs, ω-3 FAs and C18:1 *trans* total FAs with lower levels of SFAs (C16:0 and C18:0) compared with the milk of goats fed cottonseeds. The composition of kid muscle fats had the same profile of FA composition especially the ω-3 and C18:1 *trans* FAs and ω-6:ω-3 ratio, but the differences were less marked. Mele *et al.* (2007) added soybean oil to lactating goats' diets, which modified the FA composition of intramuscular lipids in suckling kids. The proportions of vaccinic acid (C18:1 *trans*-11) and rumenic acid (C18:2 *cis*-9, *trans*-11) increased from 0.20 and 0.26% to 1.15 and 0.81%, respectively, when the concentration of SFAs, and C18:1 and ω-3 FAs decreased. Camparra *et al.* (2007) also obtained interesting results with goats reared indoors or on pasture. The proportions of PUFAs, particularly ω-3 FAs in muscle fats, were higher and the percentage of C16:0 FAs was lower in kids of grazing dams than in kids with dams of the indoor group. The results from kids treated by adding linseeds to dam diets (Nudda *et al.*, 2005) or raised with grazing dams (Camparra *et al.*, 2007) have been confirmed by studies in sheep (Delmotte *et al.*, 2007).

Similarly, Martin *et al.* (1999) found that including protected fat (calcium soaps of FAs) in the dams' diet modified the FA

**Table 15.3.** Effects of dam diet on milk and kid muscle fatty acids (from Nudda *et al.*, 2005).

| Fatty acid (mg/100 mg fatty acid methyl esters) | Dam milk[a] | | Kid muscle[b] | |
|---|---|---|---|---|
| | Dam diet[c] supplemented with: | | Dam diet[c] supplemented with: | |
| | Cottonseed | Linseed | Cottonseed | Linseed |
| C4:0–C14:0 | 19.1 | 19.1 | 4.6 | 3.6 |
| C16:0 | 25.9c | 18.7d | 20.7 | 19.5 |
| C18:0 | 21.8c | 16.8d | 12.8 | 13.3 |
| C18:1 *trans*-11 (VA) | 0.8c | 7.4d | 0.9a | 1.5b |
| C18:1 total *trans* | 1.9c | 9.2d | 1.6a | 2.3b |
| C18:1 *cis*-9 | 22.0 | 20.9 | 30.2 | 27.6 |
| C18:2ω-6 | 3.3 | 4.1 | 11.6 | 12.7 |
| C18:3ω-3 | 0.2e | 1.8f | 0.8e | 1.3f |
| C18:2 *cis*-9, *trans*-11 (RA) | 0.3c | 2.2d | 0.6 | 0.9 |
| CLA total | 0.7c | 2.9d | 1.3 | 1.4 |
| C20:4ω-6 | 0.2 | 0.2 | 0.7 | 7.4 |
| C20:5ω-3 (EPA) | 0.05 | 0.08 | 0.9a | 1.3b |
| C22:6ω-3 (DHA) | 0.07 | 0.06 | 0.8 | 1.3 |
| Total ω-3 | 0.3e | 1.9f | 2.5c | 4.0d |
| Total ω-6 | 3.5 | 4.4 | 19.1 | 21.1 |
| ω-6:ω-3 ratio | 11.4e | 2.3f | 7.6 | 5.4 |

VA, vaccenic acid; RA, rumenic acid; EPA, eicosapentaenoic acid; DHA, docosahexaenoic acid.
Significant differences: on the same line, two values with different letters are significantly different at $P < 0.10$ (a, b); $P < 0.05$ (c, d) or $P < 0.01$ (e, f).
[a]Composition of milk 5 weeks after the introduction of linseeds or cottonseeds into the diet.
[b]Longissimus dorsi lipids.
[c]The dams' diet of 1.2 kg dry matter concentrate/day + hay ad libitum was supplemented with 160 g cottonseed/day or 90 g linseed/day.

composition of kid perirenal fat. The effect of milk composition appears to be more marked on subcutaneous and intramuscular fats than on internal fats. Such effects depend on the nature and composition of protected FAs. The supply of protected PUFAs to dams resulted in a higher concentration of PUFAs, especially ω-3 FAs (C20:5, C22:5 and C22:6), but lower percentages of C18:0 in kid inter- and intramuscular fats (Fernandez-Navarro *et al.*, 2005). The results were more marked with protected PUFAs than with calcium soaps.

Kids can be fed milk replacers in which animal or vegetal fats replace milk fats. This feeding system is used to save kids when dams produce insufficient milk, and in intensive milk systems when goat milk is sold at high prices. In this case, the FA composition of kid fats closely reflects the composition of the milk replacer fat. Yeom *et al.* (2002) increased the C18:2 and C18:3 content in perirenal adipose tissues from 5.8 to 11.3% and from 0.1 to 0.8%, respectively, when they raised these FAs from 6.7 to 12.4% and from 0.5 to 1.4% in milk replacers. They observed that the transfer of C18:3 FAs from milk to adipose tissue was more difficult than the transfer of C18:2 FAs.

When the composition of goat milk and milk replacer fats are similar, the FA composition of kid fats does not differ, except for the minor FAs (Rojas *et al.*, 1994).

Generally, the level of milk intake has a limited effect on the FA composition of kid fats. Sauvant *et al.* (1979) observed a significant increase in C18:0 and C18:1 contents in several internal fats and muscle lipids when milk intake in kids increased from 0.75 to 1.5 kg/day during the 35 days after birth.

Some feeding programmes consisted of early weaning with access to dry feeds from 5–6 weeks after birth. These programmes allow economical savings as goat milk is sold at high prices and milk replacers are expensive. In such cases, the biohydrogenation and isomerization of dietary FAs took place very early in the rumen. The content of C18:0 and sometimes C18:1 FAs tended to increase in the kid muscle lipids and fats, while C14:0 FAs decreased. This difference in FA composition varied according to the nature of the dietary fats before and after weaning and the nature of fibre sources in the post-weaning diet (Sauvant *et al.*, 1979; Nitsan *et al.*, 1987).

### 15.6.2 Post-weaning period

Growth performance and FA composition of meat and fats in kids is linked to the nature of the diet, particularly its energy density, which depends on the nature and nutritive value of roughages (cultivated or uncultivated forages, stage of vegetation, fresh pasture or conserved forages, treated or untreated forages, rangeland vegetation) and the nature and proportion of grain or other concentrate feeds in the diet.

Information on the effect of the concentrate levels or forage:concentrate ratio in the diet on the FA composition of meat and fats is clearly lacking in kids compared with lambs. Bas *et al.* (2005) offered concentrate to young male goats grazing in an extensive rangeland (an Argan tree forest) in Morocco (Table 15.4). There was no effect of concentrate feeding on the FA composition of the longissimus dorsi muscle lipids and perirenal adipose tissue except for a lower concentration of C14:0 FAs in muscle. The FA composition did not differ in goats grazed in the Argan forest with and without concentrate supplementation. With diets rich in fibre sources or forages, particularly diets composed exclusively of roughages, cellulolytic fermentations, biohydrogenation and isomerization of USFAs are very active in the rumen. With diets rich in concentrate feeds, especially cereals, the degradability of starch is intense. This reduces the biohydrogenation of FAs in the rumen and increases the production of minor FAs (Morand-Fehr, 1981; Doreau and Ferlay, 1994; Sauvant and Bas, 2001).

Bas and Morand-Fehr (2000) carried out a meta-analysis on lambs from 21 studies with some diets based only on pasture and others on pasture plus concentrate. There was no significant difference in the FA composition of muscle and internal or subcutaneous fats except for a lower concentration of C18:3 FAs when concentrate was added to the grazing diet. In the same analysis, there were reports on more intensive diets (57 observations of roughage plus concentrate, and 161 observations with concentrate alone). The percentages of C18:1 and C18:2 FAs were higher with concentrate only and the percentages of C14:0 and C16:0 FAs were higher with roughage plus concentrate. Some experimental work in this area has indicated similar findings with kids (P. Morand-Fehr and J. Hervieu, unpublished).

Two feeding systems are usually used during kid growth: an indoor system based on concentrate with a limited supply of hay or straw, and a grazing system based on grazing rangeland or natural or cultivated pastures. Bas *et al.* (2005) compared the FA composition of muscle lipids and intermuscular and internal fats in kids reared under an extensive rangeland system with those raised under an indoor system based on concentrate alone in Morocco, where the most popular method for producing kid and lamb meat is the indoor system. The grazing system resulted in a decrease in the cholesterol content in meat, an increase in the concentration of C18:0, C18:2ω-6 and C18:3ω-6 FAs, SFAs and PUFAs, and a reduction in the concentrations of C16:0 and C18:1 *cis* and the ω-6:ω-3 ratio (Table 15.4). The differences were more remarkable in muscle than in intermuscular and internal fats. Rhee *et al.* (2000) obtained similar results for young goats by comparing two diets based on rangeland grazing in Texas or on grain sorghum (67%). Araba *et al.* (2008) observed similar results for Timahdite lambs in Morocco using the same experimental design as that of Bas *et al.* (2005) with kids. Nürnberg *et al.* (1998) reported comparable results obtained

**Table 15.4.** Effect of the feeding system (indoors versus Argan tree forest grazing) on fatty acid composition of muscle lipids and fat (mg/100 g) in local Moroccan young goats (from Bas *et al.*, 2005).

| | Longissimus dorsi muscle tissue | | | Perirenal adipose tissue | | |
|---|---|---|---|---|---|---|
| Fat composition | Indoors, concentrate alone | Grazing + concentrate | Grazing alone | Indoors concentrate alone | Grazing + concentrate | Grazing alone |
| Cholesterol | 64 | 58 | 54 | | | |
| C14:0 | 1.5a | 1.5a | 2.0b | 2.8 | 2.2 | 2.4 |
| C16:0 | 18.4a | 16.1b | 16.8b | 27.4a | 22.5b | 23.0b |
| C18:0 | 14.0a | 17.3b | 17.4b | 36.0a | 39.3ab | 39.8b |
| C18:1 *cis* | 49.9a | 32.0b | 32.7b | 20.2a | 16.1b | 14.0b |
| C18:2ω-6 | 4.1a | 7.5b | 7.4b | 1.6a | 2.3b | 2.5b |
| C18:3ω-3 | 0.04a | 0.12b | 0.11b | 0.07a | 0.55b | 0.59b |
| C20:5ω-3 | 0.15a | 0.37ab | 0.48b | | | |
| C22:5ω-3 | 0.44a | 1.68b | 1.43b | | | |
| C22:6ω-3 | 0.17a | 0.14a | 0.30b | | | |
| SFAs | 33.9a | 35.2ab | 36.5b | 66.4 | 64.5 | 65.6 |
| MUFAs | 52.6a | 38.5b | 39.5b | 26.0 | 24.7 | 22.9 |
| PUFAs | 8.0a | 16.9b | 15.9b | 1.9a | 3.4b | 3.8b |
| ω-6 FAs | 4.9a | 7.7ab | 10.6b | | | |
| ω-3 FAs | 0.9a | 4.3b | 4.0b | | | |
| ω-6:ω-3 ratio | 8.2a | 2.9b | 2.9b | 26.7a | 5.5b | 5.0b |

SFAs, Saturated fatty acids; MUFAs, monounsaturated fatty acids; PUFAs, polyunsaturated fatty acids; FAs, fatty acids.
Values within each group of tissue, on the same line, with different letters are significantly different ($P < 0.05$).

for sheep and steers except for ω-2 FAs, which tended to decrease with grazing.

Another interesting observation by Vasta *et al.* (2007) described the effects on lambs of condensed tannins, which are frequently found in extensive rangeland vegetation. Condensed tannins may form complexes with proteins in the rumen and inhibit ruminal microorganisms responsible for dietary FA biohydrogenation in the rumen. These findings strongly confirm that the presence of condensed tannins reduces ruminal biohydrogenation and the content of C18:2 *cis*-9, *trans*-11 and C18:1 *trans*-11 FAs and CLAs. This probably explains the results obtained with some extensive feeding systems. Indeed, some contradictory results that have been reported may have been because the vegetation was rich in condensed tannins.

As ω-3 FAs and CLAs have been shown to have significant beneficial effects on human health, several authors have attempted to improve the content of these FAs in the meat and fats of beef and lambs by adding specific lipid sources to diets (e.g. fish oil, linseeds, PUFA mixtures) or by adopting feeding strategies such as grazing (Ender *et al.*, 1997; Nürnberg *et al.*, 1998; Wood *et al.*, 2005). Webb *et al.* (2005) indicated that similar results can be obtained in goats. Thus, from all of these results, grazing can be considered a good feeding system for increasing the concentration of these acids in kid meat and fats.

Marinova *et al.* (2005) fed Bulgarian White kids the same concentrate mixture with or without fish oil at a concentration of 2.5%. Intermuscular fat in the loin and intramuscular fat in the longissimus dorsi muscle increased slightly. However, the authors did not give an account of the FA composition of the meat and adipose tissues. Atti *et al.* (2007) added 5 or 10% fish oil to a concentrate used for feeding lambs but did not observe any effect on the FA composition of subcutaneous adipose tissues. In particular, there was no significant increase in ω-3 FAs and CLAs, although there was a decrease in C18:2 FAs and USFAs.

Pienak-Lendzion *et al.* (2006) added 10% linseeds to the diet of White Improved goat kids. There were significant changes in the FA profile of the intramuscular fat of the longissimus dorsi muscle, which generally corresponded to higher percentages of MUFAs and PUFAs, especially C18:1, C18:2 and C20:4 FAs, but not ω-3 FAs.

Studies in goats are limited and the results have been somewhat variable. However, several reports on sheep and cattle have shown that the concentrations of PUFAs, particularly CLAs and ω-3 FAs, can be increased by adding feeds rich in PUFAs in concentrates, such as linseeds, fish oil and PUFA mixtures, under intensive feeding conditions (Wood *et al.*, 2005). It is likely that similar results could be obtained in kids.

## 15.7 Conclusion

The FA composition of meat and fats in goat kids is rather similar to that in lambs when compared at the same stage of fattening or adiposity. Generally, more SFAs and fewer USFAs are observed in lambs, as growth and fattening rates are higher in lambs than in kids. Furthermore, kids are frequently reared under more extensive conditions. The proportions of structural lipids such as phospholipids, which are richer in PUFAs and lower in SFAs, are higher in kid adipose tissues. However, as in other ruminants, the FA composition of kid fat lipids varies greatly according to the anatomical site of the adipose tissue. Generally, muscle lipids are more unsaturated and are richer in PUFAs and lower in SFAs than subcutaneous adipose tissues, especially internal fats. Subcutaneous fats, however, are more sensitive to factors modifying FA composition, particularly dietary factors.

Information on the effect of factors modifying the FA composition of meats and fats in kids is more limited than in sheep and cattle. The major factors influencing the composition of meat lipids and fats in kids are the dietary factors. However, often it is not possible to evaluate their specific effects because of interference from other factors such as genotype, sex or stage of growth.

During the last few years, there have been significant scientific advances with regard to improvements in the dietetic quality of kid meat. Indeed, during the milk-feeding period, feeding systems based on grazing of dams and supplementation of dams' diets with linseeds, fish oil and PUFA mixtures can increase the proportions of favourable FAs such as ω-3 FAs, PUFAs and CLAs and decrease C18:1 and C16:0 FAs and SFA in dams' milk and, to a lesser extent, in the meat lipids and fats of suckling kids. During post-weaning periods, similar feeding systems and methods applied directly to kids may have similar effects on meat and fats, but the results are scanty and variable in kids.

The lipids of meat and intra- and intermuscular and subcutaneous adipose tissues play an essential role in human nutrition as they provide essential FAs (C18:2, C18:3 *cis* and C20:4 *cis*) and CLAs likely to reduce the risks of cardiovascular diseases and cancers, as well as increasing liposoluble vitamins. Moreover, PUFAs can improve the taste, juiciness and tenderness of meat. Consequently, the quantities and composition of fats can improve or decrease the dietetic, nutritional, rheological and technological qualities of meat.

In the future, progress in the commercialization of goat meat is likely to occur only if its quality, particularly dietetic quality aspects, are markedly improved. Special systems for rearing young goats may influence the FA composition of their meat and fat depots. For instance, reports on goats reared in a Moroccan Argan tree forest (Bas *et al.*, 2005) indicated that this feeding system reduced the proportion of fats in meat and increased the proportions of the favourable FAs such as CLAs, the ω-6:ω-3 ratio to about 5 and increased the cholesterol content in meat. Such results allow optimism for the future production and consumption of goat meat.

## References


Araba, A., Morand-Fehr, P., El Aïch, A., Bouarour, M., Schmidely, P. and Berthelot, V. (2008) Outlook for producing Timahdite lamb with improved dietetic quality reared on Central Middle Atlas (Morocco) rangelands. In: *Proceedings of the 10th EAAP CIHEAM Mediterranean Symposium*, Corte, France, 6–8 November 2008, pp. 186–192.

Atti, N., Mahouachi, M. and Rouissi, H. (2007) Effect of fish meal in lamb diets on growth performances, carcass characteristics and subcutaneous fatty acid composition. *Options Méditerranéennes* 74, 57–61.

Banskalieva, V., Sahlu, T. and Goetsch, A.L. (2000) Fatty acid composition of goat muscles and fat depots: a review. *Small Ruminant Research* 37, 255–268.

Bas, P. and Morand-Fehr, P. (2000) Effect of nutritional factors on fatty acid composition of lamb fat deposits. *Livestock Production Science* 64, 61–79.

Bas, P., Morand-Fehr, P., van Quackebeke, E. and Cazes, J.P. (1980) Etude du caractère mou des gras de couverture de certaines carcasses d'agneaux et de chevreaux. In: *31st Annual Meeting of the European Association of Animal Production*, Munich, Germany, 1–4 September 1980.

Bas, P., Hervieu, J. and Morand-Fehr, P. (1981) Facteurs influençant la composition des graisses chez le chevreau de boucherie: incidence sur la qualité des gras de carcasses. In: *7th International Conference on Goat Production*, Tours, France, 12–15 May 1981, pp. 90–100.

Bas, P., Morand-Fehr, P., Rouzeau, A. and Hervieu, J. (1987) Evolution de la composition des tissus adipeux du chevreau male à 4, 6, 8 semaines. *Reproduction Nutrition Development* 27, 313–314.

Bas, P., Morand-Fehr, P. and Schmidely, P. (1991) Weaning: a critical period for young kids. In: Morand-Fehr, P. (ed.) *Goat Nutrition*. Pudoc, Wageningen, the Netherlands, pp. 271–283.

Bas, P., Galouin, F. and Morand-Fehr, P. (1992) Changes in lipid content, fatty acid composition and lipoprotein lipase activity in dry goat omental adipose tissue according to tissue site. *Lipids* 26, 470–473.

Bas, P., Rouzeau, A. and Morand-Fehr, P. (1996) Changes in the content of branched-chain fatty acids of the adipose tissue from different sites in growing goats. In: *Proceedings of the 6th International Conference on Goats*, Beijing, PR China, 6–11 May 1996, Vol. 2. International Academic Publishers, p. 68.

Bas, P., Dahbi, E., El Aïch, A., Morand-Fehr, P. and Araba, A. (2005) Effect of feeding on fatty acid composition of muscles and adipose tissues in young goats raised in the Argan tree forest of Morocco. *Meat Science* 71, 317–326.

Bauchart, D. (2006) Meat fatty acids in ruminants. In: *Meeting of the Association Française des Techniciens de l'Alimentation et des Productions Animales*, 13 October 2006.

Bauchart, D., Chantelot, F. and Gaudemer, G. (2008) Qualités nutritionnelles de la viande et des abats chez le bovin: données récentes sur les principaux constituants d'intérêt nutritionnel. *Cahiers de Nutrition et de Diététique* 43, 5429–5439.

Beriain, M.J., Bas, P., Purroy, A. and Treacher, T. (2000) Effect of nutrition on lamb meat quality. *Cahiers Options Méditerranéennes Série A* 52, 75–86.

Beserra, F.J., Madruga, M.S., Leite, A.M., Da Silva, E.M.C. and Maia, E.L. (2004) Effects of age at slaughter on chemical composition of meat from Moxoto goats and their crosses. *Small Ruminant Research* 55, 177–181.

Camparra, P., Foti, F., Scerra, M., Cilione, C., Vottari, G., Galofaro, V., Sinatra, M.C. and Scerra, V. (2007) Influence of feeding system on fatty acid composition of suckling lambs. *Options Méditerranéennes Série A* 74, 95–99.

Casey, N.H., van Niekerk, W.A. and Webb, E.C. (2003) Goat meat. In: Caballero, B., Trugo, L. and Finglass, P. (eds) *Encyclopedia of Food Science and Nutrition.* Academic Press, London, UK, pp. 2937–2944.

Delmotte, C., Rondia, P., Raes, K., Dehareng, F. and Decruyenaere, V. (2007) Omega 3 and CLA naturally enhanced levels of animal products: effects of grass and linseed supplementation on the fatty acid composition of lamb meat and sheep milk. *Options Mediterranéennes Série A* 74, 41–48.

Dhanda, J.S., Taylor, D.G. and Murray, P.J. (2003) Carcass composition and fatty acid profiles of adipose tissue of male goats: effect of genotype and live weight at slaughter. *Small Ruminant Research* 50, 67–74.

Doreau, M. and Ferlay, A. (1994) Digestion and utilization of fatty acids by ruminants. *Animal Feed Science and Technology* 45, 379–396.

Ender, K., Paptein, H.J., Nürnberg, K. and Wegner, J. (1997) Muscle and fat related characteristics of grazing steers and lambs in extensive systems. In: *Proceedings of an EU Workshop on the Effect of Extensification of Animal Performance and Product Quality*, Melle, Gontrode, Belgium, 14–16 May 1997, pp. 229–232.

Enser, M. (2000) Producing meat for healthy eating. In: *Proceedings of the 46th International Congress Meat Science and Technology*, Buenos Aires, Argentina, 27 August–1 September 2000, pp. 124–129.

Fernandez-Navarro, J.R., Carmona Lopez, F.D., Gil Estremera, F., Sanz Sampelayo, M.R. and Boza, J. (2005) Kid goat adipose deposits composition in response to maternal intake of a protected fat rich in PUFAs. *Options Méditerranéennes Série A* 67, 205–208.

Ha, J.K. and Lindsay, R.C. (1990) Distribution of volatile branched chain fatty acids in perinephric fats of various red meat species. *Lebensmittel – Wissenschaft and Technologie* 23, 433–440.

Ha, J.K. and Lindsay, R.C. (1991) Volatile alkylphenols and thiophenols in species-related characterizing flavours of red meats. *Journal of Food Science* 56, 1197–1202.

Hausman, P. and Snell, H. (2000) Influence of keeping method (indoors vs. biotope) on the meat performance of goat kids of different genotypes. 2. Meat quality and composition of fatty acids. *Zuchtungskunde* 72, 308–318.

Johnson, D.D., Eastridge, J.S., Neubauer, D.R. and McGowan, C.H. (1995) Effect of sex class on nutrient content of meat from young goat. *Journal of Animal Science* 73, 296–301.

Lecerf, J.M. (2008) Acides gras et maladies cardio-vasculaires. *Cholé doc* 110, 1–4.

Ledoux, M. (2006) Les acides linoléiques conjugués: présence dans les aliments et propriétés physiologiques. *Cholé doc* 94, 1–4.

Legrand, P. (2008) Intérêt nutritionnel des principaux acides gras des lipides du lait. *Cholé doc* 105, 1–4.

Madruga, M.S., Souza, J.G., Narain, N., Beserra, F.J. and Biscontini, T.M.B. (2000a) Effect of slaughter age on fat components of the mestiço goat meat. In: *7th International Conference on Goats*, Tours, France, 15–21 May 2000, p. 842.

Madruga, M.S., Arruda, S.G.B., Narain, N. and Zouza, J.G. (2000b) Castration and slaughter age effects on panel assessment and aroma compounds of the mestiço goat meat. *Meat Science* 56, 117–125.

Madsen, A., Jakobsen, K. and Mortensen, H.P. (1992) Influence of dietary fat on carcass fat quality: a review. *Acta Agriclturae Scandinavica, Section A – Animal Science* 42, 220–225.

Mahgoub, O., Khan, A.J., Al-Maqbaly, R.S., Al-Sabahi, J.N., Annamalai, K. and Al-Sakry, N.M. (2002) Fatty acid composition of muscle and fat tissues of Omani Jebel Akhdar goats of different sexes and weights. *Meat Science* 61, 381–387.

Mandrefini, M., Massari, M., Cavani, C. and Falashini, A.F. (1988) Carcass characteristics of male Alpine kids slaughtered at different weights. *Small Ruminant Research* 1, 49–58.

Marchello, J.A., Cramer, D.A. and Miller, L.G. (1967) Effect of ambient temperature on certain ovine fat characteristics. *Journal of Animal Science* 26, 294–297.

Marinova, P., Banskalieva, V. and Tzvetkova, V. (2005) Body and carcass composition and meat quality of kids fed fish oil supplemented diets. *Options Méditerranéennes Série A* 67, 151–156.

Martin, B., Priolo, A., Valvo, M.A., Micol, D. and Coulon, J.B. (2005) Effects of grass feeding on milk, cheese and meat sensory properties. *Options Méditerranéennes Série A* 57, 213–223.

Martin, L., Rodriguez, P., Rota, A., Rojas, A., Pascual, M.R., Paton, D. and Tovar, J. (1999) Effect of protected fat supplementation to lactating goats on growth and fatty acid composition of perirenal fat in goat kids. *Animal Science* 68, 195–200.

Matsuoka, A., Furokawa, N. and Takahashi, T. (1997) Carcass traits and chemical composition of meat in male and female goats. *Journal of Agricultural Science* 42, 127–135.

Mele, M., Serra, A., La Comba, F., Buccioni, A., Conte, G. and Sechieri, P. (2007) Effect of the inclusion of soyabean oil in the diet of dairy goats on meat fatty acids composition of their suckling kids. *Options Méditerranéennes Série A* 74, 177–182.

Melton, S.L. (1990) Effects of feeds on flavor of red meat: a review. *Journal of Animal Science* 68, 4421–4435.

Molénat, G. and Thériez, M. (1973) Influence du mode d'élevage sur la qualité de la carcasse de l'agneau de bergerie. Etude des effets de l'allaitement artificiel et des régimes entièrement condensés sur la qualité des dépôts adipeux de couverture. *Annales de Zootechnie* 22, 279–293.

Moody, W.G. (1983) Beef flavor: a review. *Food Technology* 37, 227–238.

Morand-Fehr, P. (1981) Growth. In: Gall, C. (ed.) *Goat Production*. Academic Press, New York, pp. 253–283.

Morand-Fehr, P., Havrevoll, O., Bas, P., Colomer-Rocher, P., Falagan, A., Sanz-Sampelayo, M.R., Sauvant, D. and Treacher, T.T. (1991) Influence of feeding and rearing methods on the quality of young goat carcasses. In: Morand-Fehr, P. (ed.) *Goat Nutrition*. Pudoc, Wageningen, the Netherlands, pp. 292–303.

Nitsan, Z., Carasso, Y., Zoref, Z. and Nir, I. (1987) Effect of diet on fatty acid profile of adipose tissues and muscle fat of kids. *Annales de Zootechnie* 36, 339–341.

Nudda, A., Battacone, G., Fancellu, S. and Pulina, G. (2005) The transfer of conjugated linoleic acid and vaccenic acid from milk to meat in goats. *Italian Journal of Animal Science* 4, 395–397.

Nürnberg, K., Grumbach, S., Papstein, H.J., Mottles, H.D., Ender, K. and Nürnberg, G. (1996) Fatty acid composition of lamb meat. *Fett/Lipid* 98, 77–80.

Nürnberg, K., Wegner, J. and Ender, K. (1998) Factors influencing fat composition in muscle and adipose tissue of farm animals. *Livestock Production Science* 56, 145–156.

Pienak-Lendzion, K., Niedziolka, R., Borkowska, T. and Horoszewicz, E. (2006) Effect of linseeds supplement in mixtures on chemical composition and fatty acids profile in muscular tissue of male kids. *Archiv für Tierzucht* 49, 244–248.

Raes, K., de Smet, S. and Demeyer, D. (2004) Effect of dietary fatty acids on incorporation of long chain polyunsaturated fatty acids and conjugated linoleic acid in lamb, beef and pork: a review. *Animal Feed Science and Technology* 113, 199–221.

Ralph, W. (1989) Making lamb more desirable. *The Shepherd* February, 9.

Rhee, K.S., Waldron, D.F., Ziprin, Y.A. and Rhee, K.C. (2000) Fatty acid composition of goat diets vs intramuscular fat. *Meat Science* 54, 313–318.

Rojas, A., Lopez-Bote, C., Rota, A., Martin, L., Rodriguez, P.L. and Tovar, J.J. (1994) Fatty acid composition of Verata goat kids fed either goat milk or commercial milk replacer. *Small Ruminant Research* 14, 61–66.

Sauvant, D. and Bas, P. (2001) La digestion des lipides chez le ruminant. *INRA Productions Animales* 14, 303–310.

Sauvant, D., Bas, P. and Morand-Fehr, P. (1979) Production de chevreaux lourds: II. Influence du niveau d'ingestion du lait et du sevrage sur les performances et la composition des tissus adipeux. *Annales de Zootechnie* 28, 73–92.

Sheradin, R., Hoffman, L.C. and Ferreira, A.V. (2003) Meat quality of Boer kids and Mutton Merino lambs. 1. Commercial yields and chemical composition. *Animal Science* 76, 63–71.

Sikora, J. and Borys, B. (2006) Lipid profile of intermuscular fat in kids fattened to 60, 90, 180 days of age. *Archiv für Tierzucht* 49 (Special issue), 193–200.

Suzuki, J. and Bailey, M.E. (1985) Direct sampling capillary GLC analysis of flavor volatiles from ovine fat. *Journal of Agriculture and Food Chemistry* 33, 343–347.

Todaro, M., Corrao, A., Alicata, M.L., Schinelli, R., Giaccone, P. and Priolo, A. (2004) Effects of litter size and sex on meat quality traits of kid meat. *Small Ruminant Research* 54, 191–196.

Tshabalala, P.A., Strydom, P.E., Webb, E.C. and de Kock, H.L. (2003) Meat quality of designated South African indigenous goat and sheep breeds. *Meat Science* 65, 563–570.

Tsuneishi, E., Shiba, N. and Matsuzaki, M. (2001) Influence of nutritional status and anatomical location in conjugated linoleic acid (CLA) in ruminants (goat) adipose tissues and medulla ossium flava. *Animal Science Journal* 72, 218–222.

Vasta, V., Lanza, M., Pennisi, P., Bella, M. and Priolo, A. (2007) Effect of dietary condensed tannins of lamb inter-muscular fatty acids. *Options Méditerranéennes Série A* 74, 35–39.

Webb, E.C. and Casey, N.H. (1995) Genetic differences in fatty acid composition of subcutaneous adipose tissues in Dorper and SA Mutton Merino wethers at different live weights. *Small Ruminant Research* 18, 81–88.

Webb, E.C., Casey, N.H. and Simela, L. (2005) Goat meat quality. *Small Ruminant Research* 60, 153–166.

Werdi Pratiwi, N.M., Murray, P.J. and Taylor, D.G. (2004) The fatty acid composition of muscles and adipose tissues from entire and castrated male Boer goats raised in *Australia. Animal Science* 79, 221–229.

Wood, J.D., Enser, M., Fisher, A.V., Nute, G.R., Whittington, F.M. and Richarson, R.I. (2005) Effect of diets on fatty acids and meat quality. *Options Méditerranéennes Série A* 67, 133–141.

Yeom, K.H., van Trierum, G., Hovenier, R., Schellingerhout, A.B., Lee, K.W. and Beynen, A.C. (2002) Fatty acid composition of adipose tissues in goat kids fed milk replacers with different contents of $\alpha$-linolenic and linoleic acids. *Small Ruminant Research* 43, 15–22.

Zygoyiannis, D., Staniataris, C. and Katsaounis, N. (1985) The melting point, iodine value, fatty acid composition and softness index of carcass fat in three different breeds of suckled lambs in Greece. *Journal of Agricultural Science* 104, 360–365.

Zygoyiannis, D., Kufidis, D., Katsaounis, N. and Philips, P. (1992) Fatty acid composition of carcass fat of indigenous (*Capra prisca*) suckled Greek kids and milk of their does. *Small Ruminant Research* 8, 83–95.

# Index

Page numbers in bold type refer to figures and tables.


abomasum 163–164
Africa: production environments 72–73
age
effects on carcass characteristics **204**
estimated using dentition 222–223
and fatty acids composition of meat 251
hormones 202
and meat ash content 263
of puberty **121**
ageing: of meat 310–312
Algeria 73
amino acids: in goat meat 297, **298**
Animal Genetic Resource Information (FAO) 70
animals, live: evaluation *see* evaluation: of live goats
anoestrus, post-partum 131–133
antinutrients 174–175, 350
effects reduced by polyethylene glycol 184
Argentina 73
ash: content in goat meat
comparison with other species **295**
data for various breeds, sexes and ages **262**
effect of age 263
effect of breed 262–263
effect of cooking 264
effect of diet 264
effect of litter size 263–264
effect of sex 263
reflects total mineral content 261–262
Asia: production environments 71–72
Australia
cross-breeding 89, 92
feral goats 16

Bangladesh: cross-breeding 87
Batina goats 46
Bermuda grass 252
bile 164
Blakewell, Robert 70
blocks, multi-nutrient 186–187
factors affecting intake 188
body-condition scoring *see* scoring, body-condition
Boer goat 37, 95, 97
bone
growth and development 199, 243
proportion in carcass 244, 246, **247**
weights of individual bones **245**
breeding *see also* reproduction
benefit to producers 69
composite populations 92, **93**, **94**, 95, **96**, 97
cross-breeding
3- and 4-breed crosses 107–108, 152
backcrosses 78–79, 85
benefits 76–77
challenges 77–78
disease control 78
performance of cross-breeds **22**, 35, **80–84**, **90–9**, **152**
recombination loss 77

breeding *continued*
cross-breeding *continued*
rotational crosses 85, **86**
specific breed crosses 85, 87–89, 87–92
use of modelling 92
genetic improvement programmes 105–107
paucity of research 53
selection *see* selection
breeds
ash and mineral content of meat 262–263, 272
characteristics and definition 28
concept of breed 70–71
evaluation 74–76
fatty acid composition 251, 345–346
growth and gain efficiency **28**
indigenous
Boer goats 37, 95, 97, **203**
classification 71
Creole goats 42–44
effects of age and sex **204**
Omani native meat goats 44–47
performance versus cross-bred goats **22**, 35, **80–84**, **152**
replacement with exotic breeds 33
West African Dwarf 36
linear body measurements 279–280
litter sizes 143–144
oestrus cycle length **125**
ovulation rates 127
tissue distribution 238, 243, 244–246
unreported or unknown 73–74
world statistics
list of breeds by specialty, climate and country **68**
number of breeds by country **55–60**
bride price 20
browsing 165
bruising 226–227

cacti: for dietary supplementation 174, 252
calcium: in goat meat 265–266
Canada: cross-breeding 89
carbohydrates 167, 168
carcasses: characteristics
and body-condition scores 220, **221**
body size 34–35, 36
carcass indices and cuts 37, 39–41, 43–44, 328–330
carcass weight related to live body weight *see under* weight, live
classification and grading 206
Creole goats 43–44
dressing percentage *see* dressing percentage
effect of milk replacers 328–330
effects of age and sex **204**
effects of pre-slaughter conditioning **206**
linear measurements
influencing factors 282–284
method and definitions 281–282, **283**
Omani native meat goats 44–47
tissue partitioning *see* tissues: distribution in carcass
tropical versus temperate breeds 37
cardiovascular disease: in humans 339–340
China 72, 87
cholesterol: in goat meat 301
chorionic gonadotropin, equine 145
climate 26, 29
cloprostenol 142, 143
cold shortening 307–308
colour: of meat 305–306
communities: communal production 20–21
concentrates, feed 10, 28, **255**, 339
conservatism 24
consumers
attitudes to goat meat 11
extent of goat meat consumption 277, 278
conversion, feed
heritability **101**
in kids 326
copper
for control of gastrointestinal parasites 176
in goat meat 268–269
Creole goats 42–44
cross-breeding *see under* breeding
selection *see* selection
cryopreservation: sperm 136

decoquinate 182
deficiencies
minerals 180–181
protein 180
vitamins 171–172, 181
development *see* growth and development
Dhofari goats 45–47
diagnosis: pregnancy 148–150, **151**
diet *see* feeds; forages; nutrition
diseases 29
control during cross-breeding 78
in humans 339–340
domestication: goats 1, 15–16, 69–70
effect on goat form and functions 2–3
dressing percentage 212–213, 277, 284–286
duodenum 164

economy
livestock as wealth 20
and small-scale production 25
education 24

efficiency, production 34
Egypt: cross-breeding 79, 88
embryos
  cryopreservation and transfer 147–148
  mortality 128–129
  rate of growth and development 198–199
  survival correlated with nutritional status 189
energy: from nutrients 172–173
environment
  effect on mortality 97
  impact of goats 3
  production environments
    Africa and Middle East 72–73
    Americas 73
    Asia 71–72
Europe: predominance of dairy production 4, 6
evaluation: of live goats
  body-condition scoring *see* scoring, body-condition
  body weight *see* weight, live
  estimating age using dentition 222–223
  linear body measurements
    accuracy and sources of error 279
    definitions **279**
    differences between breeds 279–280
    locations of measurement 278–279
  purpose 210
  ultrasound scanning 223–224
export slaughter intervals 225
extension programmes 24

FAO (Food and Agricultural Organization)
  Animal Genetic Resource Information 70
  goat population statistics 4–6
fat
  content in goat meat 28, 105
    comparison with other species 295–296
    influencing factors 295–296
  dietary 170
  distribution in carcass *see under* tissues: distribution in carcass
  growth and development of fatty tissue 201
  scoring 220–221
fatty acids: in goat meat
  comparison with other species **299**, **341**, 342
  desirable characteristics 250–251, 297–298
  effect of age 251, 344–345
  effect of breed 251, 345–346
  effect of diet
    milk and milk replacers 334, 346–348
    post-weaning period 252–257, 348, **349**, 350 299–300
  effect of sex 251, 345
  effect of weight 251
  health implications for humans 338–339
  omega-3 and conjugated linoleic acids 342, **343**, 346–347, 350
  and organoleptic meat quality 339–340
  tissue distribution **300**, 340
  variations due to sampling location 344
feeds *see also* supplementation, dietary
  additives 182, 184
  concentrates 10, 28, **255**, 339
  forages *see* forages
  international nomenclature **182**
  milk replacers for kids *see* milk replacers
  mineral content **183**
  nutrients in feeds 165–166
    carbohydrates 167, 168
    fats 170
    fibre 167–168
    minerals 170–171
    protein 168–170
    vitamins 171–172
  types 182
fermentation, microbial 163
fibre, dietary 167–168
fitness 36
flavour 339
fleeces 226
fluorogestone actetate (FGA) 141–142
flushing 189
follicles, ovarian 123–125
forages 26
  effect of nutritional supplements on intake 185–186
  and fatty acid composition of meat 252–254
  as nutritional constraints 173–174
  survival advantage of foraging 1
France 73
freezing 306–307

gastrointestinal tract 162–164
  nitrogen utilization **169**
  parasites 175–176
genetic improvement programmes 105–107
gestation 129–131
girth, heart 214, **215**
Global Animal Genetic Data Bank 70
glycogen, muscle 201, 294
glycolysis 294
glycoproteins, pregnancy-associated (PAGs) 129–130
grasslands 22
growth and development
  birth size correlates with survival 199
  bone 199, 243
  changes in fatty acid composition 344–345

growth and development *continued*
fatty tissue 201
of goat kids on dairy farms 325–327
growth factors 200–201
influencing factors
age and sex 201–202, 203, **204**
genetics and selection 202–203
muscle 199–201
nutritional manipulation 203–205, **206**
principles and definitions 197–198
rate 198–199
skeletal system 199
typical growth pattern **198**
guajilo 175

heart girth 214, **215**
hormones
effects of age 202
in male reproductive physiology 133
in oestrus cycle 123, 125, 139
synchronization protocols 140–143
during pregnancy 129
in pseudopregnancy 131
steroid immunization 144–145
horns 134–135
hypothalamus 122

identification: of live animals 227
immunization, steroid 144–145
India: cross-breeding 79, 87
infrastructure
and small-scale production 23–24
insemination, artificial 136, 146–147
inserts, intravaginal 141–142
insulin-like growth factors 200–201
iron: in goat meat 268, 296

Kenya 1, 87
kids: on dairy farms
management methods 324
use of milk replacers *see* milk replacers

litters, size of
breed differences 143–144
environmental influences 144
improved by steroid immunization 144–145

magnesium: in goat meat 267–268
manganese: in goat meat 270
marketing
preparation of animals 224–227, **228**
and small-scale production 25
markets
contribution of goat to meat market 6–9
traditional subsistence 11–12
measurements: live goats *see* evaluation: of live goats
medicine, traditional
Chinese herbal medicine: potential use in production 13
medicinal properties of goat meat 2
Mediterranean region 72–73
melatonin 122–123
used in out-of-season breeding 138–139
used in puberty induction 137
melengestrol acetate 143
metyl acetoxy progesterone (MAP) 141–142
Mexico 73, 89
microchips 227
microorganisms: in the digestive tract 163
milk replacers
compared to milk 324–325
effect on body weight, carcass yield and offal 327–328
effect on fatty acid composition 346–348
and goat kid growth 325–327
growth curves 325–326
minerals
dietary 170–171
content in feeds **183**
deficiencies and requirements 180–181
in goat meat
comparison with other meat **269**, 270, **271**
influencing factors 271–273
levels found 264–265
macroelements 265–268
methodology 261
microelements 268–270
overview 260–261, 301–302
tissue distribution 270–271
modelling: to develop cross-breeding strategies 92
monensin 182, 184
Morocco 73
mortality 10
and body-condition score 221–222
effect of environment 97
of embryos 128–129
mouth 162
muscle
cold shortening 307–308
composition 297, **300**
distribution in carcass *see under* tissues: distribution in carcass
distribution of minerals 271
effect of milk replacers 332–334

effects of fibre types on meat quality 312–314
eye muscle depth 223–224
growth and development 199–201
myofibrillar fragmentation index 305
pH 303
post-mortem changes 294–295
effect of electrical stimulation 310
shear-force 304–305
structure **201**, **293**, 294, **313**
tissue characteristics 293
myofibrillar fragmentation index 305
myogenesis 200–201

Nigeria 72
nitrogen: use by ruminants **169**
nomads 21–23, 69, 72
Norway 92
nutrition
constraints
forage availability and quality 173–174
internal parasites 175–176
plant secondary metabolites 174–175
dietary supplementation *see* supplementation, dietary
effect on meat ash and mineral content 264, **271**, 272–273
effect on reproduction 128, 134, 189
factors of influence 161–162
feeding habits 164–165
feeds *see* feeds
milk replacers *see* milk replacers
requirements of goats
energy 177, 180
minerals 180–181
overview 176–177, **178–179**
protein 180
vitamins 181
source of energy 172–173
used for manipulation of growth 203–205, **206**
water intake 166–167

oesophagus 162
oestrus cycle
endocrine events 123, 125
follicular dynamics 123–125
oestrus expression 125–126
synchronization protocols 140–143
Oman
cross-breeding 87–88
native meat goats 44–47
organic *see under* production
organoleptic properties 339–340

Ovsynch 143
ovulation
oestrus cycle
endocrine events 123, 125
follicular dynamics 123–125
ovulation rate and reproductive wastage 126–129
seasonality 121–123, 127
timing 126

parasites: gastrointestinal tract 175–176
pH: of muscle tissue 303
effect of electrical stimulation 310
Philippines 72
phosphorus: in goat meat 266–267
photoperiodic treatment 135–136
placenta 129
polyethylene glycol 184, 253
population, goat 4–6, **55–60**
potassium: in goat meat 267
pregnancy 129–131
diagnosis 148–150, **151**
pregnancy-associated glycoproteins (PAGs) 129–130
prices 12
processing: of meat 314–316
production
based on concentrate feeding 10, 28
commercial versus non-commercial 2, 9, 15
communal 20–21
economic and political influences 3
evaluation of goats *see* evaluation: of live goats
extensive versus intensive 19–20, 26–28
increased by manipulating reproduction
puberty induction 137–138
integration with crop farming 72
integration with fish culture 72
meat within milk production system 10, 324
and natural resources 25–30
nomadic 21–23, 69, 72
organic 12–13
pastoral 1
primitive herding 23
production environments *see under* environment
small-scale systems: problems
economics 25
infrastructure 23–24
land tenure 24–25
marketing 25
people 24
social systems 24
subsistence 18–19

production *continued*
  system types 10–11, 17
    characteristics of indigenous systems 21
    definition of a system 16–17
  tethering 20
  transhumance 23, 72, 73
  world statistics **9**, 53–54, **55–60**, 60, **61–67**, **75**
progestogens: in oestrus synchronization 141–143
prostaglandin F2α 143
protein
  digestibility coefficient 296
  in feed 168–170
  in goat meat **295**, **296**
  importance of goat meat as a source 17
  requirements of goats 180
  supplementation 26, 176, 186–187
pseudopregnancy 131
puberty 119–120, **121**
  age **121**
  induction 137–138
  and nutritional status 189

recombination loss 77
refrigeration 306
reproduction
  assisted reproductive technologies
    artificial insemination 136, 146–147
    embryo transfer 147–148
    rationale 145–146
  buck management 136–137
  buck sexual behaviour 135
  economic importance 29
  effect of nutrition 128, 134, 189
  factors affecting buck reproductive capacity 134–135
  genetic improvement of reproductive traits 151–152
  gestation 129–131
  litter size *see* litters, size of
  oestrus cycle *see* oestrus cycle
  out-of-season breeding
    buck effect 139–140
    exogenous melatonin 138–139
  ovulation *see* ovulation
  post-partum anoestrus 131–133
  reproductive losses 128–129
    and body-condition score 222
  seasonality 120–123, 133–134
    photoperiodic treatment of bucks 135–136
research 3–4
reticulo-rumen 163
ribs, long: fat scoring 220–221
ribs, short: body-condition scoring 217–220
rigor mortis 294–295
ruminants, small: economic importance 2
rumination 162

Sahel 72
saliva 162
satellite cells 199–200
scales, weighing 213–214, **215**
scanning, ultrasound 223–224
scoring, body-condition
  fat scoring long ribs 220–221
  and mortality risk 221–222
  purpose 216
  and reproductive performance 222
  short ribs
    interpretation **217**
    method 216, 218
    relationship to carcass characteristics 220, **221**
    relationship to live weight 218–220
    reliability 218
seasonality: of reproduction 120–123, 133–134
  photoperiodic treatment 135–136
selection
  accelerated by minimizing generation interval 105
  advanced methodologies 109–110
  benefits 97–98
  concept 97
  effects on growth and development 202–203
  estimates of genetic parameters 98, **99**, **102–103**
    feed conversion **101**
    weight 99, **100**, 101, 104
  genetic improvement programmes 105–107
    for reproductive traits 151–152
  for meat quality 105
  predicting performance of 3- and 4-breed crosses 107–108
selectivity 164
selenium: in goat meat 270
*Sericea lespedeza* 252–254
shear-force: of meat 304–305
shearing 226
shortening, cold 307–308
sodium: in goat meat 267
soil: effect on farming systems 26
Spain
  cross-breeding 88–89
specifications, marketing 225
sperm
  cryopreservation 136, 146
  effect of photoperiodic treatment 136
  seasonal variations 133–134

Sri Lanka
    cross-breeding 79
stomach 162–163
stress, pre-slaughter 201
suckling, restricted 326–327
sulla 253
supplementation, dietary
    with cacti 174, 252
    effect on fatty acid composition 350
    effect on forage intake 185–186
    and fatty acid composition of meat 251–252
    hand-crafted supplement blocks 186–187
    protein 26, 176
    reducing intake variation 187–188
    supplements
        ingredient characteristics 185
        nutritional profile 184–185
sustainability 18
synchronization: of oestrus cycle 140–143

Taiwan: cross-breeding 85
tannins
    antinutrients 174–175, 350
        effects reduced by polyethylene glycol 184
    in parasite control 176
tapes, heart girth 214
teeth 162
    estimating age using dentition 222–223
tenderness: of meat 304–305
    effect of ageing 310–312
    effect of electrical stimulation 310
tenure, land 24–25
tethering 20
Thailand: cross-breeding 79
Tibet 72
tissues: distribution in carcass 35, 41–42, **44**, **46**, **233**, 286–288
    bone
        boneless meat yield 213, **214**
        effect of breed and size 244, 246
        effect of sex 246
        proportion of bone 244–246, **247**
    effect of milk replacers 330, **331**
    fat 239–240, **241**
        effect of body size and breed 243
        effect of sex 242–243, 286–287
    influencing factors 232
    muscle 232, **234**, **235–236**, 237
        effect of body weight 238
        effect of breed type 238
        effect of sex 238–239
tongue 162
transhumance 23, 72, 73
transport: of live animals 226–227
tropics: reasons for failure of livestock enterprises 33
Tunisia 73

ultrasound scanning 223–224
United Kingdom: cross-breeding 88
United States 73
    cross-breeding 79, 88
urea 186
    toxicity 100
uterus: involution 130–131

vegetation: effect on farming systems 26
vitamins
    dietary 171–172
        supplementation 186
    in goat meat 302

water
    intake 166–167
    retention in meat 303–304
weaning 225
weight, live
    birth weight: effect on growth 326
    and carcass fat **211**
        dressing percentage 212–213
    and carcass weight 37, **38**, **39**, **44**, **46**, 211–212, 280–281
        boneless meat yield 213, **214**
        prediction of carcass weight **41**, 210–211
    effect of diet on gain **27**
    effect of milk replacers 327–328
    heritability 99, **100**, 101, 104
    measurement and estimation
        errors 214–216
        heart girth tapes 214, **215**
        livestock scales 213–214, **215**
    and relative growth coefficient of body components **212**
West African Dwarf 36
withholding periods 225

zinc: in goat meat 269–270